# Bioanalytical Chemistry

# Bioanalytical Chemistry

Editor: Jean Boynton

NY RESEARCH PRESS

New York

Published by NY Research Press
118-35 Queens Blvd., Suite 400,
Forest Hills, NY 11375, USA
www.nyresearchpress.com

Bioanalytical Chemistry
Edited by Jean Boynton

International Standard Book Number: 978-1-63238-670-0 (Hardback)

**Cataloging-in-Publication Data**

Bioanalytical chemistry / edited by Jean Boynton.
p. cm.
Includes bibliographical references and index.
ISBN 978-1-63238-670-0
1. Analytic biochemistry. 2. Chemistry, Analytic. I. Boynton, Jean.
QP519.7 .B56 2019
572.36--dc23

# Contents

# Preface

The world is advancing at a fast pace like never before. Therefore, the need is to keep up with the latest developments. This book was an idea that came to fruition when the specialists in the area realized the need to coordinate together and document essential themes in the subject. That's when I was requested to be the editor. Editing this book has been an honour as it brings together diverse authors researching on different streams of the field. The book collates essential materials contributed by veterans in the area which can be utilized by students and researchers alike.

The quantitative measurement of biotics and xenobiotics in biological systems falls under the discipline of bioanalytical chemistry. This is important for the purposes of pharmacokinetics, bioequivalence, exposure-response and toxicokinetics. It is also crucial for forensic investigations, anti-doping testing and for the analysis of drugs with an influence on the environment. Some of the techniques used in bioanalytical studies are chromatographic methods, mass spectrometry, nuclear magnetic resonance, electrophoresis, etc. The topics included in this book on bioanalytical chemistry are of the utmost significance and bound to provide incredible insights to readers. It includes some of the vital pieces of work being conducted across the world, on various topics related to bioanalysis. This book aims to serve as a resource guide for students and experts alike and contribute to the growth of the discipline.

Each chapter is a sole-standing publication that reflects each author's interpretation. Thus, the book displays a multi-facetted picture of our current understanding of application, resources and aspects of the field. I would like to thank the contributors of this book and my family for their endless support.

Editor

# Advances and Changes in the Techniques of Multi-dimensional and Comprehensive Chromatography and when Coupled with Mass Spectrometry

**Peter Baugh J\***

*The BMSS, C/O 23, Priory Road, Sale, M33 2BU, England, UK*

**Abstract**

This review covers the principles of two dimensional gas and liquid chromatography (2DGC and 2DLC) and briefly introduces the theory accounting for the increase in separation resulting from a greater peak capacity than for the one dimensional (1D) mode. The advance in the techniques from multi-dimensional to comprehensive chromatography is discussed. The more recent development of multi-dimensional chromatography ion mobility mass spectrometry receives a mention to highlight the added dimension of molecular size and shape (molecular collision cross section) as an enabling tool for increasing component separation and peak capacity. Although both the techniques of 2DGC and LC are described the focus is on the environmental and food applications of 2DGC, principally when coupled to mass spectrometry, with TOFMS and HRMS as prime examples.

**Keywords:** Mass spectrometry; Chromatography; Pre-columns; Column chromatography

## Background and Introduction

### Advent of two dimensional techniques

**Planar chromatography:** The history of two dimensional chromatographic techniques stems from humble beginnings and an inheritance from the early development of paper chromatography (pc) and, subsequently, thin layer chromatography (tlc), involving liquid mobile and solid stationary phases (MP and SP). Essentially, a solvent is selected to separate components of a mixture in one direction, e.g., in ascending pc or tlc, followed by drying and application of a second solvent after rotating the paper or thin layer plate through 90°, which allows a greater degree of separation and less peak overlap, as illustrated in Figure 1.

Initially, a cellulosic-based SP was employed in pc and silica based SP in tlc. The early work in tlc, focused on the use of the normal phase mode (polar SP, non-polar MP) followed later by the reversed phase mode where the modification of the silica SP was carried out to inactivate polar sites by the introduction of a range of moieties to render the SP, non-polar ($CH_3$) and gradually more polar ($NH_2$), depending on the application. The selection of MP is dependent on the range of polarities of the components to be separated. In the first dimension, a non-polar MP may be employed to progress less polar analytes, followed by a more polar MP applied in the second dimension to separate the more polar analytes. The combined technique allows a greater degree of separation of a multi-component mixture than for the one dimensional technique and, for example, can be employed as a purification method for analytes, which are present only in small quantities, applying radioautography subsequently for the detection of radioisotopically-labelled pure analytes and for chemical detection of metabolites produced by biochemical processes.

**Column chromatography-GC and LC:** These planar chromatographic techniques have progressed to gas and liquid chromatography, and when linked to mass spectrometry, firstly, involving one dimensional GC or LC modes (1DGC or 1DLC, respectively) but, importantly, for consideration in this review, embracing 2DGC and LC modes (liquid adsorption column chromatography by gravity can be traced back to the work of Mikhail

Tswett Physikalisch-chemische Studien über das Chlorophyll. Die Adsorptionen (Physical-chemical studies of chlorophyll. Adsorption.), Berichte der Deutschen botanischen Gesellschaft. Tsvet coins the term "chromatography").

Giddings has stated that 1D separation process is statistically limited in case of mixtures exceeding 50-60 compounds [1]. More recently Multidimensional gas chromatography (MDGC) was developed as a technique following the ideas and proposals of several groups and Giddings in 1995 evaluated sample dimensionality as a predictor of order- disorder in component peak distribution in multidimensional separation [1]. Comprehensive two dimensional gas chromatography: the recent evolution and current trends, have been reviewed by Brinkman et al. and Tranchida et al. [2]. Interpretation of comprehensive two-dimensional gas chromatography data using advanced chemometrics has also received recent attention [3].

It is worth noting that guard or pre-columns (short and unconditioned) have long been a feature in analysis using 1DGC and LC systems, primarily for prolonging the lifetime of analytical columns, see Figure 2. The employment of an HPLC guard column (cartridge) installed in front of an analytical column, provides protection from strongly retained impurities, thus prolonging the life of the analytical column. Guard columns are identical to analytical columns except these columns are shorter and typically not tested in order to keep the cost low. Understanding the significant factors that affect guard column performance can assist in protecting analytical column and save in cost by extending analytical column lifetime. Similarly a The first work on 2D gas chromatography appeared several decades ago

\***Corresponding author:** Peter Baugh J, Environmental and Food Analysis Special Interest Group Leader, The BMSS, C/O 23, Priory Road, Sale, M33 2BU, England, UK, E-mail: peterbaugh682@hotmail.com

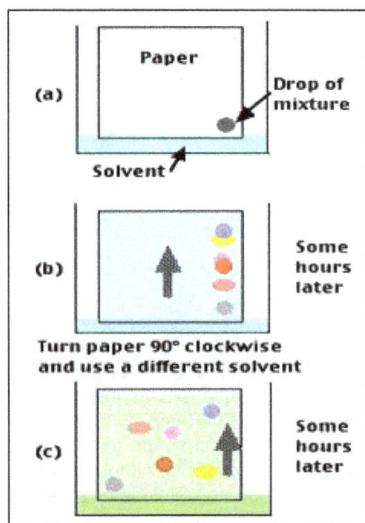

**Figure 1:** Illustration of two dimensional ascending paper chromatography.

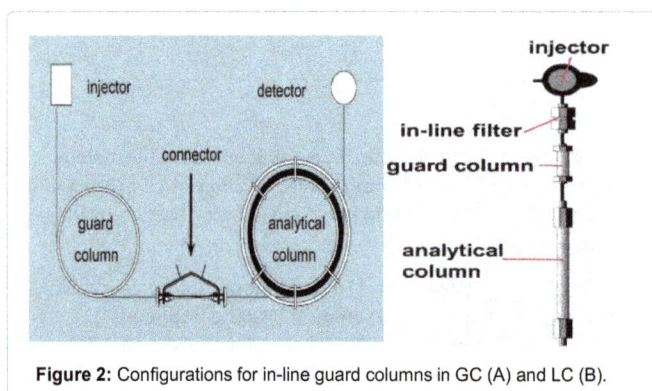

**Figure 2:** Configurations for in-line guard columns in GC (A) and LC (B).

[4] and concerned two dimensional gas chromatography in which a low polarity packed (or now capillary for GC) column was placed in sequence with a more polar column separated by an interface splitter, having now been superseded by a more high tech. modulator device (2DGC, see section 3.21). The principle of heart cutting from the first column has been applied in this respect to focus on a group of analytes selected from the first column, which can be separated on the second column possessing a suitable polarity [5]. MDGC fits this description but has the limitation of incompleteness, holistically (Note: 2DGC can be considered generic embracing MDGC or GC-GC and comprehensive GC or GC × GC).

**Comprehensive 2DGC:** Capillary GC with added feature of 2D offers the capability of higher peak capacities than other chromatographic techniques: today, some 100-150 peaks can be separated in one run in one dimension and a considerably greater number with the second dimension. With this in mind, if highly complex samples are to be effectively studied; one way to improve the separation power is to couple, through an interface, two independent columns. The practicability of so-called multidimensional gas chromatography (MDGC or GC-GC), has convincingly been demonstrated for a variety of sample types [5]. The main limitation of the MDGC approach is that its application almost invariably is of the heart-cutting type: one or, at best, a few small fractions eluting from the first column are selected for

further separation on the second column. In other words, the approach is successful in a target-analysis situation. When, on the other hand, across-the-board screening of an entire sample is required, MDGC is much too time- consuming and complicated, and a comprehensive, i.e., a GC × GC, approach has to be used. In this case, the entire first-column (or, first dimension) eluate, cut into small adjacent fractions to maintain the first-dimension resolution, is subjected to further analysis on the second (second-dimension) column.

The analytical benefits of comprehensive two-dimensional gas chromatography (GC × GC) methods have been exploited and emphasized. The unexpectedly complex guard column can be employed for GC.

Nature of many real-world samples amenable to GC analysis has been elucidated through the enhanced resolving power of GC × GC. For the effective operation of GC × GC, a range of fundamental devices, called modulators, are employed to enable continuous heart-cutting and re-injection, without which GC × GC analysis cannot be achieved. GC × GC modulation describing the advantages and the disadvantages of the most significant thermal and pneumatic modulators has been reviewed and an update of the current position regarding the state of art modulators is described in Section 3.2. Figure 3 illustrates the comparison of conventional MDGC and comprehensive GC × GC.

Summarizing, comprehensive 2DGC techniques involve the complete 1D and 2D profiling, i.e., separation of all analytes from column 1 and 2. This aspect of GC × GC, with respect to classical multidimensional chromatography, is that the entire sample is subjected to two distinct analytical separations. The resulting enhanced separating capacity makes this approach a prime choice for highly complex mixtures and when coupled with mass spectrometry.

**2DLC:** The principle of 2DLC is similar for liquid chromatography in that two columns can be selected to reflect the differences in polarities of the components of a complex mixture of target analytes and to optimize the separation. Initially, the 2DLC was hampered by extremely long analysis times although fast high temperature analysis can now be conducted in the second LC column, to vastly reduce the second phase elution time. For 2DLC the result of a comprehensive two-dimensional separation can be envisaged as a series of many (for example, 100) fast second-dimension chromatograms. These 2D

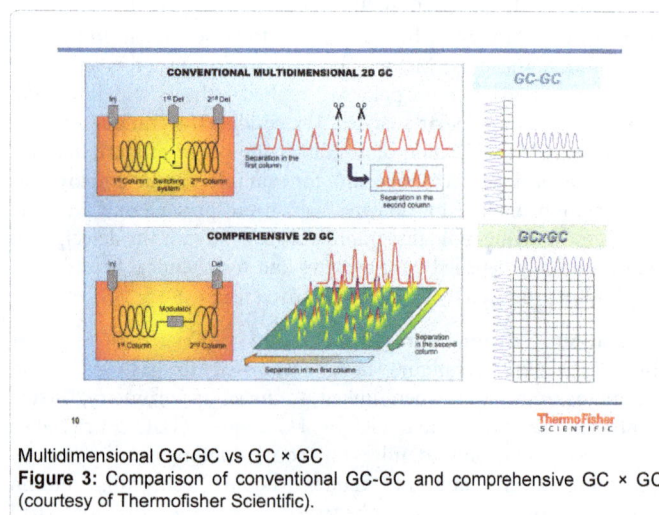

Multidimensional GC-GC vs GC × GC
**Figure 3:** Comparison of conventional GC-GC and comprehensive GC × GC (courtesy of Thermofisher Scientific).

chromatograms usually are combined into a data matrix, which then can be displayed as a contour plot.

**Special features of Multi-Dimensional Chromatography mass spectrometry recently developed:** While the major developments and emphasis in multi-dimensional techniques have involved conventional 2DGC and LC and when coupled with mass spectrometry, with TOFMS as one of the prime instruments, a further variation has come through the interfacing of a travelling wave ion mobility (TWIM) separation component to the mass spectrometer of the combined instrumentation, LC/TWIM-MS (for example, see Waters SYNAPT G2-Si High Definition MS System), which also enables an increase in peak capacity over 1D chromatography akin to the front end 2D mode. TWIM allows access to this additional dimension of separation based on molecular size and shape through the acquisition/computation of molecular collision cross sectional data. A full account of this development is outside the scope of this review [6].

## Theory of 2D Chromatography

### Peak capacity considerations- 1D and 2D

The theory relating to the relative peak capacities of 1D and 2D is well-documented and provides a basis for the advantages in separation of the 2D technique. Peak capacity determines the number of peaks of a finite width that can be separated within a specified time window.

On coupling 2 columns together, in sequence, there is a considerable increase in the theoretical and experimental peak capacity because of the enhanced separation achieved in the second column, with peaks from the first column being separated with a greater efficiency within a particular time window, which eliminate or minimize peak overlap.

The peak capacity-n, for 2DGC or LC is the product of the peak capacities for column 1 and 2, $n_1{}^{1D}$ and $n_2{}^{1D}$, respectively, and not the sum. In an ideal state the increase is an order of magnitude greater.

Thus, $\qquad n^{2D}=n_1{}^{1D}\times n_2{}^{1D}$ $\qquad\qquad$ (1)

Another consideration for the application of the 2D mode is the analysis time. For 2DGC the total analysis time is simply the sum of the time to elute the components of interest from the first column plus the time for elution and detection of all of the selected components from the second column. If heart cutting is not to be applied, the total analysis time would be the sum of the elution times for all analytes from the first and second columns to the point of final detection.

As mentioned, previously, long analysis times have been a feature of 2DLC for isocratic and gradient elution operations but with the advent of high temperature LC analysis times can be dramatically decreased with analysis times as low as 20 sec for elution from column 2.

Models for peak separation in 2D chromatography have been proposed including the Davis- Giddings Statistical Model of Overlap, SMO, which provides the theoretical basis for requiring 2D separations when considering the analysis of multi-component mixtures. The 1D mode is frequently exhibits too low a peak capacity and the 2D is a more efficient mode generating peak capacities that are of an order of magnitude greater than for 1D. Extensive details of the theory behind SMO are provided elsewhere [7]. Consideration of SMO in applications of 2D chromatography provides a basis for maximizing the separation of analytes of particular interest. The model predicts that the 1D mode is inadequate for the separation of complex mixtures and on average the maximum fraction of the peak capacity that can be identified as peaks is about 37%, equivalent to the separation of 100 analytes

using an ID separation having a peak capacity of 100, on average the maximum number of peaks observed will be 37. Also the maximum fraction of peak capacity that can be recognized as single component constituent peaks is about 18%. For more than 10-20 analytes, the prediction of the SMO are that many observed peaks will comprise two or three overlapping single constituent peaks. Guichon's work shows a better ability to deal with highly saturated chromatograms, i.e., when the number of analytes in the sample exceeds the peak capacity of the method. Davis has corrected his model of overlap in 1D separations to more accurately predict the number of observed peaks in highly saturated chromatograms. Other work involving a model of the size distribution based on data from the separation of plant extracts has shown that an increase in range of concentrations requires an increased peak capacity to obtain accurate analysis (important where the dynamic range of concentrations can be excessive)

In brief, SMO indicates that extremely high peak capacities are needed to separate complex samples with a large number of analytes creating difficulties with the limitations of 1DLC and emphasizes the need for 2D separations to generate greater peak capacities. However, relative to 1D separations the number of peaks predicted in 2D separations can be complicated by the fact that the ratio of peak widths in the direction of the two axes of the separation has an effect on the efficiency of 2D separations in producing peaks at given peak capacity and number of components. This problem has been treated in a 2D model of overlap and corrected to account for several complicating factors.

Murphy et al. [7] have considered a theoretical criterion for realizing the ideal 2D peak capacity which provides information as to how often the 1D effluent must be sampled so that the resolution gained in the 1D is not partially lost due to the sampling process, which is required to carry out separation effectively in the second dimension. Essentially the effluent must be sampled at least three to four over 8σ width of a 1D peak to avoid significant loss of 2D resolution between a pair of peaks when the 1D separation contributes significantly to the overall resolution (M-S-F sampling criterion).

## Instrumental Considerations

### 2DGC and 2DLC

The coupling of two capillary columns in sequence originally was designed for heart-cut analysis of a selected analyte group eluted from the first column using a splitter to divert from or allow analyte transfer to the second column. Over the past several decades with the added capability of modulators and when coupled with mass spectrometry, the applications of the techniques of 2D chromatography have grown, immensely. The types of modulato 2DGC are described in section 4.2.1.

For 2DLC the chromatography involves two columns, respectively for slow and fast analysis, connected by an interface device evolved from the transfer loop/multi-port valve combination, which has been, and is a feature of 1DGC and LC. This mode, suitably designed, with flow switching, is primarily employed for configuring 2DLC, as described in section 4.2.2.

### Interfaces for 2DGC and LC

There is more complication in, and a greater variety of interface devices for 2DGC than for 2DLC and a range of modulators having been designed [8].

**Modulators for 2DGC:** The introduction of a modulator for the transfer of analytes from column 1 to column 2 enables a greater separation and efficiency. Basically the introduction of a modulator has several requirements involving electronic control of the interface, transfer of analytes, and computer software for generation of the 1D and 2D chromatograms and, frequently 3D profiles by signal processing (see section 4.3). A number of designs of modulator (Cryogenic modulator; Flow modulator; Heart-cutting; In-line valve system; Longitudinally-modulated cryogenic system; Out-line valve system; Pneumatic modulator; Re-injection; Thermal modulator (for a further description refer to http://www.leco.com.au/files/sepscigcxgc209-184. pdf) have been employed for a range of applications. The modulator is a piece of hardware that transfers effluent from the exit of the primary column to the head of the secondary column as a repetitive series of pulses. The commercially available units have primarily employed a thermal modulation strategy that required the consumption of large amounts of liquid cryogen coolant and gaseous working fluid. Thermal modulation produces optimal resolution but also sacrifices some of the simplicity of conventional GC. Valve-based modulation has received less attention than thermal modulation. However, valve-based modulators use straightforward, low-cost designs and do not require additional consumables. Two main classes of valve-based modulators have been developed: sub- sampling and differential flow. Sub-sampling modulators generate pulses by briefly diverting small amounts of primary column effluent to the head of the secondary column. While easy to put into practice, sub-sampling modulators lead to decreased sensitivity and must employ short modulation periods. Differential flow modulators sample all of the primary column effluent, which means that sensitivity is not sacrificed and larger modulation periods can be used. However, these devices also use high secondary flows that lead to elevated column pressures and limit the use of direct mass spectrometric detection.

Figure 4 illustrates the general schematic diagrams for a simple fluidic modulator [9] depicting, A. the 2DGC arrangement, B and C. the fill and flush sequences, respectively.

The employment of the wide variety of modulators can be highlighted by applications in many areas. The state of the art and one of the most favoured modulators is the AGILENT capillary flow technology (CFT) device (https://www.chem.agilent.com/cag/prod/GC/2DGC_amj2_05_02_07D1a.pdf; http://www.agilent.com/en-us/products/gas-chromatography/gc-gc-ms-technologies/capillary-flow-technology/gc-x-gc). The Agilent GC × GC CFT Flow.

Modulator incorporates a proprietary deactivated flow channel and integrated gas supply connections providing unprecedented robustness that is easy to use and maintenance free. CFT Flow Modulator is not limited by sample volatility providing efficient collection and injection of solutes across the volatility range. The CFT has advantages as follows:

1. No cryogenic-cooling needed to re-focus, providing significant savings to the lab

2. Collects the material from the first column, dividing a peak into multiple cuts

3. Focuses the material collected from each cut into a narrow band

4. Introduces the bands sequentially into the second column

Thermofisher have focused on 2DGC and a range of modulators for HRMS featuring the Dual Stage Thermal Modulator:- Dual-Jet LCO$_2$ (Figure 5) Quad-Jet LN$_2$ and Loop Type; Flow Modulator:- Valves system and CFT device (Cavagnino, An Introduction and Overview of Comprehensive Gas Chromatography (GC × GC): New Trace1300GC was introduced in 2012).

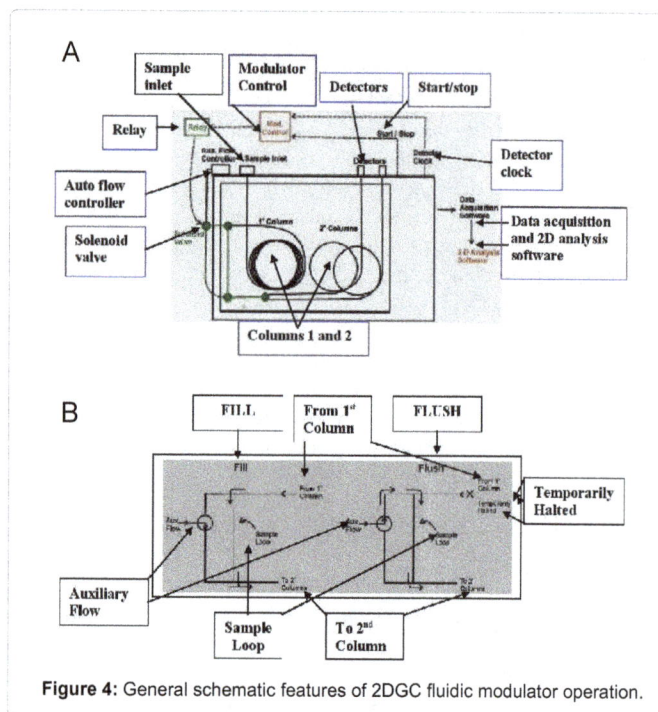

**Figure 4:** General schematic features of 2DGC fluidic modulator operation.

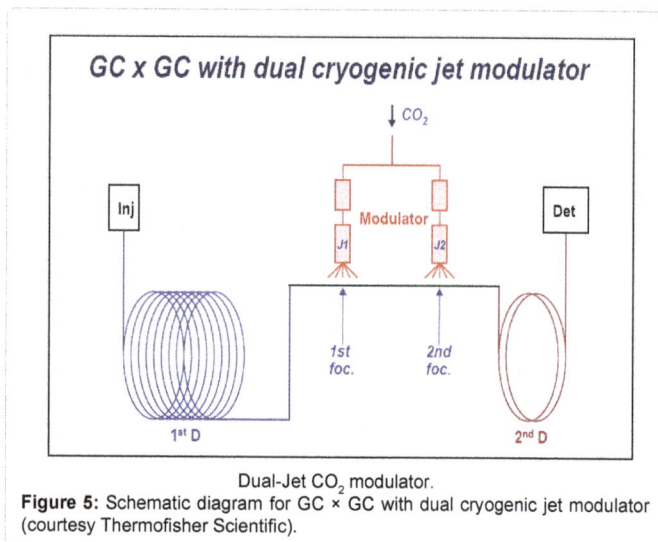

Dual-Jet CO$_2$ modulator.
**Figure 5:** Schematic diagram for GC × GC with dual cryogenic jet modulator (courtesy Thermofisher Scientific).

**Loop and flow switching interface for 2DLC:** An overview of comprehensive 2DLC has been presented by Majors and Shoenmakers in LC-GC North America [10].

Multidimensional chromatography uses a combination of several chromatography techniques, separation modes, and columns to separate multiple components. It achieves significantly higher separation than normal one-dimensional chromatography. Various separation modes and the corresponding mobile phases can be selected for HPLC, and the diverse permutations available suggest the possibility of achieving a degree of selectivity not possible using one-dimensional separation alone. Multidimensional chromatography includes an offline technique in which the eluent is temporarily fractioned (collected) in a suitable position and then part or all of it is injected into the next column, and an online technique in which part or all of the eluent is introduced into the column via the sample loop for automatic analysis. Offline

Advances and Changes in the Techniques of Multi-dimensional and Comprehensive Chromatography...

5

multidimensional chromatography requires additional handling, such as elution by solvent and concentration of the temporarily fractioned (collected) eluent.

Figure 6 shows the system configuration for comprehensive two-dimensional LC. This system permits gradient elution at both the first- and second-dimension stages. The sample solution is injected by the autosampler and separated in the first-dimension column. The sample flow rate is generally low in the first-dimension column (such as 50 μL/min) to allow separation over a long period of time. The eluent from the first column is sent to the loop of the flow-switching valve. When the loop becomes full (after two minutes at a 50 μL/min flow rate with the 100 μL loop used in this example), the valve operates to send the eluent for separation in the second-dimension column. As the flow-switching valve operates every two minutes in the example shown here, analysis in the second-dimension column (including column equilibration time for gradient elution) must be completed within two minutes.

A UHPLC (Ultra high performance liquid chromatography) column has been adopted in recent years as the second-dimension column in order to achieve a sufficiently high degree of separation in about one minute.

### Contour, surface plot generation and GC × GC software

Sophisticated software has been developed to enable the profiling of 2DGC analysis and when coupled with mass spectrometry. Figure 7 illustrates the plot generation and employment of computed stacked modulation slices in 3D profiles.

One of the favoured software facilities for 2DGC is GC Image which is used by a range of companies. GC × GC Software is multi-featured package for visualizing, analyzing, and processing data produced by comprehensive two-dimensional gas chromatography. The GC × GC Software package includes three programs:

**1. GC Image:** for visualizing, analyzing, and reporting on individual GC × GC chromatograms.

**2. GC Project:** for managing projects with multiple GC × GC runs/injections.

**3. Image Investigator:** for automated and interactive multivariate analyses.

Zoex Corporation and Shimadzu Corporation employ GC Image GC × GC Edition software for their thermal modulation GC × GC systems and GC Image software for the Comprehensive GC × GC System, respectively. Agilent Technologies favours GC Image GC × GC Edition software for its Capillary Flow Technology (CFT) GC × GC systems (7890A GC). LECO's pioneering efforts in GC × GC data processing have resulted in one of the most comprehensive software package available, Chroma TOF Software for data mining and generating the 3D chromatograms and profiles.

GC × GC Software Images are illustrated in Figure 8 to highlight the advantages in 2D contour, 3D surface and pca plots that can be produced.

## Applications of 2DGC and LC, and When Coupled to Mass Spectrometry

### 2DGC

The applications of 2DGC are many and varied and total coverage is not possible. Only a snapshot of the vast amount of work and

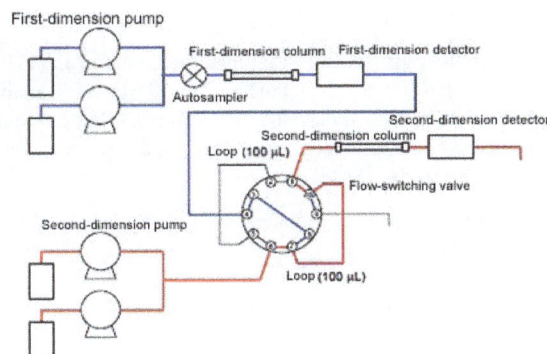

**Figure 6:** The system configuration for comprehensive two-dimensional LC (http://www.shimadzu.com/an/hplc/support/lib/lctalk/2dlc.html).

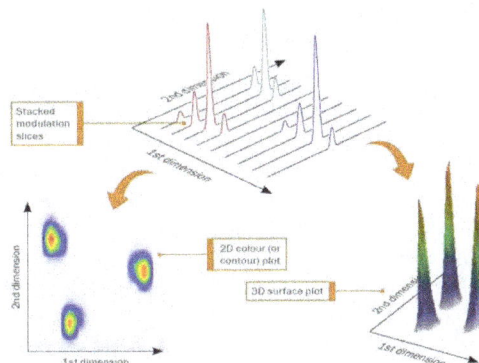

**Figure 7:** 2D Contour and 3D Surface plots for comprehensive GC × GC-TOFMS indicating the application of stacked modulation slices in 3D profiling, (by kind permission of Laura McGregor, Markes International, Llantrisant, South Wales, UK).

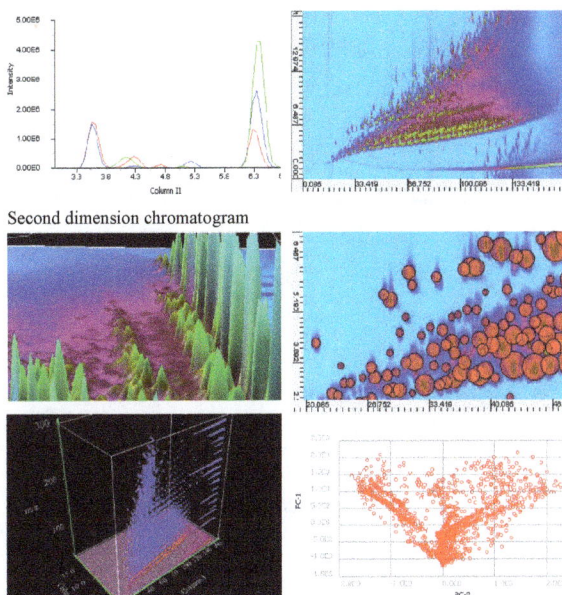

**Figure 8:** Generation of 2D contour and 3D surface plots using GC × GC software, also featuring a pca plot.

applications will be presented here to demonstrate the types of the analytical activity involved.

As an excellent introductory example, Brinkman et al. [11] in the application of 2DGC with detection by TOFMS to trace analysis of flavour compounds in food describe the advantages and requirements for the successful employment of 2DGC and detection by mass spectrometry in this field as follows:

To solve the separation problems relating to analysis by 1DGC, two-dimensional heart- cut-type 2DGC is frequently used as a more powerful alternative. However, this 2D technique is a less than ideal solution because of the limitation of the analysis to a few discrete target regions of the chromatogram and the considerable increase in analysis time which, even then, occurs. In addition, 2DGC requires sophisticated instrumentation and experienced analysts Comprehensive two-dimensional gas chromatography (GC × GC) is a new and extremely useful technique to enhance separation of analytes of interest from each other and/or the matrix background. In the past few years, GC × GC has been shown to provide the capability to considerably improve the analysis of complex samples. In GC

× GC, two independent GC separations are applied to an entire sample. The sample is first separated on a high-resolution capillary GC column under programmed-temperature conditions. Very small fractions of the effluent of this column are continuously focused in a so-called cryogenic modulator and, next, re-injected very rapidly onto a second GC column. The column is short and narrow to enable very rapid separations; the separation in this column must be finished before the next first-column fraction starts to elute. The speed of the second column is so high that it is effectively operated under isothermal conditions. To properly record very narrow peaks, with widths of typically 60-600 ms at the baseline, some 10 data points are

needed. This means that the data acquisition rate should be 50 Hz or, in other words, that a time-of-flight-mass spectrometer (TOF-MS) has to be used. The coupling of 2DGC with a TOF-MS has been reported in several recent studies, which demonstrated this technique to be a most powerful for the identification and quantification of trace-level analytes in complex mixtures. The aim of this study was to show the potential of 2DGC–TOF-MS for the trace-level determination of flavour compounds in food extracts.

In 2013 Cajka reviewed the applications of gas chromatography time of-flight mass spectrometry in food and environmental analysis [12].

Table 1A and 1B provide a summary of the wide ranging and varied applications of 2DGC (published primarily in J Chromatogr A). In addition, there are many papers dealing with the instrumental and modulator design features, and the operational procedures for different types of analyses and chromatographic parameters as summarized in Table 1C. Selected references from the recent past (2003-2015) are highlighted to exemplify the applications (Table 1).

## 2DLC

A brief introduction only is given to the applications of 2DLC. The work of several groups highlights the advances in the applications of 2DLC coupled to mass spectrometry [32,33] and an outline of an example application is given in the paragraph following.

Off-line two-dimensional liquid chromatography with tandem mass spectrometry detection (2D-LC/MS-MS) has been used to separate a set of metabolomic species. Water-soluble metabolites were extracted from Escherichia coli and Saccharomyces cerevisae cultures and were analyzed using strong cation exchange (SCX)-hydrophilic interaction chromatography (HILIC). Metabolite mixtures are well-suited for

| A. Summary of Applications - References [11-24] | | |
|---|---|---|
| Flavour Compounds in food analysis [11] Food and Environmental analysis [12] | Organic aerosols [13] | Persistent and Bioaccumulative Contaminants in Marine Environments [14] |
| Flame retardants and plasticizers [15] | Fatty acid methylesters [16] | Shale oil [17] |
| 18. Virgin olive oil [18] | Accumulated hydrocarbons [19] | PCBs, polychloro compounds [20] |
| Sewage treatment plant effluents [21] | Mineral oils in foods [22] | Naphthenic acids [23] |
| Isomeric species [24] | | |
| **B. Applications not cited** | | |
| Aromatic sulphur compounds | | Selective extraction |
| Biodegradation patterns | Non-target 2DGC | Steroids |
| Biodiesel | OPs and esters | Surfactants |
| Closely eluting compounds | Paleoenvironmental analysis | Speciation of nitrogen compounds |
| Environmental chlorinate | Polybrominateddiphenylethers | Terpanes |
| Fatty acids and sterols | Recurrent oil sheens | Volatile compounds |
| Halogenated species | Rocket propulsion | |
| Musk fragances | Screening of oil sources | |
| **C. Instrumental aspects, operational procedures, theoretical, chromatographic and analytical parameters: Selected References [25-31]** | | |
| Mass spectral matching [25] | Comparative FID and MS [26] | Tile based Fisher ratio [27] |
| Evaluation of QTOF [28] | Evaluation of near theoretical maximum peak capacity gain [29] | Mixed mode separation in the first dimension of LC × LC [30] |
| Retention time prediction [31] | | |

**Table 1:** Summary of the Titles and Topics of Comprehensive 2DGC Investigations.

multidimensional chromatography as the range of components varies widely with respect to polarity and chemical makeup. Some currently used methods employ two different separations for the detection of positively and negatively ionized metabolites by mass spectrometry. Here, a single set of chromatographic conditions were developed for both ionization modes allowing detection of a total of 141 extracted metabolite species, with an overall peak capacity of ca. 2500. It was shown that a single two-dimensional separation method is sufficient and practical when a pair or more of uni-dimensional separations are used in metabolomics immediately [34-36].

## Summary of Pending Special Issue

Comprehensive two-dimensional gas chromatography (GC × GC) has become a well- established and valuable technique for separating and analysing complex mixtures. GC × GC has been applied in many fields where conventional one-dimensional (1D) separations were traditionally used for volatile and semi-volatile analysis, including flavour and fragrance, environmental studies, metabolomics, pharmaceuticals, petroleum products, and forensic science. Compared to 1D GC, GC×GC provides enhanced sensitivity and superior separation power due to increased peak capacity. An additional dimension of information can be added by employing detection systems, such as, time-of- flight mass spectrometers (TOFMS). The ongoing evolution of GC × GC has included advances in modulation, data handling and application development. Improved quantitative data analysis is necessary to move the use of GC × GC from research laboratories to more widespread use in commercial laboratories. Discussion at recent scientific meetings has focused on reaching a near-theoretical peak capacity and the implementation of high resolution TOFMS (A special issue with a call for manuscripts, April 2016, is to be published in J Chromatography. Guest Editor is **Prof. Dr. Shari Forbes**, Centre for Forensic Science, University of Technology, Sydney, PO Box 123, Broadway, NSW 2007, Australia). This Special Issue invites contributions that highlight the latest research and advancements in GC × GC technology, and demonstrate the range of applications and fields for which its use is particularly beneficial).

## Acknowledgements

The author wishes to acknowledge the contributions of the following scientific colleagues from the gas and liquid chromatography and mass spectrometry communities.

Laura McGregor, Markes International Ltd., Llantrisant, South Wales, UK for comprehensive collection of scientific paper pdfs on 2DGC and pdfs from associated scientific researchers and selected application notes, for example: https://www.markes.com/Resources/Mass-Spec-application-notes/Select-eV.aspx Markes International Application Note 524 January (2014) Analysis of complex petrochemicals by GC × GC–TOF MS with Select-eV variable-energy electron ionization.

- Daniela Cavagnino, Thermofisher Scientific, Milan, Italy for diagrams and power point presentation.

- Keith Hall and colleagues for information about modulators.

- Ken Brady, AGILENT, Stockport, UK, for useful dialogue on comprehensive chromatography.

## References

1. Giddings JC (1995) Sample dimensionality: a predictor of order-disorder in component peak distribution in multidimensional separation. J Chromatogr A 703: 3-15.

2. Adahchour M, Beens J, Brinkman UA (2008) Recent developments in the application of comprehensive two-dimensional gas chromatography. J Chromatogr A 1186: 67-108.

3. Tranchida PQ, Franchina FA, Dugo P (2016) Comprehensive two-dimensional gas chromatography-mass spectrometry: Recent evolution and current trends. Mass Spectrom Rev 35: 524-534.

4. Pierce KM, Hoggard JC, Mohler RE, Synovec RE (2008) Recent advancements in comprehensive two-dimensional separations with chemometrics. J Chromatogr A 1184: 341-352.

5. Zeng Z, Li J, Hugel MH, Xu G, Marriott PJ (2014) Interpretation of comprehensive two-dimensional gas chromatography data using advanced chemometrics. Trends in Anal Chem 53: 150-166.

6. Mondello L, Mostafa A, Górecki T, Tranchida PQ (2011) History, Evolution, and Optimization Aspects of Comprehensive Two-Dimensional Gas Chromatography. John Wiley & Sons.

7. Blomberg J, Schoenmakers PJ, Beens J, Tijssen R (1997) Comprehensive two- dimensional gas chromatography (GC×GC) and its applicability to the characterization of complex petrochemical) mixtures. J High Res Chromatog 20: 539-544.

8. Valentine SJ, Kulchania M, Srebalus Barnes CA, Clemmer DE (2001) Multidimensional separations of complex peptide mixtures: a combined high performance liquid chromatography/ion mobility/time of-flight mass spectrometry approach. Int J Mass Spectrum 212: 97.

9. Stoll DR, Li X, Wang X, Carr PW, Porter SE, et al. (2007) Fast, comprehensive two-dimensional liquid chromatography. J Chromatogr A 1168: 3-43.

10. Tranchida PQ, Purcaro G, Dugo P, Mondello L, Purcaro G (2011) Modulators for comprehensive two dimensional gas chromatography. Trends in Analytical Chemistry 30: 1437-1461.

11. Seeley JV, Micyus NJ, McCurry JD, Seeley SK (2006) Comprehensive two-dimensional gas chromatography with simple fluidic modulator.

12. Majors RE, Schoenmakers P (2008) LCxLC: Comprehensive Two- Dimential liquids Chromatography. LCGC North America 26: 600-608.

13. Adahchour M, van Stee LL, Beens J, Vreuls RJ, Batenburg MA, et al. (2003) Comprehensive two-dimensional gas chromatography with time-of-flight mass spectrometric detection for the trace analysis of flavour compounds in food. J Chromatogr A 1019: 157-172.

14. Cajka TS (2013) Gas chromatography time of-flight mass spectrometry in food and environmental analysis. Comprehensive Analytical Chemistry 61: 271-302.

15. Goldstein AH, Worton DR, Williams BJ, Hering SV, Kreisberg NM, et al. (2008) Thermal desorption comprehensive two-dimensional gas chromatography for in-situ measurements of organic aerosols. J Chromatogr A 1186: 340-347.

16. Hoh E, Dodder NG, Lehotay SJ, Pangallo KC, Reddy CM, et al. (2012) Non-targeted Comprehensive Two-Dimensional Gas Chromatography/Time-of-Flight Mass Spectrometry Method and Software for Inventorying Persistent and Bioaccumulative Contaminants in Marine Environments. Environ Sci Technol 46: 8001-8008.

17. Ballesteros-Gomez A, de Boer J, Leonards PE (2013) Novel analytical methods for flame retardants and plasticizers based on gas chromatography, comprehensive two-dimensional gas chromatography, and direct probe coupled to atmospheric pressure chemical ionization-high resolution time-of-flight-mass spectrometry. Anal Chem 85: 9572-9580.

18. Nosheen A, Mitrevski B, Bano A, Marriott PJ (2013) Fast comprehensive two-dimensional gas chromatography method for fatty acid methyl ester separation and quantification using dual ionic liquid columns. J Chromatogr A 1312: 118-123.

19. Amer MW, Mitrevski B, Jackson WR, Chaffee AL, Marriott PJ (2014) Multidimensional and comprehensive two-dimensional gas chromatography of dichloromethane soluble products from a high sulfur Jordanian oil shale. Talanta 120: 55-63.

20. Purcaro G, Cordero C, Liberto E, Bicchi C, Conte LS (2014) Toward a definition of blueprint of virgin olive oil by comprehensive two-dimensional gas chromatography. J Chromatogr A 1334: 101-111.

21. Biedermann M, Barp L, Kornauth C, Würger T, Rudas M, et al. (2015) Mineral oil in human tissues, Part II: Characterization of the accumulated hydrocarbons by comprehensive two-dimensional gas chromatography. Science of the Total Environment 506: 644-655.

22. Muscalu AM, Edwards M, Górecki G, Reiner EJ (2015) Evaluation of a single-stage consumable-free modulator for comprehensive two-dimensional

gas chromatography: Analysis of polychlorinated biphenyls, organochlorine pesticides and chlorobenzenes. J Chomatogr A, p: 13.

23. Ouyang X, Leonard P, Legler J, van der Oost R, de Boer J, et al. (2015) Comprehensive two-dimensional liquid chromatography coupled to high resolution time of flight mass spectrometry for chemical characterization of sewage treatment plant effluents. J Chromatogr A 1380: 139-145.

24. Biedermann M, Grob K (2015) Comprehensive two-dimensional gas chromatography for characterizing mineral oils in foods and distinguishing them from synthetic hydrocarbons. J Chromatogr A 1375: 146-153.

25. Swigert JP, Lee C, Wong DCL, White R, Scarlett AG, et al. (2015) Aquatic hazard assessment of a commercial sample of naphthenic acids. Chemosphere 124: 1-9.

26. Alam MS, Stark C, Harrison RM (2016) Using Variable Ionization Energy Time-of-Flight Mass Spectrometry with Comprehensive GC×GC To Identify Isomeric Species. Anal Chem 88: 4211-4220.

27. Koo I, Kim S, Zhang X (2013) Comparative analysis of mass spectral matching-based compound identification in gas chromatography-mass spectrometry. J Chromatogr A 1298: 132-138.

28. KrupÄ Ãk J, Gorovenko R, SpÃ¡nik I, Sandra P, Armstrong DW (2013) Flow-modulated comprehensive two-dimensional gas chromatography with simultaneous flame ionization and quadrupole mass spectrometric detection. J Chromatogr A 1280: 104-111.

29. Marney LC, Siegler WC, Parsons BA, Hoggard JC, Wright BW, et al. (2013) Tile-based Fisher-ratio software for improved feature selection analysis of comprehensive two-dimensional gas chromatography -time-of-fl ight mass spectrometry data. Talanta 112: 887-895.

30. Mitrevski B, Marriott PJ (2014) Evaluation of quadrupole-time-of-flight mass spectrometry in comprehensive two-dimensional gas chromatography. J Chromatogr A 1362: 262-269.

31. Klee MS, Cochran J, Merrick M, Blumberg LM (2015) Evaluation of conditions of comprehensive two-dimensional gas chromatography that yield a near-theoretical maximum in peak capacity gain. J Chromatogr A 1383: 151-159.

32. Li D, Dück R, Schmitz OJ (2014) The advantage of mixed-mode separation in the first dimension of comprehensive two-dimensional liquid-chromatography. J Chromatogr A 1358: 128-135.

33. Barcaru A, Anroedh-Sampat A, Janssen HG, Vivo-Truyols G (2014) Retention time prediction in temperature-programmed, comprehensive´ two-dimensional gas chromatography: modelling and error assessment, Journal of Chromatography A 1368: 190-198.

34. Davis JM, Stoll DR (2014) Likelihood of total resolution in liquid chromatography: evaluation of one-dimensional, comprehensive two-dimensional, and selective comprehensive two-dimensional liquid chromatography. J Chromatogr A 1360: 128-142.

35. Stoll DR, Wang X, Carr PW (2008) Comparison of the practical resolving power of one- and two-dimensional high-performance liquid chromatography analysis of metabolomic samples. Analytical Chemistry 80: 268-278.

36. Fairchild JN, Horvath K, Gooding JR, Campagna SR, Guiochon G (2010) Two-dimensional liquid chromatography/mass spectrometry/mass spectrometry separation of water-soluble metabolites. J Chromatogr A 1217: 8161-8166.

# 2

# Features of N-Glycosylation of Immunoglobulins from Knockout Pig Models

Marjorie Buist[1], Emy Komatsu[1], Paul G Lopez[1], Lauren Girard[1], Edward Bodnar[1], Apolline Salama[2,3], David H Sachs[4], Cesare Galli[5,6,7,8], Andrea Perota[5], Sophie Conchon[2], Jean-Paul Judor[2], Jean-Paul Concordet[9], Giovanna Lazzari[5,6], Jean-Paul Soulillou[2,8] and Hélène Perreault[1]*

[1]Department of Chemistry, University of Manitoba, Canada
[2]INSERM UMR 10-64, Institut de Transplantation Urology Nephrology (ITUN), Université de Nantes, France
[3]Xenothera, Nantes, France
[4]Massachusetts General Hospital, Harvard University, Cambridge, MA, USA
[5]Avantea Laboratory of Reproductive Technologies, Cremona, Italy
[6]Avantea Foundation, Cremona, Italy
[7]Department of Veterinary Medical Sciences, University of Bologna, Ozzano Emilia, Italy
[8]Translink Framework Program (FP7), Padova, Italy
[9]Université Paris Descartes, Paris, France

## Abstract

For the first time, the N-glycosylation patterns of immunoglobulin G (IgGs) isolated from the serum of two varieties of knockout pigs (lacking N-glycolylneuraminic acid (Neu5Gc) and/or α 1,3 galactose) were examined for the presence of potential glycan xenoantigens and compared to N-glycosylation patterns obtained for wild-type (WT) pig IgGs. Glycopeptide analysis was chosen over glycan release, as protein-A eluates from pig serum may contain IgA and IgM as shown previously. The experiments focused on the analysis of tryptic glycopeptides EEQFNSTYR and AEQFNSTYR from IgGs, and excluded IgA and IgM, in which N-glycosylated peptides have different sequences and masses. WT pig IgG glycopeptides showed the presence of N-glycolylneuraminic acid (Neu5Gc) and absence of N-acetylneuraminic acid (Neu5Ac). Released glycans from the protein-A eluate, however, showed the presence of both types of sialic acids, allowing Neu5Ac to be attributed to IgA and/or IgM. The WT IgG samples also showed the presence of glycans that could by composition have been α-galactosylated, but treatments with α- and β-galactosidases produced inconclusive results as to the linkage nature of the terminal Gal residues. Single knockout (α-Gal transferase) pig IgG was shown to contain Neu5Gc residues, and there was a definite absence of α-Gal. Double knockout pigs (DKO for α-Gal transferase and cytidine monophosphate-A-acetylneuraminic acid hydroxylase (CMAH)) showed the definite absence of α-Gal and Neu5Gc. Instead of the latter, Neu5Ac residues were observed. Further investigation into the sialylation patterns of WT and DKO pig IgGs consisted of esterifying the glycopeptides to allow the detection and differentiation of α-2,3 and α-2,6 sialic acid-galactose linkages. Fucosylation levels were also compared between IgG species.

**Keywords:** Immunoglobulin; Knockout pig; Xenoantigen; MALDI-ToF-MS; Sialylation; Galactosylation

## Introduction

Animal biological products offer possible clinical opportunities such as xenotransplantation in order to remedy to the frequent shortage of human organs. In the clinical arena, animal derived engineered tissues such as tendons [1], scaffolds [2], heart valves [3] or even polyclonal IgGs [4] have been designed and used. Although pigs have been considered as candidates of choice for these purposes, several immunological challenges still create obstacles to the grafting processes and cause potential concerns where animal derived products are used in humans. Indeed, a strong antibody-mediated response shortens the lifetime of xenografts [5], and this is also observed following grafting of engineered pig skin [6] or infusion of foreign immunoglobulins (IgGs) [7,8]. Genetically engineered donor pigs have thus been designed in order to eliminate the expression of xenogenic antigens α1,3 galactose (αGal) [9] and/or N-glycolylneuraminic acid (Neu5Gc) [10], found in glycoproteins and other glycoconjugates expressed in wild-type animals and considered as major xenoantigenic barriers [10]. A recent report showed that representative glycoproteins from wild-type pigs, IgGs, contain Neu5Gc, although no αGal was detected with the methods used [11].

Information specific to these IgGs is important, as clinical applications of xenotransplantation are not restricted to using organs or tissues but could also concern the use of specific molecules such as IgGs. It is expected that modifying the glycans on these antibodies can reduce the immunogenicity of polyclonal IgGs [8] which still can modify the

course of Ebola infection in guinea pigs in passive immunotherapy [4]. Indeed, by knocking-out both the genes responsible for the expression of α-galactosyltransferase (GT) and cytidine monophosphate-A-acetylneuraminic acid hydroxylase (CMAH), the latter's function being to add a glycolyl to N-acetylneuraminic acid (Neu5Ac) [12], it is highly likely that these IgGs will have a much lower immunogenic potential [10], despite the fact that they can still prolong survival of EBOV-infected guinea pigs as well as decrease the extent of EBOV replication in these animals [4].

Single α-galactosyltransferase knock-out (GTKO) pig models have also been used in different studies (reviewed in [13]). As there is still antibody-mediated xenograft rejection of organs from GTKO pig organs in non-human primates, this rejection can be directed to non-Gal pig proteins and carbohydrate antigens, a situation likely predictable if such grafts would have been done in humans. The analysis of glycans from pig GTKO tissues did not result in the identification

*Corresponding author: Hélène Perreault, Department of Chemistry, University of Manitoba, 144 Dysart Road Winnipeg, Manitoba Canada R3T
E-mail: 2N2Helene.Perreault@umanitoba.ca

of new antigens, however high levels of *N*-glycolylneuraminic acid (Neu5Gc) were detected in glycoproteins and glycolipids [13].

The purpose of this article is to characterize the *N*-glycosylation of IgGs isolated from the serum of double KO pigs (DKO, GT+CMAH KO) and GTKO pigs. This has not been reported before. Mass spectrometry is used as the main detection method, after extensive sample preparation as discussed herein. Although Burlak and co-workers profiled the serum *N*-glycome of these same two types of KO pigs [14], information specific to IgGs is also of primary importance given their potential utilisation in clinical applications, as mentioned above [4]. For whole serum *N*-glycomes, it was highlighted that DKO pig glycoproteins had more mannosylated, xylosylated, core-fucosylated and truncated *N*-glycans than domestic WT pigs [14]. Reflection of these findings on IgGs isolated from pig serum may bring further insight on the immunogenic properties of these modified antibodies.

## Experimental

### Materials

Wild-type pigs of the strain Landrace × Large White of 25 kg were used (EARL du Pont Romain, Surzur, France) and were housed at the Large Animal Facility of the INSERM UMR1064 (Nantes University Hospital, France, agreement number: D44011). Whole blood was sampled, and serum was stored at -80°C until use for purification. Blood samples (100 mL) were also obtained from male GTKO and DKO pigs (14 months, weighing about 102 kg) [8]. All animal procedures were approved by the local ethic committee and carried out in accordance to DGL 116/92 for the Italian regulation. Trypsin Gold (MS Grade) was purchased from Promega (Madison, WI). Ammonium bicarbonate, trifluoroacetic acid (TFA), 2,5-dihydroxybenzoic acid (DHB), dithiothreitol (DTT) and iodiacetamide (IA) were purchased from Sigma-Aldrich (St. Louis, MO). Acetic acid (AcOH) was purchased from Fisher Scientific (Ontario, Canada), acetonitrile (ACN) was bought from EMD Millipore (Dermstadt, Germany) and ethanol (EtOH) was purchased from Commercial Alcohols (Ontario, Canada). C-18 cartridges were purchased from Phenomenex (Torrance, CA). Galactosidase α1-3,4,6 was obtained from New England Biolabs (Whitby, ON) and galactosidase β from bovine testes was obtained from Sigma-Aldrich.

**Purification of porcine IgG:** IgGs were purified from blood on a Protein-A column (high performance Sepharose™ (GE Healthcare)) using a low pressure chromatograph and 280 nm UV detection. The chromatograph allowed to record pH and conductivity. The IgGs were eluted with a solution of 0.1 M citric acid (pH 3), followed by immediate pH neutralization of the eluate to pH 7 - 7.4 with a solution of 1 M TRIS at pH 8. IgGs were then dialyzed against PBS 1X and their amount was assessed by UV spectrophotometry at 280 nm. In each case the final IgG amount was about 800 mg.

**Tryptic digestion of IgGs:** Digestion was conducted without prior reduction and alkylation. The Ab (75 µg) was reconstituted in 200 µL of 50 mM ammonium bicarbonate. Trypsin (1.5 µg) was added and digestion proceeded at 37°C for ~18 h.

**HPLC fractionation of tryptic digestion mixtures:** The digestion mixtures were injected on a Synergi C18 polar column (Phenomenex, Torrance, CA), and eluted with a gradient from 0 to 30% of acetonitrile in water at a flow rate of 0.25 mL/min. The HPLC system used was a Waters1525 binary pump equipped with a Waters 2707 autosampler and a Waters 2998 photodiode array detector. Fractions (2-min each)

were collected, dried and resuspended in 3:7 ACN-water with 0.1% TFA for MALDI-MS analysis.

**Alpha- and beta galactosidase digestions:** For α1-3,4,6 galactosidase digestion, glycopeptides from HPLC fractionation (250 ng) were dried and resuspended in 15 µL of water, to which were added 3 µL of sodium phosphate buffer at pH 6, 3 µL of 1 mg/mL bovine serum albumin solution, and 3 µL of the enzyme solution provided by the manufacturer. The mixture was incubated for 17 h at 37°C, and digestion products were cleaned by HPLC using the method described above. For digestion with β-galactosidase, glycopeptides (250 ng) were resuspended in 13 µL of water, to which were added 12 µL of 100 mM sodium citrate/phosphate buffer at pH 4 and 6 µL of the enzyme solution provided by the manufacturer. The incubation time was 17 h at 37°C. Digestion products were cleaned by HPLC.

**PNGase digestion:** This procedure was initiated by the reduction and alkylation of 100 µg of antibody protein-A eluate. The protein sample was reconstituted in 10 µL of 50 mM ammonium bicarbonate buffer (pH 7.8). Dithiothreitol (DTT, 50 µL, 10 mM in buffer) was added and reduction was allowed to proceed at 50-60°C for 1 h. After cooling, iodoacetamide (IA, 40 µL, 10 mM in water) was added and the mixture was stored in the dark for 30 min for alkylation. The sample was then cleaned and desalted on a C18 cartridge, which was conditioned with 5 × 1 mL of ACN, then 5 × 1 mL of $H_2O$. The protein sample was loaded and desalted with 5 × 1 mL of $H_2O$. The protein was then eluted with 1.5 mL of 50:50 ACN:$H_2O$+0.1% TFA, collected in an Eppendorf tube, and reconstituted in 40 µL of buffer. For PNGase F digestion, the enzyme (4 µL, at 10 units/µL) was added, and digestion proceeded 37°C for 18 h. Following the digest, the glycans were separated from the deglycosylated protein on a C18 cartridge. The column was conditioned with 5 × 1 mL of ACN+0.1% TFA and then with 5 × 1 mL of $H_2O$+0.1% TFA. The sample was loaded and glycans were eluted with 3 mL of $H_2O$+0.1% TFA and collected in two fractions. The glycan samples were resuspended in 20 µL of $H_2O$.

**Preparation of samples for MALDI-MS analysis:** Fractions in 0.1% TFA in 30:70 ACN:water mixed with DHB matrix saturated solution in the same solvent at a 1:1 ratio. This mixture (1 µL) was spotted onto the stainless steel target and allowed to dry. It was estimated that for 100 µg of antibody, a maximum of 1.8 µg of glycopeptides was obtained, divided in 10 fractions, giving an average of ca. 0.18 µg per fraction.

**Mass spectrometry:** All MS analyses were performed on an UltrafleXtreme™ mass spectrometer (Bruker Daltonics, Bremen, Germany) equipped with LID-LIFT™ technology for tandem MS (MS/MS) experiments. The following peptide calibration mixture was used (American Peptide Company, Vista, Ca): Bradykinin(1-7) 757.3992; angiotensin II 1046.542; angiotensin I 1296.685; substance P 1346.735; bombesin 1619.822; ACTH clip (1-17) 2093.086; ACTH clip (18-39) 2465.198; ACTH (1-39) 4539.267, where ACTH=adrenocorticotropic hormone and numbers are calculated *m/z* values of [M+H]⁺ ions. For sample preparation of this calibrant, 1 µL of a solution containing 0.125 µg/µL was mixed in the MALDI target with the sample volume of DHB matrix saturated solution. All experiments presented in this report were conducted in positive ionization reflector mode. As this study focuses on qualitative profiling, the mass accuracy was of the order of 150-200 ppm.

## Results and Discussion

A previous study of IgG heavy chain glycosylation in WT pigs [11] revealed the presence of Neu5Gc on Fc glycans, whereas no α-galactosylated residues were identified, using a glycopeptide-based approach. The Asn-297 site tends to be frequently designated as the only glycosylation site in IgGs, however in as least 10-15% of polyclonal antibodies the variable region is also glycosylated [15]. Some studies have shown even higher glycosylation rates in the variable portion of polyclonal IgGs with up to to 30-40% [16,17]. Plomp et al. recently reported the detection of O-glycans in the hinge region of human IgG3 [18], and previous IgG glycopeptide studies (e.g., Ref. [11,19]) have found evidence of N-glycosylation in the variable region. Particular to pig IgG, no O-glycosylation was found in a 2014 study [20], and no detectable O- glycopeptides were listed in heavy chain tryptic digestion products [11]. As the N-linked glycans of serum proteins are thought to reflect organism-wide glycosylation patterns [14], the present study focuses on the characterization of only the Fc N-glycans of two KO pig models as compared to WT.

### Glycoproteomic approach

**Study of N-glycopeptides from the Fc portion:** Figure 1 shows a comparison between WT, GKO and DKO IgG N-glycans measured using glycopeptide profiling. The polyclonal nature of these IgG samples reflects itself in the observation of two main peptide chains bearing the N-glycans, EAQFNSTYR (1) and EEQFNSTYR (2), for which all equivalent glycoforms (letters referring to Figure 2) are separated by 58 m/z units, i.e., the difference between aspartic acid E and alanine A. These chains had been identified and sequenced in a previous report [11]. The relative abundances of these two chains will not be discussed, as it was shown that within the same race of WT pigs, different animals produced different ratios of EAQFNSTYR (subtype IgG6a) and EEQFNSTYR (subtypes IgG1a, IgG1b, IgG2a, IgG4a, IgG4b, IgG5a, IgG6b) in their IgG [11,21]. IgG subtypes and the deficiency/proficiency of, as well as their immunogenic and binding properties have been well characterized in humans [15], but not in pigs. Two other isobaric amino acid sequences could have been expected in this sample, i.e., EGQFNSTYR for subtype IgG2b and EEQFNSSYR for IgG 5b [21], however they were not observed.

In Figure 1a, the main glycoforms observed are peptides bearing biantennary fucosylated glycans labelled 1D, 2D, 1F, 2F, 1G and 2G, with the number referring to the peptide chain and the letters, to the structures of Figure 2. Some minor glycoforms include xenoantigen Neu5Gc: 1J-2J, 1L-2L, and 2O. The latter (2O) could potentially include αGal, as well as glycoform 2H. These structures had not been reported in a first study [11] most possibly due to sample preparation procedures. The IgG chains had been first separated on gel followed by in-gel tryptic digestion, whereas in the present study the whole protein-A IgG eluate was directly subjected to tryptic digestion in solution. This method allowed to recover more material and hence the observation of these signals. However, repeated digestions with α-galactosidase did not modify species 2H and 2O. With β-galactosidase, both 2H and 2O were partially hydrolyzed to lower glycoforms, but the same situation was observed for peptides bearing F and G. Figure S1 shows the spectra before and after galactosidase treatments. These experiments with exoglycosidases therefore remained inconclusive. Other minor glycoforms found in Figure 1a are glycopeptides with truncated glycans A and B, and afucosylated glycans C and E.

Figure 1b shows the N-glycoforms obtained from the IgG of a GTKO pig. There is no significant variation in ions observed, for instance 1E and 2E are not as abundant as in Figure 1a, and species containing Neu5Gc are also in lower abundance. Peptides with glycans H and O were not detected.

The last spectrum (Figure 1c) corresponds to the tryptic N glycopeptides from the IgG of DKO pigs. A third peptide is also observed with sequence PKEEQFNSTYR (3), i.e., formed from a tryptic cleavage before proline, which is not common but has been reported [22]. Peaks corresponding to glycoforms of this peptide appear at m/z 2843 and 3005 and are labelled 3D and 3F. Of interest was that in the absence of Neu5Gc, replacement by Neu5Ac verified that no CMAH enzyme was available to convert Neu5Ac into Neu5Gc. This validated the DKO model, and indicated that no potentially diet derived Neu5Gc were present on the IgG from a classical pig diet. The spectrum emphasizes the sialylation of glycoforms, in this case 1I-2I, 1K-2K and 2M when referring to Figure 2.

Figure 3 compares the tandem mass spectra obtained from the [M+H]+ precursor ions of glycoforms 2K and 2L. The prevalent feature of each spectrum is the loss of sialic acid (-307 for Neu5Gc, -291 for Neu5Ac). Otherwise, characteristic P+84 ions (where P=bare peptide) give access to the mass of the peptide [23] and thus to the mass of glycan component. These MS/MS spectra help confirm the presence of NeuGc in WT pig IgG and of NeuAc in DKO pig IgG. It was not possible to perform MS/MS for the GTKO pig IgG sample due to weakness of the signal.

Although Burlak et al. observed both Neu5Ac and Neu5Gc in the glycome of domestic pigs, the present study and previous work [11] show that it is not the case for IgG by itself. Burlak's study also reported mostly Neu5Ac in DKO pigs with one instance of Neu5Gc being present [14]. In this study of glycopeptides, IgG from DKO pigs is shown to contain Neu5Ac exclusively. In humans, it has been shown that anti-Neu5Gc antibody is produced and that anti- Neu5Gc antibody response is induced after WT pig tissue grafts [13]. There is still uncertainty however about the pathogenicity of the human anti-Neu5Gc antibody xenografts using pig organs. Non-human primates used as models for such transplantations have shown anti-Gal immunity but, as expected, no production of anti-Neu5Gc antibodies [13]. In these models, post-transplantation antibody induction is rather directed to pig endothelial cells proteins and to a glycan due to the pig B4GALNT2 gene [24]. If this type of glycan is to be found in pig IgG, it would be on an O-site, in the hinge or Fab region, which were not specifically studied in this report. A more detailed study of Fab glycans, using papain to cleave the antibodies into two distinct portions [25] is underway and will be the object of a future manuscript.

Levels of fucosylation were compared in triplicate between WT and DKO samples. They were determined to be 14 ± 4% (WT) and 9 ± 1% (DKO), which could be determinant in explaining the observed higher toxicity dependence of DKO pig IgG vs WT [26,27]. On the other hand, DKO pig IgG has shown a higher complement (C1q) binding than WT pig IgG using Biacore™ technology [27] which, according to the template of interactions proposed by Butler et al. [21], could relate to a higher IgG3 content in DKO samples. This could not be determined in this current glycopeptide study, however experiments using quantitative markers are underway to determine IgG3 in WT and DKO IgG samples.

There was thus no major variation in the predominant glycan structures observed in IgGs from WT, GTKO and DKO pigs in that they were mostly biantennary, core-fucosylated oligosaccharides. No significant increased abundance of mannosylated species was detected, as signaled by Burlak et al. in the glycome of DKO pig serum [14]. High mannose species (Man)$^5$, monitored at m/z 2331 and 2389 for Peptides 1 and 2, were not observed in Figure 1a, but close inspection of an exploded view of the spectrum in (c) revealed a trace presence of these

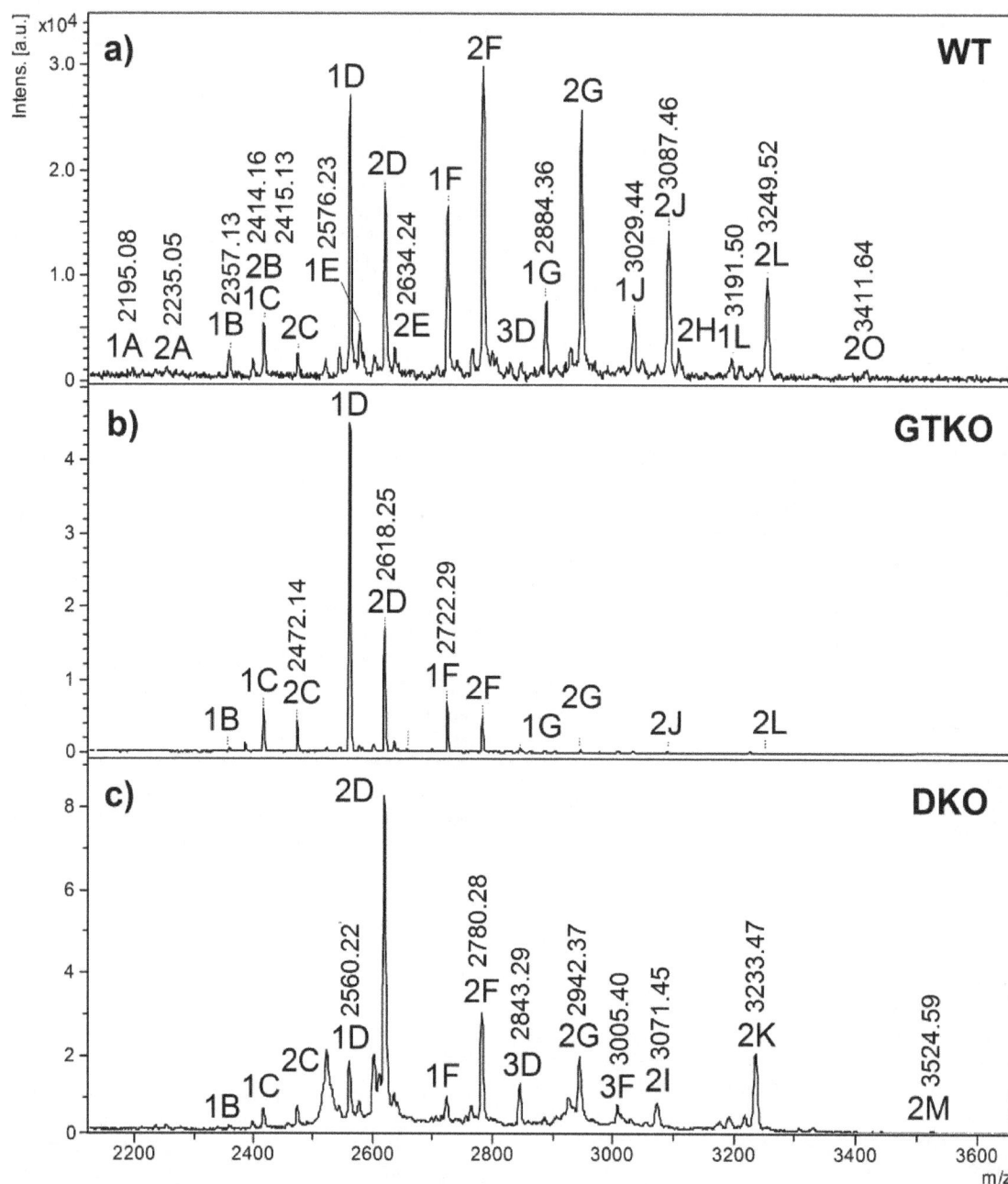

**Figure 1:** Linear, positive mode MALDI-TOF-MS *N*-glycopeptide profiles for wild-type pig IgG (a), GTKO pig IgG (b), and DKO pig IgG (c). Peptide numbering: 1: EAQFNSTYR; 2: EEQFNSTYR; 3: PKEEQFNSTYR. Letters refer to glycans shown in Figure 2.

glycoforms. Glycoforms Man6 to Man9 were not observed at all in all spectra of Figure 1. As for truncated glycan forms noted by Burlak et al. in GTKO and DKO glycomes [14], they were not detected in this study, nor were xylose residues.

Another interesting aspect of terminal sialic acids is the linkage patterns to distal galactoses in the α2,3- or α2,6- position [28]. Sialylation has often been shown to enhance the half-lifes of glycoproteins, as sialic acid caps terminal galactose residues that are otherwise recognized by hepatic asialoglycoprotein receptors (ASGPR) [29]. It has been suggested that α 2,3 sialylation provides better half-life extension, as ASGPR recognizes Sia α 2,6Gal and Sia α 2,6GalNAc residues as well

as Gal and GalNAc moieties [30,31]. In IgG samples, esterification of sialylated glycopeptides has allowed to differentiate these types of linkages [32,33], and this method was applied to asialylated pig IgG glycopeptides in a previous study [11], although the main purpose was to establish that peptides chains were EEQFNSTYR and EAQFNSTYR, E being reactive to esterification but not A. Glycopeptides from WT and DKO porcine IgG were subjected to this reaction and results are shown in Figure 4. Non-sialylated glycopeptides corresponding to Sequence 1 (EAQF<u>N</u>STYR) saw their masses go up by 10 units (+28 for ethyl substitution, -18 for loss of $H_2O$) [11,32]. Those corresponding to Sequence 2, EEQFNSTYR, showed an increment of 38 units (+(2

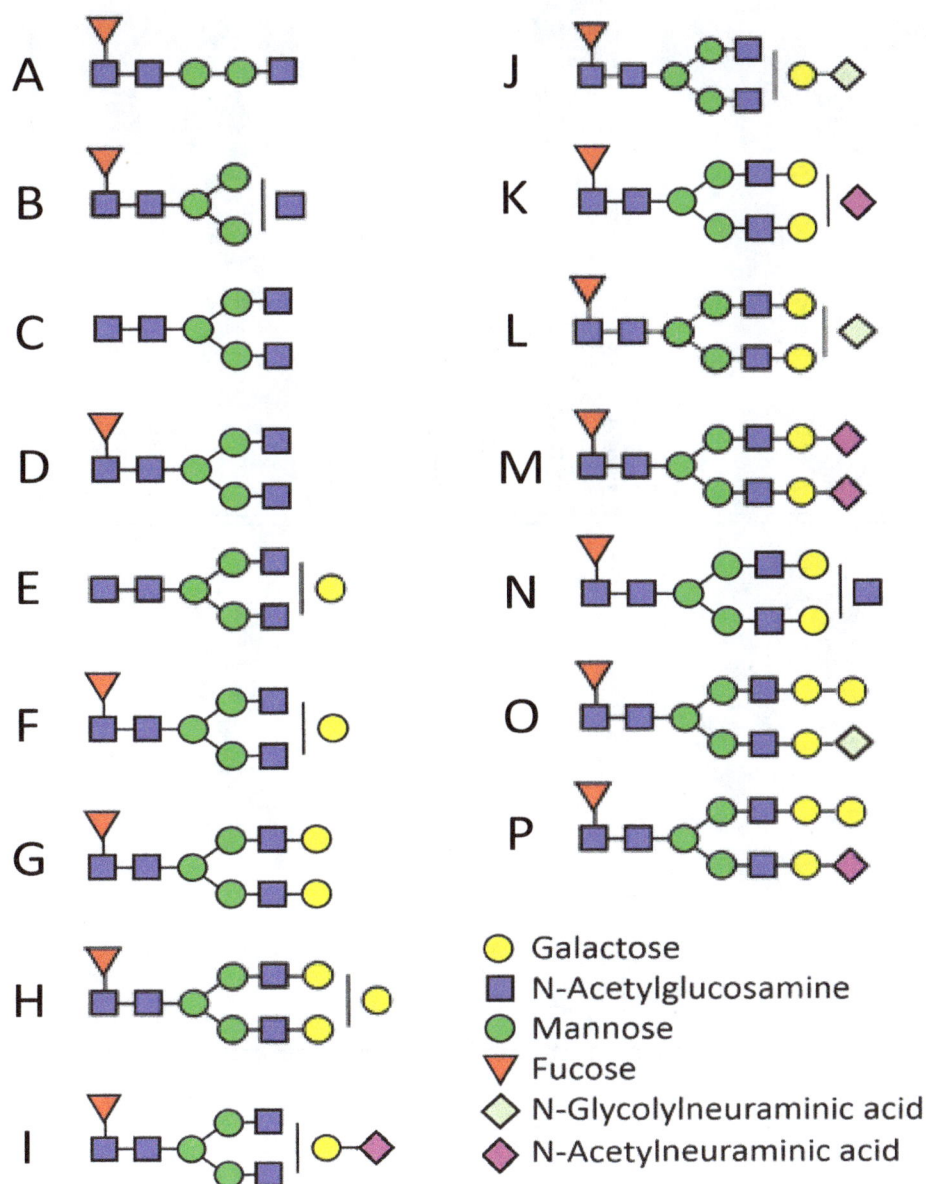

**Figure 2:** *N*-glycans detected on tryptic digest glycoforms of pig IgGs discussed in this study. Letters are used as labels in Figures 1, 4 and 5.

× 28)-18) [11,32]. Sialylated glycopeptides showed either an extra increment of 28 (335 instead of 307 for Neu5Gc, 319 instead of 291 for Neu5Ac) due to esterification of the carboxyl group or a loss of $H_2O$ (273 instead of 291 for Neu5Ac, 289 instead of 307 for Neu5Gc) due to lactonization. These mass differences are indicated by red numbers on the spectra. Some studies have pointed out that lactonization (-18) occurs for Sia α 2,3Gal linkages, while esterification takes place for Sia α 2,6Gal linkages, based on methods relying on methyl- and ethyl-esterification of carboxyl groups in glycans released from glycoprotein [28,32-35]. From this approach, used here for the first time with demonstrated lactonization of NeuGc, it can be observed in Figure 4 that IgGs from WT and DKO pigs exhibit both α 2,3 and α 2,6 linkages, for Neu5Gc (WT) and Neu5Ac (DKO). It is possible to assign semi-quantitative figures to the abundances of each type by comparing peak areas. Overall for WT the α 2,6 to α 2,3 ratio was 1.7 and for DKO, 1.3,

showing a small predominance of α 2,6 linkages. The half-life of the DKO pig IgG has been measured as normal (170 h) vs. WT IgG [27], which is consistent with these MS results.

These types of linkages were found to be important with respect to cell infectivities of human parainfluenza virus type 1 and type 3: Type 1 recognized only α2,3 sialic acid linkages as viral receptors, while type 3 recognized both α2,3 and α2,6 sialic acid linkages [36]. As the terminal and only (negatively) charged residue in Fc glycans, it has been suggested that sialic acids have the most effect on the Fc domain structure [15,37].

**Glycomic approach**

**Study of N-glycans released from protein-A eluate:** At the glycomic level, mass spectra of native glycans detached from full

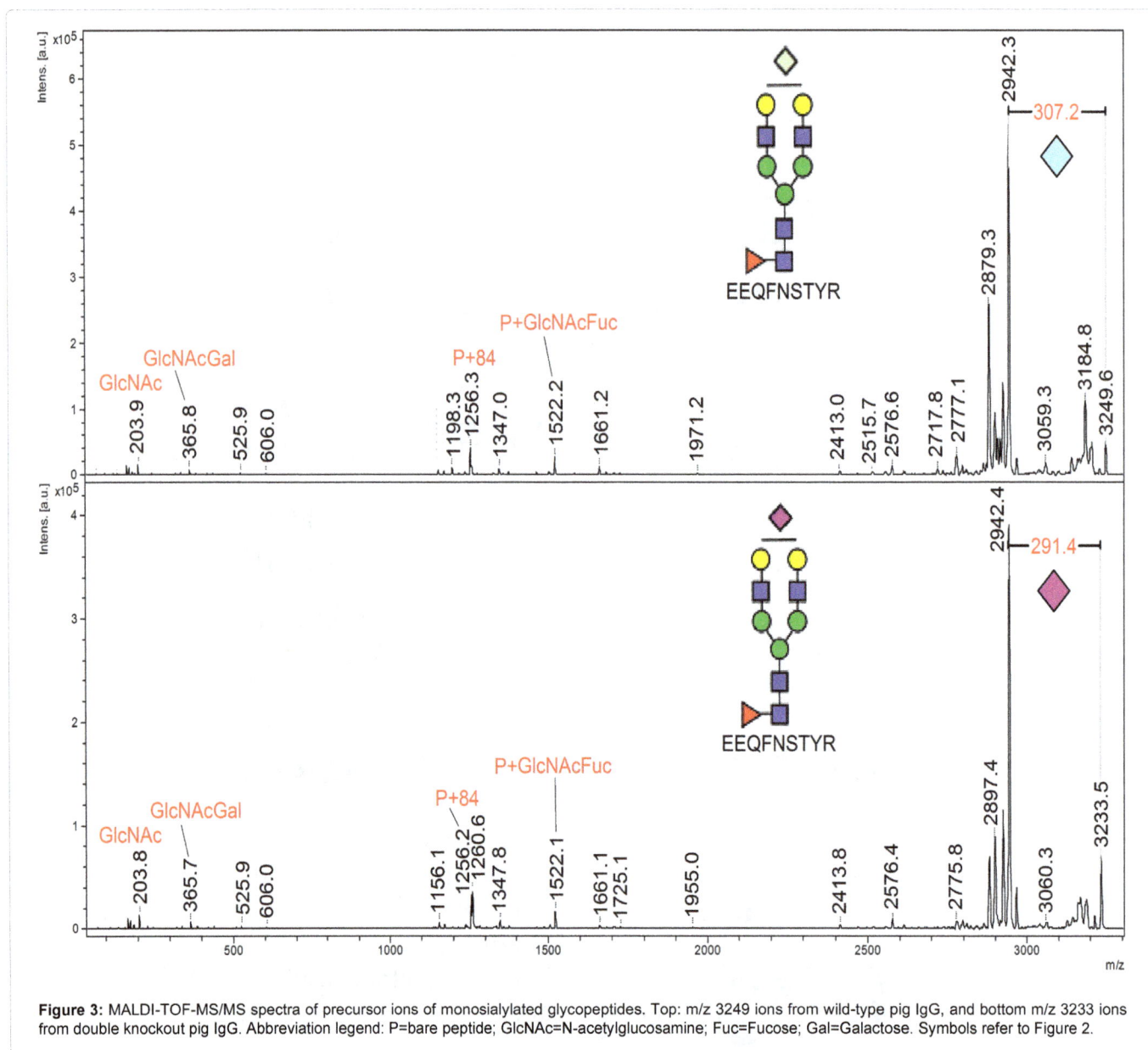

**Figure 3:** MALDI-TOF-MS/MS spectra of precursor ions of monosialylated glycopeptides. Top: m/z 3249 ions from wild-type pig IgG, and bottom m/z 3233 ions from double knockout pig IgG. Abbreviation legend: P=bare peptide; GlcNAc=N-acetylglucosamine; Fuc=Fucose; Gal=Galactose. Symbols refer to Figure 2.

antibodies (WT and DKO) with PNGaseF are shown in Figure 5. These glycans were detached from the whole protein-A eluate, which also contains IgM and IgA [11], and thus are not uniquely related to the heavy chains of IgG. In Figure 5, neutral glycans are detected as $[M+Na]^+$ ions, whereas monosialylated species produce $[M-H+2Na]^+$ signals, and disialylated, $[M-2H+3Na]^+$ [38]. The observation of glycans containing Neu5Ac in the WT (m/z difference of 313), not detected on specific IgG glycopeptides from the same antibody sample (Figure 1a), suggests that these glycans are foreign to the heavy chain of IgG. A study by Marco-Ramell et al. reported the presence of Neu5Ac in IgA glycans, but also in IgG [20], possibly released from the IgG Fab light chain portion. Fab vs. Fc glycans in the same antibody encompass some differences, especially in increased levels of bisection, galactosylation, and thereby sialylation [39,40]. For instance disialylation is rarely observed in Fc glycans but more frequently in Fab glycans. Reduced levels of fucosylation have also been observed from Fab to Fc glycans

[39,40]. Comparing the glycans detected in Figure 5 with heavy chain-specific glycoforms of Figure 1 leads to the observation of disialylated (di-NeuAc, species M) glycans in the WT sample of Figure 5, which were not present in Figure 1. Glycans M were present in both Figures 1 and 5 in the case of the DKO sample. Significant changes in the WT sample between relative abundances of glycoforms/glycans suggest that a significant proportion of released glycans came from either the Fab domain, or IgM/IgA. There was more consistency for the DKO samples analyzed in Figures 1 and 5.

MS/MS spectra were recorded to verify the presence of both Neu5Ac and Neu5Gc in the sample, as Figure S2 indicates. The mass selection tool of the UltraFleXtreme instrument was not able to completely isolate each of the m/z 2212 and 2228 precursor ions, resulting in overlapping MS/MS spectra which however show the respective loss of 313 (Neu5Ac+Na) and 329 (Neu5Gc+Na) from these two precursor masses.

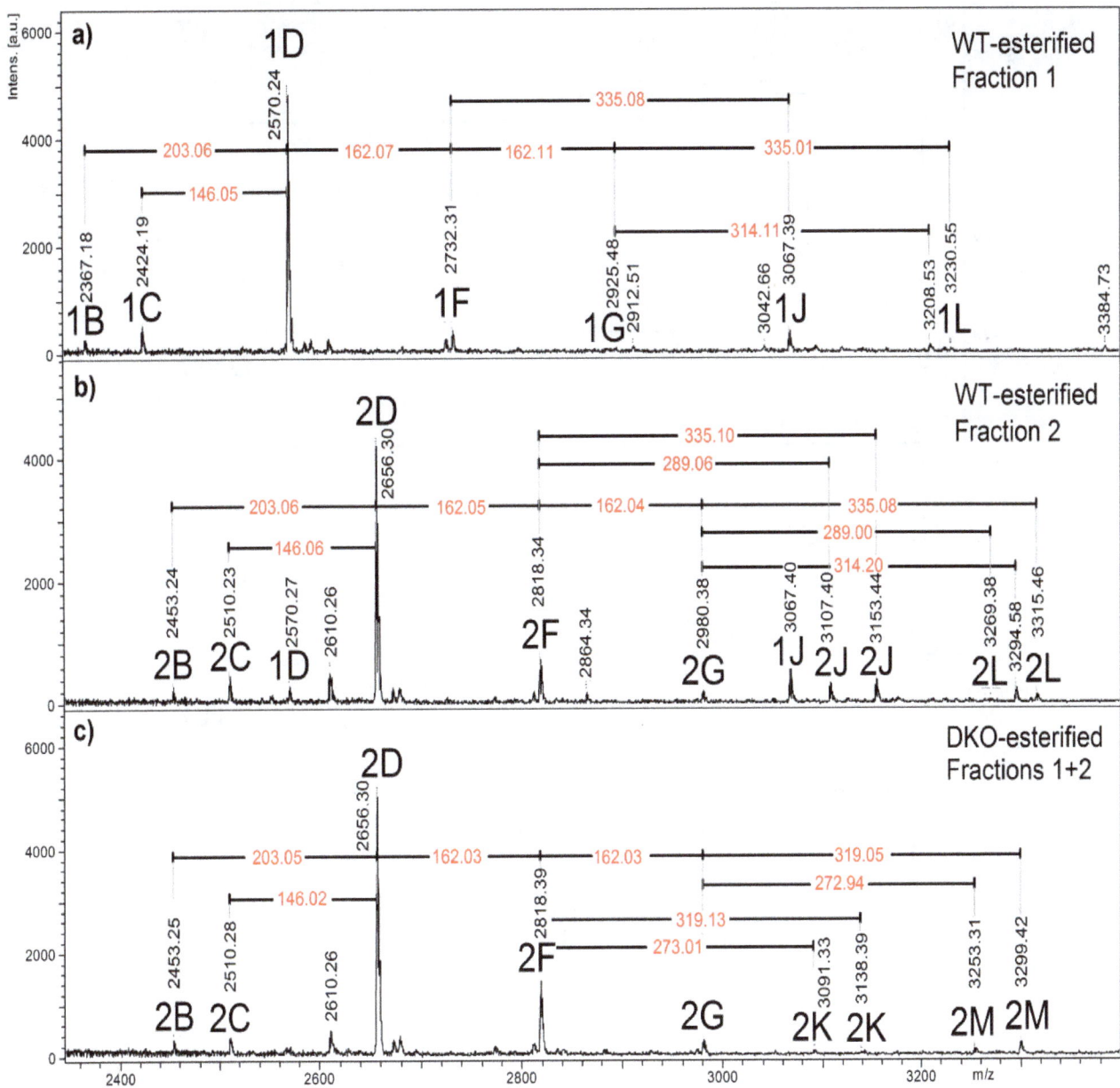

**Figure 4:** MALDI-TOF-MS spectra of ethyl-esterified tryptic glycopeptide HPLC fractions from a) and b) WT pig IgG and c) from DKO pig IgG. Peak labels refer to Figure 2. Numbers in red refer to mass differences between glycoforms: 162=hexose, 203=N-acetylhexosamine, 273=lactonized NeuAc, 289=lactonized NeuGc, 319=esterified NeuAc, 335=esterified NeuGc, 314=Undetermined.

Overall, results from the PNGase experiment help to emphasize the importance of glycosylation site specificity of analysis when studying glycopeptides. Where porcine IgG species contain variations of the Fc tryptic sequence EEQFNSTYR, IgA and IgM produce different tryptic peptides, according to UniprotKB accession numbers K7ZJP7 (IgM) and K7ZRK0 (IgA). The constant portion of porcine IgM has four consensus sequences for N-glycosylation which would be included in tryptic peptides ESLNISWTR and TSIVFSEIYANGTFGAR and two other longer peptides, whereas IgA's conserved part would produce a peptide of sequence LAGKPTHVNVSVVMAEAEGICY.

## Conclusions

This report gathers important information specific to polyclonal IgGs from KO pig species and makes the parallel with IgG from WT pigs. Modified glycans on these antibodies can modulate their immunogenicity in passive immunotherapy treatments. Knocking-out the gene responsible for αGal resulted in IgGs where in N-glycosylation no α Gal residues were found, and where Neu5Gc residues were present. Knocking-out both the genes for αGal and Neu5Gc produced IgGs in which no αGal is detected, and where Neu5Ac residues were detected. In WT and DKO pig IgG, an esterification reaction allowed to determine

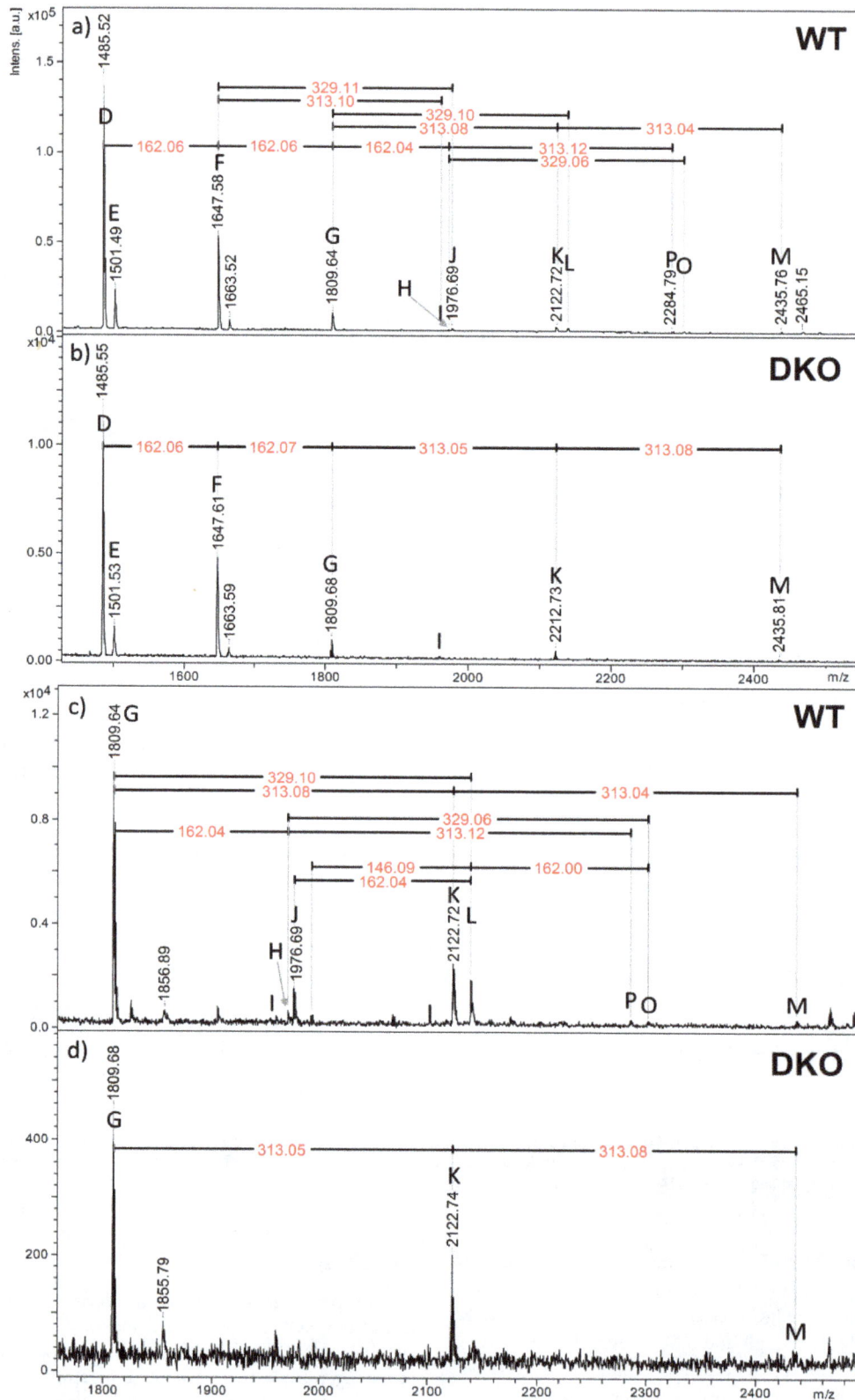

**Figure 5:** MALDI-MS spectra of glycans released from protein-A eluate of porcine serum. a) WT; c) detailed view of WT glycans; b) DKO; d) detailed view of DKO glycans. Neutral glycans are observed as [M+Na]$^+$ ions, sialylated glycans as [M-nH+(n+1)Na]$^+$, where n=number of sialic acids. Numbers in red: 146=fucose; 162=hexose; 313=NeuAc+Na; 329=NeuGc+Na.**e 1.**

the binding pattern of sialic acids, which showed a greater proportion of α2,6 binding over α2,3, which could have important biological implications. The fucosylation level was also higher in the WT than in DKO IgG, which could be significant for toxicity dependence levels. Finally, this study emphasizes the specificity of using a glycoproteomic rather than strictly glycomic approach in the characterization of these antibodies, as *N*-glycans released from whole protein-A eluates showed compositions varying from those measured on IgG *N*-glycopeptides. Future work will focus on Fab *N*- and *O*-glycans.

## Acknowledgments

This work was carried out under the European Union Seventh Framework Programme collaborative Project Translink (Grant agreement No. 603049). The authors thank the Natural Sciences and Engineering Research Council of Canada (NSERC, RGPIN/170241-2011) and the Canadian Foundation for Innovation (Grant no. 23391) for funding this research. We also thank Bernard Martinet (Inserm 1064, Nantes) for preparing the protein A purification of pig IgGs.

## References

1. Na JY, Song K, Kim S, Lee HB, Kim JK, et al. (2014) Evaluation of porcine xenograft in collateral ligament reconstruction in beagle dogs. Res Vet Sci 97: 605-610.

2. Hussein KH, Park KM, Kim HM, Teotia PK, Ghim JH, et al. (2015) Construction of a biocompatible decellularized porcine hepatic lobe for liver bioengineering. The International journal of artificial organs 38: 96-104.

3. Manji RA, Lee W, Cooper DK (2015) Xenograft bioprosthetic heart valves: Past, present and future. International journal of surgery (London, England) 23: 280-284.

4. Reynard O, Jacquot F, Evanno G, Mai HL, Salama S, et al. (2016) Anti-EBOV GP IgGs Lacking a1-3-Galactose and Neu5Gc Prolong Survival and Decrease Blood Viral Load in EBOV-infected Guinea Pigs. PLoS One 11: e0156775.

5. Vadori M, Cozzi E (2014) Immunological challenges and therapies in xenotransplantation. Cold Spring Harb Perspect Med 4: a015578.

6. Scobie L, Padler-Karavani V, Le Bas-Bernardet S, Crossan C, Blaha J, et al.(2013) Long-term IgG response to porcine Neu5Gc antigens without transmission of PERV in burn patients treated with porcine skin xenografts. Journal of immunology 191: 2907-2915.

7. Couvrat-Desvergnes G, Salama A, Le Berre L, Evanno G, Viklicky O, et al. (2015) Rabbit antithymocyte globulin-induced serum sickness disease and human kidney graft survival. J Clin Invest 125: 4655-4665.

8. GlobalSurg Collaborative (2016) Mortality of emergency abdominal surgery in high-, middle- and low-income countries. Br J Surg 103: 971-988.

9. Galili U (2005) The alpha-gal epitope and the anti-Gal antibody in xenotransplantation and in cancer immunotherapy. Immunol Cell Biol 83: 674-686.

10. Salama A, Evanno G, Harb J, Soulillou JP (2015) Potential deleterious role of anti-Neu5Gc antibodies in xenotransplantation. Xenotransplantation 22: 85-94.

11. Lopez PG, Girard L, Buist M, de Oliveira AG, Bodnar E, et al. (2016) Characterization of N-glycosylation and amino acid sequence features of immunoglobulins from swine. Glycoconj J 33: 79-91.

12. Lutz AJ, Li P, Estrada JL, Sidner RA, Chihara RK, et al. (2013) Double knockout pigs deficient in N- glycolylneuraminic acid and galactose alpha-1,3-galactose reduce the humoral barrier to xenotransplantation. Xenotransplantation 20: 27-35.

13. Byrne GW, McGregor CG, Breimer ME (2015) Recent investigations into pig antigen and anti-pig antibody expression. International journal of surgery (London, England) 23: 223-228.

14. Burlak C, Bern M, Brito AE, Isailovic D, Wang ZY, et al. (2013) N-linked glycan profiling of GGTA1/CMAH knockout pigs identifies new potential carbohydrate xenoantigens. Xenotransplantation 20: 277-291.

15. Vidarsson G, Dekkers G, Rispens T (2014) IgG subclasses and allotypes: from structure to effector functions. Front Immunol 5: 520.

16. Dwek RA, Lellouch AC, Wormald MR (1995) Glycobiology: 'the function of sugar in the IgG molecule'. J Anat 187: 279-292.

17. Wormald MR, Rudd PM, Harvey DJ, Chang SC, Scragg IG, et al. (1997) Variations in oligosaccharide-protein interactions in immunoglobulin G determine the site-specific glycosylation profiles and modulate the dynamic motion of the Fc oligosaccharides. Biochemistry 36: 1370-1380.

18. Plomp R, Dekkers G, Rombouts Y, Visser R, Koeleman CA, et al. (2015) Hinge-Region O-Glycosylation of Human Immunoglobulin G3 (IgG3). Mol Cell Proteomics 14: 1373-1384.

19. Lattová E, Kapková P, Krokhin O, Perreault H (2006) Method for investigation of oligosaccharides from glycopeptides: direct determination of glycosylation sites in proteins. Anal Chem 78: 2977-2984.

20. Marco-Ramell A, Miller I, Nobauer K, Moginger U, Segales J, et al. (2014) Proteomics on porcine haptoglobin and IgG/IgA show protein species distribution and glycosylation pattern to remain similar in PCV2-SD infection. Journal of proteomics 101: 205-216.

21. Butler JE, Wertz N, Deschacht N, Kacskovics I (2009) Porcine IgG: structure, genetics, and evolution. Immunogenetics 61: 209-230.

22. Rodriguez J, Gupta N, Smith RD, Pevzner PA (2008) Does trypsin cut before proline. J Proteome Res 7: 300-305.

23. Krokhin O, Ens W, Standing KG, Wilkins J, Perreault H (2004) Site-specific N-glycosylation analysis: matrix-assisted laser desorption/ionization quadrupole-quadrupole time-of-flight tandem mass spectral signatures for recognition and identification of glycopeptides. Rapid communications in mass spectrometry: RCM 18: 2020-2030.

24. Byrne GW, Du Z, Stalboerger P, Kogelberg H, McGregor CG (2014) Cloning and expression of porcine beta1,4 N-acetylgalactosaminyl transferase encoding a new xenoreactive antigen. Xenotransplantation 21: 543-554.

25. https://www.thermofisher.com/ca/en/home/life-science/protein-biology/protein-biology-learning-center/protein-biology-resource-library/pierce-protein-methods/antibody-fragmentation.html

26. Kanda Y, Yamada T, Mori K, Okazaki A, Inoue M, et al. (2007) Comparison of biological activity among nonfucosylated therapeutic IgG1 antibodies with three different N-linked Fc oligosaccharides: the high-mannose, hybrid, and complex types. Glycobiology 17: 104-118.

27. Salama A, Conchon S, Perota A, Martinet B, Judor JP et al., (2015) Immune phenotype and IgG characteristics of Neu5Gc and alpha-1-3-gal double knock-out pigs. Xenotransplantation 22 (S1):S2-S47 no. 348.

28. Wheeler SF, Domann P, Harvey DJ (2009) Derivatization of sialic acids for stabilization in matrix-assisted laser desorption/ionization mass spectrometry and concomitant differentiation of alpha(2 --> 3)- and alpha(2 --> 6)-isomers. Rapid communications in mass spectrometry: RCM 23: 303-312.

29. Morell AG, Gregoriadis G, Scheinberg IH, Hickman J, Ashwell G (1971) The role of sialic acid in determining the survival of glycoproteins in the circulation. J Biol Chem 246: 1461-1467.

30. Park EI, Mi Y, Unverzagt C, Gabius HJ, Baenziger JU (2005) The asialoglycoprotein receptor clears glycoconjugates terminating with sialic acid alpha 2,6 GalNAc. Proc Natl Acad Sci USA 102: 17125-17129.

31. Unverzagt C, André S, Seifert J, Kojima S, Fink C, et al. (2002) Structure-activity profiles of complex biantennary glycans with core fucosylation and with/without additional alpha 2,3/alpha 2,6 sialylation: synthesis of neoglycoproteins and their properties in lectin assays, cell binding, and organ uptake. J Med Chem 45: 478-491.

32. Gomes de Oliveira AG, Roy R, Raymond C, Bodnar ED, Venkata ST, et al. (2015) A systematic study of glycopeptide esterification for the semi-quantitative determination of sialylation in antibodies. Rapid Commun Mass Spectrom 29: 1817-1826.

33. de Haan N, Reiding KR, Haberger M, Reusch D, Falck D, et al. (2015) Linkage-specific sialic acid derivatization for MALDI-TOF-MS profiling of IgG glycopeptides. Anal Chem 87: 8284-8291.

34. Harvey DJ (2005) Fragmentation of negative ions from carbohydrates: part 3. Fragmentation of hybrid and complex N-linked glycans. J Am Soc Mass Spectrom 16: 647-659.

35. Reiding KR, Blank D, Kuijper DM, Deelder AM, Wuhrer M (2014) High-throughput profiling of protein N-glycosylation by MALDI-TOF-MS employing linkage-specific sialic acid esterification. Analytical chemistry 86: 5784-5793.

36. Fukushima K, Takahashi T, Ito S, Takaguchi M, Takano M, et al. (2014) Terminal sialic acid linkages determine different cell infectivities of human parainfluenza virus type 1 and type 3. Virology 464-465: 424-431.

37. Sondermann P, Pincetic A, Maamary J, Lammens K, Ravetch JV, et al. (2013) General mechanism for modulating immunoglobulin effector function. Proceedings of the National Academy of Sciences of the United States of America 110: 9868-9872.

38. Snovida SI, Chen VC, Krokhin O, Perreault H (2006) Isolation and identification of sialylated glycopeptides from bovine alpha1-acid glycoprotein by off-line capillary electrophoresis MALDI- TOF mass spectrometry. Analytical chemistry 78: 6556-6563.

39. Zauner G, Selman MH, Bondt A, Rombouts Y, Blank D, et al. (2013) Glycoproteomic analysis of antibodies. Molecular & cellular proteomics: MCP 12: 856-865.

40. Bondt A, Rombouts Y, Selman MH, Hensbergen PJ, Reiding KR, et al. (2014) Immunoglobulin G (IgG) Fab glycosylation analysis using a new mass spectrometric high-throughput profiling method reveals pregnancy-associated changes. Molecular & cellular proteomics: MCP 13: 3029-3039.

# Chemometrical Evaluation of Metoprolol Tartarate Enantiomers Separation Applying Conventional Achiral Chromatography

Ivkovic B[1]*, Karljikovic-Rajic K[2], Vujic Z[1] and Ibric S[3]

[1]Department of Pharmaceutical Chemistry and Drug Analysis, Faculty of Pharmacy, University of Belgrade, Serbia
[2]Department of Analytical Chemistry, Faculty of Pharmacy, University of Belgrade, Serbia
[3]Department of Pharmaceutical Technology and Cosmetology, University of Belgrade, Serbia

## Abstract

In this paper, the separation of metoprolol tartarate enantiomers using conventional achiral chromatography was followed employing experimental design $2^4$. Design was useful tool in evaluation of chromatographic behavior of enantiomers. For the separation, Supelcosil $LC_{18}$ 4.6 mm × 250 mm, 5 m particle size column was used. UV detection was performed at 275 nm. Because of the large number of factors which can influence chromatographic separation, in preliminary study *"one factor at a time"* method was applied to define values of important factors. As outputs, retention factor, selectivity factor and resolution factors were chosen. Chromatographic behavior of investigated enantiomers was affected by acetonitrile content, chiral modificator content in the mobile phase and flow rate the most, demonstrated by obtained linear models. On the basis of results from screening of experiment, factors with strong influence on the separation, were analyzed in method optimization applying response surface methodology (RSM). The appropriate region of chromatographic behavior of metoprolol tartarate enantiomers was defined using three-D graphs and analysis of variance. The methodology proposed represents an efficient approach in resolving the problem of searching for optimal HPLC chromatographic conditions via experimental design.

**Keywords:** Metoprolol tartarate; Enantiomers; RP-HPLC; Experimental design

## Introduction

Metoprolol tartarate ((R,S)–3[4–(2–methoxyethyl) phenoxy)–1–(isopropylamino) propan-2-ol] tartarate) is a $\beta_1$-adrenoreceptor antagonist ($\beta_1$-blocker) widely used in the treatment of hypertension, angina pectoris and cardiac dysrhythmias [1,2].

Most β-adrenergic antagonists (β-blockers) are therapeutically used as a racemic mixture and their enantiomers demonstrate different pharmacodynamic and pharmacokinetic properties [1,2]. The both enantiomers are reducing blood pressure, but the therapeutic effects of the (S)-enantiomer is about 100 times stronger than (R)-enantiomer. The increasing demands for the production of enantiomerically pure drugs have led to enantioselective separations becoming one of the most important analytical tasks. In this context, enantioselective HPLC is one of the most powerful and widely employed separation techniques, both for analytical and preparative purposes, as well as for research in pharmaceutical and biomedical analysis.

Separation of $\beta_1$-blocker enantiomers included conventional achiral chromatography with pre-column derivatisation [3,4]. Using different chiral stationary phases or chemically different chiral mobile phase additives, β-blockers has been recently separated [5-8]. Also, the spectrophotometric method for analysis of β-cyclodextrin/metoprolol tartarate inclusion complex was investigated [9].

The aim of our investigation was analysis of metoprolol tartarate enantiomers using achiral stationary phase with β-cyclodextrin as chiral additive in the mobile phase. The experimental design was applied as the best way to define chromatographic behavior of enantiomers. Analysis employed *"one factor at a time"* method for preliminary study, full factorial design for screening of experiment and response surface methodology (RSM) for method optimization. A literature search showed many experimental design applications in analytical method development and validation, especially in the area of separation science. Experimental design has been used for separation optimization [10-12] and for validation in RP-HPLC method [13,14]. It was used for robustness testing in RP-HPLC method [15,16] and capillary electrophoresis [17]. The methodology proposed in this paper represents novel, efficient and easily attainable approach in resolving metoprolol tartarate enantiomers using conventional achiral chromatography.

## Experimental

### Reagents and samples

All reagents used were of an analytical grade. Methanol-gradient grade (*Merck*, Darmstadt, Germany), acetonitrile (*Merck*, Darmstadt, Germany), water-HPLC grade, triethylamine (TEA) (*Merck*, Darmstadt, Germany). Beta cyclodextrine (-CD) was obtained from *Cyclolab*, Hungary. Presolol' tablets (containing 100 mg of metoprolol taratarat) were manufactured by Hemofarm d.o.o., SCG. Working standard of metoprolol tartarate was obtained from *Hemofarm* d.o.o., SCG.

### Chromatographic conditions

The chromatographic system Hewlett Packard 1100 (*Agilent*, Technologies) consisted of a HP 1100 pump, HP 1100 UV-VIS detector and HP ChemStation integrator. Separations were performed on a Supelcosil $LC_{18}$ 4.6 mm × 250 mm, 5 μm particle size column.

*Corresponding author: Branka Ivkovic, Department of Pharmaceutical Chemistry and Drug Analysis, Faculty of Pharmacy, University of Belgrade, Vojvode Stepe 450, 11221 Belgrade-Kumodraz, PO Box 146, Serbia, E-mail: blucic@pharmacy.bg.ac.rs

UV detection was performed at 275 nm. The samples were introduced through a Rheodyne injector valve with a 20 μL sample loop.

## Buffer solution

Buffer solution was prepared by adding of 2 ml of TEA to 600 ml of HPLC water, pH was adjusted in the range from 2 to 6 with glacial acetic acid.

## Solution for equlibration and storage of column

Solution was prepared in concentration of 3 mM of β-CD in water.

## Results and Discussion

In preliminary investigations, influence of different chromatographic factors on separation of metoprolol tartarate enantiomers was analyzed. As the separation was performed employing conventional achiral chromatography, beta cyclodextrine (β-CD) was added to the mobile phase as chiral modificator. Mobile phases consisted of acetonitrile and β-CD in triethylamine/glacial acetic acid buffer in different ratios. Column was stored and conditioned with solution of 3 mM of β-CD in water. In chromarography many factors can influence separation e.g., content of organic modifier in mobile phase, pH of the mobile phase, column temperature, flow rate, concentration of solute etc. In the first step of our study, pH of the mobile phase and temperature were defined using "one factor at a time" method. pH of the mobile phase (2.5; 3.0; 4.0; 5.0 and 6.0) was changed and other factors were kept at constant level. Obtained chromatograms demonstrated the best separation of enantiomers at pH 3.0 and in following investigations it remained constant. Secondly, temperature was analyzed on two levels 30°C and 40°C. The accepted separation was at 35°C column temperature.

In the second step, for the screening of experiment, full factorial design $2^4$ was chosen. Full factorial designs at two levels are mainly used for screening, that is, to determine the influence of a number of effects on a response and to eliminate those that are not significant [18]. Selected factors and their "low" (−) and "high" (+) levels are presented in Table 1. Matrix of the experiment is given in Table 2.

As outputs (Y) capacity factors ($k_1$ ($Y_1$)-retention factor for enantiomer R (+), $k_2$ ($Y_2$)-retention factor for enantiomer S (-)), selectivity factor (α–$Y_3$) and resolution factor (Rs–$Y_4$) were analyzed. The obtained results are presented in Table 3.

In experimental design for the evaluation of influence of investigated factors, on measured response, mathematical model was applied. Often form of a mathemathical model is:

$$y = b_0 + b_1 x_1 + b_2 x_2 + b_3 x_3 + \cdots + b_{N-1} x_{N-1} + b_N x_N + b_{12} x_1 x_2 + b_{13} x_1 x_3 + b_{23} x_2 x_3 + \cdots + b_{(N-1)N} x_{N-1} x_N \quad \text{(Eq. 1)}$$

where y presents the estimate response, $b_0$, is the average experimental response, the coefficients $b_1$ to $b_N$ are the estimated effects of the factors considered and the extend to which these terms affect the performance of the method is called main effect. The coefficients $b_{12}$ to $b_{(N-1)N}$ are called the interaction terms. We can see that the factorial design provides information about the importance of interaction between the factors [18]. The calculating coefficients of mathematical models for outputs are presented in Table 4.

The results showed that acetonitrile content and concentration of β-CD had the biggest influence on retention factors ($Y_1$ and $Y_2$). The flow rate and content of β-CD influenced selectivity factor ($Y_3$) and resolution ($Y_4$) the most. Concentration of metoprolol tartarate had

negligible influence on analyzed outputs and in further investigations it was kept constant.

In the third step of method optimization, three factors (content of acetonitrile, content of -CD and flow rate) were analyzed in 22 experiments. Matrix of experiment for optimization is presented in Table 5. Experimental data for outputs are presented in Table 6.

On the basis of the results, outputs $Y_1$ and $Y_4$ were chosen to analyze separation. The results for others outputs gave bad coefficient of determination ($\leq 0.5$) and they fitted badly in the obtained model. Those results could be explained with different characters of

| | Factors | Levels of the investigation | |
|---|---|---|---|
| | | (-) | (+) |
| $X_1$ | Acetonitrile (%) | 9.1 | 14.5 |
| $X_2$ | β–CD (mM) | 5.81 | 8.81 |
| $X_3$ | Flow rate (mlmin⁻¹) | 1 | 1.5 |
| $X_4$ | Concentration of metoprolol tartarate (mlL⁻¹) | 0.25 | 0.50 |

(-) and (+) are "low" and "high" levels

**Table 1:** Factor and their levels.

| Number of experiments | $X_1$ | $X_2$ | $X_3$ | $X_4$ |
|---|---|---|---|---|
| 1 | - | - | - | - |
| 2 | + | - | - | - |
| 3 | - | + | - | - |
| 4 | + | + | - | - |
| 5 | - | - | + | - |
| 6 | + | - | + | - |
| 7 | - | + | + | - |
| 8 | + | + | + | - |
| 9 | - | - | - | + |
| 10 | + | - | - | + |
| 11 | - | + | - | + |
| 12 | + | + | - | + |
| 13 | - | - | + | + |
| 14 | + | - | + | + |
| 15 | - | + | + | + |
| 16 | + | + | + | + |

**Table 2:** Matrix for screening of experiment.

| Number of exp | $k_1$ ($Y_1$) | $k_2$ ($Y_2$) | α ($Y_3$) | Rs ($Y_4$) |
|---|---|---|---|---|
| 1 | 1.03 | 1.17 | 1.13 | 0.96 |
| 2 | 0.75 | 0.85 | 1.14 | 0.94 |
| 3 | 1.03 | 1.17 | 1.14 | 0.93 |
| 4 | 0.68 | 0.77 | 1.13 | 0.93 |
| 5 | 1.03 | 1.18 | 1.15 | 0.96 |
| 6 | 0.73 | 0.84 | 1.16 | 0.94 |
| 7 | 1.03 | 1.18 | 1.15 | 0.96 |
| 8 | 0.70 | 0.80 | 1.14 | 0.94 |
| 9 | 1.03 | 1.17 | 1.13 | 0.98 |
| 10 | 0.75 | 0.85 | 1.14 | 0.94 |
| 11 | 1.03 | 1.17 | 1.14 | 0.93 |
| 12 | 0.68 | 0.77 | 1.13 | 0.93 |
| 13 | 1.03 | 1.18 | 1.15 | 0.95 |
| 14 | 0.73 | 0.84 | 1.16 | 0.94 |
| 15 | 1.03 | 1.18 | 1.15 | 0.97 |
| 16 | 0.70 | 0.80 | 1.14 | 0.95 |

**Table 3:** Results for outputs.

| Number of exp | $k_1$ | $k_2$ | $\alpha$ | Rs |
|---|---|---|---|---|
| $b_1$ | -0.315 | -0.36 | 0 | -0.01625 |
| $b_2$ | -0.025 | -0.03 | -0.005 | -0.00875 |
| $b_1b_2$ | -0.025 | -0.03 | -0.01 | 0.00625 |
| $b_3$ | 0.00000 | 0.01000 | 0.01500 | 0.00875 |
| $b_1b_3$ | -4.16334E-17 | -1.38778E-17 | 0 | -0.00125 |
| $b_2b_3$ | 0.01 | 0.01 | -0.005 | 0.01625 |
| $b_1b_2b_3$ | 0.01 | 0.01 | 0 | -0.00875 |
| $b_4$ | 0 | 5.55112E-17 | -3.60822E-16 | 0.00375 |
| $b_1b_4$ | 1.38778E-17 | -6.93889E-17 | 0 | -0.00125 |
| $b_2b_4$ | 0 | 0 | -2.77556E-17 | 0.00125 |
| $b_1b_2b_4$ | 0.01 | 0.01 | 0 | -0.00875 |
| $b_3b_4$ | 0 | 5.55112E-17 | -2.77556E-17 | -0.00125 |
| $b_1b_3b_4$ | 1.38778E-17 | -1.38778E-17 | 0 | 0.00375 |
| $b_2b_3b_4$ | 0 | 0 | 2.77556E-17 | 0.00625 |
| $b_1b_2b_3b_4$ | 0 | -1.38778E-17 | 0 | -0.00375 |
| $b_0$ | 0.8725 | 0.995 | 1.1425 | 0.946875 |

**Table 4:** Coefficients for mathematical models.

| Number of exp. | $X_1$ | $X_2$ | $X_3$ |
|---|---|---|---|
| 1 | - | - | - |
| 2 | 0 | - | - |
| 3 | + | - | - |
| 4 | - | 0 | - |
| 5 | 0 | 0 | - |
| 6 | + | 0 | - |
| 7 | - | + | - |
| 8 | 0 | + | - |
| 9 | + | + | - |
| 10 | - | - | + |
| 11 | 0 | - | + |
| 12 | + | - | + |
| 13 | - | 0 | + |
| 14 | 0 | 0 | + |
| 15 | + | 0 | + |
| 16 | - | + | + |
| 17 | 0 | + | + |
| 18 | + | + | + |
| 19 | 0 | 0 | - |
| 20 | 0 | 0 | - |
| 21 | 0 | 0 | + |
| 22 | 0 | 0 | + |

**Table 5:** Matrix of experiment for optimization.

metoprolol tartarate enantiomers. The results for analysis of variance concerning influence of acetonitrile content and β-CD content, as the most important inputs, are presented in Table 7 for output $Y_1$ and in Table 8 for output $Y_4$.

Coefficients of determination ($R^2$) and results for factor Fisher value (F) demonstrated good fitting of obtained results in mathematical model. Suitable three-D graphs are presented in Figure 1 for output $Y_1$ and in Figure 2 for output $Y_4$.

Obtained three-D graphs gave information about influence of acetonitrile content and chiral modificator concentration on metoprolol tartarate enantiomers separation. The connection between influence of the factors and outputs can be presented with second order polynoms. The obtained polynoms are presented as Equation 2 and Equation 3.

$$Y_1 = -3.585 + 0.059x_1 + 1.288x_2 - 0.004x_1^2 - 0.003x_1x_2 - 0.086x_2^2 \qquad \text{(Eq. 2)}$$

where is $Y_1$-is retention factor of enantiomer R (+)

$$Y_4 = 6.918 - 0.52x_1 - 0.875x_2 + 0.022x_1^2 + 0.001x_1x_2 + 0.059x_2^2 \qquad \text{(Eq. 3)}$$

where is $Y_4$-is resolution factor

As it could be seen from the Figure 1, retention factor of the first enantiomer has a higher value for higher content of β-CD and higher content of acetonitrile. Also, strong influence of acetonitrile content is obvious and drastical decrease of retention factor was observed. On the other hand, resolution factor is under strong influence of both investigated factors. It is clear that both factors must be carefully set in order to achieve acceptable separation of enantiomers investigated.

According to the presented results the best separation of metoprolol tartarate enantiomers can be achieved with mobile phase:

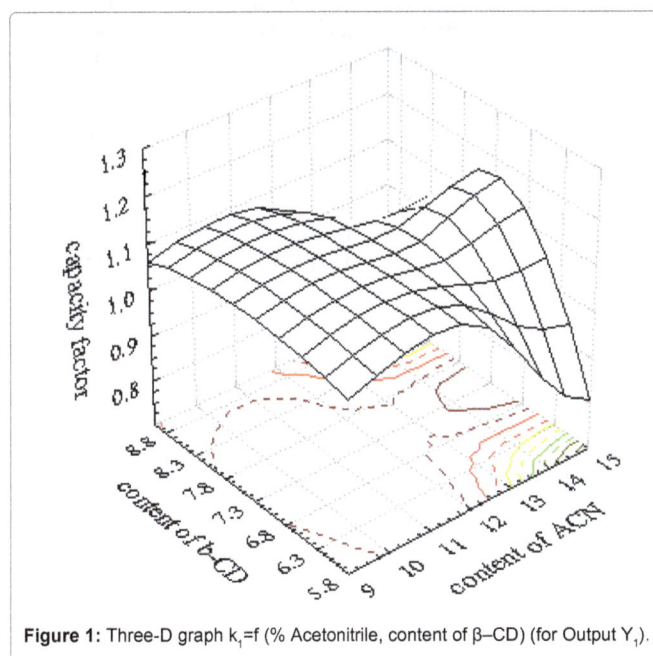

**Figure 1:** Three-D graph $k_1$=f (% Acetonitrile, content of β–CD) (for Output $Y_1$).

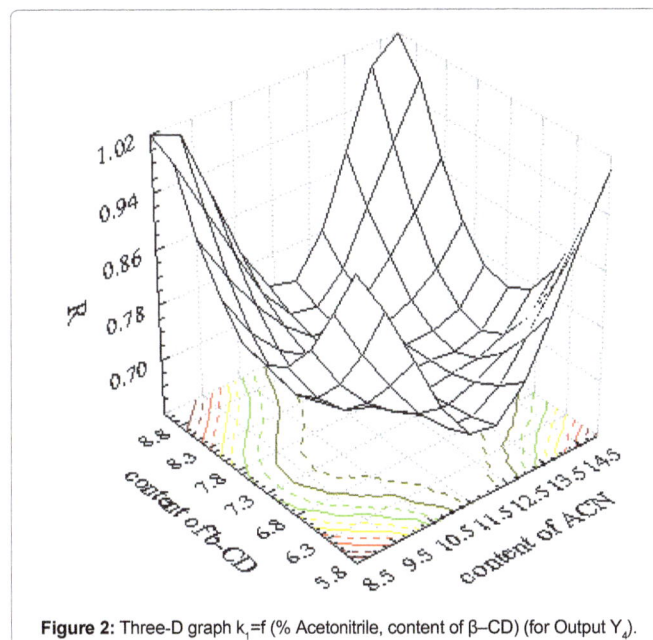

**Figure 2:** Three-D graph $k_1$=f (% Acetonitrile, content of β–CD) (for Output $Y_4$).

acetonitrile-5.87 mM β-CD (14.3:85.7), pH of the mobile phase adjusted to 3.0 and flow rate 1.5 mlmin$^{-1}$ at 35°C column temperature. Under these conditions the value of the resolution factor is 0.98. The representative chromatograms are presented in Figure 3.

## Conclusion

Applying experimental design it was possible to achieve optimal separation of metoprolol tartarate enantiomers performing a relatively small number of experiments. In this paper full factorial design $2^4$ was used for experiment screening. Chromatographic behavior of investigated enantiomers was affected by acetonitrile content, chiral modificator content in the mobile phase and flow rate the most, which was demonstrated by obtained linear models. After experimental screening, RSM was used for optimization of RP–HPLC method and optimal chromatographic conditions were settled. The proposed methodology represents an efficient and easily attainable approach in resolving the problem of searching for optimal HPLC chromatographic conditions via experimental design.

| Source of variation | Sum of squares | d.f. | Mean square | F-ratio | F-tab |
|---|---|---|---|---|---|
| $X_1$ | 0.00187500 | 1 | 0.00187500 | 0.42 | $F_{1,17}=1.45$ |
| $X_2$ | 0.00020833 | 1 | 0.00020833 | 0.05 | |
| $X_1^2$ | 0.20423045 | 1 | 0.20423045 | 46.19 | |
| $X_2^2$ | 0.08965974 | 1 | 0.08965974 | 20.28 | $F_{4,17}=2.96$ |
| Model | 0.29597352 | 4 | 0.0739934 | 16.7335 | $F_{14.3}=8.71$ |
| Error | 0.0751719 | 17 | 0.00442188 | | |
| Lack of fit | 0.05171 | 14 | 0.004083 | 0.6805 | |
| Purely experimental uncertainty | 0.01800 | 3 | 0.00600 | | |
| Total (corr) | 0.44086 | 21 | | | |
| $R^2$ | | | 0.7974 | | |

**Table 8:** Analysis of variance for variables and for the full regression for resolution factor (Rs).

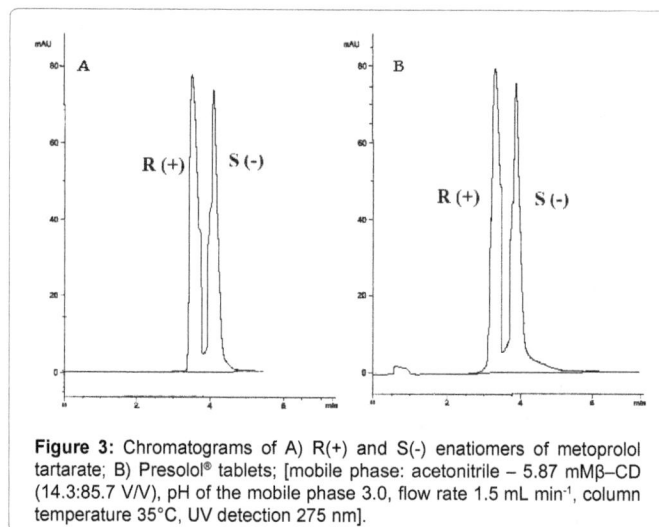

**Figure 3:** Chromatograms of A) R(+) and S(-) enatiomers of metoprolol tartarate; B) Presolol® tablets; [mobile phase: acetonitrile – 5.87 mMβ–CD (14.3:85.7 V/V), pH of the mobile phase 3.0, flow rate 1.5 mL min$^{-1}$, column temperature 35°C, UV detection 275 nm].

### Acknowledgements

Our work was supported by the Scientific Research Grants from the Ministry of Science and Technology OI172041 (Serbia).

| Number of exp | $k_1$ ($Y_1$) | $k_2$ ($Y_2$) | α ($Y_3$) | Rs ($Y_4$) |
|---|---|---|---|---|
| 1 | 1.03 | 1.17 | 1.13 | 0.96 |
| 2 | 1.05 | 1.18 | 1.12 | 0.68 |
| 3 | 0.75 | 0.85 | 1.14 | 0.94 |
| 4 | 1.31 | 1.48 | 1.13 | 0.72 |
| 5 | 1.05 | 1.19 | 1.13 | 0.67 |
| 6 | 1.14 | 1.28 | 1.12 | 0.68 |
| 7 | 1.03 | 1.17 | 1.14 | 0.93 |
| 8 | 1.05 | 1.18 | 1.13 | 0.67 |
| 9 | 0.68 | 0.77 | 1.13 | 0.93 |
| 10 | 1.03 | 1.18 | 1.15 | 0.96 |
| 11 | 1.1 | 1.23 | 1.12 | 0.67 |
| 12 | 0.73 | 0.84 | 1.16 | 0.94 |
| 13 | 1.32 | 1.51 | 1.14 | 0.72 |
| 14 | 1.09 | 1.23 | 1.13 | 0.67 |
| 15 | 1.15 | 1.31 | 1.14 | 0.67 |
| 16 | 1.03 | 1.18 | 1.15 | 0.96 |
| 17 | 1.03 | 1.17 | 1.14 | 0.67 |
| 18 | 0.7 | 0.8 | 1.14 | 0.94 |
| 19 | 1.04 | 1.2 | 1.13 | 0.66 |
| 20 | 1.05 | 1.19 | 1.13 | 0.67 |
| 21 | 1.1 | 1.22 | 1.13 | 0.67 |
| 22 | 1.09 | 1.23 | 1.13 | 0.66 |

**Table 6:** Results of experiment for optimization.

| Source of variation | Sum of squares | d.f. | Mean square | F-ratio | F-tab |
|---|---|---|---|---|---|
| $X_1$ | 0.21333333 | 1 | 0.2133333 | 22.15 | $F_{1,18}=1.14$ |
| $X_2$ | 0.00240833 | 1 | 0.0024083 | 0.25 | |
| $X_1^2$ | 0.41423037 | 1 | 0.41423037 | 21.15 | |
| $X_2^2$ | 0.21781833 | 1 | 0.2178183 | 22.61 | $F_{3,18}=3.16$ |
| Model | 0.43356000 | 3 | 0.1445184 | 15.0029 | $F_{15.3}=8.70$ |
| Error | 0.173390 | 18 | 0.00963278 | | |
| Lack of fit | 0.05539 | 15 | 0.00369 | 5.594 | |
| Purely experimental uncertainty | 0.01800 | 3 | 0.00066 | | |
| Total (corr) | 0.59814 | 21 | | | |
| $R^2$ | | | 0.7143 | | |

**Table 7:** Analysis of variance for variables and for the full regression for retention factor ($k'_1$).

### References

1. Gisvold O, Delgado JN, Remers WA (1998) Wilson and Gisvold's Textbook of Organic Medicinal and Pharmaceutical Chemistry. Lippincott-Raven, Philadelphia, New York, pp: 498-503.

2. Foye WO, Lemke TL, Williams DA (1995) Principles of Medicinal Chemistry. Williams and Wilkins, Baltimore, pp: 359-379.

3. Kleidernigg OP, Posch K, Lindner W (1996) Synthesis and application of a new isothiocyanate as a chiral derivatizing agent for the indirect resolution of chiral amino alcohols and amines. J Chromatogr A 729: 33-42.

4. Olsen L, Bronnumhansen K, Helboe P, Jorgensen GH, Kryger S (1993) Chiral separations of ß-blocking drug substances using derivatization with chiral reagents and normal-phase high-performance liquid chromatography. J Chromatogr 636: 231-241.

5. Hedeland M, Isaksson R, Pettersson C (1998) Cellobiohydrolase I as a chiral additivein capillary electrophoresis and liquid chromatography. J Chromatogr A 807: 297-305.

6. Aboulenein HY, Abou-Basha LI (1996) HPLC Separation of Nadolol and Enantiomers on Chiralcel OD Column. J Liq Chromatogr Relat Tehnol 19: 383-392.

7. Piperaki S, Tsantilikakoulidou A, Parissi-Polou M (1995) Solvent selectivity in chiral chromatography using a ß-cyclodextrin-bonded phase. Chirality 7: 257-266.

8.  Egginger G, Lindner W, Vandebosch C, Massart DL (1993) Enantioselective bioanalysis of beta-blocking agents: Focus on atenolol, betaxolol, carvedilol, metoprolol, pindolol, propranolol and sotalol. Biomed Chromatogr 7: 277-295.

9.  Radulovic DM, Karljikovic-Rajic KD, Lucic BM, Vujic ZB (2001) A preliminary study of ß-cyclodextrin/metoprolol tartrate inclusion complex for potential enantiomeric separation. J Pharm Biomed Anal 24: 871-876.

10. Araujo P (2000) A new high performance liquid chromatography multifactor methodology for systematic and simultaneous optimisation of the gradient solvent system and the instrumental/experimental variables. Trends Anal Chem 19: 524-529.

11. Gennaro MC, Marengo E, Gianotti V, Angioi S (2002) Simultaneous reversed-phase high-performance liquid chromatographic separation of mono-, di-and trichloroanilines through a gradient elution optimised by experimental design. J Chromatogr A 945: 287-292.

12. Medenica M, Jancic B, Ivanovic D, Malenovic A (2004) Experimental design in reversed-phase high-performance liquid chromatographic analysis of imatinib mesylate and its impurity. J Chromatogr A 1031: 243-248.

13. Ye C, Liu J, Ren F, Okafo N (2000) Design of experiment and data analysis by JMP (SAS institute) in analytical method validation. J Pharm Biomed Anal 23: 581-589.

14. Ficarra R, Ficarra P, Tommasini S, Melardi S, Calabro ML, et al. (2000) Validation of a LC method for the analysis of zafirlukast in a pharmaceutical formulation. J Pharm Biomed Anal 23: 169-174.

15. Vander Heyden Y, Nijhuis A, Smeyers-Verbeke J, Vandeginste BG, Massart DL (2001) Guidance for robustness/ruggedness tests in method validation. J Pharm Biomed Anal 24: 723-753.

16. Hund E, Vander Heyden Y, Haustein M, Massart DL, Smeyers-Verbeke J (2000) Robustness testing of a reversed-phase high-performance liquid chromatographic assay: comparison of fractional and asymmetrical factorial designs. J Chromatogr A 874: 167-185.

17. Fabre H (1996) Robustness testing in liquid chromatography and capillary electrophoresis. J Pharm Biomed Anal 14: 1125-1132.

18. Deming SN, Morgan SL (1993) Experimental design: a chemometrical approach. Elsevier, Amsterdam, pp: 317.

# Determination of Specific Activity and Tellurium (IV) Concentration in Non-Carrier Added Iodine-124 Solutions, Via UV-Vis Spectroscopy

**Atilio I Anzellotti\*, Robert Ylimaki and Alexander Yordanov**

*IBA Molecular, Virginia Commonwealth University, Richmond, USA*

### Abstract

In this work, two fast and highly sensitive analytical methods, based on UV-VIS spectroscopy, were adapted for its use on the determination of specific activity (SA) and Tellurium (IV) concentration, in $^{124}I^-$ solutions produced by the $^{124}Te(p,n)^{124}I$ reaction. The solutions analyzed showed high specific activity (209 - 216 Ci/mg) and Te(IV) contents below the regulatory acceptable limit (1 ppm).

**Keywords:** Iodine-124; Tellurium (IV); HPLC; UV-Vis spectroscopy; Specific activity

## Introduction

Positron Emission Tomography (PET) is a highly sensitive imaging modality that provides functional information about pathologically relevant processes. The procedure generally involves the injection of a positron-emitting radiotracer and further detection/image reconstruction of the gamma rays produced upon electron-positron annihilation [1,2]. Currently the most used radionuclides for PET applications have relative short half-lives, for instance $^{11}C$ (20.4 min half-life), $^{13}N$ (9.97 min) and $^{18}F$ (110 min). The use of radiotracers with longer half-lives involves a range of potential benefits, such as an increased time available for synthesis and quality control, imaging of long metabolic/physiologic processes, etc. [3,4].

Iodine-124 ($^{124}I^-$) is an attractive isotope of iodine due to its complex radioactive decay scheme and convenient half-life (4.18 d). There are several modes for the production of Iodine-124, but the nuclear reaction $^{124}Te(p,n)^{124}I$ offers the highest purity (with only few other radionuclidic impurities, <0.1%) and also the most convenient one, since the maximum cross-section for the reaction is below 14 MeV placing it within the capacity of most available cyclotrons [5,6].

With the increasing use of positron emission tomography (PET) in nuclear medicine, medical oncology, pharmacokinetics and drug metabolism, $^{124}I$-labeled radiopharmaceuticals could be most useful for PET imaging [7-9]. Furthermore, the 4.18 d half-life would permit their use in PET facilities far away from the radionuclide production centers (satellite concept). Sodium [$^{124}I$] iodide is already reportedly used for diagnosis and treatment planning in thyroid disease [10,11]. In addition to the growing number of [$^{124}I$]-labeled small molecules studied in recent years, notably [$^{124}I$]*m*-iodobenzylguanidine ([$^{124}I$]MIBG) [12] and [$^{124}I$]2'-fluoro-2'-deoxy-5'-iodo-1β-D-arabinofuranosyl-uracil ([$^{124}I$]FIAU) [13]. There is also a strong interest in the high selectivity/sensitivity features from [$^{124}I$]-labeled antibodies [14,15]. Limited availability of this radionuclide so far has been a hindrance to its wider development and clinical use, for example its commercially available from just a couple of producers worldwide (IBA Molecular, Perkin Elmer, etc.).

There are two key quality attributes that need to be determined in $^{124}I^-$ solutions to be used in radiotracer synthesis:

1) Specific Activity (SA), it is defined as the radioactivity per unit mass of a radionuclide or a radiotracer. In practice, the radioactivity and mass of a sample are determined by different methods in order to get the final SA expressed as a ratio, where the radioactivity is expressed in Curies (Ci) or Becquerels (Bq), and the mass can be expressed in grams or moles.

2) Amount of tellurium present in the form Te(IV), potentially coming from the staring material (enriched Tellurium oxide), it can potentially interfere with labelling reactions and also represents a chemical impurity (heavy metal), so it needs to be taken into account in an eventual clinical application.

In our case, the SA is obtained from the radioactive measurement of $^{124}I^-$ using a dose calibrator or another appropriate instrument; however the measurement of the often small amount of matter (non-radioactive iodine, ng to fg range) requires very sensitive methods of detection. For this purpose spectrophotometric methods are preferred due to its simplicity and low-cost, but the extinction coefficient for the analyte of interest (I$^-$) is not appropriate for the working range intended and therefore the sensitivity needs to be increased. To this end we use a derivatization process in which the iodine is reacted *in-situ* with 2-naphthol or β-naphthol using either Chloramine T (C-T) or Iodogen as oxidation reagents. The resulting product 1-iodo-2-naphthol (Figure 1A) has a very large extinction coefficient, thanks to the 2-naphthol moiety and since the reaction is quantitative, it can be used to measure the concentration of iodine in an indirect mode.

Another important characteristic of $^{124}I^-$ solutions to be used as raw material in radiotracer development would be a minimal Te(IV) content, due to chemical and biological reasons (*vide supra*). We adapted a highly sensitive UV-VIS method to evaluate the Te(IV) levels in $^{124}I^-$ solutions by the formation of an association complex with sodium tungstate in the presence of poly(vinyl alcohol) [16].

---

**\*Corresponding author:** Atilio I. Anzellotti, IBA Molecular, Virginia Commonwealth University, Richmond, VA 23298, USA
E-mail: aanzellotti@abt-mi.com

## Experimental

### Instruments

A Waters 515 HPLC system equipped with an Alltech 460 UV-vis detector and radioactivity detector (Bioscan, model FC-3300), was used for determination of specific activity in $^{124}I^-$ batches. The column used was an Alltech lichrosorb RP-18 5 μ (length of 250 mm, ID 4.6 mm). For Te(IV) determination was used a Genesys 10 UV-VIS spectrometer (Thermo Spectronic) with multi-wavelength capability, the wavelength of interest was set to 580 nm.

### Materials and reagents

All solvents and reagents (highest purity grade) were purchased from Sigma-Aldrich and used without further purification. 1-iodo-2-naphthol (used for the calibration curve) was synthesized from 2-naphthol and iodine monochloride in dichlorometane according to Figure 1B. The solvent was evaporated and the product purified by gradient elution flash chromatography (hexane/acetone). The final characterization was done by Elemental analysis (C,H,N) and $^1$H-NMR spectroscopy, for the latter observing shifts between 0.10-0.14 ppm compared to free 2-naphthol; solvent residual peak $D_2O$ at 4.69 ppm was taken as reference: 7.92 (1H, d), 7.74 (1H, d), 7.73 (1H, d), 7.54 (1H, d), 7.37 (1H, d) and 7.25 (1H, d). HPLC analysis revealed that the product was 98% pure and eluted in a different retention time compared to the starting material. Tellurium dioxide 99.9995% (Sigma Aldrich, part # 435902) was used in the calibration curve for Te(IV) determination.

$^{124}I^-$ solution batches were prepared by the $^{124}Te(p,n)^{124}I$ nuclear reaction using the COSTIS Compact Solid Target Irradiation System (IBA S.A., Louvain-La-Neuve, Belgium). $^{124}I^-$ was separated from the target using the Reetz GmBH (Berlin, Germany) quartz thermochromatographic furnace for radio-iodine recovery from solid targets [16]. Typically 5 μL of the concentrated solution are diluted to 50 μL with sodium hydroxide 0.02M, this volume is enough to perform SA and Te(IV) determination in duplicate (if needed), since only 20 μL are needed for SA determination and 5 μL for Te(IV) determination.

SA and Te(IV) determination were evaluated under the international guidelines for method validation [17,18]. Data analysis suite from Microsoft Excel (Microsoft office professional plus 2010, ver. 14.0.71635000 32 bit) was used to obtain calibration data including LOQ and LOD values. Resolution ($R_s$) between peaks was calculated according to the following formula:

$$R_s=1.18[(t_{RA}-t_{RB})/(WA_{1/2h}+WB_{1/2h})],$$

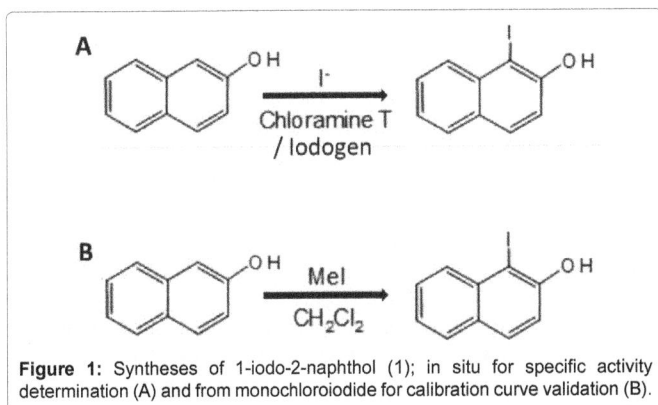

**Figure 1:** Syntheses of 1-iodo-2-naphthol (1); in situ for specific activity determination (A) and from monochloroiodide for calibration curve validation (B).

Where, $R_s$=resolution

t=elution time for species A or B

W=width of the peak at half height for A or B

### Specific activity determination

Solutions for calibration curve using the standard 1-iodo-2-naphthol were prepared by dilution of a 1 mg/mL solution in methanol/water mixture 1:1. Blank injections were performed between data points to account for residual 1-iodo-2-naphthol in the column. Injection volume was 20 μL, typical volume and radioactivity in a sample was 50 μL and 5.0-1.0 mCi, respectively.

Linearity was evaluated based on the linear regression coefficient value ($R^2 \geq 0.99$) obtained for average data points performed along seven different $^{124}I^-$ values (i.e., 500, 250, 150, 100, 50, 20 and 3 ng, for a 500 to 3 ng working range). Every data point was performed in triplicate. Repeatability was evaluated through relative standard deviation (RSD) calculations at three different $^{124}I^-$ values (high, mid and low section of the working range, i.e., 500, 150 and 3 ng). Every data point was performed six times LOQ and LOD were calculated based on the residual standard deviation (σ) and slope (S) of the calibration curve obtained for the linearity section.

Mobile phase used was methanol 70% in water and the flow was adjusted to 1 ml/min. UV-detector was set at 254 nm. System suitability test for the HPLC system in reverse-phase conditions was performed at the beginning of each day using a mixture of different ketones (Waters Co; part # WAT042887). A criteria of resolution ($R_s$=4.0) was established for the propiophenone and butyrophenone peaks using the same method for SA. 1 was produced *in situ* using 10 μL of a chloramine-T solution (1 mg/mL) and 10 μL of concentrated trifluoroacetic acid, or alternatively by addition of 25 μL of Iodogen solution (0.5 mg/mL) in ethanol, both in the presence of excess 2-naphthol. Dilution was accounted for in the iodine concentration calculation.

### Determination of Te(IV) content

This method was scale-down from the literature in order to work with 4-5 μL aliquots with an activity <0.5 mCi, the samples were diluted up to 1.5 mL prior to measurement giving concentrations in the range 0.5-4.0 ng/mL. Briefly to 500 μL of the diluted sample in water it was added 120 μL of $Na_2WO_4$ solution (1 μg/mL in sulfuric acid 2M); in addition to 60 μL of poly(vynil alcohol) solution (1 mg/mL). 120 μL of nile blue solution (0.5 mg/mL) were added within 5 minutes of measurement. Disposable cuvettes with 100 L capacity on the path length were used for UV-VIS measurement, and fresh solutions for the calibration curve were prepared before measuring each new batch of $^{124}I^-$. Absorption was measured at 580 nm.

## Results and Discussion

### Specific activity determination

$^{124}I^-$ solutions prepared by the $^{124}Te(p,n)^{124}I$ nuclear reaction are considered non-carrier added (nca), in principle this means that no iodine was added to the matrix and we could achieve the theoretical value for maximum specific activity in iodine, which is 251.6 Ci/mg or 31,200 Ci/mmol [19]. In practice this value is never achieved but it is useful to establish a limit for the detection of the method. Given the conditions of the method and the amount of sample tested the minimum amount of iodine that can be found is in the range of 116 to 23 ng of $^{124}I^-$.

We adapted an analytical approach that has been used for SA determination in solutions of [123]I- and [131]I-, and modified the procedure to customize the quality control in non-carrier added [124]I- [20]. In order to assess the potential of this derivatization method for [124]I- solutions, the reference compound 1-iodo-2-naphthol (1) was first prepared and characterized by [1]H-NMR spectroscopy and HPLC. The linearity, accuracy and repeatability of the method was tested using solutions in the concentration range of 22.5-0.15 μg/mL or actual 450 to 3.0 ng analyzed. From the results obtained in Table 1 it can be seen that the method was linear in the tested range, with a limit of quantification (LOQ) of 9.3 ng and limit of detection (LOD) of 2.6 ng. The method was linear, accurate (recovery % 90-110, n=9) and repeatable (Relative Standard Deviation <5%, n=18) in the working range, which was appropriate for the iodine concentrations expected.

After the method was validated with the compound 1, additional calibration curves were obtained in which 1 was produced *in situ* via the derivatization reaction of 2-naphthol, from known amounts of potassium iodine and either oxidation agent chloramine-T (C-T) or Iodogen (IG) (Figure 2). Usually the latter is considered milder since it does not need the presence of strong acids such as trifluoroacetic acid. A comparison between the linearity values and working ranges for all the calibration curves found is presented in Table 1. It can be observed that all three methods are equivalent, in terms of response obtained vs. concentration of iodine, independently of compound 1 being prepared *in situ*.

Optimization for the derivatization reaction in the SA method was made for the following variables: time of injection after mixing (2 to 15 min), amount of oxidation agent used (5 to 25 μg, total mass) and temperature (25°C to 45°C). Iodine was introduced as potassium iodine. The derivatization reaction to produce 1 *in situ* was fast, the time was set to 5 minutes after mixing, for both oxidation agents, by comparing the peak areas obtained to the reference compound 1. In the case of chloramine-T the time for injection mixing was important since it was observed that the peak for 1 was not stable with time, eventually decreasing in favor of the specie 1-chloro-2-naphthol (2) which is also formed in excess given the concentration of C-T and presence of trifluoroacetic acid (see structure of both oxidation agents used in Figure 2). No significant change in the peak for 1 was observed up to 15 minutes after mixing when IG was used as an oxidant, and negligible amounts of 2 were also noted, possibly due to the milder nature of this reagent and the absence of trifluoroacetic acid in the matrix mixture. Finally temperature was found to not modify results obtained in the range tested, 25°C to 45°C, as a default 25°C was chosen due to convenience. Blanks for both methods without potassium iodide as a source of iodide did not produce any peak in the window for compound 1.

In a typical chromatogram the peak for the specie of interest 1 was in a very consistent window, determined from specificity experiments (n=6), eluting between 11.0 to 11.3 min. The elution of this peak using the reference compound was generally seen later compared to the production of 1 *in-situ* for both C-T and IG methods, possibly because of the absence of organic solvents in the matrix, although for all cases the peak for 1 remained inside the aforementioned window. For chromatograms obtained using dilutions of the reference compound there was only one peak present for 1, for the other methods peaks identified with unreacted 2-naphthol and other negligible and unidentified peak was detected.

As seen in Figure 3, the chromatogram for the chloramine-T method exhibited the greater amount of peaks, although given the

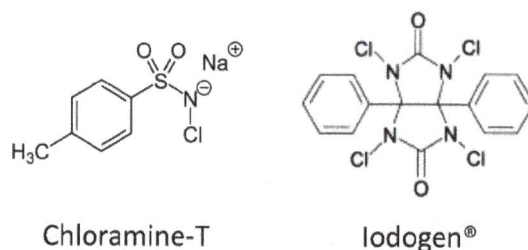

**Figure 2:** Comparison between the structures of the oxidation reagents used, chloramine-T at left and Iodogen® at right

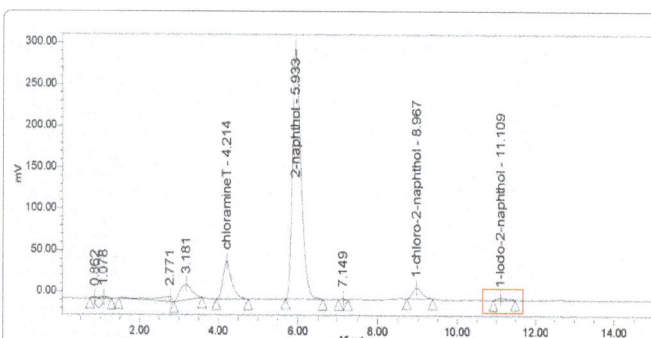

**Figure 3:** Representative chromatogram (using Chloramine-T) showing the chemical species in solution after 5 min of mixing, the signal for 1 is highlighted.

| Method | Slope | Intercept | Correlation coefficient | Working range |
|---|---|---|---|---|
| 1 | 13,154 | 12,205 | 0.9997 | 3-450 ng |
| C-T | 14,797 | 2,295 | 0.999 | 3-500 ng |
| IG | 13,572 | 16,862 | 0.993 | 3-500 ng |

**Table 1:** Comparison between the linearity parameters found in the different SA determination methods. 1-iodo-2-naphthol (1), Chloramine-T (C-T) and Iodogen® (IG).

| Quality Attribute | Chloramine-T | Iodogen® |
|---|---|---|
| Specificity | Specific | Specific |
| Linearity | R²=0.999 | R²=0.993 |
| Range | 3-500 ng | 3-500 ng |
| LOQ | 9 ng | 6 ng |
| LOD | 4 ng | 3 ng |
| Accuracy | 90-110% | 95-105% |
| Precision (Repeatability) | RSD<20% | RSD<10% |
| Intermediate Precision | RSD<25% | RSD<10% |

RSD: relative standard deviation

**Table 2:** Comparison between the qualities attributes found in the validation of SA determination using chloramine-T and Iodogen®.

resolution with the peak of interest (Resolution of 1 and 2 was $R_s = 3$) this did not represented an issue.

In Table 2 the main differences in quality attributes validated for the chloramine-T and Iodogen® methods are summarized, in general both methods comply with general requirements for analytical methods in terms of specificity, linearity and range. The Iodogen® method exhibits lower limits of quantification and detection and more importantly it showed to be more repeatable and accurate than the chloramine-T method. These differences could be due to the instability of 1 after mixing. Specifically Robustness was evaluated fin both methods by changes in flow rate (10%) and mobile phase composition (10%), these changes caused the peak for 1 to go out of the specified window obtained from specificity experiments. It is therefore demonstrated that such variations in either mobile phase composition or flow would impact the result of the method. An appropriate generic System Suitability Test was used at the beginning and end of day of experiments to assure consistency of results.

After validation of both methods for SA determination, twelve different batches of $^{124}$I⁻ solutions prepared by the $^{124}$Te(p,n)$^{124}$I nuclear reaction, were analyzed in parallel. In Figure 4 it can be observed a plot for the SA determined on each batch using both analytical methods. The results for iodine concentration found were within the working range obtained for both methods, with values between 15 and 25 ng. The resulting averages for SA values were 216 ± 13 Ci/mg for IG and 209 ± 12 Ci/mg for C-T, further comparison using unpaired t-test indicated that the statistical difference between both methods is not significant (P=0.1843). SA values found using chloramine-T as an oxidant were usually lower compared to the Iodogen® method (Figure 4), however this variability could be attributed to matrix effects since the calibration curve was performed in water and $^{124}$I⁻ solutions have a matrix of sodium hydroxide 0.02M.

The fact that the average values of SA obtained are below the expected maximum theoretical value means that not all the Iodine atoms in the sample are radioactive. It is not surprising to obtain lower values since the theoretical value represents an upper limit in the range and also "cold" or non-radioactive iodine in the system is ubiquitous. The values obtained are notwithstanding very high, 86% (IG) and 83% (C-T) of the theoretical value of 251.6 Ci/mg, and demonstrate a valid raw material for use in the manufacture of radiotracers.

## Determination of Te(IV) content

The determination of Te(IV) in $^{124}$I⁻ solution batches was possible due to the formation of an ion association complex Te(IV)-WO$_4$-nile blue in acidic solutions stabilized with poly(vinyl alcohol). We have scaled down this method in order to be used for radioactive samples and prepared a calibration curve using standard solutions of TeO$_2$. The method was validated in the range 3.10 to 0.70 ng/mL since the change in color depending on Te(IV) concentration can be followed visually (Figure 5). The resulting values for the quality attributes evaluated are summarized in Table 3.

Briefly the method was found to be linear in the expected working range, and under matrix effects coming from the basic $^{124}$I⁻ solutions. The typical specification for Te(IV) content as a heavy metal is less than 1 μg/mL or 1 ppm. Twelve $^{124}$I⁻ batches were tested using this procedure and the average concentration found was 0.40 ± 0.12 ppm, thus confirming that the Te(IV) content present in $^{124}$I⁻ batches produced by the $^{124}$Te(p,n)$^{124}$I reaction was within specifications. The color change was monitored via UV-vis at 580 nm.

## Conclusion

As a conclusion, two highly sensitive methods for SA and Te(IV) determination in $^{124}$I⁻ batches were validated and applied successfully. In addition, $^{124}$I⁻ batches produced by the COSTIS target and the Reetz GmBH quartz thermochromatographic furnace possess high SA and low Te(IV) concentration. Given the small amount of sample needed for the SA and Te(IV) determination methods a workflow was developed in which only a small amount of $^{124}$I⁻ solution was needed (<20 μL) to perform both quality control tests. This is important since given the typical activity concentration present on these solutions (ca. 1 mCi/

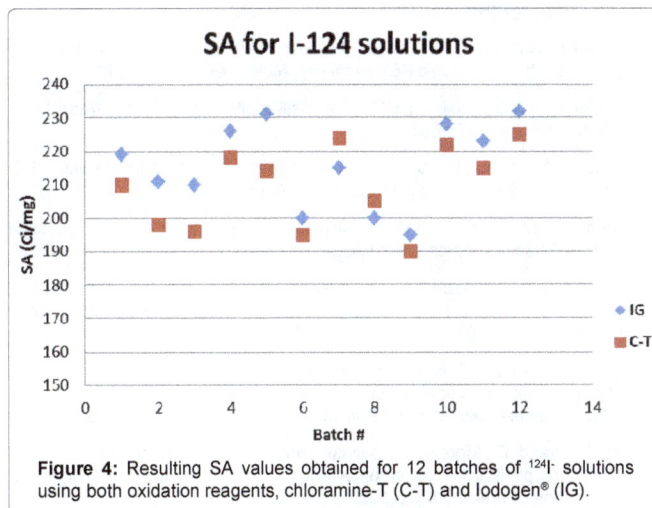

**Figure 4:** Resulting SA values obtained for 12 batches of $^{124}$I⁻ solutions using both oxidation reagents, chloramine-T (C-T) and Iodogen® (IG).

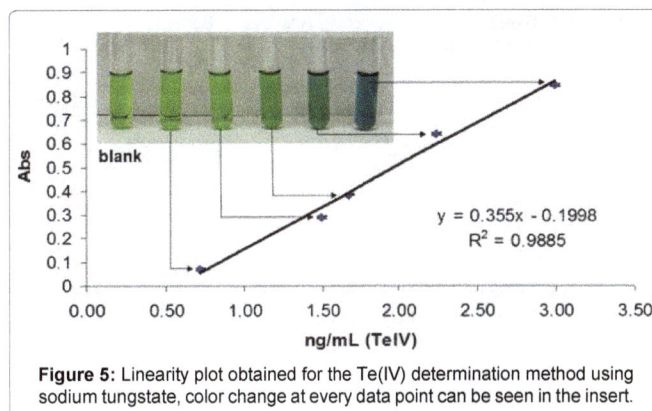

**Figure 5:** Linearity plot obtained for the Te(IV) determination method using sodium tungstate, color change at every data point can be seen in the insert.

| Quality Attribute | Result |
|---|---|
| Specificity | Specific for Te(IV) |
| Linearity | R²=0.9885 |
| Range | 0.70-3.1 ng/mL |
| LOQ | 1 ng/mL |
| LOD | 0.7 ng/mL |
| Accuracy | 90-110% |
| Precision (Repeatability) | RSD<15% |
| Intermediate Precision | RSD<20% |

**Table 3:** Summary of quality attributes obtained from the validation of the method for Te(IV) using sodium tungstate.

mL) the exposure to quality control personnel can be high. Decrease in radiation exposure is of main importance when considering that $^{124}$I⁻ solutions also exhibit high energy gamma radiation (0.6 and 1.7 MeV) and have the potential for airborne contamination.

## References

1. Mittra E, Quon A (2009) Positron emission tomography/computed tomography: the current technology and applications. Radiol Clin North Am 47: 147-160.

2. Miller PW, Long NJ, Vilar R, Gee AD (2008) Synthesis of 11C, 18F, 15O, and 13N radiolabels for positron emission tomography. Angew Chem Int Ed Engl 47: 8998-9033.

3. Rice SL, Roney CA, Daumar P, Lewis JS (2011) The next generation of positron emission tomography radiopharmaceuticals in oncology. Semin Nucl Med 41: 265-282.

4. Glaser M, Luthra SK, Brady F (2003) Applications of positron-emitting halogens

in PET oncology (Review). Int J Oncol 22: 253-267.

5.  Koehler L, Gagnon K, McQuarrie S, Wuest F (2010) Iodine-124: a promising positron emitter for organic PET chemistry. Molecules 15: 2686-2718.

6.  Fonslet J, Koziorowski J (2013) Dry distillation of radioiodine from TeO2 targets. Appl Sci 3: 675-683.

7.  Lubberink M, Herzog H (2011) Quantitative imaging of 124I and 86Y with PET. Eur J Nucl Med Mol Imaging 38 Suppl 1: S10-18.

8.  Braghirolli AM, Waissmann W, da Silva JB, dos Santos GR (2014) Production of iodine-124 and its applications in nuclear medicine. Appl Radiat Isot 90: 138-148.

9.  Cascini GL, Niccoli Asabella A, Notaristefano A, Restuccia A, Ferrari C, et al. (2014) 124 Iodine: a longer-life positron emitter isotope-new opportunities in molecular imaging. Biomed Res Int 2014: 672094.

10. Sgouros G, Hobbs RF, Atkins FB, Van Nostrand D, Ladenson PW, et al. (2011) Three-dimensional radiobiological dosimetry (3D-RD) with 124I PET for 131I therapy of thyroid cancer. Eur J Nucl Med Mol Imaging 38 Suppl 1: S41-47.

11. Van Nostrand D, Moreau S, Bandaru VV, Atkins F, Chennupati S, et al. (2010) 124I positron emission tomography versus 131I planar imaging in the identification of residual thyroid tissue and/or metastasis in patients who have well-differentiated thyroid cancer. Thyroid 20: 879-883.

12. Moroz MA, Serganova I, Zanzonico P, Ageyeva L, Beresten T, et al. (2007) Imaging hNET reporter gene expression with 124I-MIBG. J Nucl Med 48: 827-836.

13. Simões MV, Miyagawa M, Reder S, Städele C, Haubner R, et al. (2005) Myocardial kinetics of reporter probe 124I-FIAU in isolated perfused rat hearts after in vivo adenoviral transfer of herpes simplex virus type 1 thymidine kinase reporter gene. J Nucl Med 46: 98-105.

14. Kraeber-Bodéré F, Rousseau C, Bodet-Milin C, Mathieu C, Guérard F et al. (2015) Tumor immunotargeting using innovative radionuclides. Int J Mol Sci 16: 3932-3954.

15. van Dongen GA, Visser GW, Lub-de Hooge MN, de Vries EG, Perk LR (2007) Immuno-PET: a navigator in monoclonal antibody development and applications. Oncologist 12: 1379-1389.

16. Qiu-e C, Zhide H, Zubi L, Jialin W, Qiheng, X (1998) Highly sensitive spectrophotometric determination of trace amounts of tellurium(IV) with the tungstate-basic dyes-poly(vinyl alcohol) system. Analyst 123: 695-698.

17. Alvarenga L, Ferreira D, Altekruse D, Menezes JC, Lochmann D (2008) Tablet identification using near-infrared spectroscopy (NIRS) for pharmaceutical quality control. J Pharm Biomed Anal 48: 62-69.

18. FDA (2015) Analytical Procedures and Methods Validation for Drugs and Biologics. Guidance for Industry. Food and Drug Administration.

19. Wilbur DS (1992) Radiohalogenation of proteins: an overview of radionuclides, labeling methods, and reagents for conjugate labeling. Bioconjug Chem 3: 433-470.

20. Kloster G, Laufer P (1983) Determination of specific activity of radiohalide preparations (75Br, 77Br, 123I, 131I) by HPLC-UV detection following chemical derivatization to 1-halonaphthol-2. J Labelled Comp Radiopharm 20: 1305-1315.

# 5

# Evaluation of Bio-Layer Interferometric Biosensors for Label-Free Rapid Detection of Norovirus Using Virus Like Particles

Xiuli Dong[1], Jessica J. Broglie [1], Yongan Tang[2] and Liju Yang[1]*

[1]Biomanufacturing Research Institute and Technology Enterprise (BRITE) and Department of Pharmaceutical Sciences, North Carolina Central University, Durham, NC 27707, USA
[2]Department of Mathematics and Physics, North Carolina Central University, Durham, NC 27707, USA

## Abstract

This study evaluated the label-free bio-layer interferometric (BLI) biosensor for the detection of norovirus (NoV) using two types of virus like particles (VLPs) that represent human NoV GI.1 and GII.4. To construct biosensors for NoV GI.1 and GII.4 detection, the commercial AMC sensors, on which anti-mouse Fc-specific antibodies were pre-immobilized on the surfaces, were further bound with the capture antibodies mAb3901 and mAb NS14, respectively, by using the Blitz system. The kinetics of immobilization of capture antibodies on the AMC sensors demonstrated that mAb3901 and mAb NS14 reached saturated binding phase almost at the same time (~415 s). The optimal concentration of capture antibodies for immobilization was 15 µg/mL for both mAb3901 and mAb NS14. The AMC sensors loaded more mAb NS14 than mAb3901 at the same binding condition. The biosensors constructed by immobilization of the capture antibodies at their optimal concentration showed tight binding interactions with their respective GI.1 VLPs and GII.4 VLPs, with the affinity constant of $6.01 \times 10^{-7}$ M and $2.01 \times 10^{-7}$ M, respectively. For both biosensors, the VLPs binding rates were linearly increased with the increase of VLP concentrations. These biosensors were able to detect GI.1 or GII.4 VLPs at the concentration of 5 µg/mL in PBS, and showed intense and stable binding interactions at VLP concentration of 10 µg/mL and above. The mAb NS14-immoblized biosensors for GII.4 VLP detection were more sensitive than the mAb3901-immoblized biosensors for GI.1 VLP detection. This detection technique was label-free, easy, rapid (2 min), and accurate, requiring a very small sample volume (4 µL).

**Keywords:** Bio-Layer interferometry; Biosensor; Rapid detection; Virus like particles; Novovirus

## Introduction

Human Norovirus (NoV) is the most common cause of nonbacterial, acute gastroenteritis outbreaks worldwide [1,2], accounting for more than 21 million illnesses and hospitalizations, and at least 570 deaths in the United States each year (Centers for Disease control and Prevention, 2013). NoVs are a genetically diverse group of single-stranded RNA, non-enveloped viruses in the Calicivirdae family. NoVs are classified into six genogroups (GI to GIV) and further subclassified into genotypes based on their capsid sequence [2]. Most NoVs that infect humans belong to genogroups GI and GII [3]. NoV GI.1 is the first isolated genotype and is considered the prototype virus of the genus, whereas NoV GII.4 is currently the most frequently detected genotype in humans [4,5]. NoV is extremely contagious and affects people of all ages with a low infectious dose of 18 particles or less [5]. The transmission of NoV occurs directly through person to person (62-84% of all reported outbreaks) and indirectly via contaminated water and food [6]. NoV aerosols are formed during vomiting and toilet flushing when vomit or diarrhea is present. It is estimated that as many as 30 million virus particles are released in a single episode of vomiting Infection may develop after eating food or breathing air near an episode of vomiting, even if it is cleaned up. NoV shedding can be detected many weeks after infection symptoms have subsided [7].

The main challenges hindering work with human NoVs are that they cannot be cultivated *in vitro* and there is no animal model for their propagation. Consequently, surrogate viruses, which are morphologically similar and cultivable, are widely used to mimic human NoV behavior [8]. NoV virus-like particles (VLPs) are also used as model viral systems in research. VLPs are formed by the expression of the major capsid protein (ORF2) in baculovirus [9] and Venezuelan equine encephalitis virus [10]. Each VLP is ~38 nm in diameter and has repeating arch-like surface features. These arches

are formed by 90 dimers of a single capsid protein and contain both a shell and protruding (P) domain. The former houses the capsid's N-terminus, consisting of 225 residues of the 530 amino acid (aa) sequence [11], while the latter forms the top (P2 domain) and body (P1 domain) of each arch-like structure. The P1 and P2 domains contain the C-terminus and the central regions of the amino acid sequence, respectively [12]. These VLPs do not contain genomic RNA and are replication deficient, however their morphologies are nearly identical to native virus particles. Recent studies have proved that VLPs could be used to understand the role of immunological factors on the evolution and emergence of new strains [13,14]. NoV VLPs have been successfully expressed using several expression systems. The characteristics of NoV VLPs make them appropriate models for NoVs in biological assays to answer human NoV-specific questions, and for the development of detection and inactivation methods for human NoV [15-17].

Various technologies have been investigated for detection of NoVs in clinical and environmental samples. These technologies mainly include electron microscopy techniques, molecular detection techniques, and immunological techniques. Diagnostic electron microscopies (DEM) have been widely accepted as a diagnostic method since 1980s for being able to visualize virus particles and other

*Corresponding author: Liju Yang, Biomanufacturing Research Institute and Technology Enterprise (BRITE) and Department of Pharmaceutical Sciences, North Carolina Central University, Durham, NC 27707, USA
E-mail: lyang@nccu.edu

pathogens including bacteria and parasites. However, the high costs of DEM equipment and the need of experienced staff for the operation are hampering its usage and timely renewal [17]. The molecular detection techniques, such as polymerase chain reaction (PCR) and nonisotopic detection methods, have had the greatest impact on the clinical virology laboratory. Their low detection limits are ideal for screening the low viral loads common to contaminated foods and environmental samples [18] but the detection sensitivity and specificity are largely affected by the efficacy of the concentration, purification, and reaction conditions. The typical immunological techniques, enzyme-linked immunosorbent assays (ELISAs), are generally easy to perform without the need for sophisticated equipment. However, their need for high viral loads limits the assays' application mostly in clinical settings [15,17,19].

Biosensor technologies represent a class of analytical methods/ devices that combine the high selectivity from the bio-recognition molecules and the sensitivity for quantification measurement from the transducers, and offer rapid detection, easy to use, and the possibility of miniaturization advantages over typical instrumental analysis [20,21]. The label-free bio-layer interferometry (BLI) technology-based biosensors have gained popularity in recent years as a reliable method for analyzing biomolecular interactions such as protein-protein interactions, protein-liposome interactions, and others [22,23]. BLI technology is based on the monitoring of the interference pattern of white light reflected from two surfaces: a layer of immobilized protein on the fiber optic biosensor tip and an internal reference layer [24]. The binding of specific molecules in sample solutions to the biosensor tip causes a shift in the interference pattern. This shift can be monitored in real time, and allows rapid identification, quantitation, and characterization of proteins and other biomolecules in a very small sample volume (4 μL). The BLI-based biosensor can be a disposable sensor made from a biocompatible matrix that is uniform, non-denaturing and minimizes non-specific binding. Only molecules that bind directly to the biosensor surface are detected, providing exceptional specificity for individual applications. In addition, this optical-based measurement can minimize interferences from colored samples and has the potential for testing clinical samples while overcoming the issues from difficult sample matrix [25]. The objective of this study was to evaluate a bio-layer interferometry (BLI)-based biosensor platform for the detection of NoV using GI.1 and GII.4 VLPs. In this study, anti-mouse IgG Fc Capture (AMC) sensors were used for immobilizing anti-GI.1 and anti-GII.4 VLP antibodies as the capturing antibodies for detection of NoV GI.1 and GII.4 VLPs, respectively. The binding kinetics between VLPs and the capture antibodies on the biosensors were examined in real time and the detection sensitivities of the biosensors to GI.1 and GII.4 VLPs were evaluated.

## Materials and Methods

### VLPs, antibodies and chemicals

Stock solutions of GI.1 VLPs, GII.4 VLPs, monoclonal anti-GI.1 VLP antibody 3901 (mAb 3901), and monoclonal anti-GII.4 antibody NS14 (mAb NS14) were obtained from Dr. Robert Atmar's laboratory at the Baylor College of Medicine (Houston, TX). Phosphate buffered saline (PBS), pH 7.4, was prepared in-house from a 1X (0.01 M) PBS recipe (Cold Spring Harbor Protocols) using NaCl, KCl, $Na_2HPO_4$, and $KH_2PO_4$, which were all purchased from Fisher Scientific.

### Biosensor construction and detection procedures

To construct the biosensors for VLP detection, anti-mouse IgG Fc capture (AMC) sensors were used (Pall FortéBio Corp., Menlo Park,

CA). The AMC sensors were pre-immobilized with a layer of high-affinity antibody against the Fc portion of mouse IgG (mIgG) on their surfaces, which can be used for further immobilizing mIgG or other Fc-containing ligands to produce a stable surface for specific capturing of target molecules. Figure 1A shows the image of the AMC sensor and the illustration of antibody immobilization steps. The AMC sensor is a needle-shaped sensor with approximately 1 cm in length but with a flat tip of approximately 1 mm in diameter. The actual sensing surface is the surface of its flat tip. For detection of GI.1 and GII.4 VLPs, monoclonal antibodies mAb3901 (specific to GI.1) and mAb NS 14 (specific to GII.4) at various concentrations in 1 × PBS buffer were freshly prepared and used to immobilize on AMC sensors, respectively. The immobilization step was carried out using the BLItz instrument (Fortébio Inc., Menlo Park, CA) by immersing the AMC biosensors in 200 μL antibody solutions with constant shaking (1000 rpm) at room temperature for 420 s. The immobilization curves were recorded in real time using the BLItz instrument with software BLItz Pro (version 1.2.0.49, Fortébio Inc., Menlo Park, CA). To optimize the antibody concentration for immobilization, the sensors were immobilized with different concentrations of mAb3901 or mAb NS14. The binding rates of the resulting sensors to GI.1 or GII.4 VLPs at the concentration of 10 μg/mL were measured and compared.

Once the optimal antibody concentration for immobilization was selected, the detection of GI.1 or GII.4 VLPs by the resulting biosensors was performed by measuring the binding rates of GI.1 or GII.4 VLPs at various concentrations in PBS buffer to the antibody-immobilized biosensors under constant shaking (1000 rpm) for 120 s. Binding affinity between the immobilized antibodies on the sensors and the target VLPs were determined by analyzing the binding kinetic curves using the software BLItz Pro. Linear response ranges of the biosensors to GI.1 and GII.4 VLPs were demonstrated.

## Results and Discussion

### The typical sensogram

The typical sensogram for the stepwise antibody immobilization and the detection of VLPs is shown in Figure 1B. The sensogram showed the real-time signal of the sensor in response to the binding of mAb3901 and mAb NS14 antibodies to the AMC sensor surfaces, which include a quick initial binding phase and a slow to non-increase phase toward the saturation of antibody binding. This was followed by the wash step in which free antibodies and non-specific bindings were washed away, and the final VLP binding step in which each strain of VLPs showed their individual characteristic binding curves. These binding curves can be analyzed to determine the kinetics of antibody immobilization and the VLP binding, and the binding affinities between mAb3901 or mAb NS14 to the AMC sensors, and between VLPs to mAb3901-immobilized sensor or mAb NS14-immobilized sensors. The binding rate extracted from these binding curves can be used as the detection signals for VLPs detection, as demonstrated in the following sections.

### Kinetics of antibody immobilization on AMC biosensors

The first step to construct the biosensor was the immobilization of mouse anti-NoV VLP antibodies (mAb3901 or mAb NS14) onto the AMC sensors through the binding of their Fc region to the pre-existing anti-mouse IgG antibodies on the surface. Figure 2A and 2B shows the representative binding curves of antibody mAb3901 and mAb NS14 at two different concentrations (15 μg/mL and 25 μg/mL) onto the AMC sensors that resulted from three replicates, respectively. In both cases, 1 × PBS

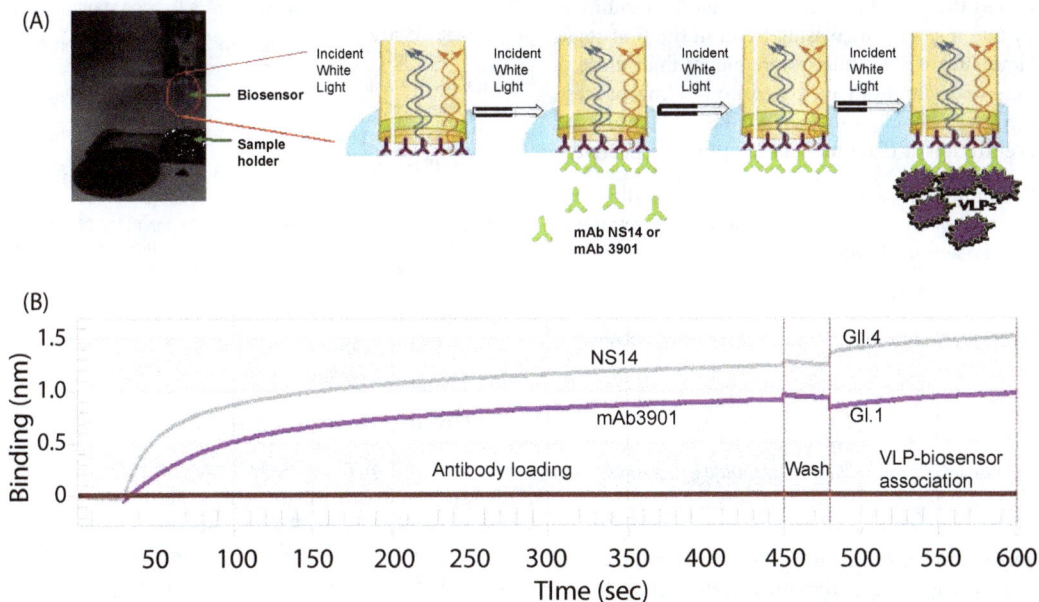

**Figure 1:** (A) The image of the AMC sensor and the illustration of the steps for antibody immobilization and VLP capture onto the BLI biosensor. (B) The typical stepwise sensograms of the BLI biosensors to GI.1 and GII.4 VLPs during antibody immobilization and the detection of VLPs.

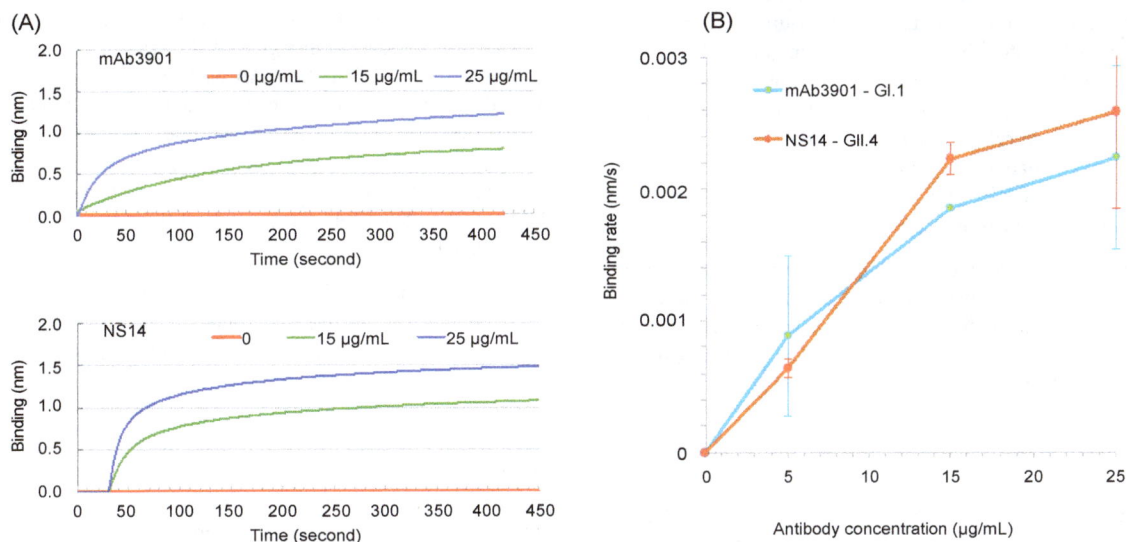

**Figure 2:** (A) The binding curves of mAb3901 and mAb NS14 antibodies onto the AMC biosensors. Each line was for a representative sample selected from three replicates. (B) The effect antibody concentration during the antibody immobilization step on the binding of VLPs to the resulting antibody-immobilized sensors.

without antibody was used a control. As shown in both Figures 2A and 2B, a higher binding rate was observed at a higher antibody concentration (25 µg/mL) during the binding of mAb3901 or mAb NS14 to the AMC sensor than that at a lower antibody concentration (15 µg/mL), which is common as the binding event should follow the law of mass action. At higher concentrations, more antibody molecules diffuse to reach and bind to the anti-mouse IgG molecules on the surface of the AMC sensor. Looking at the entire binding processes for both mAb3901 and mAb NS14, the binding curves presented similar patterns, where at the first 20 s, the binding of both mAb3901 and mAb NS14 to the AMC sensors increased immediately and rapidly, showing exponential

increasing binding rates, followed by a slower binding stage, and a final saturated stage. Closer examination of the individual binding curves provided detailed kinetics of mAb3901 and mAb NS14 immobilization onto the AMC sensor. For mAb 3901, the binding signal level at the first 20 s were 0.099 nm and 0.242 nm, at the concentration of 15 µg/mL and 25 µg/mL, respectively. From 20 s to 125 s, the binding of mAb3901 at both concentrations showed linear increasing trends, and from 125 s to 415 s, the binding event showed a slower increasing trend and started the stationary phase at 415s. Between 415 s to 420 s, the binding event reached the saturation stage and had the highest binding signal level of 0.795 nm and 1.22 nm at the concentration of 15 µg/mL and 25 µg/mL, respectively. The binding of mAb NS14 to the AMC sensors showed a similar overall trend as that of mAb3901, except that the binding of mAb NS14 antibody had a steeper increasing trend in the first stage. The binding signal levels for NS14 binding were higher than those of mAb3901 at the same concentrations in the entire binding process. At saturation, the mAb NS14 binding signal levels were 1.075 nm and 1.477 nm at the concentration of 15 μg/mL and 25 μg/mL, respectively.

Looking into the binding process of mAb3901 or NS14 antibodies to anti-mouse IgG on the AMC sensor surface, it is a reversible reaction that can be expressed as reaction (1) and (2):

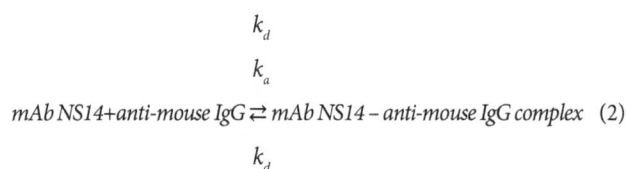

$$k_a$$

$$mAb3901 + anti\text{-}mouse\ IgG \rightleftharpoons mAb3901 - anti\text{-}mouse\ IgG\ complex \quad (1)$$

$$k_d$$

$$k_a$$

$$mAb\ NS14 + anti\text{-}mouse\ IgG \rightleftharpoons mAb\ NS14 - anti\text{-}mouse\ IgG\ complex \quad (2)$$

$$k_d$$

The kinetic patterns of mAb3901 or mAb NS14 binding follow the law of mass action. At the beginning, as the concentration of mAb3901 or NS14 antibody near the AMC surface was high, the reaction was predominant on the forward reaction to form the complex formation; As time elapsed, the concentration of complex on the AMC sensor surface increased and the available binding sites (anti-mouse IgG) for mAb3901 or mAb NS14 gradually decreased, leading to the gradual slowdown of the forward reaction until equilibrium was reached. The binding of molecules onto a biosensor surface might be characterized as a fractal system along with its different complexities, including heterogeneities on the surface and in solution, diffusion-coupled reactions, and time-varying adsorption (or binding), and even dissociation rate coefficient. In general, the molecule concentration in solution has a mild effect on the fractal dimension for binding compared to those reactions in solutions.

The kinetics of the antibody binding can be used for estimating the dissociation constants between mAb3901 or mAb NS14 antibodies and anti-mouse IgG antibodies on the AMC surfaces. According to reaction (1), the binding affinity constant between mAb3901 or NS14 antibody and the anti-mouse IgG antibody can be expressed by eq. (1):

$$K_A = \frac{1}{K_D} = \frac{[complex]}{[mAb3901\ or\ mAb\ NS14] \times [anti-mouse\ IgG]} = \frac{k_a}{k_d} \quad (1)$$

Whereas; $k_a$, $k_d$, $K_A$, $K_D$ represent the association rate constant, the dissociation rate constant, the equilibrium associate constant, and the equilibrium dissociation constant, for reaction (1) or (2); square brackets represent chemical concentrations. The association rate constant, $k_a$, describes how fast molecules bind to the sensor surface and the dissociation rate constant, $k_d$, describes how fast complexes fall apart. The equilibrium associate constant, $K_A$, is considered as the affinity constant, it is time independent and indicates the complex strength, which is termed the binding strength. Therefore, the affinity determines how much complex is formed at equilibrium [26,27]. Using the software BLItz Pro (version 1.2.0.49) to analyze the binding kinetics, Table 1 shows the kinetic constants, including $k_a$, $k_d$, and $K_D$ of both mAb3901 and NS14 binding to anti-mouse IgG on AMC sensors. The equilibrium dissociate constants ($K_D$) for mAb3901 and mAb NS14 binding to anti-mouse IgG on AMC sensors were $2.45 \times 10^{-8}$ M and $1.71 \times 10^{-8}$ M, respectively. This observation indicated that mAb NS14 antibodies had a slightly higher affinity to anti-mouse IgG on the AMC

| Complex | Association rate constant ($k_a$, M⁻¹s⁻¹) | Dissociation rate constant ($k_d$, s⁻¹) | Binding affinity constant ($K_D$, M) |
|---|---|---|---|
| mAb3901-anti-mouse IgG | $6.09 \times 10^4$ | $1.49 \times 10^{-3}$ | $2.45 \times 10^{-8}$ |
| mAb NS14-anti-mouse IgG | $5.02 \times 10^4$ | $8.60 \times 10^{-4}$ | $1.71 \times 10^{-8}$ |
| mAb3901-GI.1 VLP | $1.21 \times 10^4$ | $7.29 \times 10^{-3}$ | $6.01 \times 10^{-7}$ |
| mAb NS14-GII.4 VLP | $4.57 \times 10^4$ | $9.20 \times 10^{-3}$ | $2.01 \times 10^{-7}$ |

**Table 1:** Summary of kinetic constants for mAb3901 and mAb NS14 antibody binding on AMC sensors and GI.1 and GII.4 VLPs binding on the mAb3901-immobilized and mAb NS14-immobilized biosensors.

biosensors than mAb3901, which was observed on the binding curves, and that at the same concentration, mAb NS14 antibodies bound more onto the AMC biosensors than mAb3901.

## Antibody concentration effects on VLP bindings

The optimal antibody concentration for immobilization was determined based on the sensor response to the binding of 10 μg/mL VLP GI.1 or VLP GII.4 on the resulting biosensors that were immobilized with mAb3901 or mAb NS14 at concentrations of 0, 5, 15, and 25 μg/mL. Figure 2B shows the response signals of the resulting sensors at 120 s binding time in the binding of VLP GI.1 and GII.4, respectively. Both sensors showed an increased signal in binding rates with the increasing antibody concentration from 5 μg/mL to 25 μg/mL used in the immobilization step. For the mAb3901-immobilized sensors for GI.1 VLPs detection, the concentration of mAb3901 used in the immobilization step increased from 5 to 15 μg/mL and to 25 μg/mL, and the detection signal of the resulting sensor increased by ~250% and 13.1%. For the mAb NS14-immobilized sensor for GII.4 VLP detection, the sensor signal increased by ~110% and 21%, when the concentration of mAb NS14 antibody used in the immobilization increased from 5 to 15 μg/mL and to 25 μg/mL, respectively. Therefore, for both sensors, antibody concentration at 15 μg/mL (mAb3901 or NS14) was selected for construction of the sensors for GI.1 and GII.4 VLP detection.

It was also noted that the sensors immobilized with the same antibody concentration and for the detection of the same concentration of VLPs, the resulting binding signals of the sensors for GII.4 VLPs were higher than those of sensors for GI.1 VLPs. At the antibody concentration of 15 μg/mL, the binding rates of mAb3901-immobolized sensors to GI.1 VLPs and mAb NS14-immobolized sensors to GII.4 VLPs were 0.0019 nm/s and 0.0022 nm/s, respectively. The binding rates of mAb NS14-immobolized sensors to GII.4 VLPs was ~15% higher than that of the mAb3901-immobilized sensors to GI.1 VLPs at the same concentrations. This was most likely related to the nature of the binding affinity between the capture antibody and the target VLPs. Antibodies are large and extremely flexible molecules that are able to adopt a wide range of conformations. The antigen-antibody binding process is closely related to the internal dynamics of the IgG [27]. Antibody mAb3901 was specific for GI viruses, and its epitope mapped to NoV amino acids 454 to 520, and specifically to E472, which forms a salt bridge with K514 [28]. mAb NS14 antibody specifically recognized GII viruses, and its epitope mapped to GII.4 amino acids 473 to 495. Although a number of conserved amino acids showed at the epitope of GI and GII antibody binding domains from various tested VLPs, only alanine was completely conserved within this domain [29]. The different amino acid sequence between GI and GII epitopes and the difference in conformation between mAb3901 and mAb NS14 antibody could be the factors that affect the VLP-antibody binding rate, except for those factors that are commonly observed to affect immunoassay performance, including antigen concentration, quality of immune-components solid phase, fluid phase (pH, ionic strength) [26], and reaction temperature.

## Binding kinetics of VLPs onto the antibody-immobilized biosensors

Figures 3A and 3B show the binding curves of GI.1 and GII.4 VLPs at various concentrations to the mAb3901 and mAb NS14 antibody-immobilized sensors, respectively. On both sensors, the binding curves of all VLP concentrations presented an almost linear binding phase in the first ten seconds, and after 120 s the binding signal slowed down significantly. Therefore, the binding curves were recorded up to 120 s for both sensors and for all VLP concentrations. The binding signal level at 120 s read from the instrumental graph (reading on y-axe) could be used as the binding signal to evaluate the sensor performance for VLP detection. For the mAb3901-immobolized sensors to detect GI.1 VLPs, the binding signal at 120 s increased with increasing concentrations of VLPs, with the binding signal of 0.05 nm, 0.14 nm, and 0.24 nm at the GI.1 concentration of 5 μg/mL, 10 μg/mL, and 20 μg/mL, respectively. Similar increasing binding signals at 120 s were observed for the mAb NS14-immobolized sensors to detect GII.4 VLPs. The binding signals were 0.01 nm, 0.19 nm and 0.34 nm at the GII.4 concentration of 5 μg/mL, 10 μg/mL, and 20 μg/mL, respectively. Both sensors demonstrated the increasing trend in binding signal with increasing concentrations of VLPs. However, low VLP concentration (5 μg/mL) hardly generated detectable signals. Between the two sensors, the mAb NS14-immobilized sensors for GII.4 VLPs generated higher binding signals at detectable VLPs concentrations. For example, the binding signal of GII.4 VLPs at 20 μg/mL to the NS14-immobilized sensor at 120 s was 41.67% higher than the binding signal of GI.1 VLPs at the same concentration to its sensor.

The binding rate (the slope of the binding curve) was obtained by the instrumental software and used to evaluate its linear relationship with VLP concentrations for quantitative analysis. Figure 3C shows the relationship between the binding rate and the VLP concentration for the two sensors for GI.1 and GII.4 VLPs detection. It indicated that both sensors had low binding rates when VLP concentrations were lower than 10 μg/mL. However, there was a linear correlation between the binding rate and VLP concentration when VLP concentrations were in the range of 10 μg/mL to 20 μg/mL for both sensors, with linear regression equations: Binding Rate=0.0004 [GII.4 VLPs] -0.0035 for GII.4 VLPs, and Binding Rate=0.0002 [GI.1 VLPs] – 0.0014 for GI.1 VLPs.

In view of the linear relationships between the two sensors for GI.1 and GII.4 VLPs, the mAb NS 14-immobilized sensors for GII.4 VLPs were more sensitive than the mAb3901-immobilized sensors for GI.1 VLPs. For example, at VLP concentration of 20 μg/mL, the binding rates of GI.1 and GII.4 VLPs to their sensors were 0.0021 and 0.0046 nm/s, respectively. GII.4 VLPs' binding rate was 54.34% higher than that of GI.1 VLPs. Again, the difference in sensitivity between the two sensors was most likely related to the nature of the binding of mAb3901 antibody to GI.1 VLPs and mAb NS14 antibody to GII.4 VLPs, the amount of capture antibodies immobilized on the sensors, as well as other factors as discussed above.

The binding events between mAb3901 antibodies and GI.1 VLPs and between mAb NS14 antibodies and GII.4 VLPs to form the antibody-VLP complex followed the similar pattern as equations (1) and (2), and the kinetic parameters can be expressed and analyzed in the similar way as equation (3). Again, by analyzing the binding curves using the software BLItz Pro (1:1 fitting), the kinetic parameters for the binding between mAb3901 and GI.1 VLPs and between the NS14 antibody and GII.4 VLPs were obtained (Table 1). The dissociation constants ($K_D$) for GI.1 VLP - mAb3901 complexes on the biosensor, and GII.4 VLP – mAb NS14 complexes on the biosensor, were $6.01 \times 10^{-7}$ M and $2.01 \times 10^{-7}$ M, respectively, which indicated that the affinity between mAb NS14 and GII.4 VLP was higher than that between mAb3901 and GI.1 VLP. In the process of VLP detection using the resulting

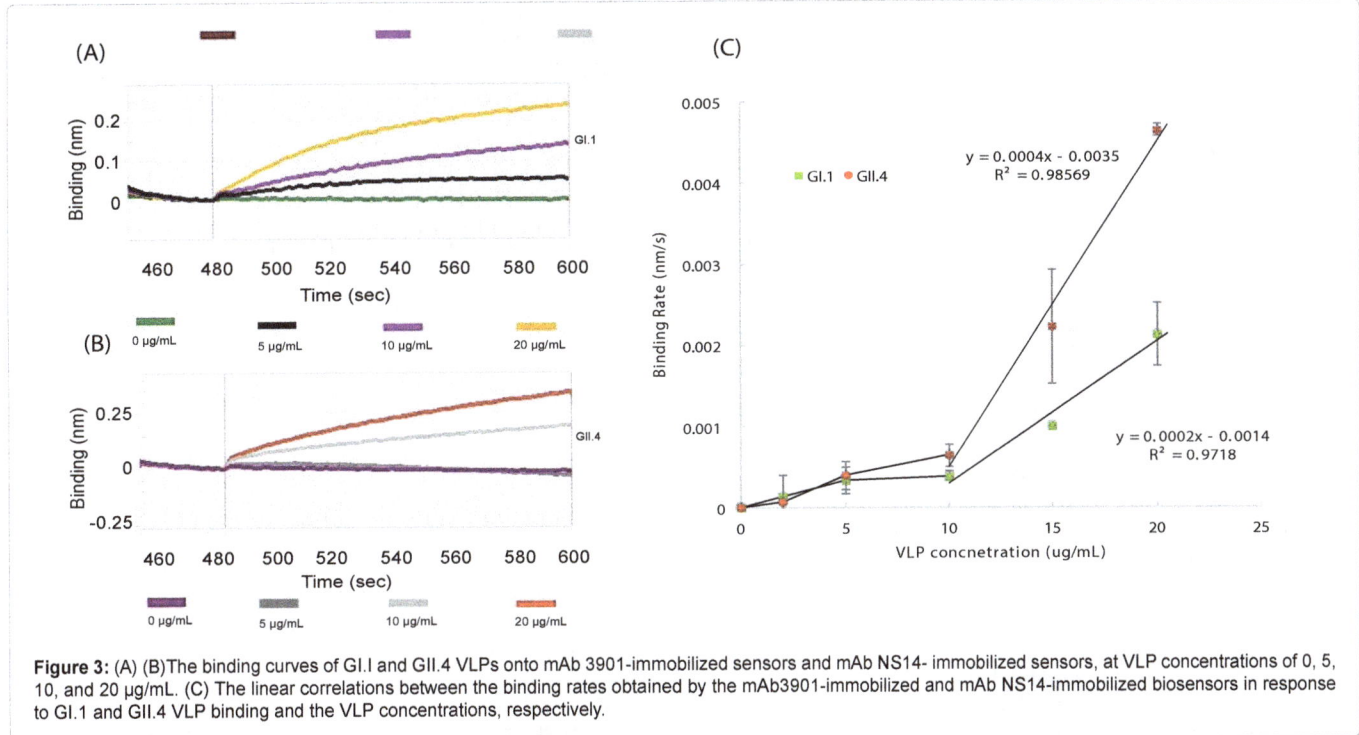

**Figure 3:** (A) (B)The binding curves of GI.I and GII.4 VLPs onto mAb 3901-immobilized sensors and mAb NS14- immobilized sensors, at VLP concentrations of 0, 5, 10, and 20 μg/mL. (C) The linear correlations between the binding rates obtained by the mAb3901-immobilized and mAb NS14-immobilized biosensors in response to GI.1 and GII.4 VLP binding and the VLP concentrations, respectively.

biosensors at the given conditions, lower $K_D$ value or higher antibody concentration would improve the VLP-antibody complex formation at a given VLP concentration. The results also demonstrated that mAb NS14-immobilized sensors had higher antibody immobilization on the sensors compared to mAb3901-immoblized sensors at the same conditions. This explained the observation that GII.4 VLPs had a higher binding value than GI.1 VLPs to their respective sensors, and the more sensitive detection of GII.4 VLPs by the NS14-immobilized sensors than the mAb3901 immobilized sensor for GI.1 VLPs detection.

The BLI biosensors developed in this study were quick and easy to use, conducted real-time detection, and did not require isolation of the virus, genetic material, or any other complicated procedures associated with sample preparation. They only needed a 4 µL sample, and the test procedure was simply dipping and reading. The detection time was 2 min, which was much faster than the other reported biosensors for virus detection. For instance, Nidzworski et al. [30] developed universal immunosensors by using electrochemical impedance spectroscopy and direct attachment of antibodies to the gold electrode for the detection of all serotypes of the influenza A virus, andtheir sensors need about 30 min to finish analysis. Xu et al. [31] developed a piezoelectric diaphragm-based immunoassay chip to simultaneously detect HBV and α-detoprotein antibodies. The total assay time was less than 2 h. Wang et al. [32] evaluated the performance of a newly developed impedance biosensor for avian influenza virus detection with the detection time less than 1 h. Though the detection limit of this BLI-based sensor at its current condition was higher than commonly used ELISA-based methods reported in our previous study [15] and by others [33] it presents both challenges and opportunities for further development of a more sensitive method for Human Nov detection. Alternatively, with the combination of an effective concentrating method, it could possibly provide a strategy for rapidly screening suspect foods and environmental samples for norovirus, or to meet the growing need for rapid detection of HuNoV for in clinical and point-of-care settings, contributing to the reduction, prevention, and eventual eradication of norovirus-derived outbreaks [34,35].

## Conclusions

This study demonstrated the proof-of-concept to use interferometry-based biosensors for the detection of Norovirus VLPs. The sensors was able to detect the binding of NoV VLPs in real time. The binding signal was detectable at the VLP concentration of 5 µg/mL, however, the linear correlations between the sensor signal and the VLP concentration were demonstrated in the VLP concentration range of 10 g/mL to 20 µg/mL for both sensors for detection of GI.1 and GII.4 VLPs. In the antibody immobilization step, the nature of the binding between the anti-mouse IgG on the AMC sensor and mAb3901 antibody or mAb NS14 antibody affected the efficiency of antibody immobilization on the sensor, and further in the detection step, the nature of the binding between the capture antibody and the target VLPs affected the capture efficiency and the detection sensitivity of VLPs. The analysis of the kinetics of binding curves enabled the estimation of the affinity (dissociation constants) between capture antibodies and anti-mouse IgG antibody on the sensor, and between the capture antibodies and VLPs. This was in agreement with the observation that the sensors for the detection of GII.4 VLPs had higher sensitivity than the sensors for the detection of GI.1 VLPs. This BLI-based sensor detection technique was a label-free, easy-to-use, rapid (2 min) detection method, requiring only 4 µL sample volume. The results demonstrated the potential application prospect of BLI-based sensor for detection of NoV.

## Acknowledgements

This study was supported by USDA-NIFA Agriculture and Food Research Initiative Competitive Grant No. 2011-68003-30395. The authors acknowledge Dr. Robert L. Atmar's lab at Baylor College of Medicine (BCM) to provide all VLPs samples and antibodies to VLPs, also acknowledge the technical discussion provided by Dr. Robert L. Atmar at BCM and Dr. Lee-Ann Jaykus at North Carolina State University.

## References

1. Patel MM, Hall AJ, Vinje J, Parashar UD (2009) Noroviruses: a comprehensive review. J Clin Virol 44: 1-8.

2. Zheng DP, Ando T, Fankhauser RL, Beard RS, Glass RI, et al. (2006) Norovirus classification and proposed strain nomenclature. Virology 346: 312-323.

3. Vinje J, Green J, Lewis DC, Gallimore CI, Brown DW, et al. (2000) Genetic polymorphism across regions of the three open reading frames of Norwalk-like viruses. Arch Virol 145: 223-241.

4. Glass RI, Parashar UD, Estes MK (2009) Norovirus gastroenteritis. N Engl J Med 361: 1776-1785.

5. Teunis PF, Moe CL, Liu P, Miller SE, Lindesmith L, et al. (2008) Norwalk virus: how infectious is it. J Med Virol 80: 1468-1476.

6. Moore MD, Goulter RM, Jaykus LA (2015) Human norovirus as a foodborne pathogen: challenges and developments. Annu Rev Food Sci Technol 6: 411-433.

7. Atmar RL, Opekun AR, Gilger MA, Estes MK, Crawford SE, et al. (2008) Norwalk virus shedding after experimental human infection. Emerg Infect Dis 14: 1553-1557.

8. Tung-Thompson G, Libera DA, Koch KL, de Los Reyes FL, Jaykus LA (2015) Aerosolization of a Human Norovirus Surrogate, Bacteriophage MS2, during Simulated Vomiting. PLoS One 10: e0134277.

9. Jiang X, Wang M, Graham DY, Estes MK (1992) Expression, self-assembly, and antigenicity of the Norwalk virus capsid protein. J Virol 66: 6527-6532.

10. Baric RS, Yount B, Lindesmith L, Harrington PR, Greene SR, et al. (2002) Expression and self-assembly of norwalk virus capsid protein from venezuelan equine encephalitis virus replicons. J Virol 76: 3023-3030.

11. Prasad BV, Rothnagel R, Jiang X, Estes MK (1994) Three-dimensional structure of baculovirus-expressed Norwalk virus capsids. Journal of virology 68: 5117-5125.

12. Hale A, Mattick K, Lewis D, Estes M, Jiang X, et al. (2000) Distinct epidemiological patterns of Norwalk-like virus infection. Journal of medical virology 62: 99-103.

13. Donaldson EF, Lindesmith LC, Lobue AD, Baric RS (2010) Viral shape-shifting: norovirus evasion of the human immune system. Nat Rev Microbiol 8: 231-241.

14. Lindesmith LC, Donaldson EF, Baric RS (2011) Norovirus GII.4 strain antigenic variation. J Virol 85: 231-242.

15. Jessica Jenkins B, Matthew DM, Jaykus J (2014) Design and Evaluation of Three Immuno-based Assays for Rapid Detection of Human Norovirus Virus-like Particles. Journal of Analytical & Bioanalytical Techniques 5.

16. Escudero-Abarca BI, Rawsthorne H, Goulter RM, Suh SH, Jaykus LA (2014) Molecular methods used to estimate thermal inactivation of a prototype human norovirus: More heat resistant than previously believed. Food Microbiology 41: 91-95.

17. Gentile M, Gelderblom HR (2014) Electron microscopy in rapid viral diagnosis: an update. New Microbiol 37: 403-422.

18. Knight A, Li D, Uyttendaele M, Jaykus LA (2013) A critical review of methods for detecting human noroviruses and predicting their infectivity. Crit Rev Microbiol 39: 295-309.

19. Hirneisen KA, Kniel KE (2012) Comparison of ELISA attachment and infectivity assays for murine norovirus. J Virol Methods 186: 14-20.

20. Hong SA, Kwon J, Kim D, Yang S (2015) A rapid, sensitive and selective electrochemical biosensor with concanavalin A for the preemptive detection of norovirus. Biosens Bioelectron 64: 338-344.

21. Mandal SS, Navratna V, Sharma P, Gopal B, Bhattacharyya AJ (2014) Titania nanotube-modified screen printed carbon electrodes enhance the sensitivity in the electrochemical detection of proteins. Bioelectrochemistry 98: 46-52.

22. Levina A, Lay PA (2014) Influence of an anti-metastatic ruthenium(iii) prodrug on extracellular protein-protein interactions: studies by bio-layer interferometry. Inorganic Chemistry Frontiers 1: 44-48.

23. Wallner J, Kuhleitner M, Brunner N, Lhota G, Vorauer-Uhl K (2014) Application of the log-normal model for long term high affinity antibody/antigen interactions using Bio-Layer Interferometry. Journal of Mathematical Chemistry 52: 575-587.

24. Groner M, Ng T, Wang W, Udit AK (2015) Bio-layer interferometry of a multivalent sulfated virus nanoparticle with heparin-like anticoagulant activity. Anal Bioanal Chem 407: 5843-5847.

25. Auer S, Koho T, Uusi-Kerttula H, Vesikari T, Blazevic V, et al. (2015) Rapid and sensitive detection of norovirus antibodies in human serum with a biolayer interferometry biosensor. Sensors and Actuators B-Chemical 221: 507-514.

26. Goldblatt D, van Etten L, van Milligen FJ, Aalberse RC, Turner MW (1993) The role of pH in modified ELISA procedures used for the estimation of functional antibody affinity. J Immunol Methods 166: 281-285.

27. Galanti M, Fanelli D, Piazza F (2016) Conformation-controlled binding kinetics of antibodies. Sci Rep 6: 18976.

28. Parker TD, Kitamoto N, Tanaka T, Hutson AM, Estes MK (2005) Identification of Genogroup I and Genogroup II broadly reactive epitopes on the norovirus capsid. J Virol 79: 7402-7409.

29. Crawford SE, Ajami N, Parker TD, Kitamoto N, Natori K, et al. (2015) Mapping broadly reactive norovirus genogroup I and II monoclonal antibodies. Clin Vaccine Immunol 22: 168-177.

30. Nidzworski D, Pranszke P, Grudniewska M, Krol E, Gromadzka B (2014) Universal biosensor for detection of influenza virus. Biosens Bioelectron 59: 239-242.

31. Ting X, Jianmin M, Zhihong W, Ling Y, Li L, et al. (2011) Micro-piezoelectric immunoassay chip for simultaneous detection of Hepatitis B virus and α-fetoprotein. Sensors and Actuators B: Chemical 151: 370-376.

32. Wang R, Lin J, Lassiter K, Srinivasan B, Lin L, et al. (2011) Evaluation study of a portable impedance biosensor for detection of avian influenza virus. J Virol Methods 178: 52-58.

33. Kele B, Lengyel G, Deak J (2011) Comparison of an ELISA and two reverse transcription polymerase chain reaction methods for norovirus detection. Diagn Microbiol Infect Dis 70: 475-478.

34. Lou F, Huang P, Neetoo H, Gurtler JB, Niemira BA, et al. (2012) High-pressure inactivation of human norovirus virus-like particles provides evidence that the capsid of human norovirus is highly pressure resistant. Appl Environ Microbiol 78: 5320-5327.

35. Sadana A, Vo-Dinh T (2001) A kinetic analysis using fractals of cellular analyte-receptor binding and dissociation. Biotechnol Appl Biochem 33: 17-28.

# 6

# A New Optical Sensor for Selective Monitoring of Nickel Ion Based on A Hydrazone Derivative Immobilized on the Triacetyl Cellulose Membrane

**Kamal Alizadeh\* and Nasim Abbasi Rad**

*Department of Chemistry, Lorestan University, Khorramabad, Iran*

## Abstract

A new highly selective optical sensor was prepared by de-esterification of triacetyl cellulose transparent film and chemical immobilization of 1-acenaphthoquinone 1-thiosemicarbazone (**L**) on it. The absorbance variation of immobilized 1-acenaphthoquinone 1-thiosemicarbazone on hydrolyzed cellulose acetate film of upon addition of $1.5 \times 10^{-5}$ mol $L^{-1}$ aqueous solutions of $Zn^{2+}$, $Pb^{2+}$, $K^+$, $Cu^{2+}$, $Ag^+$, $Ni^2$, $Cd^{2+}$, $Ca^{2+}$, $CrO_4^{2-}$, $Hg^{2+}$, $Co^{2+}$, $Mn^{2+}$, $Cr^{3+}$, $S_2O_3^{2-}$, $Mg^{2+}$, $Na^+$, $Al^{3+}$, $Tl^+$ and $Fe^{3+}$ indicated a substantiality much larger variation for the Nickel ion in compare to other studied ions. Consequently, the new hydrazone derivative **L** possesses a high selectivity towards this metal ion. Influences of various experimental parameters on $Ni^{2+}$ sensing, including the reaction time, the solution pH and the concentration of reagents were studied. A linear relationship was observed between the variance in membrane absorbance$(\Delta A)$ at 337 nm and $Ni^{2+}$ concentrations in a range from $5.01 \times 10^{-10}$ to $2.04 \times 10^{-5}$ mol $L^{-1}$ with a detection limit $(3\sigma)$ of $1.00 \times 10^{-10}$ mol $L^{-1}$. No significant interference from 100 times concentrations of a number of potentially interfering ions was detected for the nickel ion determination. The sensor showed a good durability and short response time with no evidence of reagent leaching. The optical sensor was successfully applied to the determination of nickel in real water samples.

**Keywords:** Optical sensor; Nickel ion; Triacetyl cellulose membrane; Hydrazone derivative; Spectrophotometric

## Introduction

In the recent years, pollution of the environment by heavy metals has received considerable attention. Nickel is a moderate toxic element compared to other transition metals. However, it is known that inhalation of nickel and its compounds can lead to serious problems, including respiratory system cancer. Moreover, nickel can cause a disorder known as nickel-eczema [1,2]. Nickel is an excellent alloying metal in steel industry and is the metal component of the enzyme urease and as such is considered to be essential to plants and some domestic animals [3]. This metal normally occurs at very low level in the environment, so sensitive methods are needed to detect it in most environmental samples. Thus, the development of simple methods for selective determination of nickel in trace amounts in different matrices is critical.

Optical sensors have found great interest in recent years as they have many uses in clinical analysis, environmental analysis, and process control [4]. Optical chemical sensors (optodes), are usually based on acid–base indicators, which can be adsorbed on the surface of support materials [5-7]. Several different support materials including lipophilic polymers and plasticizers, hydrophilic polymers, ionic polymers, sol-gel glass and molecularly imprinted polymers have been used for preparation of optical sensors [5-8]. Covalently immobilized dyes, in contrast to the physically adsorbed or entrapped dyes, do not suffer from leaching or hysteresis and exhibit long lifetimes [8].

Optodes are simple and selective tools for the determination of heavy metal ions that have been extensively developed in recent years. Optodes are generally used in combination with inexpensive spectrophotometic or spectroflourometic techniques to provide simple and fast determination methods with enhanced selectivity and low detection limits [9-13].

Fabrication of membrane optical sensors have been reported for many cations including $Ca^{2+}$, $Na^+$, $K^+$, $Ni^{2+}$, $Pb^{2+}$, $Hg^{2+}$ and $Cu^{2+}$ [14-22]. Use of transparent triacetyl cellulose and agarose membranes as supports for preparation of covalently immobilized optical sensors for some ions determination were reported by different laboratories. It was shown that these membranes can be easily manufactured and simply activated and functionalized with an ionophore. In construction of optical sensors, ionophores play an important role. The compounds contain some donor atoms have been frequently used as ionophores in construction of membrane sensors because of their ability to form stable complexes with transition metal ions. They produce remarkable selectivity, sensitivity and stability for a specific ion [23-26].

In the present study, a hydrazone derivative ligand 1-acenaphthoquinone 1-thiosemicarbazone, L (Figure 1) [27,28], is covalently immobilized on a triacetyl cellulose membrane to be used as an effective ionophore with N and O donor atoms for construction of a selective optical sensor for the spectrophotometric determination of $Ni^{2+}$ in aqueous solutions. The studied compound as an ionophore is the family of hydrazone which is a kind of asymmetric Schiff's base. Schiff's bases (also called azomethines or imines) are functional groups with the general formula of $R_1R_2C=N-R_3$. Schiff's base can be divided into two groups; symmetric and asymmetric Schiff's base. Hydrazones are the members of the asymmetric Schiff's bases. Hydrazones are a class of organic compounds with the general structure of $R_2C=NNR_2$ which are related to ketones and aldehydes by the replacement of the oxygen with $NNR_2$ functional group. These compounds are commonly

**\*Corresponding author:** Kamal Alizadeh, Department of Chemistry, Lorestan University, Khorramabad, Iran
E-mail: Alizadehkam@yahoo.com (or) Alizadeh.k@lu.ac.ir

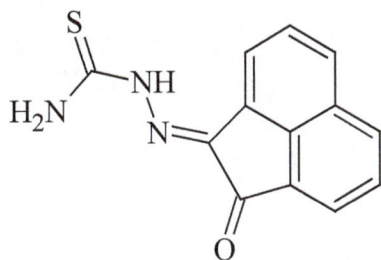

**Figure 1:** Chemical structure of 1-acenaphthoquinone 1-thiosemicarbazone.

formed through the reaction of hydrazine on ketones or aldehydes [29-31].

## Experimental Section

### Materials and instruments

All reagents were of analytical reagent grade and were used as received. Deionized double-distilled water was used throughout and test solutions were buffered in a 0.02 mol $L^{-1}$ solution of acetic acid/sodium acetate and pH adjusted with dropwise addition of 1 mol $L^{-1}$ solution of HCl or NaOH. The Schiff's base **L**, with the chemical name of 1-acenaphthoquinone 1-thiosemicarbazone, was synthesized and purified using a previously reported method [27,28].

A Jenway (USA) model 3020 pH meter with a combined glass electrode was used after calibration against standard Merck buffers for pH determinations. A Shimadzu (Japan) model 1650PC double-beam spectrophotometer was used for running the electronic absorption spectra (controlled to ± 0.1°C). A home-made polyacrilamide holder was used for holding triacetyl cellulose membranes inside the quartz cells of the spectrophotometer. A totally glass Fisons (UK) double distiller was used for preparation of doubly distilled water.

### Procedures

A method described elsewhere was used for the preparation of transparent triacetyl cellulose membranes. They were produced from waste photographic film tapes, which were previously treated with commercial sodium hypochlorite for several seconds in order to remove the colored gelationous layers [26,32]. The triacetyl cellulose transparent film was hydrolyzed in order to deesterify the acetyl groups and to increase the porosity of the membrane by treating the membrane into 0.10 M NaOH solution for 24 h. The films were treated with a solution of 0.007 g of the compound **L**, in 10 ml ethylene diamine for 2 min at ambient temperature. Afterwards, they were washed with water for the removal of ethylene diamine and the loosely trapped indicator. The prepared membranes were kept under water, when not in use. A 1 cm × 2 cm peace of the fabricated membranes sensor was cut and mounted in a polyacrylamide holder and placed inside the quartz cell of the spectrophotometer. The cell was then used as usual for the absorbance determinations. All the measurements on the transparent triacetyl cellulose membrane were performed in aqueous medium.

The recommended procedure was applied to the determination of nickel in several real water samples collected from the west of Iran, Kermanshah. The pH was adjusted to 6.0 before analysis, without any further treatment. For analysis, about 2.5 ml of the samples were transferred into a 1 cm quartz cell equipped with the membrane sensor. The absorbance's were then measured at 337 nm and subtracted from an absorbance reading for a buffer solution at the same wavelength. The $Ni^{2+}$ concentration was then derived using an ordinary calibration curve method.

## Results and Discussion

### Preliminary studies

In preliminary studies, we recorded the absorbance variations followed by absorbance readings at maximum wavelength of immobilized 1-acenaphthoquinone 1-thiosemicarbazone on hydrolyzed cellulose acetate into a quartz cell of the spectrometer. They occur upon addition of $1.5 \times 10^{-5}$ mol $L^{-1}$ aqueous solutions of $Zn^{2+}$, $Pb^{2+}$, $K^+$, $Cu^{2+}$, $Ag^+$, $Ni^2$, $Cd^{2+}$, $Ca^{2+}$, $CrO_4^{2-}$, $Hg^{2+}$, $Co^{2+}$, $Mn^{2+}$, $Cr^{3+}$, $S_2O_3^{2-}$, $Mg^{2+}$, $Na^+$, $Al^{3+}$, $Tl^+$ and $Fe^{3+}$ which was obtained after equilibration at pH 6. According to the shape reported in Figure 2, It should be noted that the largest variation is observed for $Ni^{2+}$, whereas for the other studied ions, negligible or small variations in the absorbance maximum is observed by increasing the concentration of corresponding ions. Based on the relatively high selectivity of 1-acenaphthoquinone 1-thiosemicarbazone for $Ni^{2+}$, as was concluded from its absorbance variation, the mentioned compound, **L** was expected to possess a high selectivity towards this metal ion.

Immobilization of ligand **L** on a triacetyl cellulose transparent film changed, in some extent, its optical properties. The absorbance maximum of L showed a blue shift from 417 nm to about 337 nm upon the immobilization as is obvious from Figure 3. This can suggest that the structured conformation of the immobilized 1-acenaphthoquinone 1-thiosemicarbazone compound is less flat than that of its soluble analogue [26]. Furthermore, it is evident that the first region of spectra at about 200-300 nm for the membrane sensor in compare to the dissolved form of **L**, in methanol was disappeared too.

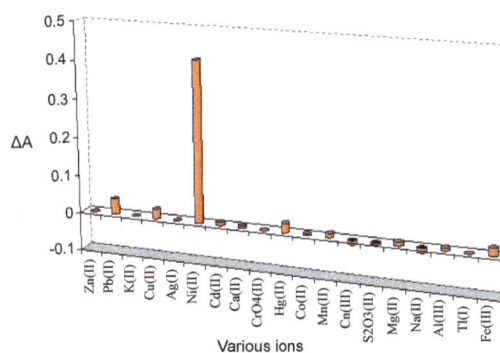

**Figure 2:** The absorbance variations of the 1-acenaphthoquinone 1-thiosemicarbazone membrane sensor at maximum wavelength for the studied ions.

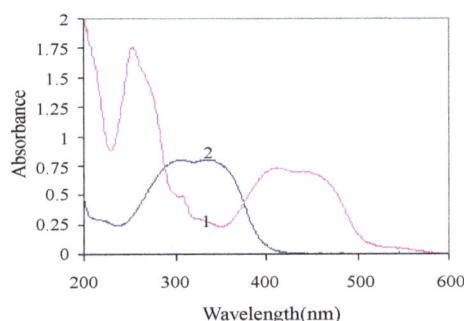

**Figure 3:** The curves 1 and 2 show the absorbance spectra of the 1-acenaphthoquinone 1-thiosemicarbazone in methanol solution ($1 \times 10^{-4}$ mol $L^{-1}$) and after immobilization on a triacetyl cellulose membrane, respectively. The blue shift observed for L (curve 2) upon immobilization on the membrane.

Figure 4 show the absorption spectra of immobilized 1-acenaphthoquinone 1-thiosemicarbazone on hydrolyzed cellulose acetate which was obtained after equilibration at pH 6.0 containing different concentrations of $Ni^{2+}$. The spectral characteristic of this optical sensor indicate maxima at 337 nm. It is evident that the membrane absorbance at 337 nm decrease by increasing $Ni^{2+}$ concentration as a result of the complex formation in the optode. During the titration, no measurable spectral shift was observed, which is typical for an absorption process involving a strong complex formation [26].

## Effect of pH on the sensor response

The response characteristic of the prepared membrane sensor was highly dependent to pH. Since variation of pH changed the absorbance of both the free and complexed forms of the immobilized **L**, for the study of the effect of pH absorbance differences (ΔA) before and after addition of $Ni^{2+}$ was followed in a pH range of 4 to 10. As shown in Figure 5, the change in absorbance increased rapidly by changing the pH from 4 to about 5.5, while it was decreased at pH values higher than 6.5. The diminished response at the low pH region may be explained by the extraction of $H^+$ from the test solution into the membrane, via protonation of the donor atoms of **L**, resulting in an expected change in the formation of a $Ni^{2+}$-**L** complex. On the other hand the reduced optical response of the proposed sensor due to a possible of $Ni^{2+}$ hydrolysis in higher pH values. Thus, a pH of 6.0 was considered as optimum and used for further studies [33].

## Calibration curve of the sensor

The dynamic working ranges for the proposed membrane sensor was studied by stepwise addition of $Ni^{2+}$ to a series of test solutions followed by the absorbance difference monitoring at 337 nm. It was found that the absorbance decreased continuously by increasing the Ni(II) concentration and the membrane was saturated when the $Ni^{2+}$ concentration exceeded $10^{-4}$ mol $L^{-1}$. Under the specified experimental conditions, the calibration curve in a logarithmic scale for $Ni^{2+}$ was linear from $5.01 \times 10^{-10}$ to $2.04 \times 10^{-5}$ mol $L^{-1}$. According to the definition of IUPAC, the limit of detection (LOD, 3σ) of this method was $1.00 \times 10^{-10}$ mol $L^{-1}$ which is sufficiently low for $Ni^{2+}$ monitoring in environmental samples [10,34,35]. The regression equation for the calibration curve shown in Figure 6 was ΔA=0.097 × Log ($Ni^{2+}$)+0.925 with a correlation coefficient ($R^2$) of 0.994.

## The sensor response time

The response time of the $Ni^{2+}$ sensor was calculated by plotting the absorbance as a function of time at two levels of $Ni^{2+}$ concentrations. As shown in Figure 7 the profile of the response of $Ni^{2+}$ optical sensor at 337 nm with time, the absorbance gets to 95% of the steady state signal in about 1 min. In general the response time decreased by increasing the analyte concentration. This may be explained by the fact that at a higher analyte concentration the rate of its diffusion in the membrane phase may be increased [10,34].

## Effect of interfering ions

Perhaps the most important characteristics of an ion-selective optode are its selectivity, which reflects its relative response for primary ion over diverse ions present in solution. Thus, the influence of several potentially interfering ions on the response behavior of the membrane sensor was studied. To investigate the selectivity of the proposed $Ni^{2+}$ optical membrane sensor, the absorbance of a fixed concentration of nickel ion, at $1.0 \times 10^{-8}$ mol $L^{-1}$ level, in a solution of pH=6.0 was recorded before ($\Delta A_o$) and after (ΔA) addition of some

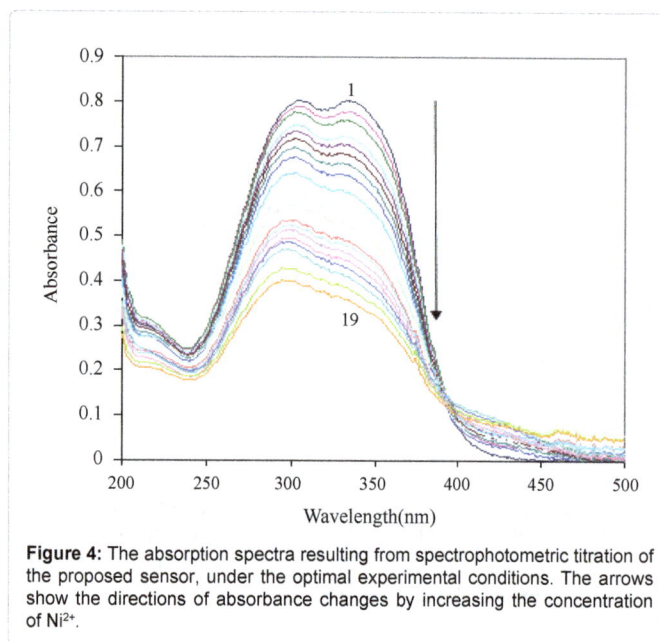

**Figure 4:** The absorption spectra resulting from spectrophotometric titration of the proposed sensor, under the optimal experimental conditions. The arrows show the directions of absorbance changes by increasing the concentration of $Ni^{2+}$.

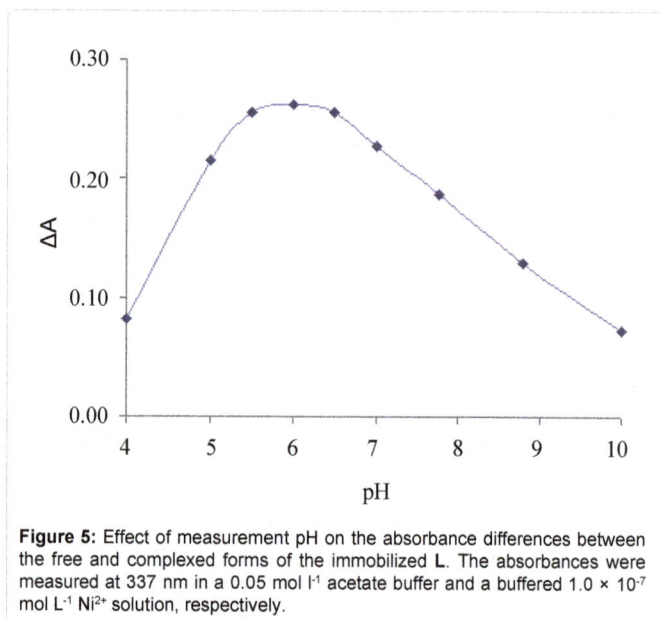

**Figure 5:** Effect of measurement pH on the absorbance differences between the free and complexed forms of the immobilized **L**. The absorbances were measured at 337 nm in a 0.05 mol $l^{-1}$ acetate buffer and a buffered $1.0 \times 10^{-7}$ mol $L^{-1}$ $Ni^{2+}$ solution, respectively.

potentially interfering ions such as $Ca^{2+}$, $Mn^{2+}$, $Al^{3+}$, $Zn^{2+}$, $Cd^{2+}$, $Pb^{2+}$, $Ag^+$, $Hg^{2+}$, $Co^{2+}$, $Cu^{2+}$, $Cr^{3+}$, $Mg^{2+}$, $Na^+$, $K^+$, $Tl^+$, $Fe^{3+}$, $Li^+$, $Ba^{2+}$ and $Ce^{3+}$ at concentrations up to 100 times of the analyte ion. The resulting relative error is defined as RE(%)=[$(\Delta A-\Delta A_o)/\Delta A_o$] × 100. The results of the selectivity studies are summarized in Figure 8. The data clearly indicate that, for all the studied metal ions, the relative error is up to 4%, which demonstrated that the studied interfering ions with a concentration of at least 100 times of $Ni^{2+}$ ion, have no significant effect on the analytical signal [21,34,35].

## Regeneration and reproducibility of the sensor

Multiple usage of an optical sensor is feasible if the sensor can be easily regenerated and give reproducible responses. Different compounds, EDTA and $SCN^-$ solutions with different concentrations were tested for regeneration of the membrane sensor and desorption of

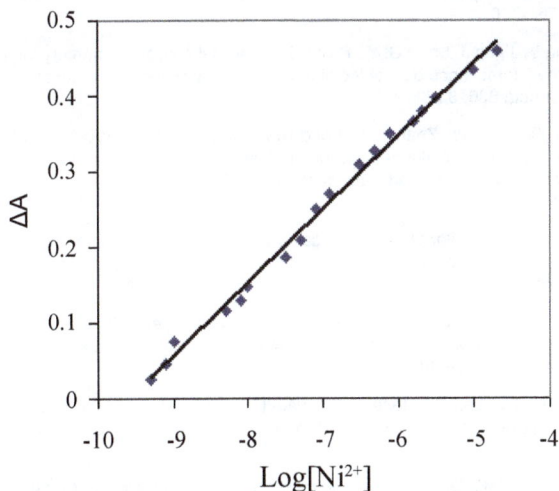

**Figure 6:** The logarithmic scale calibration curve for the 1-acenaphthoquinone 1-thiosemicarbazone membrane sensor at 337 nm.

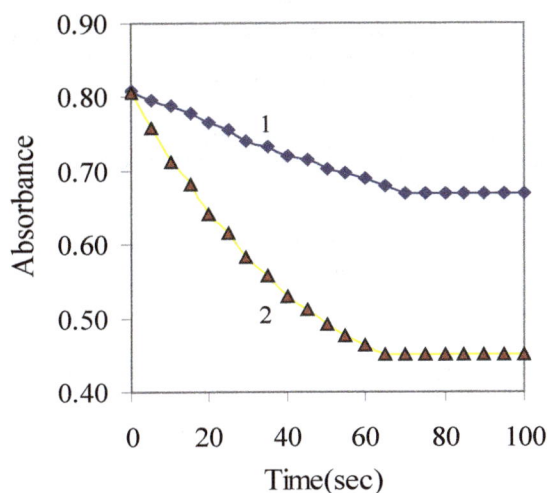

**Figure 7:** Plot of absorption vs. time of the sensor at 337 nm for two $Ni^{2+}$ concentration levels of $5.0 \times 10^{-9}$ and $1.0 \times 10^{-6}$ mol $L^{-1}$ according to curves 1 and 2, respectively.

$Ni^{2+}$ from it, see Figure 9. The best reagent was an EDTA solution with a concentration of 0.1 mol $L^{-1}$ or higher, which can efficiently remove any adsorbed $Ni^{2+}$ from the membrane and returns its absorbance to its initial value for membrane ($\Delta A \approx 0$) in less than about three minute.

The reproducibility of the sensor response was tested by its multiple usages for $Ni^{2+}$ monitoring in test solutions at two concentration levels of $1.0 \times 10^{-9}$ and $1.0 \times 10^{-7}$ mol $L^{-1}$. After each absorbance reading, the membrane was cleaned by 0.1 mol $L^{-1}$ EDTA solution, pure water and a 0.05 mol $L^{-1}$ acetate buffer solution, respectively. As shown in Figure 10, good reproducibilities were obtained at both $Ni^{2+}$ concentration levels. The corresponding RSD values were 1.45% and 0.83%, respectively [21,34,35].

### Lifetime and stability

The life time of the membrane sensor was tested over a period of 4 months during which the membranes were stored in water [5,36-

38]. The mean absorbances of the membranes at 337 nm were found to be 0.801 ($\pm$ 0.025) and 0.805 ($\pm$ 0.020), before and after this period, respectively. Hence, the membranes are stable within this period with a minimum life time of 4 months. Also no evidence of the ionophore leaching or signal drift was observed during multiple usages of the membrane.

Additionally, the short-term stability of the optode membrane was investigated by monitoring its absorbance values during its contact with a $1.0 \ 10^{-7}$ M solution of $Ni^{2+}$ at pH 6.0 over a period of 6 hours. From the absorbance measurements in 30 min intervals (n=2), it was found that the response was almost unchanged with only 1.0% increase in absorbance at tested wavelength after the 6 hours monitoring [34].

### Application

The proposed $Ni^{2+}$-selective optical sensor was used for the determination of $Ni^{2+}$ in two natural water samples. The data given in Table 1 show a good agreement between the measured values using the proposed method and those obtained by the atomic absorption spectrometry (AAS) laboratory at Razi University, Kermanshah, Iran. It may be concluded that the proposed membrane sensor is selective to $Ni^{2+}$ and may be used for monitoring of this ion in real water samples.

### Conclusion

The $Ni^{2+}$ optical sensor that is prepared on the basis of 1-acenaphthoquinone 1-thiosemicarbazone in this study shows very good selectivity for $Ni^{2+}$ over other common metal ions. The proposed sensor showed very favorable optical properties for its use as an optical sensor, such as high selectivity, adequate life time, fast and reproducible regeneration, low cost and simple fabrication and handling. The sensor

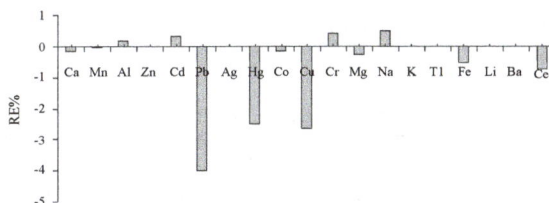

**Figure 8:** Interferences from different metal ions to the spectrophotometric determination of $Ni^{2+}$ ion for the 1-acenaphthoquinone 1-thiosemicarbazone membrane sensor at pH 6. Concentration of $Ni^{2+}$ and the diverse ions were $1.0 \times 10^{-8}$ and $1.0 \times 10^{-6}$ mol $L^{-1}$, respectively.

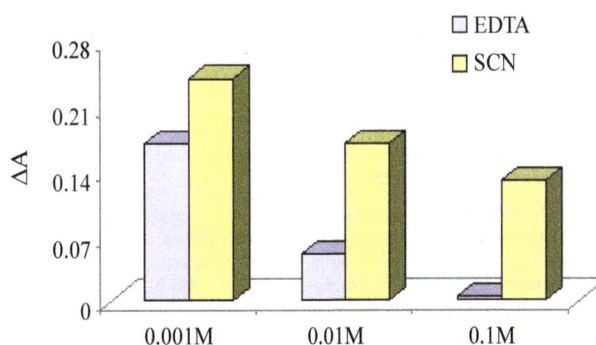

**Figure 9:** Regeneration of the membrane in the presence of different concentrations of EDTA and $SCN^{-}$. Experimental conditions: wavelength, 337 nm; measurement pH was 6.0.

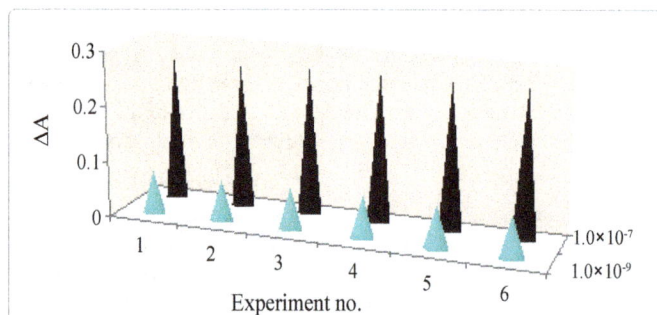

**Figure 10:** Reproducibility of the optical sensor signal at two Ni$^{2+}$ concentration levels of $1.0 \times 10^{-8}$ and $1.0 \times 10^{-6}$ mol L$^{-1}$.

| Samples | Ni$^{2+}$ (mol L$^{-1}$)$^a$ | |
| --- | --- | --- |
| | Measured value (Triacetyl cellulose membrane) | Reference value (AAS) |
| **Gharahsoo River water** | 1.68 (± 0.22) × 10$^{-7}$ | 1.73 ( ± 0.23) × 10$^{-7}$ |
| **Spring water** | 1.21 (± 0.19) × 10$^{-7}$ | 1.25 ( ± 0.20) × 10$^{-7}$ |

$^a$values in the parentheses are standard deviation based on the three replicate analyses

**Table 1:** Application of the membrane sensor for the determination of Ni$^{2+}$ in different real samples.

can be regenerated readily with an EDTA solution and demonstrated a long life time with the possibility of multiple uses for environmental monitoring of Ni$^{2+}$. Due to the advantages of the proposed method with respect to previously reported ones it may be used as an alternative method for Ni$^{2+}$ determination over a range of $5.01 \times 10^{-10}$ to $2.04 \times 10^{-5}$ mol L$^{-1}$ values without any significant interference from other metal ions.

## Acknowledgements

The authors thank the Lorestan University for supporting this study.

## References

1. Ferreira SLC, Santos WNL dos, Lemos VA (2001) On-line preconcentration system for nickel determination in food samples by flame atomic absorption spectrometry. Anal Chim Acta 445: 145-151.

2. Underwood EJ (1977) Trace elements in human and Animal Nutrition. Fourth edn. Academic Press, New York, USA.

3. Hu NL, Gao HW, Zhang B, Zhan GQ (2005) Simultaneous determination of cobalt and nickel in wastewater with 2-hydroxyl-5-benzeneazoformoamithiozone by spectral correction technique. J Chin Chem Soc 52: 1145-1152.

4. Wu MH, Lin JL, Wang J, Cui Z, Cui Z (2009) Development of high throughput optical sensor array for on-line pH monitoring in micro-scale cell culture environment. Biomed Microdevices 11: 265-273.

5. Mohr GJ, Wolfbeis OS (1994) Optical sensors for a wide range of pH based on azo dyes immobilized on a novel support. Anal Chim Acta 292: 41-48.

6. Wang E, Chow KF, Kwan V, Chin T, Wong C, et al. (2003) Fast and long term optical sensors for pH based on sol-gels. Anal Chim Acta 495: 45-50.

7. Safavi A, Bagheri M (2003) Novel optical pH sensor for high and low pH values. Sens Actuators B 90: 143-150.

8. Alizadeh K, Nemati H, Zohrevand S, Hashemi P, Kakanejadifard A, et al. (2013) Selective dispersive liquid-liquid microextraction and preconcentration of Ni(II) into a micro droplet followed by ETAAS determination using a yellow Schiff's base bisazanyl derivative. Mater Sci Eng C 33: 916-922.

9. Amini MK, Khezri B, Firooz AR (2008) Development of a highly sensitive and selective optical chemical sensor for batch and flow-through determination of mercury ion. Sens Actuators B 131: 470-478.

10. Hashemi P, Abolghasemi MM, Alizadeh K, Afzari Zarjani R (2008) A calmagite immobilized agarose membrane optical sensor for selective monitoring of Cu2+. Sens Actuators B 129: 332-338.

11. Dridi C, Ben Ali M (2008) Electrical and optical study on modified thiacalix(4) arene sensing molecules: Application to Hg2+ ion detection. Mater Sci Eng C 28: 765-770.

12. Yang Y, Jiang J, Shen G, Yu R (2009) An optical sensor for mercury ion based on the fluorescence quenching of tetra(p-dimethylaminophenyl)porphyrin. Anal Chim Acta 636: 83-88.

13. Han ZX, Luo HY, Zhang XB, Kong RM, Shen GL, et al. (2009) A ratiometric chemosensor for fluorescent determination of Hg(2+) based on a new porphyrin-quinoline dyad. Spectrochim Acta A Mol Biomol Spectrosc 72: 1084-1088.

14. Hisamoto H, Watanabe K, Nakagawa E, Siswanta D, Shichi Y, et al. (1994) Flow-through type calcium ion selective optodes based on novel neutral ionophores and a lipophilic anionic dye. Anal Chim Acta 299: 179-187.

15. O'Neill S, Conway S, Twellmeyer J, Egan O, Nolan K, et al. (1999) Ion-selective optode membranes using 9-(4-diethylamino-2-octadecanoatestyryl)-acridine acidochromic dye. Anal Chim Acta 398: 1-11.

16. Tóth K, Thu Lan BT, Jeney J, Horváth M, Bitter I, et al. (1994) Chromogenic calix[4]arene as ionophore for potentiometric and optical sensors. Talanta 41: 1041-1049.

17. Shortreed MR, Dourado S, Kopelman R (1997) Development of a fluorescent optical potassium-selective ion sensor with ratiometric response for intracellular applications. Sens Actuators B 38-39: 8-12.

18. Ensafi A, Bakhshi M (2003) A stable optical film sensor based on immobilization of 2-amino-1-cyclopentene-1-dithiocarboxylic acid on acetyl cellulose membrane for Ni(II) determination. Sens Actuators B 96: 435-440.

19. Shamsipur M, Poursaberi T, Karami AR, Hosseini M, Momeni A, et al. (2004) Development of a new fluorimetric bulk optode membrane based on 2,5-thiophenylbis(5-tert-butyl-1,3-benzexazole) for nickel(II) ions. Analytica Chimica Acta 501: 55-60.

20. Alizadeh N, Moemeni A, Shamsipur M (2002) Poly(vinyl chloride)-membrane ion-selective bulk optode based on 1,10-dibenzyl-1,10-diaza-18-crown-6 and 1-(2-pyridylazo)-2-naphthol for Cu2+ and Pb2+ ions. Anal Chim Acta 464: 187-196.

21. Shamsipur M, Hosseini M, Alizadeh K, Alizadeh N, Yari A, et al. (2005) Novel fluorimetric bulk optode membrane based on a dansylamidopropyl pendant arm derivative of 1-aza-4,10-dithia-7-oxacyclododecane ([12]aneNS2O) for selective subnanomolar detection of Hg(II) ions. Anal Chim Acta 533: 17-24.

22. Kim SH, Han SK, Park SH, Lee SM, Mi Lee S, et al. (1999) Use of squarylium dyes as a sensing molecule in optical sensors for the detection of metal ions. Dyes and Pigments 41: 221-226.

23. Ganjali MR, Poursaberi T, Hajiagha-Babaei L, Rouhani S, Yousefi M, et al. (2001) Highly selective and sensitive copper(II) membrane coated graphite electrode based on a recently synthesized Schiff's base. Anal Chim Acta 440: 81-87.

24. Gholivand MB, Ahmadi F, Rafiee E (2006) A novel Al(III)-selective electrochemical sensor based on N,N'-bis(salicylidene)-l,2-phenylenediamine complexes. Electroanalysis 18: 1620-1626.

25. Ganjali MR, Rezapour M, Norouzi P, Salavati-Niasari M (2005) A new pentadentate S-N Schiffs-Base as a novel ionophore in construction of a novel Gd(III) membrane sensor. Electroanalysis 17: 2032-2036.

26. Alizadeh K, Rezaei B, Khazaeli E (2014) A new triazene-1-oxide derivative, immobilized on the triacetyl cellulose membrane as an optical Ni2+ sensor. Sens Actuators B 193: 267-272.

27. Hahn R, Herrmann WA, Artus GRJ, Kleine M (1995) Biologically relevant metal coordination compounds: MoVIO2 and nickel(II) complexes with tridentate aromatic Schiff bases. Polyhedron 14: 2953-2960.

28. Su X, Aprahamian I (2014) Hydrazone-based switches, metallo-assemblies and sensors. Chem Soc Rev 43: 1963-1981.

29. Alizadeh K, Seyyedi S, Avanes A, Ganjali MR, Faridbod F (2011) Concentration and Temperature Effects on the Electronic Absorption Spectra of 1-pyridinyl-2-methylene-benzenecarbohydrazonic Acid Following Solvatochromic Studies. Acta Chim Slov 58: 251-255.

30. Faridbod F, Ganjali MR, Dinarvand R, Norouzi P, Riahi S (2008) Schiff's bases and crown ethers as supramolecular sensing materials in the construction of potentiometric membrane sensors. Sensors 8: 1645-1703.

31. Alizadeh K, Ghiasvand AR, Borzoei M, Zohrevand S, Rezaei B, et al. (2009) Experimental and computational study on the aqueous acidity constants of some new aminobenzoic acid compounds. J Mol Liq 149: 60-65.

32. Noroozifar M, Khorasani Motlagh M, Taheri A, Zare Dorabei R (2008) Diphenylthiocarbazone immobilized on the triacetyl cellulose membrane as an optical silver sensor. Turk J Chem 32: 249-257.

33. Hashemi P, Hosseini M, Zargoosh K, Alizadeh K (2011) High sensitive optode for selective determination of Ni2+ based on the covalently immobilized thionine in agarose membrane. Sens Actuators B 153: 24-28.

34. Alizadeh K, Parooi R, Hashemi P, Rezaei B, Ganjali MR (2011) A new Schiff's base ligand immobilized agarose membrane optical sensor for selective monitoring of mercury ion. J Hazard Mater 186: 1794-1800.

35. Shamsipur M, Alizadeh K, Hosseini M, Caltagirone C, Lippolis V (2006) A selective optode membrane for silver ion based on fluorescence quenching of the dansylamidopropyl pendant arm derivative of 1-aza-4,7,10-trithiacyclododecane ([12]aneNS3). Sens Actuators B 113: 892-899.

36. Ahmad M, Hamzah H, Sufliza Marsom E (1998) Development of an Hg(II) fibre-optic sensor for aqueous environmental monitoring. Talanta 47: 275-283.

37. Kilian K, Pyrzyaska K (2003) Spectrophotometric study of Cd(II), Pb(II), Hg(II) and Zn(II) complexes with 5,10,15,20-tetrakis(4-carboxylphenyl)porphyrin. Talanta 60: 669-678.

38. Shamsipur M, Shirmardi Dezaki A, Akhond M, Sharghi H, Paziraee Z, et al. (2009) Novel PVC-membrane potentiometric sensors based on a recently synthesized sulfur-containing macrocyclic diamide for Cd2+ ion. Application to flow-injection potentiometry. J Hazard Mater 172: 566-573.

# 7

# Growth Rate and Morphology of a Single Calcium Carbonate Crystal on Polysulfone Film Measured with Time Lapse Raman Micro Spectroscopy

**Barbara Liszka M[1,2]\*, Aufried Lenferink TM[1] and Cees Otto[1]\***

[1]Department of Medical Cell Biophysics, MIRA Institute, University of Twente, Enschede, The Netherlands
[2]Wetsus, European Centre of Excellence for Sustainable Water Technology, Leeuwarden, The Netherlands

### Abstract

The growth of single, self- nucleated calcium carbonate crystals on a polysulfone (PSU) film was investigated with high resolution, time lapse Raman imaging. The Raman images were acquired on the interface of the polymer with the crystal. The growth of crystals could thus be followed in time. PSU is a polymer that is used as a membrane material in water cleaning technology. The intensity of the Raman band at the position of 1086 cm$^{-1}$, which is due to the symmetric stretching of the C-O bonds in the carbonate group of calcite was used to translate the number of $CO_3^{2-}$ ions in a crystal to the growth in time. The growth rate of single crystals of calcium carbonate on a surface was obtained from successive Raman images. We are presenting for the first time time-lapse Raman images of single crystal growth as a direct method to determine a crystal growth rate on an industrially relevant membrane material, like polysulfone.

**Keywords:** Crystal growth rate; Calcium carbonate; Raman imaging; Polymer surface; Time lapse images

**Abbreviations:** PSU: Polysulfone; SS1: Solution made by mixing of 4 mM NaHCO$_3$ and 8 mM of CaCl$_2$; SS2: Solution made by mixing of 10 mM NaHCO$_3$ and 10 mM of Calcium chloride (CaCl$_2$); ROI: Region of Interest.

## Introduction

The phenomenon of calcium carbonate crystal growth from aqueous solutions on surfaces widely occurs in systems where carbonate and bicarbonate ions are present, such as in domestic systems, waste and drinkable water treatment systems or industrial apparatus where water is used. Crystal growth leads to the formation of mineral scale which reduces the performance of membrane materials in equipment. The scale development is affected by factors such as: pH, super-saturation index, temperature, water composition etc.

The process of scaling and crystal growth includes the following stages:

1) The induction period involve nucleation and crystal growth.

2) The mineralizing crystals and others particles transportation from the bulk and its adhesion to the surface,

3) Ageing of crystals at a surface, for instance due to recrystallization and dehydration.

Other components present in a mineralizing solution could increase or decrease the crystal growth rate by adhesion to crystal surfaces.

The abovementioned physico-chemical factors play an important role in crystal growth on a surface and hence in the performance over time of membrane materials. An understanding of the kinetics governing formation and growth of calcium carbonate crystals on surface is important to gain the ability to predict, control and direct or stop this process. Many studies have been performed of calcium carbonate crystal growth on a macroscopic scale using indirect methods that monitor changes of solution chemistry [1]. These studies have revealed a dependence of growth kinetics upon parameters such as pH, supersaturation ratio, ionic strength or temperature [2-4]. Atomic Force Microscopy (AFM) has been extensively used to study

mechanisms and growth rates of single crystals from solution [5-7], including calcite crystals [1,8-14]. The high resolution of AFM can visualize monomolecular steps on atomically flat crystal surfaces. It has been also observed that the AFM tip can influence the growth rate under supersaturated conditions [9]. Vertical scanning interferometry is an alternative approach to study growth rates and morphology of single barite crystals [5]. In these approaches the mineralization was studied after seeding. Therefore a measured growth rate is dependent on e.g., seed preparation method. In another study, cryo-electron tomography was used to investigate self-nucleated, template controlled growth of CaCO$_3$ crystals from the solution–phase [15]. Early crystallization events of a few nanometer were observed.

Raman micro-spectroscopy is a well-established method to study crystal morphology as it provides information about the chemical composition and crystal polymorphism [16-20]. The method enables investigations under ambient conditions in an aqueous environment. Raman micro-spectroscopy is a non-invasive, contact-free method that does not influence crystal formation. Furthermore, the method is quantitative as the Raman scattering intensity of a typical group vibration is directly related to the total number of that chemical group.

This paper presents time lapse Raman spectroscopic imaging for *in situ* study of single calcium carbonate crystals growing on a polysulfone film under diffusion-limited conditions in a stagnant solution. In time lapse Raman imaging a time-sequence of Raman images is made

---

**\*Corresponding authors:** Barbara Maria Liszka, Department of Medical Cell Biophysics, MIRA Institute, University of Twente, Enschede, The Netherlands, E-mail: b.liszka@utwente.nl

Cees Otto, Department of Medical Cell Biophysics, MIRA Institute, University of Twente, Enschede, The Netherlands, E-mail: c.otto@utwente.nl

of a region of interest. All the individual images can be analyzed simultaneously to provide physico-chemical data of the growing crystal. It will be shown below that the mass growth rate of individual calcite crystals can be obtained directly from time lapse Raman images.

## Materials and Methods

### Materials

Sodium bicarbonate ($NaHCO_3$), calcium chloride ($CaCl_2$), potassium chloride (KCl), sodium hydroxide (NaOH) was of analytical grade and obtained from Sigma Aldrich, Germany. Pellets of polysulfone (PSU) with ~ 35,000 Mw was purchased from Sigma Aldrich, Germany. Cyclopentanone was used as a solvent for polysulfone thin film production. Analytical grade cyclopentanone was purchased from Fluka Analytical. Raman grade calcium fluoride windows with a thickness of 200 μm and a diameter of 1 cm were obtained from Crystran Ltd.

### Methods

**Liquid cell:** A liquid cell was custom made from glass (DURAN® borosilicate) with a volume of 32 ml. At the top of the cell a window with a diameter of 0.8 mm was made by removing the glass by cutting. The window was sealed with 200 μm calcium fluoride disc ($CaF_2$). The disc was covered with a 4.5 ± 1.5 mm thin polysulfone film. The polymer-covered surface was mounted towards the interior of the liquid cell. The orientation of the active polysulfone surface above the mineralization solution avoided sedimentation on the surface of crystals that may have grown in solution. High resolution Raman imaging could be performed on the interface polymer- mineralization solution with a high numerical aperture (NA=0.95) objective with a working distance of 300 μm. $CaF_2$ was selected as a substrate because this material gives minimal optical contributions to the Raman scattering. The cell was placed on a microscope stage with the measurement window towards the objective. Measurements were performed in a closed liquid cell to prevent exchange of carbon dioxide from the chamber with the atmosphere. A schematic figure of the fluidic cell is presented in Figure

1A. The Raman signal was acquired by stepping the laser in x- and y directions over the region of interest. The Raman spectral information comes from a laser focal volume (c), with an ellipsoidal shape with dimensions of 1500 nm (a) by 330 nm in diameter (b), shown in Figure 1B together with a schematic representation of imaging.

The liquid cell was cleaned before use with 1M $H_2SO_4$ and rinsed thoroughly with milliQ water. The polysulfone film was prepared by dissolving polymers pellets in cyclopentanone up to a 0.4%wt concentration. The entire calcium fluoride window was wetted with a polymer solution and the substrate was accelerated to a spin speed of 4000 rpm. The total spinning time was 6 s. The thickness of the polysulfone film was determined to 4.5 ± 1.5 mm.

**Supersaturated solutions:** Two types of supersaturated solution were prepared, which will be distinguished as SS1 and SS2. In order to prepare solution SS1 two solutions of 250 mL of $NaHCO_3$ (4.0 mM) and $CaCl_2$ (8.0 mM) were prepared. The ionic strength was adjusted to 0.10 by the addition of 3.2 g KCl to the solution of $NaHCO_3$. The pH was adjusted to a value of 8.89 by the addition of 30 μL of aqueous NaOH (3.0 M) to the solution of $CaCl_2$. The final SS1 was prepared by mixing aqueous solutions of $NaHCO_3$ and $CaCl_2$ in a ratio of 1:1 to obtain the desired volume.

SS2 was prepared by mixing 250 mL of a solution of $NaHCO_3$ (10.0 mM) and 250 mL of a solution of $CaCl_2$ (10.0 mM). All solutions were prepared with Milli-Q water.

The driving force for crystals formation is the super-saturation ratio, which is defined in following equation:

$$\frac{\Delta \mu}{RT} = \vartheta \ln(S) \qquad 1.1$$

Where $R$ [$J\ mol^{-1}\ K^{-1}$] is the gas constant, $T[K]$ is the absolute temperature, $\Delta \mu$ ($J\ mol^{-1}$) is the change in chemical potential of ions in the crystal ($\mu_{ic}$) and the chemical potential of ions in the liquid phase ($\mu iL$). $\vartheta$ is the number of ions in the formula units (for $CaCO_3$ $\upsilon$=2), $S$ is super-saturation ratio.

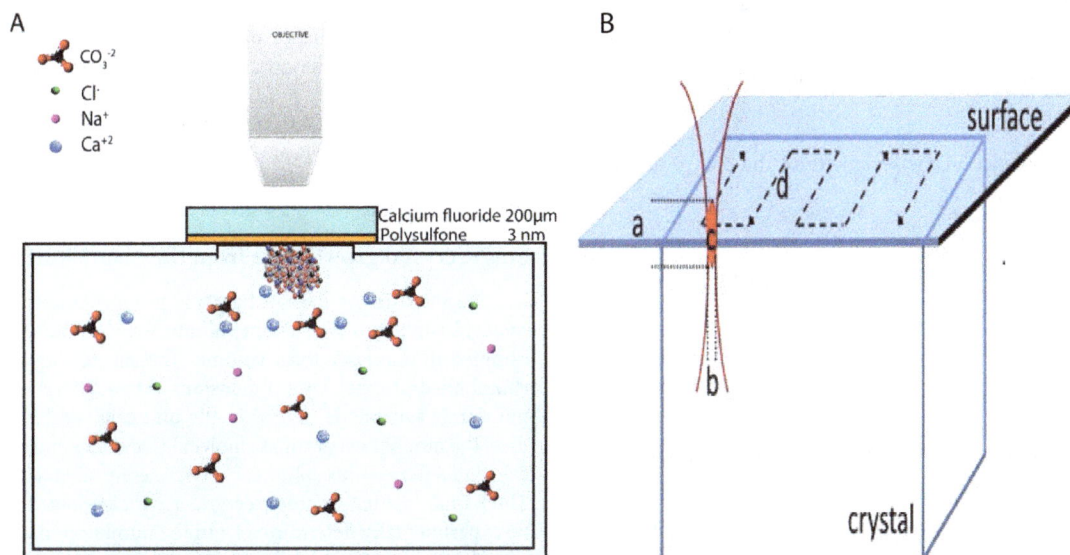

**Figure 1:** A) Schematic representation of the liquid cell. B) Schematic representation of the scanning procedure in Raman imaging: "c" is the laser focal volume with dimensions a: 1500 nm and b: 330 nm. The laser is focused on the surface and image is acquired by stepping the laser along the lines e. Vertical and horizontal dimensions in this schematic representation are not to scale.

For ionic solutions, such as $CaCO_3$, the super-saturation can best be expressed in terms of ion activities [21].

$$s = \left( \frac{IAP}{K_{sp}} \right)^{1/9} \qquad 1.2$$

Where $IAP$ is the ion activity product, $K_{sp}$ is the thermodynamic solubility product, which is $3.36 \times 10^{-9}$ for calcium carbonate in the calcite structure [22].

The super-saturation ratio and the chemical speciation of the solutions in this study were calculated in the program Visual MinteQ 3.1 from the total concentration of ions in solution ($Cl^-$, $Ca^{2+}$, $Na^+$, $H^+$, $CO_3^{2-}$, $K^+$) and the activity coefficients as computed by the program [23-25].

**Confocal Raman micro spectrometer and data analysis:** All Raman measurement were conducted with a home-built Raman spectrometer [26]. The excitation source was a Kr-ion laser (Coherent, Innova 90-K. Santa Clara, CA) with an emission wavelength of 647.09 nm. Raman spectra were acquired using an objective with 40X magnification and a NA of 0.95. A pinhole with a diameter of 15 µm functioned as the entrance aperture to the spectrograph. The spectrograph dispersed the light in the range from 645-847 nm. The Raman signal was detected by a back illuminated CCD camera with $1600 \times 200$ pixels (Newton DU-970N-BV; Andor Technology). The combined spectrograph and CCD camera recorded Raman shifts between -50 $cm^{-1}$ and 3650 $cm^{-1}$.

The series of images were acquired under following conditions for SS1 laser power 75 mW, an acquisition time 0.1 s/per pixel, Raman maps $32 \times 32$ spectra, and for SS2 laser power 80 mW, an acquisition time 0.034 s/per pixel, Raman maps $32 \times 32$ spectra. The position of a crystal was located on the polysulfone surface by bright field optical microscopy. Crystals with a diameter larger than 100 nm could be discerned by bright field microscopy. The Raman excitation laser beam was focused on the polymer surface. The Raman images were acquired with the crystal in the center of the region of interest (ROI). The position of the ROI was adjusted in between images during time lapse measurement to follow the direction of crystal growth. The size of the square ROI was adjusted to the size of the crystal. The images were taken one after another with user defined time delay between images.

The measured Raman spectra were corrected for:

1) Cosmic ray events,

2) wavelength-dependence of the sensitivity of the complete Raman spectrometer,

3) CCD camera-etaloning,

4) Absolute amplitude correction by a calibrated tungsten halogen lamp and

5) Conversion from pixels to wavenumbers by the combined output from a mercury-argon lamp and the recorded Raman spectrum of toluene.

After this procedure fully corrected Raman spectra were stored and used for further analysis.

The corrected Raman spectra were subjected to singular value decomposition (SVD) to improve the signal to noise ratio [27]. The Raman scattering intensity in each pixel of the time lapse images was determined with a custom written Labview program. The intensity was calculated for the carbonate vibrational mode at 1086 $cm^{-1}$ after subtraction of a local linear baseline between selected anchor points left and right of the band of interest.

## Results and Discussion

The onset of crystal growth was monitored by bright field microscopy and upon the observation of crystal growth event the region of interest (ROI) of the polysulfone polymer surface was monitored by time lapse Raman imaging with the crystal in the center of the ROI. The intensity of the carbonate ion symmetric stretch vibration at a position of 1086 $cm^{-1}$ was used to construct images. Each resulting Raman image shows the morphology of the crystal and the time sequence of the images shows crystal growth from a supersaturated solution (Figure 2A). The region of interest in the first four images was 4.5 µm × 4.5 µm. The step size (x, y) of the laser scanner was 146 nm. This step size is in accordance with the Nyquist sampling frequency for optimal information retrieval. The 5th to 15th images were acquired with a step size of 397 nm to maintain a clear border between the growing crystal and the edge of the region of interest. An example of an average Raman spectrum of a crystal is presented in Figure 2B. The characteristic Raman frequencies of carbonate ion in a calcite crystal lattice occur at 1086 $cm^{-1}$: C-O symmetric stretching, 711 $cm^{-1}$: C-O antisymmetric stretching, 1436 $cm^{-1}$, 1750 $cm^{-1}$: C=O stretching and lattice related modes in positions at 156 $cm^{-1}$ and at 282 $cm^{-1}$ [17,20-28]. The Raman spectra therefore, show that the crystal on a PSU film formed in a calcite polymorphism. The presence of calcite polymorph was observed in all the Raman images of the time lapse series. The earliest image was obtained 25 minutes after mixing. No Raman bands of other calcium carbonate polymorphs, like vaterite, aragonite or amorphous calcium carbonate, could be discerned. Calcite is thermodynamically the most stable polymorph of calcium carbonate and no changes were observed in crystal polymorphism during growth. The Raman spectrum of the $4.5 \pm 1.5$ nm polysulfone polymer layer that acted as the active material for crystal growth is shown in Figure 2C. The spectrum contains characteristics PSU bands in position 1073, 1108, 1148 $cm^{-1}$. The band in the position 1086 $cm^{-1}$, characteristic for carbonate, is still visible in the spectrum of the background as weak contributions from crystal edges contribute to the background signal.

The anisotropic growth of the crystal after the first 43 minutes can be observed in the images presented in Figure 2A. After the image taken at 43 minutes, the crystal starts to grow irregularly due to rapid anisotropic growth. The positions at the crystal surface where irregular growth occurs are marked in the images by arrows. The irregularity in the crystal shape disappears during its growth and in the last image, acquired after 24 h, the crystal achieved a regular shape very familiar for calcite crystals.

### Single crystal growth rate from time lapse Raman images

The intensity of Raman bands is proportional to the number of molecules in the measurement volume, which is the volume effectively sampled by the laser focal volume. The calcite single unit cell has a rhombohedral shape with dimensions a=b=4.9896 Å, c=17.061 Å, and the number of calcite groups in the unit cell Z=6 [29]. Based on this data, the number of carbonate molecules in the microscope volume was calculated in case this volume was completely filled with calcite crystal. The Raman scattering cross section, $\sigma_R$, of carbonate ion in calcite can be experimentally determined from the number of measured photons, as follows:

$$\sigma_R = \frac{(N_c h \upsilon_R)}{N_m \frac{I_l}{A}} \qquad 2.1$$

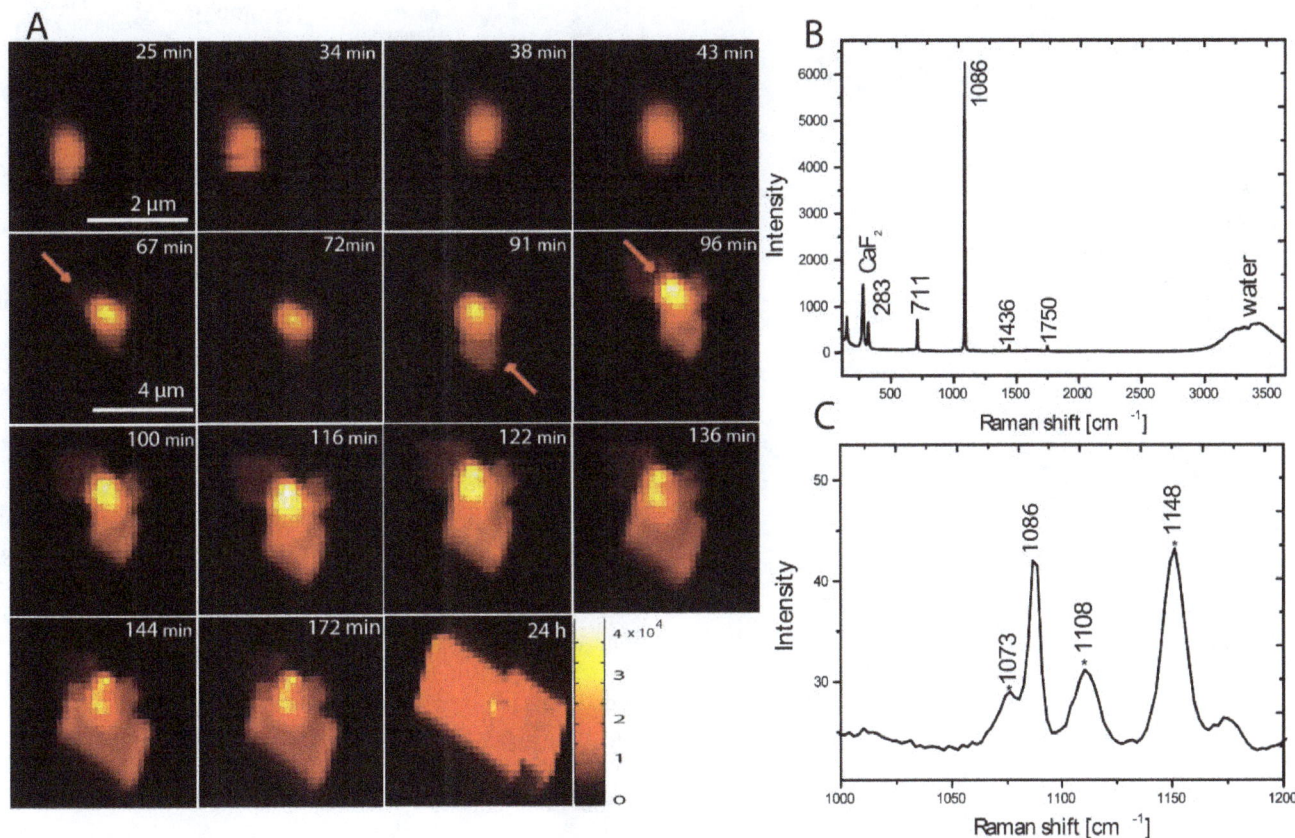

**Figure 2:** A) Time lapse Raman imaging of $CaCO_3$ crystal growth on a polysulfone film. The images are based on the intensity of band at 1086 cm⁻¹. The images were acquired with 75 mW and an acquisition time of 0.1 s/pixel. Each image contains 1024 spectra. The arrows show a position of anisotropic growth of the crystal. B) An average Raman spectrum of the calcite crystal, C) An average Raman spectrum of a 4.5 ± 1.5 nm polysulfone film.

Where $N_c$ is the number of measured counts in the integral intensity of the Raman band of the carbonate ion stretch vibration at 1086 cm⁻¹; $h = 6.63 \times 10^{-34}$ is Planck's constant; $v_r$ is the frequency [1/s] of the Raman scattered light at 1086 cm⁻¹ upon excitation with the 647.09 nm laser, $I_l$ is the laser intensity [W], A is the area of the laser beam [m²] in the focal plane of the microscope objective, $N_m$ is the number of $CO_3^{2-}$ ions in the microscope volume.

Equation (2.1) includes the measurement parameters and the set up characteristics to convert the detected number of counts to the emitted and collected photons. With the Raman cross section established, every pixel, with position coordinates (x, y), in the time lapse Raman series of images can be converted to an actual number of carbonate ions $N_m$, also when the size of a crystal is smaller than the laser focal volume.

$$N_m(x,y) = \frac{N_c(x,y)h v_R A}{\sigma_R \, I_L} \qquad 2.2$$

Figure 3A and 3B represents individual crystal growth from the super saturated solution SS1. The data were fitted to a quadratic function. Figure 3C shows crystal growth from a super saturated solution SS2. The starting concentration of solution SS2 was two and a half times higher degree of super saturation than SS1. The higher super saturation leads to a more rapid crystal growth, which was modeled by a sigmodal function, which represents full growth (until ions depletion) during this measurement time measurement time.

The growth rate of a crystal $R_G$ was calculated from equation 2.3 [21].

$$R_G = \frac{1}{S}\frac{dn}{dt} \quad \left[ \text{mol m}^{-2}\text{s}^{-1} \right] \qquad 2.3$$

Where $n$ is the number of moles $CO_3^{2-}$ ions, $S$ is the surface area of one crystal face measured for each time point and $t$ is time.

The surface $S$ of the growing crystal was calculated from the pixels in the Raman images, using the amount of pixels of the measured crystal surface multiplied by the known pixel size. The crystal growth rate was expressed as an average of growth rates calculated per time point [30,31].

The growth rate depends on solution parameters, such as: the supersaturation index (SI), the ionic strength (I), the pH and the $[Ca^{2+}]/[CO_3^{2-}]$ ratio [8,32-35]. Table 1 compares the results from this study with results from the literature [33-35]. Experimental conditions for crystal growth rates vary widely in the literature and importantly influence observed growth rates. The conditions in the literature included in Table 1 most closely approximated the conditions of this study.

The AFM data on crystal growth rate are usually obtained at constant saturation index. In our experiment growth was studied at a starting saturation index of 1.2, which was not maintained constant during growth.

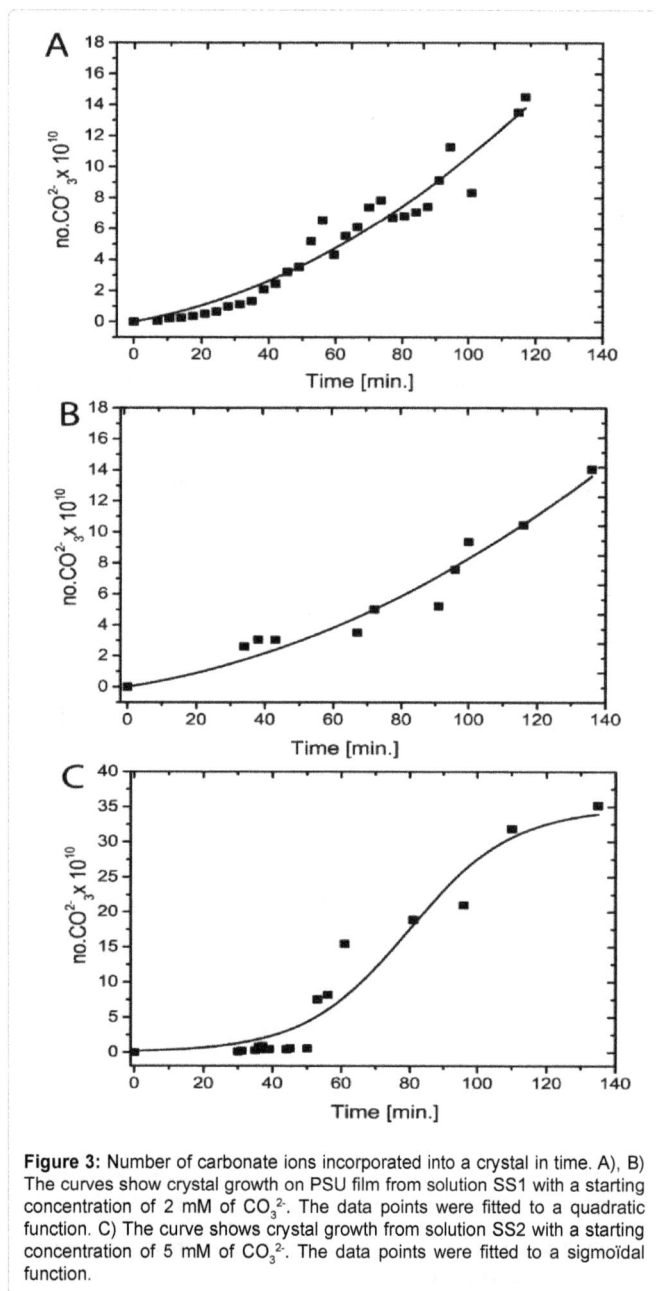

**Figure 3:** Number of carbonate ions incorporated into a crystal in time. A), B) The curves show crystal growth on PSU film from solution SS1 with a starting concentration of 2 mM of $CO_3^{2-}$. The data points were fitted to a quadratic function. C) The curve shows crystal growth from solution SS2 with a starting concentration of 5 mM of $CO_3^{2-}$. The data points were fitted to a sigmoïdal function.

| | Growth rate (µmol m⁻² s⁻¹) [References] | $[Ca^{2+}]/[CO_3^{2-}]$ | Experimental conditions | |
|---|---|---|---|---|
| This study | 2.22 | 2 | SI=1.20* | I=0.1 KCl |
| | 2.64 | 2 | SI=1.20* | I=0.1 KCl |
| Other studies | 1.25 [32,35] | 2 | SI=0.97 SI | I=0.1 KCl |
| | 0.45 [31,35] | 1 | ≈ 1.00 | I=0.1 NaCl |
| | 1.2 ± 0.4 [35] | 2 | SI=0.97 | I=0.1 KCl |

*Starting conditions

**Table 1:** Crystal growth rates.

the visualization of the number of growth events would be desirable. Therefore an integration of Raman microscopy with a fast method for event recognition, e.g., dark-field microscopy, is required.

## Conclusion

Time lapse Raman imaging was developed as a method to study *in situ* single, self-nucleated calcite crystals on the surface of an active polysulfone polymer layer. The analysis of the material-specific information in the Raman images enables the extraction of the mass growth rate. The experimentally obtained values for growth rate of calcite from time lapse Raman imaging is in good agreement with values from the literature. Differences in these values can be related to the differences in conditions prior to and during the crystallization process and are in agreement with the expected trend. Further development requires a more efficient search for growth events in spontaneous crystallization.

### Author Contributions

The manuscript was written through contributions of all authors. All authors have given approval to the final version of the manuscript.

### Funding Sources

This work is part of the research program "Spectroscopic analysis of particles in water" which is (partly) financed Fundamenteel Onderzoek der Materie (FOM), which is financially supported by the Nederlandse Organisatie voor Wetenschappelijk Onderzoek (NWO). This work was performed in the cooperation framework of Wetsus, Centre of Excellence for Sustainable Water Technology. Wetsus is co-funded by the Dutch Ministry of Economic Affairs and Ministry of Infrastructure and Environment, the European Union Regional Development Fund, the Province of Fryslân, and the Northern Netherlands Provinces.

### References

1. Teng HH, Dove PM, De Yoreo JJ (2000) Kinetics of calcite growth: Surface processes and relationships to macroscopic rate laws. Geochim Cosmochim Acta 64: 2255-2266.

2. Spanos N, Koutsoukos PG (1998) Kinetics of precipitation of calcium carbonate in alkaline pH at constant supersaturation, Spontaneous and seeded growth. J Phys Chem B 102: 6679-6684.

3. Busenberg E, Plummer LN (1989) Thermodynamics of Magnesian Calcite Solid-Solutions at 25-Degrees-C and 1-Atm Total Pressure. Geochim Cosmochim Acta 53: 1189-1208.

4. Christoffersen J, Christoffersen MR (1990) Kinetics of Spiral Growth of Calcite Crystals and Determination of the Absolute Rate-Constant. J Cryst Growth 100: 203-211.

5. Godinho JRA, Stack AG (2015) Growth Kinetics and Morphology of Barite Crystals Derived from Face-Specific Growth Rates. Cryst Growth Des 15: 2064-2071.

6. Pina CM, Becker U, Risthaus P, Bosbach D, Putnis A (1998) Molecular-scale mechanisms of crystal growth in barite. Nature 395: 483-486.

7. Higgins SR, Bosbach D, Eggleston CM, Knauss KG (2000) Kink dynamics and step growth on barium sulfate (001): A hydrothermal scanning probe microscopy study. J Phys Chem B 104: 6978-6982.

The experimentally determined growth rate (*vide supra*) is in good correspondence with experimental values obtained in the literature. Especially when considering that in this work, experiments were performed at a higher saturation index of 1.2 and a higher ratio of the calcium ion to carbonate ion concentration. Both aspects tend to increase the growth rate, which was observed. Moreover, the calculated growth rate could be an overestimate, since the focal volume of the laser will also partialkly measure the side surfaces of the growing crystal, which were not included in the calculation.

The growth rate is strongly influenced by the presence of seeds for crystallization. In this work no seeds were used and crystals were spontaneously formed at the active polymer layer. The occurrence of a growth event is rare in the case of unseeded crystallization. Also the location of a growth event is not known beforehand. An increase of

8. Bracco JN, Grantham MC, Stack AG (2012) Calcite Growth Rates As a Function of Aqueous Calcium-to-Carbonate Ratio, Saturation Index, and Inhibitor Concentration: Insight into the Mechanism of Reaction and Poisoning by Strontium. Cryst Growth Des 12: 3540-3548.

9. McEvoy AL, Stevens F, Langford SC, Dickinson JT (2006) Scanning-induced growth on single crystal calcite with an atomic force microscope. Langmuir 22: 6931-6938.

10. Gratz AJ, Hillner PE, Hansma PK (1993) Step Dynamics and Spiral Growth on Calcite. Geochim Cosmochim Acta 57: 491-495.

11. Stipp SLS, Eggleston CM, Nielsen BS (1994) Calcite Surface-Structure Observed at Microtopographic and Molecular Scales with Atomic-Force Microscopy (Afm). Geochim Cosmochim Acta 58: 3023-3033.

12. Britt DW, Hlady V (1997) In-situ atomic force microscope imaging of calcite etch pit morphology changes in undersaturated and 1-hydroxyethylidene-1,1-diphosphonic acid poisoned solutions. Langmuir 13: 1873-1876.

13. Teng HH, Dove PM, Orme CA, De Yoreo JJ (1998) Thermodynamics of calcite growth: baseline for understanding biomineral formation. Science 282: 724-727.

14. Dobson PS, Bindley LA, Macpherson JV, Unwin PR (2005) Atomic force microscopy investigation of the mechanism of calcite microcrystal growth under Kitano conditions. Langmuir 21: 1255-1260.

15. Pouget EM, Bomans PH, Goos JA, Frederik PM, de With G, et al. (2009) The initial stages of template-controlled CaCO3 formation revealed by cryo-TEM. Science 323: 1455-1458.

16. Tlili MM, Ben Amor M, Gabrielli C, Joiret S, Maurin G, et al. (2002) Characterization of CaCO3 hydrates by micro-Raman spectroscopy. J Raman Spectrosc 33: 10-16.

17. Herman RG, Bogdan CE, Sommer AJ, Simpson DR (1987) Discrimination among Carbonate Minerals by Raman-Spectroscopy Using the Laser Microprobe. Appl Spectrosc 41: 437-440.

18. Urmos J, Sharma SK, Mackenzie FT (1991) Characterization of Some Biogenic Carbonates with Raman-Spectroscopy. Am Mineral 76: 641-646.

19. Behrens G, Kuhn LT, Ubic R, Heuer AH (1995) Raman-Spectra of Vateritic Calcium-Carbonate. Spectrosc Lett 28: 983-995.

20. Gabrielli C, Jaouhari R, Joiret S, Maurin G (2000) In situ Raman spectroscopy applied to electrochemical scaling. Determination of the structure of vaterite. J Raman Spectrosc 31: 497-501.

21. Mullin JW (2001) Crystallization. Butterworth-Heinemann, Oxford, UK.

22. Lide DR (2006) Hanbook for crystallization and Physics. Taylor& Francis Group.

23. Guggenheim EA, Turgeon JC (1955) Specific Interaction of Ions. T Faraday Soc 51: 747-761.

24. Scatchard G (1936) Concentrated solutions of strong electrolytes. Chem Rev 51: 747-761.

25. Brönsted JN (1992) Studies on solubility. IV. The principle of the specific interaction of ions. J Am Chem Soc 44: 877-898.

26. Pully VV, Lenferink A, Otto C (2010) Hybrid Rayleigh, Raman and two-photon excited fluorescence spectral confocal, microscopy of living cells. J Raman Spectrosc 41: 599-608.

27. van Manen HJ, Kraan YM, Roos D, Otto C (2005) Single-cell Raman and fluorescence microscopy reveal the association of lipid bodies with phagosomes in leukocytes. Proc Natl Acad Sci USA 102: 10159-10164.

28. Porto SPS, Giordmai J, Damen TC (1966) Depolarization of Raman Scattering in Calcite. Phys Rev 147: 608.

29. Haüy RJ (1832) Traite de Mineralogie. Paris.

30. Deng Z, Habraken GJM, Peeters M, Heise A, de With G, et al. (2011) Fluorescein functionalized random amino acid copolymers in the biomimetic synthesis of $CaCO_3$. Soft Matter 7: 9685.

31. Nehrke G, Reichart GJ, Van Cappellen P, Meile C, Bijma J (2007) Dependence of calcite growth rate and Sr partitioning on solution stoichiometry: Non-Kossel crystal growth. Geochim Cosmochim Acta 71: 2240-2249.

32. Gebrehiwet TA, Redden GD, Fujita Y, Beig MS, Smith RW (2012) The Effect of the $CO_3^{2-}$ to $Ca^{2+}$ Ion activity ratio on calcite precipitation kinetics and $Sr^{2+}$ partitioning. Geochem Trans 13: 1.

33. Stack AG, Grantham MC (2010) Growth Rate of Calcite Steps As a Function of Aqueous Calcium-to-Carbonate Ratio: Independent Attachment and Detachment of Calcium and Carbonate Ions. Cryst Growth Des 10: 1409-1413.

34. Perdikouri C, Putnis CV, Kasioptas A, Putnis A (2009) An Atomic Force Microscopy Study of the Growth of a Calcite Surface as a Function of Calcium/Total Carbonate Concentration Ratio in Solution at Constant Supersaturation. Cryst Growth Des 9: 4344-4350.

35. Bracco JN, Stack AG, Steefel CI (2013) Upscaling calcite growth rates from the mesoscale to the macroscale. Environ Sci Technol 47: 7555-7562.

# 8

# Development of Microextraction by Packed Sorbent - Gas Chromatography-Mass Spectrometry Method for Quantification of Nitro-Explosives in Aqueous and Fluidic Biological Samples

Pooja Bansal[1], Gaurav G[1], Nidhi N[2], Susheela Rani[1] and Ashok Kumar Malik[1*]

[1]Department of Chemistry, Punjabi University, Patiala, Punjab, India
[2]Department of Chemistry, Atma Ram Sanatan Dharam College, New Delhi, India

## Abstract

A new method for quantification of twelve nitroaromatic compounds including 2,4,6-TNT, its metabolites and Tetryl with microextraction by packed sorbent (MEPS), followed by gas chromatography - mass spectrometric (GC-MS) detection in environmental and biological samples is developed. MEPS employ 4 mg of $C_{18}$ silica sorbent inserted into a micro-volume syringe for sample preparation. Several parameters capable of influencing the microextraction procedure viz., number of extraction cycles, washing solvent, volume of washing solvent, elution solvent, elution solvent volume and pH of matrix etc., were optimized. Helium gas was used as mobile phase during GC operation. The developed method produced satisfactory results with excellent values of coefficient of determination ($R^2 > 0.9804$) within the established calibration range. The extraction yields were satisfactory for all analytes (>89.32%) for aqueous samples and (>87.45%) for biological samples. The limits of detection values lie in the range 14 - 828 pg/mL. Due to procedural simplicity, high sensitivity and efficient resolution of all analytes, the developed method was applied successfully for quantification of nitro aromatic explosives in real aqueous and fluidic biological samples.

**Keywords:** Microextraction; Packed sorbent; MEPS; TNT; Tetryl; Nitro-explosive; GC-MS

## Introduction

2,4,6-Trinitrotoluene (TNT) is one of the most frequently used organic high explosive by the armed forces and terrorist groups around the world. Military activities like training, testing, mining, dismantling, open burning/ open detonation, and wars have caused widespread contamination of soil and water by TNT and its degradation products [1,2]. Major route of entry of TNT to surface water and thus environment is through discharge of waste streams generated during its manufacturing and processing. It is estimated that a single munition manufacturing plant can generate as much as 2000 $m^3$ of wastewater per day containing TNT and other nitro compounds [3]. TNT has been classified as a substance that is suspected of being potentially carcinogenic for humans [4-6]. It has been detected in surface and ground water samples collected in the vicinity of munitions facilities and waste disposal sites [7]. The biodegradation and photolysis of TNT introduces highly reactive compounds of greater polarity and water solubility, which may constitute even greater environmental concern than TNT itself. 4-Amino-2,6-dinitrotoluene (4-Am-2,6-DNT), 2-amino-4,6-dinitrotoluene (2-Am-4,6-DNT), 2,4-dinitrotoluene (2,4-DNT), 2,6-dinitrotoluene (2,6-DNT), are main but mildly toxic metabolites of TNT and are eliminated in urine after conjugation to acid-labile glucuronides [8-13]. Other metabolites include nitrobenzene (NB), 1,3-dinitrobenzene (1,3-DNB), 1,3,5-trinitrobenzene (1,3,5-TNB), 2-nitrotoluene (2-NT), 3-nitrotoluene (3-NT) and 4-nitrotoluene (4-NT), that are generated either during industrial manufacturing of TNT or as its degradation products. Workers of explosives manufacturing plant and social communities are regularly exposed to these hazardous nitro-organic compounds. Therefore, precise identification and quantification of TNT and its degradation products at sub ppb concentration levels in environmental and biological samples is desirable. Similarly, 2,4,6-trinitrophenyl-N-methylnitramine (Tetryl) has acute toxic effects and is found to be mutagenic in several different bacterial assays [14].

In the recent times, development of fast, accurate, precise and sensitive analytical methods have become an important issue. Miniaturization, short analysis times and consumption of lesser volumes of organic solvents are also important parameters for analytical improvement. The sample preparation is utmost necessary in order to extract, isolate and concentrate the analytes of interest from complex environmental and biological matrices since most of the analytical instruments cannot handle these matrices directly. Proteins and other interferences present in biological matrices result in lowering of performance of analytical columns. Conventional sample pre-treatment techniques such as liquid-liquid extraction (LLE) and solid-phase extraction (SPE) have inherent drawbacks like complicated and time-consuming procedures, large sample size, use of organic solvents, solvent evaporation, sample reconstitution and difficulty in automation [15]. Incorporation of various chemicals and organic solvents in large proportions results in environmental pollution, health hazards and extra operational costs for waste treatment [16,17]. Although, other preconcentration techniques such as solid phase micro extraction (SPME) and single drop micro extraction (SDME) has several advantages over LLE and SPE. But high selectivity of SPME fiber sorbents, lack of robustness, fast ageing, high cost and low recoveries poses problems in analytical measurements [18-21]. Similarly, ease of dislodgment of the micro-drop hanging from the

*Corresponding author: Ashok Kumar Malik, College of Engineering and Management, Punjabi University Neighborhood Campus, Rampura Phul-151 103, Punjab, India
E-mail: malik_chem2002@yahoo.co.uk

tip of the micro-syringe needle during the extraction process limits the rate of agitation of sample solution and practical utility of SDME technique [18,22].

In an attempt to overcome above problems, microextraction by packed sorbent (MEPS) appears to be a good alternative. This technique for sample preparation is based on the miniaturization of conventional SPE technique, using a gas-tight syringe as the extraction device containing solid packing material. This allows direct online coupling either to gas or liquid chromatographic systems without hardware modifications [23,24]. MEPS is based on multiple extractions in which fluidic sample flows through a bed of solid sorbent particles and the particle size has to be as small as possible to speed up the transfer of analytes from the sample matrix to the sorbent phase. The packed sorbent of MEPS can be reused more than 100 times for plasma or urine samples and more than 400 times for aqueous samples, whereas a conventional SPE column can be used only once [25]. The MEPS apparatus can handle volumes from 10 to 1000 μL and is more robust when compared to other microextraction devices such as SPME. This technique can be used comfortably for complex matrices such as plasma, urine, blood and organic solvents [26-32], which is not always the case with SPME. Moreover, high extraction recoveries can be achieved with this technique [33-36]. Analysis of organo nitro-explosives has been dominated so far by HPLC protocols [37,38]. But large sample volumes have been used for lowering the detection limits during HPLC analysis. An alternative approach to HPLC relies on GC, where analytes move through the column on an elevated temperature in the gas phase. For nitro-aromatic explosives, GC-MS offered two additional significant advantages over the standard HPLC methods *viz.*, lower detection limits and improved chromatographic resolution of the isomers of DNT and Am-DNT [39].

Therefore, a simple, fast, efficient and reliable analytical method has been developed by hyphenation of MEPS with GC-MS for analysis of nitro-explosives and evaluated with real samples.

## Experimental

### Instrumentation

The separations were carried on Shimadzu QP2010 Plus gas chromatography-mass spectrometer system (Shimadzu Corporation, Kyoto, Japan) equipped with a Rtx-200 column (30 m × 0.25 mm i.d. × 0.25 μm) supplied by Restek (Bellefonte, PA, USA). A personal computer equipped with the GC-MS real time analysis software was used to process the MS data. MEPS procedure was carried out by means of a BIN (Barrel Insert and Needle Assembly) containing 4 mg of solid phase silica $C_{18}$ sorbent inserted into a 250 μL gas tight syringe from SGE Analytical Science (Melbourne, VIC, Australia). Aqueous and non-aqueous solvents were filtered with 0.45 μm Nylon-6,6 membrane filters (Rankem, New Delhi, India) in a glass filtration assembly (Riviera, Mumbai, India).

### Reagents and standards

Standard solutions of TNT, Tetryl, 1,3,5-TNB, 4-Am-2,6-DNT and 2-Am-4,6-DNT were purchased from Supelco (Bellefonte, USA) in acetonitrile solution (1000 μg/mL). Other metabolites of TNT like NB, 1,3-DNB; 2,4-DNT; 2,6-DNT; isomers of nitrotoluene i.e., 2-NT, 3-NT and 4-NT were obtained from Aldrich, India in solid state. LC-grade methanol was purchased from Rankem (New Delhi, India). The purity range of all chemical standards was 97-99.9%.

## Preparation of standard solution

Separate stock solutions (1000 μg/mL) were prepared for each solid analyte in methanol. Mixtures of all analytes bearing different concentration i.e., 0.1, 1.0, 10 and 25 μg/mL were prepared in methanol from their stock solutions. All samples were wrapped in the aluminum foil to prevent photo-decomposition and later stored in deep refrigerator at -10°C. Working samples were prepared from mixture solutions by stepwise dilution as per requirement.

## Preparation of biological samples

Blood (3 mL) and urine (10 mL) specimens were obtained from five healthy volunteers. Blood samples were stored in separate glass tubes containing ethylenediaminetetraacetic acid (EDTA) as an anticoagulant and then centrifuged (within 2 h from collection) at 4000 rpm for 10 min. The supernatant (plasma) was transferred to polypropylene tubes and stored at −10°C. The plasma samples were stable over a period of six months. Urine samples were also centrifuged at same rate and the supernatant collected were stored at −10°C in glass tubes. Before use, the plasma and urine samples were thawed at room temperature and centrifuged at 4000 rpm for 5 min again. Spiked plasma and urine samples (1 mL) of desired concentrations i.e., 1, 10, 100 and 250 ng/mL were prepared from mixture solutions such that the ratio of methanol to sample volume was always 1:100. Samples were preconcentrated with MEPS and analyzed chromatographically.

## Preparation of aqueous samples

River water was obtained from Ghaggar River, Patiala (Punjab, India) in Pyrex borosilicate amber glass container. Ghaggar River originates from the Shivalik hills of Himachal Pardesh, India. The industrial and domestic waste water discharged from the towns located alongside Ghaggar or its tributaries deteriorates the quality of water. Since river water contains high concentration of particulate matter and suspended impurities therefore, sample was first filtered with Whattman filter paper (grade no. 1) and then with 0.45 μm pore size Nylon-6,6 membrane filter. It was degassed with an ultrasonic bath, and stored at 4°C. Ground water sample was obtained from a bore well dig inside Punjabi university campus, Patiala, Punjab, India in pyrex borosilicate amber glass containers previously rinsed with triply distilled water. The bore well water is pumped out at the depth of around 400 feet from ground level and is supplied to the university campus for daily needs and routine activities. The collected ground water sample was then filtered with 0.45 μm pore size Nylon-6,6 membrane filter. Spiked aqueous samples (1 mL) of desired concentrations i.e., 1, 10, 100 and 250 ng/mL were prepared from mixture solutions such that the ratio of methanol to sample volume was always 1:100. Samples were preconcentrated using MEPS and analyzed chromatographically. In another approach, variation in recoveries due to change in spiking strategy was also tested. For this, 100 mL water samples (both river and ground) were spiked (100 ng/mL) prior to filtration. Spiked aqueous samples were filtered in bulk through filtration media. Preconcentraion of 1 mL of each sample was carried out with MEPS and analyzed with GC-MS. No significant variation in recoveries was observed with this approach. This may be due weaker interaction of analytes with particulate matter, filter papers media and large volume of aqueous samples taken for filtration. So, the spiking of aqueous samples was performed after filtration in developed method.

## Micro Extraction by Packed Sorbent (MEPS) procedure

Before being used for the first time, the packed sorbent was conditioned with 100 μL of methanol followed by 100 μL of deionized water. The volumes of methanol and water drawn up every time were discarded out at an approximate flow rate of 20 µL/s (± 5 µL/s).

50 µL of plasma and urine samples were pulled and pushed through the syringe 10 times manually. Samples were drawn slowly at an approximate flow rate of 20 µL/s (± 5 µL/s) with caution to obtain good percolation between sample matrix and solid adsorbent. The sorbent was then washed with 50 µL deionized water to remove biological interferences. Thereafter, the analytes were eluted with 30 µL methanol in a vial and 1 µL of it was injected into GC instrument. After each extraction, sorbent of MEPS BIN was cleaned with 3 cycles of 100 µL methanol followed by 3 cycles of 100 µL water. The cleaning step of the sorbent also acted as the conditioning step for the next extraction cycle. The same protocol was repeated with river and ground water samples except the washing step. One packing bed was used for about 60 extractions and then it was discarded due to low analyte extraction and sorbent clogging. The whole sample preconcentration process (i.e., sampling, enrichment and elution steps) takes about 20 min. Therefore, it could be carried out simultaneously while the chromatographic separations of the prior injected sample are taking place on the GC-MS leading to fast and coordinated analytical procedure.

### GC-MS determination

The split-splitless injector was maintained at 250°C and operated in the splitless mode with the split closed for 0.60 min. Helium (>99.999% pure) was used as the carrier gas at constant linear velocity 28.1 cm/s. The GC column oven was initially set at 60°C for 1 min, increased by 12°C min⁻¹ up to 240°C and held for 9 min and then increased by 15°C min⁻¹ up to 270°C. The interface temperature was set at 290°C and ion source temperature at 200°C. The electron ionization source was set at 70 eV. A 4 min solvent cut time was allowed for all analyses. The mass range of m/z 45-600 was recorded in full scan mode. Peak identification of targets was based on the retention times and full scan spectra of the standards. The selected ion monitoring (SIM) mode was employed for quantification of ions. The characteristic ions selected for qualitative and quantitative studies are listed in Table 1. The chromatographic separation of all nitro-explosive components was achieved within 23.5 min and the total run time was 27 min.

### Method validation

Calibration curves of spiked urine, plasma and aqueous samples were prepared with MEPS-GC-MS. Five concentration values i.e., 1, 10, 50, 100 and 250 ng/mL were used for each matrix while preparing the curve. The limit of detection (LOD) values was calculated with formula

signal to noise ratio 3:1. Spiked samples of concentrations 1, 10 and 100 ng/mL were used as quality control (QC) samples to calculate the accuracy and precision, for both intraday and interday analysis. The experiments were done six times during six different days. The extraction recoveries from different types of matrices were calculated by comparing the peak areas of extracted samples with the calibration curves (Figure 1).

## Results and Discussion

### GC-MS analysis

The parameters evaluated and optimized during GC-MS analysis were injector temperature, injector mode, ion source temperature, interface temperature, column temperature gradient and carrier gas flow rate. The temperature in the ion source was varied between 120°C to 250°C. Sufficient ionization of all analytes was obtained at 200°C. Lower temperatures resulted in insufficient ionization and much higher temperatures resulted in decomposition of molecules. Running the MS in SIM mode allowed the analysis of nitroaromatic compounds according to their abundance. One risk associated with this method was the probability of presence of other compounds with the same m/z as that of target analytes (Table 1). This could lead to false positives, if the concentration of the contaminant is much higher than that of the actual value, or false negatives due to ion-quenching. The first way to identify and eliminate the false positives is to compare the retention time of peaks of analytes of real sample with those of standard sample and excluding all other peaks that deviate. Another way is to look at more than one fragment, since the fragmentation pattern is one way to identify a compound. The use of more fragments will also prevent ion quenching, which is unlikely for all fragments to occur at the same time. In an effort to increase the sensitivity, a group of ions obtained from the electron ionization were used to devise a SIM procedure.

### MEPS method development

The aim of this study was to validate MEPS as a new sample preparation technique for quantification of TNT and other nitroaromatic energetic compounds. Factors affecting the recovery with MEPS such as pulling or pushing speed, number of extraction cycles, composition and volume of washing solution, composition and volume of elution solution, pH of matrix and carryover effects were studied.

**Number of extraction cycles:** The number of extraction cycles and speed of extraction process are two important parameters which affect the retention of analytes on the sorbent. Two procedures were tested

| Analytes | Acronym | Retention time (min) | Target ion (m/z) | Qualifier ion #1 | | Qualifier ion #2 | | Qualifier ion #3 | |
|---|---|---|---|---|---|---|---|---|---|
| | | | | Ion m/z | Abundance vs. Quant. ion | Ion m/z | Abundance vs. Quant. ion | Ion m/z | Abundance vs. Quant. ion |
| Nitrobenzene | NB | 8.489 | 77 | 51 | 66.28 | 123 | 34.91 | 93 | 12.52 |
| 2-Nitrotoluene | 2-NT | 9.273 | 65 | 120 | 62.55 | 92 | 44.16 | 91 | 43.00 |
| 3-Nitrotoluene | 3-NT | 9.981 | 91 | 65 | 83.73 | 137 | 53.34 | 63 | 29.53 |
| 4-Nitrotoluene | 4-NT | 10.309 | 65 | 91 | 90.40 | 137 | 63.24 | 63 | 36.12 |
| 2,6-Dinitrotoluene | 2,6-DNT | 13.760 | 63 | 77 | 92.50 | 165 | 62.90 | 91 | 31.39 |
| 1,3-Dinitrobenzene | 1,3-DNB | 14.159 | 76 | 50 | 91.94 | 75 | 89.01 | 168 | 39.38 |
| 2,4-Dinitrotoluene | 2,4-DNT | 15.014 | 89 | 63 | 86.92 | 165 | 71.26 | 90 | 33.47 |
| 2,4,6-Trinitrotoluene | 2,4,6-TNT | 17.148 | 137 | 63 | 65.12 | 89 | 48.12 | 210 | 32.10 |
| 1,3,5-Trinitrobenzene | 1,3,5-TNB | 17.471 | 75 | 74 | 57.21 | 213 | 18.11 | 120 | 15.93 |
| 4-Amino-2,6-dinitrotoluene | 4-Am-2,6-DNT | 18.900 | 104 | 78 | 93.71 | 180 | 66.21 | 197 | 43.24 |
| 2-Amino-4,6-dinitrotoluene | 2-Am-4,6-DNT | 20.500 | 78 | 104 | 60.09 | 180 | 47.95 | 197 | 30.10 |
| Tetryl | Tetryl | 23.500 | 77 | 75 | 32.03 | 194 | 7 | 242 | 4.7 |

**Table 1:** Retention times, quantification ions and qualifier ions selected for analysis of the target analytes.

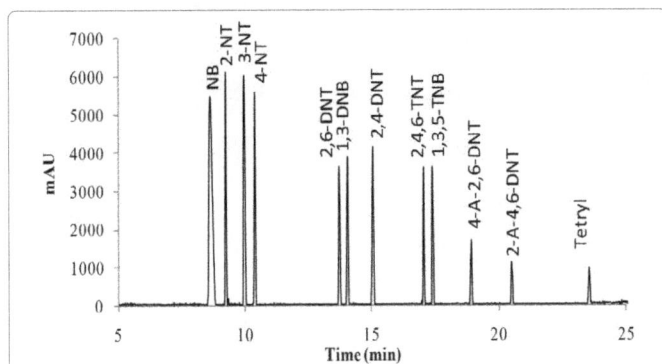

**Figure 1:** MEPS/GC-MS chromatogram of nitroaromatic compounds at concentration 1 ng/mL for NB, 2-NT, 3-NT, 4-NT, 2,6-DNT, 1,3-DNB, 2,4-DNT, 2,4,6-TNT, 1,3,5-TNB, 4-Am- 2,6-DNT, 2-Am-4,6-DNT, Tetryl under optimized conditions.

for sample loading (extraction regime). In first approach, multiple draw-eject cycle mode was tested where the drawn up sample in MEPS BIN was ejected back into the same vial from which it was taken. In second approach, draw-discard mode was evaluated in which each sample aliquot was pumped only once through the MEPS BIN and the loaded solution after extraction was discarded into waste before the next aliquot. Better results were obtained with draw-discard extraction regime. Since lesser number of strokes were required in draw-discard mode, therefore mechanical stress to the MEPS syringe plunger was reduced which consequently extends the lifetime of the MEPS syringe. Therefore, second approach was applied for further studies. Extraction cycles (5, 10, 12 and 15) were evaluated with 50 µL sample size. Best recoveries were obtained with 10 extraction cycles as shown in Figure 2a. Time taken for ten draw-discard cycles was about 10 minutes.

**Composition and volume of washing solution:** The influence of different washing solutions on absolute recoveries was investigated. Different fractions of water with methanol and acetonitrile were evaluated like pure water, water/methanol (9:1, 8:2), water/acetonitrile (9:1) and water/methanol/acetonitrile (9:1:1) for washing the MEPS BIN and removal of interferences caused by the sample matrix. Best results were obtained with pure water (50 µL). The incorporation of organic solvents resulted into the leakage of adsorbed analytes and hence lowered the absolute recoveries.

**Composition and volume of elution solvent:** The quantity of sorbent enclosed in MEPS BIN is very small, therefore desorption can be performed with a relatively small volume of solvent. The desorbed solvent could be transferred to the GC instrument either totally or partially. Generally, a GC compatible volatile solvent is required. Since retention of analytes on silica gel modified with $C_{18}$ chain is hydrophobic in nature therefore, organic solvents would be enough to disrupt the forces between the analytes and the sorbent. Five different solvents namely n-hexane, dichloromethane, toluene, acetonitrile and methanol were tried for desorption purpose. Dichloromethane, toluene and n-hexane were not able to elute all analytes efficiently. With these solvents, desorption was good for nitrotoluene isomers but were not able to elute other high polarity molecules efficiently. Methanol and acetonitrile yielded better results than other solvents. Methanol was selected as an optimized elution solvent as per Figure 2b. The effect of elution volume of methanol on analytes recovery was studied and 30 µL solvent gave the best elution result as shown in Figure 2c.

**pH of matrix:** The variation in pH can alter the solubility of analytes in matrices and consequently their retention on sorbent. It was investigated by extracting 50 µL of spiked aqueous samples (5 ng/mL) on BIN cartridge under optimized extraction conditions and

eluting the components with 30 µL methanol. The measured pH of the initial spiked aqueous sample was 7.4. The pH was varied in the range 4 to 11 and extraction recoveries were calculated. Out of all analytes, recoveries of only Am-DNTs were affected by pH change and their values decreased in acidic pH range, while recoveries of other analytes remain appreciably unaltered. Therefore, pH 7.4 was maintained for further studies.

**Carryover and matrix effect:** Compared to conventional SPE disposable cartridges, MEPS sorbent requires a detailed evaluation of carryover phenomena [40]. A thorough clean up step is required after each extraction process to eliminate the matrix effects. The carryover effect was investigated on the column by injecting three successive aliquots of a standard mixture containing all analytes at a high concentration in organic solvent into the chromatographic system. A non significant carryover effect (<0.09%) was evident. Moreover, assays were also carried out to ensure that the small quantity of sorbent phase (4 mg) in the MEPS BIN can be easily and effectively washed before the subsequent extraction to reduce the possibility of carryover. For this, MEPS was washed three times with the eluting solvent (3 × 100 µL methanol) after each extraction from all types of matrices spiked with a high concentration of the analytes. No carryover was observed and the effect was minimal. To eliminate the memory effect, three cycles of 100 µL water were used to rinse the BIN cartridge after washing with three cycles of 100 µL methanol.

## Method validation

**Selectivity:** Method selectivity is defined as non interference of impurity substances in the regions of interest in chromatogram. Under optimized MEPS and GC-MS conditions (Table 2), there were no interfering peaks in the regions of quantification for nitro explosives in blank urine, plasma and aqueous samples. GC-MS chromatograms of blank and spiked samples of river water, ground water, plasma and urine at concentration 1 ng/mL are presented (Figures 3-6).

**Limit of detection and quantification:** The MEPS followed by GC-MS analysis exhibits good linearity for calibration curves with coefficients of determination ($R^2$) greater than 0.9804 for all nitroaromatics in 1-250 ng/mL concentration range. LOD and LOQ values were calculated as three and ten times of signal to noise (S/N) ratio, respectively. The LOD values ranges from 0.014 to 0.828 ng/mL, whereas the LOQ values from 0.046 to 2.732 ng/mL for complete spectrum of analytes in all matrices as summarized in Table 3.

## Recovery

Environmental and biological samples were analyzed after spiking with nitroaromatic components at four different concentrations to validate the developed method. No major interfering peaks appeared at the retention time of the analytes during their determination in these samples. Good recoveries (91-98%) were obtained from ground and river water samples at different spiked concentrations except 2 and 4-ADNT (89-92%). The recovery percentages of nitro components from biological matrices were lesser as compared to aqueous matrices at a particular spiked concentration (Table 4). Extraction yield and precision were determined at four concentration levels, corresponding to one lowest level, one highest level and two middle point of each calibration curve. The interday precision, expressed as the relative standard deviation (RSD%), varied between 2.3% and 4.9% (n=6) while the intraday precision between 2.1% and 4.7% (n=6) (Table 4).

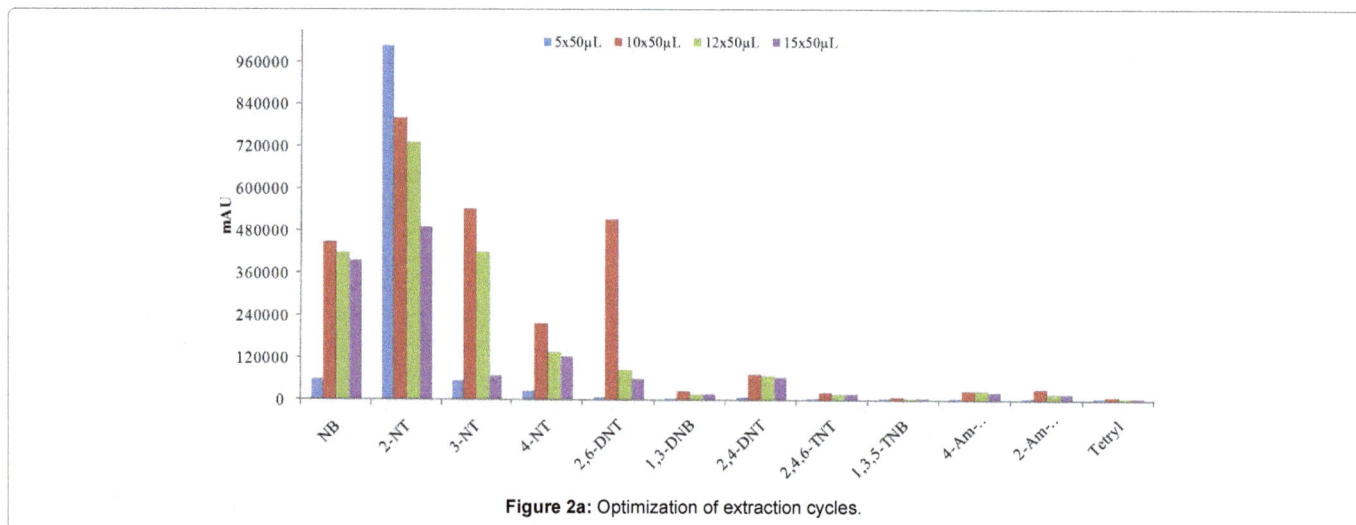

**Figure 2a:** Optimization of extraction cycles.

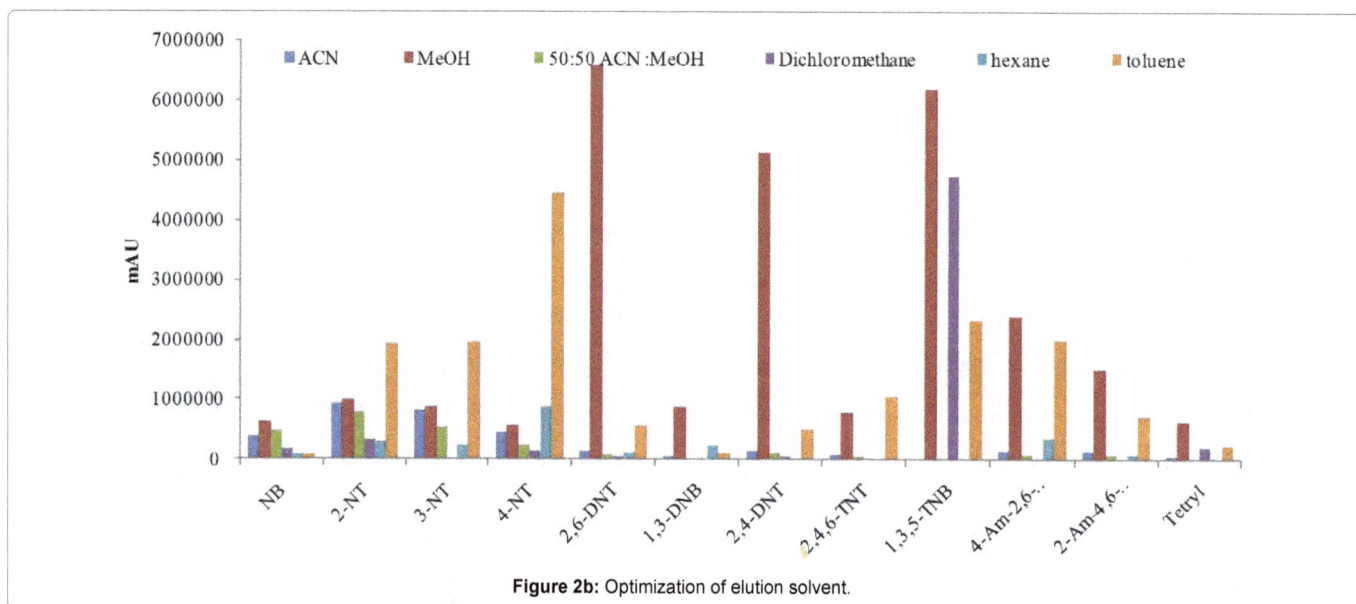

**Figure 2b:** Optimization of elution solvent.

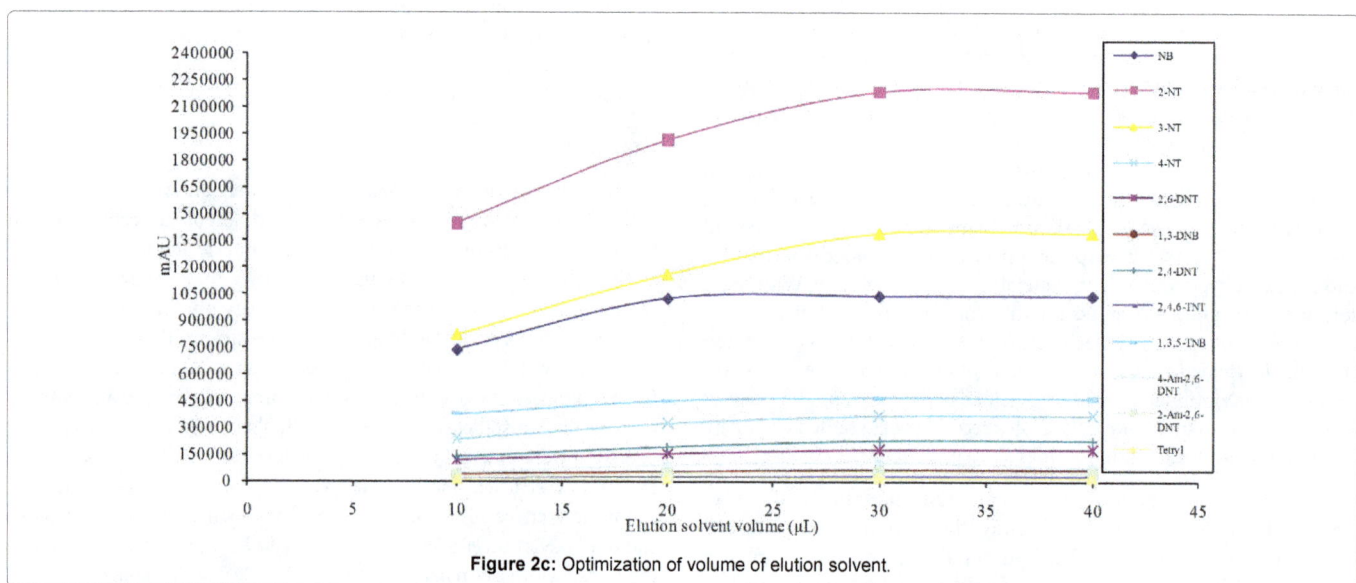

**Figure 2c:** Optimization of volume of elution solvent.

| MEPS | | Adsorbent | 4 mg solid phase silica ($C_{18}$); particle size 45 µm; 60A° porosity |
|---|---|---|---|
| GC | | Extraction mode | Draw-discard |
| | | No. of extraction cycles | 10 |
| | | Extraction volume per cycle | 50 µL |
| | | Washing solvent | Water; 50 µL |
| | | Elution solvent | Methanol; 30 µL |
| | | Pump | Shimadzu QP2010 Plus |
| | | Column | Restek Rtx-200 column; 30 m × 0.25 mm i.d. × 0.25 µm |
| MS detection | | Mobile phase | Helium gas |
| | | Injector temperature | 250°C |
| | | Flow rate | Linear velocity; 28.1 cm s$^{-1}$ |
| | | Mass range | m/z 45-600 |
| | | Quantification of ions | selected ion monitoring (SIM) mode |

Table 2: Optimized MEPS-GC-MS conditions for analysis of TNT and its metabolites.

| | | Analyte | NB | 2-NT | 3-NT | 4-NT | 2,6-DNT | 1,3-DNB | 2,4-DNT | 2,4,6-TNT | 1,3,5-TNB | 4-Am-2,6-DNT | 2-Am-4,6-DNT | Tetryl |
|---|---|---|---|---|---|---|---|---|---|---|---|---|---|---|
| Linearity range (ng/mL) | | | 1-250 | 1-250 | 1-250 | 1-250 | 1-250 | 1-250 | 1-250 | 1-250 | 1-250 | 5-250 | 5-250 | 5-250 |
| $R^2$ | | Plasma | 0.9929 | 0.9918 | 0.9889 | 0.9908 | .9865 | 0.9950 | 0.9878 | 0.9883 | 0.9847 | 0.9806 | 0.9891 | 0.9832 |
| | | Urine | 0.9915 | 0.9923 | 0.9908 | 0.9951 | 0.9900 | 0.9971 | 0.9886 | 0.9889 | 0.9869 | 0.9804 | 0.9899 | 0.9841 |
| | | River water | 0.9966 | 0.9906 | 0.9921 | 0.9982 | 0.9976 | 0.9989 | 0.9982 | 0.9902 | 0.9931 | 0.9889 | 0.9901 | 0.9935 |
| | | Tap water | 0.9971 | 0.9955 | 0.9945 | 0.9976 | 0.9879 | 0.9992 | 0.9988 | 0.9954 | 0.9981 | 0.9972 | 0.9970 | 0.9962 |
| EquationCoefficientsy = mx+c | mc | Plasma | 4767.5 | 5442.6 | 4940.2 | 4431.0 | 3082.7 | 3436.4 | 3538.3 | 2858.2 | 2814.0 | 283.1 | 171.2 | 145.4 |
| | | Urine | 3757.6 | 6036.3 | 4967.3 | 5089.6 | 3420.3 | 3463.8 | 3688.7 | 2801.5 | 2868.4 | 290.3 | 179.5 | 149.5 |
| | | River water | 4976.7 | 5531.4 | 4721.2 | 4929.9 | 3543.9 | 3364.3 | 3673.4 | 3035.9 | 3061.2 | 317.6 | 188.3 | 156.4 |
| | | Tap water | 4920.9 | 5580.3 | 5274.3 | 4843.9 | 3446.6 | 3420.9 | 3530.7 | 3316.0 | 3281.1 | 344.9 | 177.8 | 165.8 |
| | | Plasma | -1876.3 | -4755.8 | -8145.4 | 3459.9 | -958.1 | -312.3 | -1348.6 | -920.8 | 3363.6 | 150.6 | 10.4 | -272.4 |
| | | Urine | 5172.4 | -1149.6 | 5761.6 | -4972.1 | -3198.7 | -3676.9 | 3796.9 | 3503.6 | 9801.2 | -268.3 | -587.7 | 97.6 |
| | | River water | -1086.5 | -5938.3 | 4546.2 | 1025.8 | -5051.7 | -5239.5 | 1040.9 | 5161.3 | 2340.4 | -363.9 | -427.4 | 435.5 |
| | | Tap water | -8799.4 | -6399.2 | -1190.4 | -2692.5 | -4484.6 | -3295.4 | -704.93 | 7425.3 | 1319.0 | -1636.8 | 107.9 | -111.9 |
| LOD$^a$ ((ng/mL) | | Plasma | 0.025 | 0.023 | 0.024 | 0.027 | 0.039 | 0.036 | 0.035 | 0.044 | 0.042 | 0.437 | 0.715 | 0.828 |
| | | Urine | 0.023 | 0.019 | 0.020 | 0.022 | 0.032 | 0.029 | 0.028 | 0.034 | 0.034 | 0.387 | 0.604 | 0.744 |
| | | River water | 0.021 | 0.018 | 0.014 | 0.016 | 0.031 | 0.022 | 0.027 | 0.034 | 0.024 | 0.351 | 0.532 | 0.643 |
| | | Tap water | 0.015 | 0.014 | 0.018 | 0.020 | 0.024 | 0.027 | 0.021 | 0.025 | 0.029 | 0.284 | 0.438 | 0.515 |
| LOQ$^b$ (ng/mL) | | Plasma | 0.082 | 0.076 | 0.079 | 0.089 | 0.129 | 0.119 | | 0.116 | 0.145 | 0.139 | 2.359 | 2.732 |
| | | Urine | 0.076 | 0.063 | 0.066 | 0.073 | 0.106 | 0.096 | 0.092 | 0.112 | 0.112 | 1.277 | 1.993 | 2.455 |
| | | River water | 0.069 | 0.059 | 0.046 | 0.053 | 0.102 | 0.073 | 0.089 | 0.112 | 0.079 | 1.158 | 1.756 | 2.122 |
| | | Tap water | 0.049 | 0.046 | 0.059 | 0.066 | 0.079 | 0.089 | 0.069 | 0.083 | 0.096 | 0.937 | 1.445 | 1.699 |

$^a$LOD limit of detection (S/N=3): Each result is the average of six determinations; $^b$LOQ limit of quantification (S/N=10): Each result is the average of six determinations

Table 3: Linearity parameters derived from assays on spiked river water, tap water, plasma and urine samples.

## Analysis of the real sample

The validated method was applied to the analysis of underground water sample obtained from an area nearby to Terminal Ballistics Research Laboratory, Chandigarh, India, which is an explosive testing range. This range is spread over 5500 acres where explosive or ammunition are detonated in the open. This gives flexibility in operation

| Analyte | Amount added (ng/mL) | Extraction yield (%) | | | | Intraday (RSD%)[a] | | | | Interday (RSD)[a] | | | |
|---|---|---|---|---|---|---|---|---|---|---|---|---|---|
| | | Ground water | River water | Plasma | Urine | Ground water | River water | Plasma | Urine | Ground water | River water | Plasma | Urine |
| NB | 1 | 94.26 | 93.12 | 90.22 | 91.88 | 3.7 | 3.8 | 4.3 | 4.1 | 3.9 | 4.0 | 4.5 | 4.3 |
| | 10 | 95.78 | 92.65 | 91.45 | 92.12 | 3.3 | 3.9 | 4.1 | 4.1 | 3.6 | 3.8 | 4.3 | 4.1 |
| | 100 | 96.29 | 95.39 | 92.12 | 94.57 | 2.9 | 3.5 | 3.7 | 3.4 | 3.1 | 3.7 | 3.9 | 3.8 |
| | 250 | 98.08 | 96.49 | 4.66 | 96.03 | 2.1 | 2.9 | 3.4 | 3.2 | 2.4 | 3.5 | 3.6 | 3.8 |
| 2-NT | 1 | 93.18 | 91.96 | 89.45 | 90.84 | 3.8 | 3.6 | 4.0 | 3.9 | 3.9 | 3.8 | 4.2 | 3.7 |
| | 10 | 94.23 | 93.47 | 91.71 | 92.60 | 3.2 | 3.5 | 3.9 | 3.7 | 3.4 | 3.7 | 4.0 | 3.5 |
| | 100 | 96.04 | 95.19 | 92.56 | 94.91 | 3.0 | 3.2 | 3.6 | 3.7 | 3.2 | 3.6 | 3.6 | 3.1 |
| | 250 | 98.16 | 97.68 | 94.01 | 96.42 | 2.6 | 2.7 | 3.4 | 3.5 | 2.8 | 3.3 | 3.5 | 2.8 |
| 3-NT | 1 | 92.76 | 91.39 | 89.69 | 90.34 | 3.8 | 3.7 | 3.8 | 3.5 | 4.1 | 3.9 | 3.9 | 3.6 |
| | 10 | 93.92 | 92.12 | 91.26 | 91.32 | 3.4 | 3.7 | 4.0 | 3.7 | 3.6 | 3.7 | 4.2 | 3.4 |
| | 100 | 95.89 | 94.58 | 92.27 | 93.44 | 2.9 | 3.2 | 3.4 | 3.5 | 3.2 | 3.6 | 4.7 | 3.0 |
| | 250 | 98.15 | 97.65 | 94.15 | 96.79 | 2.7 | 2.8 | 3.1 | 3.2 | 3.0 | 3.3 | 3.5 | 2.9 |
| 4-NT | 1 | 93.45 | 91.24 | 89.86 | 90.06 | 3.7 | 3.9 | 4.2 | 4.1 | 3.9 | 4.2 | 4.5 | 4.5 |
| | 10 | 94.20 | 92.32 | 90.15 | 91.41 | 3.6 | 3.5 | 3.9 | 3.7 | 3.8 | 4.0 | 4.2 | 4.4 |
| | 100 | 95.89 | 93.83 | 91.60 | 92.52 | 2.9 | 3.2 | 3.6 | 3.6 | 4.2 | 3.8 | 3.9 | 3.9 |
| | 250 | 97.27 | 96.73 | 93.27 | 95.78 | 2.3 | 2.7 | 3.5 | 3.2 | 3.6 | 3.6 | 3.7 | 3.7 |
| 2,6-DNT | 1 | 94.24 | 92.52 | 88.42 | 90.04 | 3.6 | 3.6 | 3.8 | 3.9 | 3.8 | 3.8 | 4.2 | 4.1 |
| | 10 | 95.74 | 92.98 | 89.17 | 90.92 | 3.3 | 3.5 | 3.9 | 3.6 | 3.6 | 3.7 | 4.1 | 3.9 |
| | 100 | 96.98 | 93.13 | 90.62 | 92.51 | 3.2 | 3.2 | 3.6 | 3.2 | 3.2 | 3.8 | 3.8 | 3.6 |
| | 250 | 97.81 | 95.80 | 92.74 | 93.29 | 2.9 | 3.2 | 3.4 | 3.0 | 3.0 | 3.3 | 3.4 | 3.7 |
| 1,3-DNB | 1 | 94.09 | 93.65 | 90.63 | 92.56 | 3.5 | 3.9 | 4.1 | 3.9 | 4.3 | 3.8 | 4.3 | 4.2 |
| | 10 | 95.73 | 92.15 | 91.18 | 93.07 | 2.9 | 3.2 | 4.0 | 3.7 | 4.1 | 3.6 | 4.1 | 4.0 |
| | 100 | 96.46 | 94.43 | 92.57 | 94.19 | 3.4 | 3.8 | 3.5 | 3.3 | 3.7 | 3.3 | 3.9 | 3.7 |
| | 250 | 98.22 | 95.82 | 94.43 | 96.26 | 2.8 | 3.0 | 3.3 | 3.1 | 3.5 | 3.2 | 3.5 | 3.5 |
| 2,4-DNT | 1 | 94.14 | 91.08 | 88.07 | 90.09 | 2.9 | 3.2 | 3.4 | 3.2 | 3.6 | 3.7 | 3.7 | 3.5 |
| | 10 | 95.85 | 92.17 | 89.31 | 91.15 | 2.6 | 3.1 | 3.2 | 3.0 | 3.3 | 3.5 | 3.4 | 3.2 |
| | 100 | 96.39 | 94.45 | 0.56 | 92.68 | 2.5 | 2.7 | 2.7 | 3.2 | 3.0 | 3.1 | 3.0 | 3.0 |
| | 250 | 97.47 | 96.38 | 92.42 | 93.32 | 2.2 | 2.5 | 2.7 | 2.8 | 2.7 | 3.0 | 2.8 | 2.9 |
| 2,4,6-TNT | 1 | 94.08 | 91.56 | 88.58 | 89.79 | 3.8 | 3.9 | 4.2 | 4.2 | 4.0 | 4.1 | 4.5 | 4.5 |
| | 10 | 95.35 | 2.13 | 90.62 | 91.45 | 3.6 | 3.7 | 4.2 | 3.9 | 3.9 | 3.9 | 4.4 | 4.2 |
| | 100 | 96.27 | 94.09 | 91.37 | 92.29 | 3.4 | 3.5 | 3.6 | 3.5 | 3.7 | 3.8 | 3.9 | 4.0 |
| | 250 | 96.42 | 95.83 | 93.69 | 94.54 | 3.3 | 3.4 | 3.5 | 3.3 | 3.9 | 3.6 | 3.7 | 4.0 |
| 1,3,5-TNB | 1 | 92.57 | 91.53 | 88.30 | 91.48 | 2.8 | 2.6 | 2.9 | 3.4 | 2.9 | 2.9 | 3.2 | 3.6 |
| | 10 | 91.13 | 92.47 | 89.24 | 91.58 | 2.7 | 2.3 | 2.9 | 3.2 | 2.7 | 2.7 | 3.0 | 3.4 |
| | 100 | 93.87 | 93.67 | 90.85 | 93.87 | 2.5 | 2.3 | 2.6 | 2.8 | 2.4 | 2.3 | 2.8 | 2.9 |
| | 250 | 95.49 | 94.19 | 91.80 | 94.92 | 2.2 | 2.1 | 2.4 | 2.6 | 2.4 | 2.3 | 2.5 | 2.7 |
| 4-Am-2,6-DNT | 5 | 90.00 | 89.72 | 88.02 | 88.92 | 4.2 | 3.9 | 4.7 | 4.4 | 4.4 | 3.9 | 4.9 | 4.7 |
| | 10 | 91.65 | 90.06 | 88.94 | 89.86 | 4.1 | 3.5 | 4..4 | 4.1 | 4.3 | 3.7 | 4.6 | 4.4 |
| | 100 | 92.72 | 90.89 | 89.18 | 90.40 | 3.9 | 3.9 | 4.5 | 3.9 | 3.9 | 3.6 | 4.4 | 4.2 |
| | 250 | 92.90 | 91.54 | 90.47 | 91.25 | 3.6 | 3.3 | 4.3 | 3.5 | 3.4 | 3.7 | 4.4 | 4.1 |
| 2-Am-4,6-DNT | 5 | 91.04 | 89.32 | 87.45 | 88.52 | 4.1 | 4.1 | 4.3 | 4.0 | 4.2 | 4.0 | 4.5 | 4.2 |
| | 10 | 91.14 | 89.62 | 88.18 | 88.79 | 3.9 | 4.0 | 3.9 | 3.8 | 4.0 | 3.9 | 4.2 | 4.0 |
| | 100 | 91.90 | 90.46 | 89.85 | 89.47 | 3.8 | 3.8 | 36 | 3.5 | 3.9 | 3.6 | 3.9 | 3.7 |
| | 250 | 92.13 | 91.72 | 90.07 | 90.65 | 3.4 | 3.8 | 3.3 | 3.8 | 3.6 | 3.5 | 3.9 | 3.5 |
| Tetryl | 5 | 94.87 | 92.14 | 89.29 | 90.57 | 4.0 | 3.7 | 4.0 | 3.8 | 4.0 | 4.0 | 4.4 | 4.2 |
| | 10 | 95.13 | 92.85 | 90.81 | 90.62 | 3.8 | 3.5 | 3.9 | 3.9 | 4.2 | 3.8 | 4.0 | 4.0 |
| | 100 | 96.11 | 93.42 | 91.01 | 91.36 | 3.6 | 3.5 | 3.6 | 3.6 | 3.8 | 3.4 | 3.7 | 3.7 |
| | 250 | 96.77 | 94.29 | 91.56 | 92.13 | 3.3 | 3.7 | 3.4 | 3.3 | 3.6 | 3.5 | 3.8 | 3.5 |

**Table 4:** Validation parameters. [a]Each value is the mean of 6 independent assays.

and permits explosion of high calibre warheads, ammunition and large explosive charges with adequate safety measures. These activities cause the presence of explosives and other nitro components into soil and consequently their movement to underground water resources. The chromatogram of analyzed raw water sample is presented in Figure 7. The sample was found to be having 1.03 ng/mL of NB, 0.38 ng/mL of 2-NT, 0.81 ng/mL of 3-NT and 0.22 ng/mL of 2,4-DNT.

## Comparison of present method

The results of the present study were compared with other sample preparation techniques such as SPE, SPME and SDME and were in close agreement with earlier published data [7,18,37]. MEPS has been employed to preconcentrate RDX and TNT in biological and environmental samples with increased sensitivity of detection [41]. MEPS had been proved as novel sample pre-concentration technique

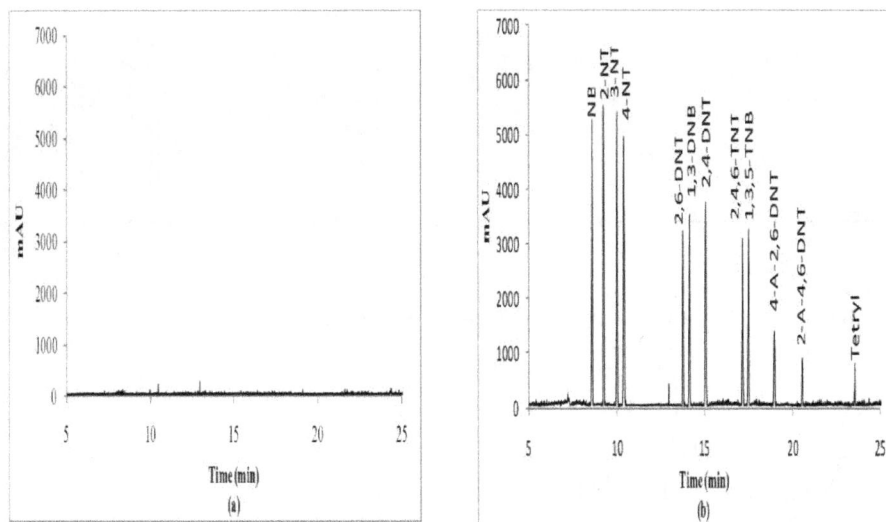

**Figure 3:** MEPS/GC-MS chromatogram of (a) blank river water (b) spiked river water (conc. same as in Figure 1).

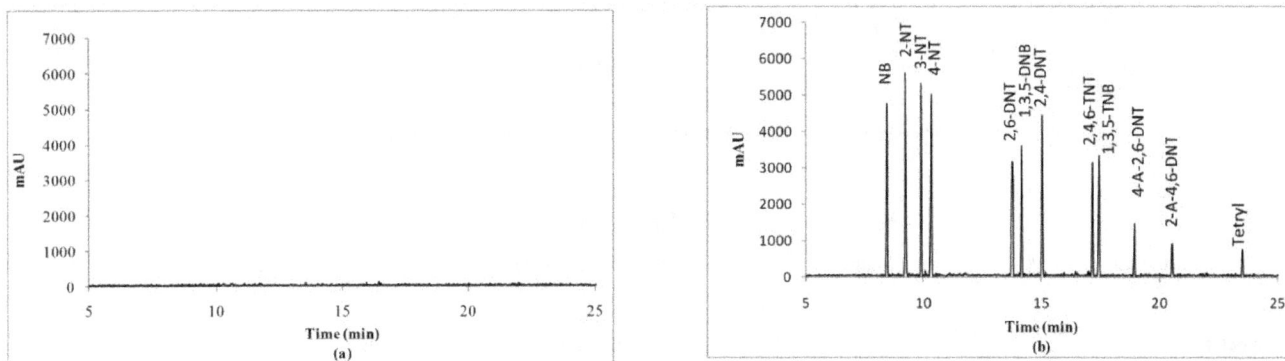

**Figure 4:** MEPS/GC-MS chromatogram of (a) blank ground water (b) spiked ground water (conc. same as in Figure 1).

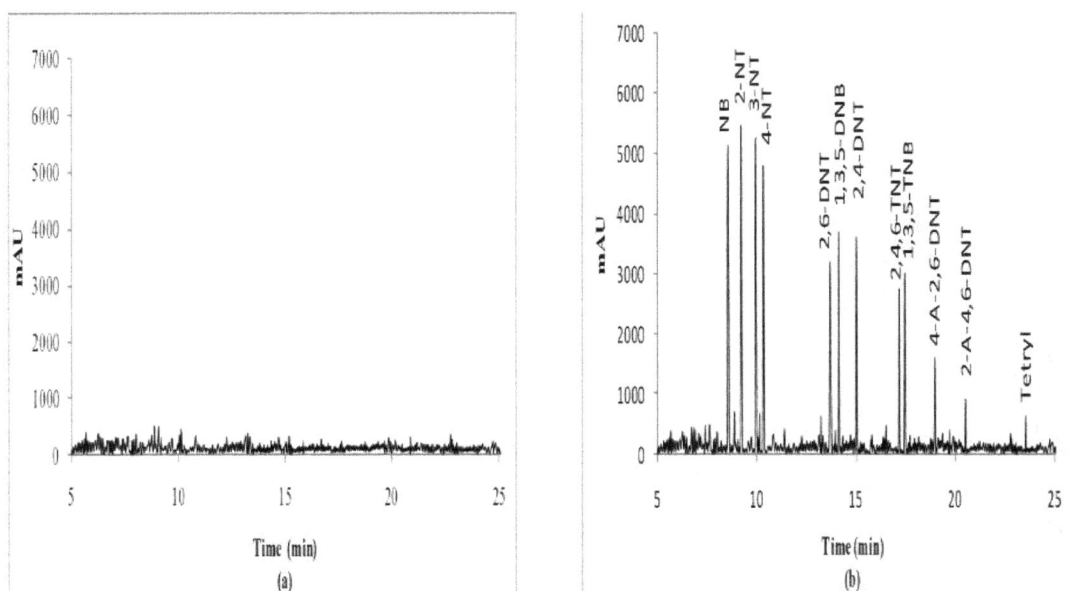

**Figure 5:** MEPS/GC-MS chromatogram of (a) blank plasma (b) spiked plasma (conc. same as in Figure 1).

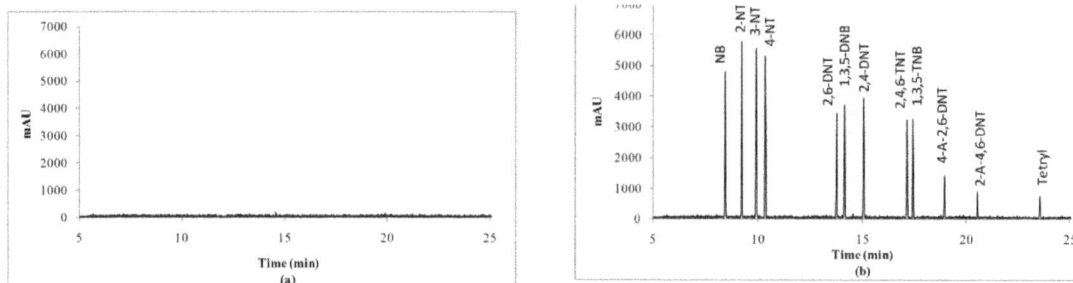

**Figure 6:** MEPS/GC-MS chromatogram of (a) blank urine (b) spiked urine (conc. same as in Figure 1).

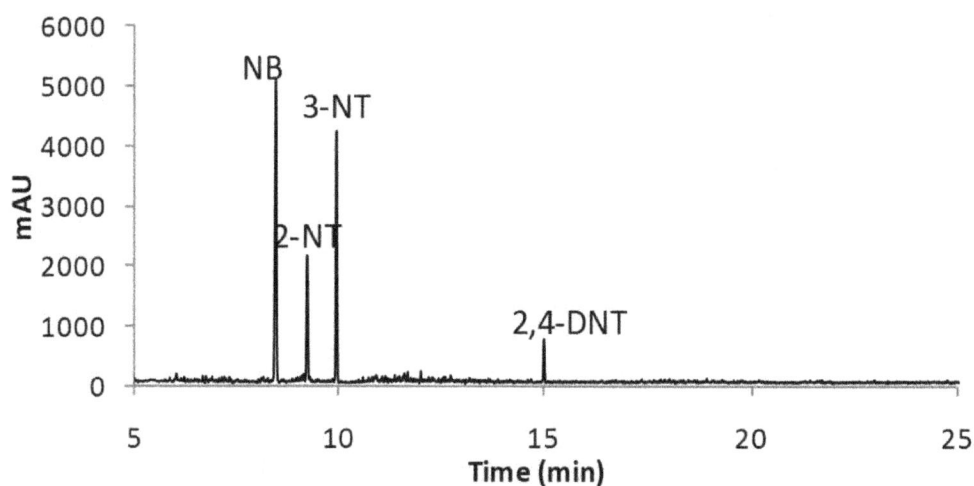

**Figure 7:** MEPS/GC-MS chromatogram of underground water sample obtained from an area nearby to explosives testing range.

| Analyte | Preconcentration technique | Chromatographic technique | LOD (ng/mL) | Recovery (%) | RSD (%) | Reference |
|---|---|---|---|---|---|---|
| HMX, RDX, 1,3,5-TNB, Tetryl, 3,4-DNT, 2,4,6-TNT, 4-Am-2,6-DNT, 2,4-DNT | SPME | HPLC-UV | 1.0 - 10.1 | 67.2 - 121.9 | 10.2 - 27.2 | 7 |
| HMX, RDX, 1,3,5-TNB, Tetryl, 3,4-DNT, 2,4,6-TNT, 4-Am-2,6-DNT, 2,4-DNT | SPE | HPLC-UV | 0.03 - 0.29 | 75.9 - 100.3 | 3.1 - 13.12 | 7 |
| 2-NT, 3-NT, 4-NT, 1,3-DNB, 2,6-DNT, 1,3,5-TNB, 2,4,6-TNT, 4-Am-2,6-DNT, 2-Am-4,6-DNT, Tetryl | SPME | GC-MS | 0.03 - 1.10 | 86.0 - 114.0 | 2.0 - 8.9 | 18 |
| 2-NT, 3-NT, 4-NT, 1,3-DNB, 2,6-DNT, 1,3,5-TNB, 2,4,6-TNT, 4-Am-2,6-DNT, 2-Am-4,6-DNT, Tetryl | SDME | GC-MS | 0.08 - 1.3 | 82.0 - 102.0 | 4.3 - 9.8 | 18 |
| 2-NT, 3-NT, 4-NT, NB, 1,3-DNB, 2,6-DNT, 1,3,5-TNB, 2,4,6-TNT, 4-Am-2,6-DNT, 2-Am-4,6-DNT, Tetryl | SPME | HPLC-UV | 0.17 - 0.92 | 87.4 - 98.8 | 2.0 - 4.8 | 37 |
| TNT, RDX | MEPS | HPLC-UV | 0.062- 0.099 | 81.2-96.3 | 3.5-5.6 | 41 |
| NB, 2-NT, 3-NT, 4-NT, 2,6-DNT, 1,3-DNB, 2,4,6-TNT, 1,3,5-TNB, 4-Am-2,6-DNT, 2-Am-4,6-DNT and Tetryl | MEPS | GC-MS | 0.014 - 0.828 | 77.5 - 99.2 | 2.3 - 4.9 | Developed method |

**Table 5:** Comparison of developed method with earlier reported methods.

in terms of extraction yields, sensitivity, selectivity and reproducibility for these two nitro organics. For analysis of TNT and its degradation products, the developed method is more rapid, simpler and robust. The developed procedure exhibited the comparable values of accuracy and precision with other reports (Table 5). The new approach offers the reduced sample preparation and analysis time as compared to previous work [42,43]. Furthermore, the developed method involves low sample volume of around 0.5 mL for ten extraction cycles (50 µL × 10), as compared to large sample volumes for other microextraction procedures.

## Conclusions

A fast, sensitive and accurate sample preparation technique has been developed and validated for the determination of selected nitroaromatic compounds in environmental aqueous and fluidic biological samples. MEPS provide an avenue for improved sample preparation to aid the speed, sensitivity, and selectivity options provided by GC-MS technique. MEPS reduced the sample preparation time and organic solvent consumption. Also, small sample volumes can be treated, in contrast with the large volumes used in conventional LLE, SPE and other procedures. Due emphasis has been given on the quantitative as well as qualitative aspects of analysis such as accuracy, precision and limits of detection for developed method. The values of extraction yields, LOQ and precision were satisfactory. The very low LOD values achieved by the method directly imply high sensitivity. The sample pre-treatment by MEPS has greatly decreased the matrix effect, which causes the hindrance in chromatographic analysis. It is highly compatible with both qualitative and quantitative aspects of separation science and allows further improvement since it provides flexibility in different parameters including type of adsorbent materials, loading environment and sample load size, etc.

### Acknowledgements

The authors are thankful to the Council of Scientific and Industrial Research, New Delhi for providing the Senior Research Fellowship No. 09/140 (0147)/2009-EMR-I to author Gaurav.

## References

1. Stucki H (2004) Toxicity and degradation of explosives. Chimia 58: 409-413.

2. Park C, Kim TH, Kim S, Kim SW (2003) Optimization for biodegradation of 2,4,6-trinitrotoluene (TNT) by Pseudomonas putida. J Bio Sci Bio Eng 95: 567-571.

3. Pereira WE, Short DL, Manigold DB, Roscio PK (1979) Isolation and characterization of TNT and its metabolites in groundwater by gas chromatograph-mass spectrometer-computer techniques. Bull Environ Cont Toxicol 21: 554-562.

4. Sabbioni G, Sepai O, Norppa H, Yan H, Hirvonen A, et al. (2007) Comparison of biomarkers in workers exposed to 2,4,6-trinitrotoluene. Biomarkers 12: 21-37.

5. Honeycutt ME, Jarvis AS, McFarland VA (1996) Cytotoxicity and mutagenicity of 2,4,6-trinitrotoluene and its metabolites. Ecotox Environ Safe 35: 282-287.

6. Yan C, Wang Y, Xia B, Li L, Zhang Y, et al. (2002) The retrospective survey of malignant tumor in weapon workers exposed to 2,4,6-trinitrotoluene. Chin J Ind Hyg Occup Dis 20: 184-188.

7. Rivera FM, Beaulieu C, Deschamps S, Paquet L, Hawari J (2004) Determination of explosives in environmental water samples by solid phase microextraction-liquid chromatography. J Chromatogr A 1048: 213-221.

8. Neuwoehner J, Schofer A, Erlenkaemper B, Steinbach K, Hund-Rinke TK (2007) Eisentraeger Oxicological characterization of 2,4,6-trinitrotoluene, its transformation products, and two nitramine explosives. Environ Toxicol Chem 26: 1090-1099.

9. Haidour AJ, Ramos L (1996) Identification of products resulting from the biological reduction of 2,4,6-trinitrotoluene, 2,4-dinitrotoluene, and 2,6-dinitrotoluene by Pseudomonassp. Environ Sci Technol 30: 2365-2370.

10. Sabbioni G, Liu YY, Yan H, Sepai O (2005) Hemoglobin adducts, urinary metabolites and health effects in 2,4,6-trinitrotoluene exposed workers. Carcinogenesis 26: 1272-1279.

11. Walsh ME (1990) US Army Cold Regions Research and Engineering Laboratory, Hanover, NH, Special Report 90.

12. Checkai RT, Major MA, Nwanguma RO, Amos CT, Phillips CT, et al. (1993) ERDEC Technical Report 135, US Army Edgewood Research, Development and Engineering Center. Aberdeen Proving Ground.

13. Maloney SW, Adrian NR, Hickey RF, Heine RL (2002) Anaerobic treatment of pink water in fluidized bed reactor containing GAC. J Hazard Mater 92: 77-88.

14. Wen-Zong W, Speciner ND, Edwards GS (1980) Mutagenic activity of tetryl, a nitroaromatic explosive, in three microbial test systems. Toxicol Lett 5: 11-17.

15. Feltes J, Levsen K, Volmer D, Spiekermann M (1990) Gas chromatographic and mass spectrometric determination of nitroaromatics in water. J Chromatogr A 518: 21-40.

16. Furton KG, Wu LM, Almirall JR (2000) Optimization of solid-phase microextraction (SPME) for the recovery of explosives from aqueous and post-explosion debris followed by gas and liquid chromatographic analysis. J Forensic Sci 45: 857-864.

17. Becanov J, Friedl Z, Simek Z (2010) Identification and determination of trinitrotoluenes and their degradation products using liquid chromatography–electrospray ionization mass spectrometry. Intern J Mass Spectrom 291: 133-139.

18. Psillakis E, Kalogerakis N (2001) Solid-phase microextraction versus single-drop microextraction for the analysis of nitroaromatic explosives in water samples. J Chromatogr A 938: 113-120.

19. Mayfield HT, Burr E, Cantrell M (2006) Analysis of explosives in soil using solid phase micro extraction and gas chromatography. Anal Letters 39: 1463-1474.

20. Wu L, JR Almirall, KG Furton (1999) An Improved interface for coupling solid-phase microextraction (SPME) to high performance liquid chromatography (HPLC) applied to the analysis of explosives. J Sep Sci 22: 279-282.

21. Barshick SA, Griest WH (1998) Trace analysis of explosives in seawater using solid-phase microextraction and gas chromatography/ion trap mass spectrometry. Anal Chem 70: 3015-3020.

22. Kin CM, Huat TG (2009) Comparison of HS-SDME with SPME and SPE for the determination of eight organochlorine and organophosphorus pesticide residues in food matrices. J Chromatogr Sci 47: 694-699.

23. Miyaguchi H, Iwata Y, Kanamori T, Tsujikawa K, Kuwayama K, et al. (2009) Rapid identification and quantification of methamphetamine and amphetamine in hair by gas chromatography/mass spectrometry coupled with micropulverized extraction, aqueous acetylation and microextraction by packed sorbent. J Chromatogr A 1216: 4063-4070.

24. Rehim MA (2010) Recent advances in microextraction by packed sorbent for bioanalysis. J Chromatogr A 1217: 2569-2580.

25. Rehim MA (2004) New trend in sample preparation: on-line microextraction in packed syringe for liquid and gas chromatography applications: I. Determination of local anaesthetics in human plasma samples using gas chromatography–mass spectrometry. J Chromatogr B 801: 317-321.

26. Vlckova H, Solichova D, Blaha M, Solich P, Novakova L (2011) Microextraction by packed sorbent as sample preparation step for atorvastatin and its metabolites in biological samples-critical evaluation. J Pharm Biomed Anal 55: 301-308.

27. Chaves AR, Leandro FZ, Carris JA, Queiroz MEC (2010) Microextraction in packed sorbent for analysis of antidepressants in human plasma by liquid chromatography and spectrophotometric detection. J Chromatogr B 878: 2123-2129.

28. Saracino MA, Tallarico K, Raggi MA (2010) Liquid chromatographic analysis of oxcarbazepine and its metabolites in plasma and saliva after a novel microextraction by packed sorbent procedure. Anal Chim Acta 661: 222-228.

29. Said R, Kamel M, Beqqali AE, Rehim MA (2010) Microextraction by packed sorbent for LC–MS/MS determination of drugs in whole blood samples. Bioanalysis 2: 197-205.

30. Beqqali AE, Kussak A, Blomberg L, Rehim MA (2007) Microextraction in packed syringe/liquid chromatography/electrospray tandem mass spectrometry for quantification of acebutolol and metoprolol in human plasma and urine samples. J Liq Chromatogr 30: 575-586.

31. Beqqali AE, Kussak A, Rehim MA (2007) Determination of dopamine and serotonin in human urine samples utilizing microextraction online with liquid chromatography/electrospray tandem mass spectrometry. J Sep Sci 30: 421-424.

32. Rehim MA, Dahlgren M, Blomberg L (2006) Quantification of ropivacaine and its major metabolites in human urine samples utilizing microextraction in a packed syringe automated with liquid chromatography-tandem mass spectrometry (MEPS-LC-MS/MS). J Sep Sci 29: 1658-1661.

33. Moreno IED, da Fonseca BM, Barroso M, Costa S, Queiroz JA, et al. (2012) Determination of piperazine-type stimulants in human urine by means of microextraction in packed sorbent and high performance liquid chromatography-diode array detection. J Pharm Biomed Anal 61: 93-99.

34. Beqqali AE, Rehim MA (2007) Quantitative analysis of methadone in human urine samples by microextraction in packed syringe-gas chromatography-mass spectrometry (MEPS-GC-MS). J Sep Sci 30: 2501-2505.

35. Rehim MA, Altun Z, Blomberg L (2004) Microextraction in packed syringe (MEPS) for liquid and gas chromatographic applications. Part II—Determination of ropivacaine and its metabolites in human plasma samples using MEPS with liquid chromatography/tandem mass spectrometry. J Mass Spectrom 39: 1488-1493.

36. Fu S, Fan J, Hashi Y, Chen Z (2012) Determination of polycyclic aromatic hydrocarbons in water samples using online microextraction by packed sorbent coupled with gas chromatography–mass spectrometry. Talanta 94: 152-157.

37. Gaurav AK, Rai PK (2009) Rapid analysis of nitro explosives using solid phase microextraction high performance liquid chromatography on reverse phase amide column with UV detection and application to analysis of aqueous samples. J Hazard Mater 172: 1652-1658.

38. Gaurav AK, Rai PK (2007) High-performance liquid chromatographic methods for the analysis of explosives. Crit Rev Anal Chem 37: 1-39.

39. Walsh ME (2001) Determination of nitroaromatic, nitramine, and nitrate ester explosives in soil by gas chromatography and an electron capture detector. Talanta 54: 427-438.

40. Moeder M, Schrader S, Winkler U, Rodil R (2010) At-line microextraction by packed sorbent-gas chromatography–mass spectrometry for the determination of UV filter and polycyclic musk compounds in water samples. J Chromatogr A 1217: 2925-2932.

41. Bansal P, Gaurav G, Nidhi AK, Malik F (2012) Rapid identification and quantification of RDX and TNT in blood and water samples utilizing microextraction in packed syringe with HPLC. Chromatogr 75: 739-745.

42. Holmgren E, Ek S, Colmsjö A (2012) Extraction of explosives from soil followed by gas chromatography–mass spectrometry analysis with negative chemical ionization. J Chromatogr A 1222: 109-115.

43. Tachon R, Pichon V, Le Borgne MB, Minet J (2008) Comparison of solid-phase extraction sorbents for sample clean-up in the analysis of organic explosives. J Chromatogr 1185: 1-8.

# 9

# Localized Electrochemical Impedance Spectroscopy Observation on Scratched Epoxy Coated Carbon Steel in Saturated Ca(OH)$_2$ with Various Chloride Concentration

**Balusamy T and Nishimura T***

*Corrosion Resistant Steel Group, Research Center for Structural Materials (RCSM), National Institute for Materials Science, Ibaraki, Tsukuba, 305-0047, Japan*

## Abstract

The *in-situ* local corrosion behavior of scratched epoxy coated carbon steel is investigated in sat. Ca(OH)$_2$ with varying concentration of Cl$^-$ ions by localized electrochemical impedance spectroscopy (LEIS). The localized corrosion process and mechanism of coated steel (scratch area) is measured by LEIS plots and 3D topographic images. The LEIS responses measured at the defect are attributed to the pore impedance with defect in the high-frequency range and an interfacial corrosion reaction in the low-frequency range of corroding steel at the base of defect within 1-10 h immersion. The continuous decrease in |Z| at the scratch is due to the higher extent of dissolution of Fe with increase of Cl$^-$ ion concentration. However, the resistance values of coated steel in sat. Ca(OH)$_2$ with each concentration of Cl$^-$ ions are not changed significantly with increase in immersion time from 1-10 h. On the other hand, LEIS Nyquist plots clearly showed that the measured impedance at high frequency is related to corrosion products formed at the defect which acts as anodic zones and the low frequency part are related to corroding of carbon steel with immersion of 1-5 days. 2D topographic images clearly showed that corrosion occurs at scratch and followed by coating degradation at scratch front as well as away from scratch due to cathodic reactions (reduction of O$_2$) leads to coating delamination. No significant change in corrosion resistance is observed for 0 and 0.0085 M/L of Cl$^-$ ions containing solution for 5 days of immersion as well as 1-10 h immersion. This is due the formation of better passive film on the steel surface (defect) in which the competition between the aggressive Cl$^-$ ions and the inhibitive OH$^-$ ions determines the rate of corrosion. A significant decrease in corrosion resistance is observed with higher concentration of Cl$^-$ ions (0.17 and 0.51 M) due to the preferential adsorption of Cl$^-$ ions at the defect site.

**Keywords:** Carbon steel; Epoxy coating; Corrosion mechanism; Localized electrochemical impedance spectroscopy (LEIS); Saturated Ca(OH)$_2$; Chlorides

## Introduction

Electrochemical impedance spectroscopy (EIS) has long been used to evaluate ability of coatings to resist corrosion of metals/alloys. The typical impedance spectra (Nyquist, Bode impedance and bode phase angle plots) being used to characterize the impedance parameters and the physical meaning of those parameters to determine the corrosion mechanism of steels under coating has been explored extensively [1,2]. However, the major limitation of this method is its inadequacy since the measured impedance values correspond to the electrochemical response of the whole electrode and it fails to reflect the averaged behavior of the macroscopic electrode having pinholes/defects. Hence, it is very difficult to extract complete quantitative information about the initial stages of corrosion and coating degradation. In recent years, verity of local electrochemical techniques have been developed to study local corrosion processes at initial stages and coating degradation at the microscopic level which includes Scanning Vibrating Electrode Technique (SVET), scanning reference electrode techniques (SRET), Scanning Kelvin Probe (SKP), Scanning Electrochemical Microscopy (SECM) and localized electrochemical impedance spectroscopy (LEIS), etc. [3-7]. LEIS measurements provide a promising alternative to investigate the coating degradation and localized corrosion of steel under coating. Moreover, the corrosion mechanism is a complex process in the scratched area of the epoxy coated carbon steel and depends on the type of metal/alloy, its chemical composition and ability to form a passive film, defect size, concentration of Cl$^-$ ions, transport phenomena and the nature of corrosion products [8-10]. In addition, LEIS could effectively separate the local impedance properties of the organic coating with defect when compared to the conventional

EIS, in which it provides comprehensive understanding on the time of initiation of local corrosion and mechanism in aqueous environment.

Zhong et al. [11-13] have studied the corrosion of steel under defected coating (~200 and 1000 μm diameter) in near-neutral pH solution using LEIS. According to them, the LEIS response was dependent on the size of the defect. It was found that the 200 μm diameter was lost due to the blocking effect, which was mainly dominated by the diffusion process. On the other hand, the blocking effect was not experienced in the 1000 μm defect due to its relatively open geometry. Jorcin et al. [14] have explored the use of LEIS mapping to assess the delamination phenomena at the steel/epoxy-vinyl primer interface in NaCl and identified that the delamination has originated from the artificial defects. Hence, LEIS is a promising alternative technique to explore the corrosion mechanism of coatings with defects at microscopic level to understand their degradation behavior as well as the localized corrosion behavior of steel in aqueous solutions.

Carbon steels are widely used as construction materials in many industries due to their low cost and ability to provide reasonably good mechanical properties besides being weldable. However, the structural

---

***Corresponding author:** Nishimura T, Corrosion Resistant Steel Group, Research Center for Structural Materials (RCSM), National Institute for Materials Science, Ibaraki, Tsukuba, 305-0047, Japan, E-mail: NISHIMURA.Toshiyasu@nims.go.jp

components fabricated using carbon steels encounter failures during service, particularly due to surface related failures such as fatigue, wear and corrosion. Deterioration of concrete structures has become a serious social problem and the deterioration caused by corrosion of the reinforcing steel is due to salt damage, since the total content of chlorides in freshly mixed concrete has been set at $0.3 \text{ kg/m}^3$ or less [15]. Epoxy based protective coatings have been applied to the steels bars for many applications such as bridges, parking structures, pavements, marine structures etc., against corrosion for many years which acts as a physical barrier layer against corrosion and cost effective [16-18]. However, in real time all polymers are permeable to corrosive species such as oxygen, water and ions which could not be isolated from the metal substrate and the environment [19,20]. Further, the corrosion appeared to develop at imperfections in the coating, especially where the disbondment had taken place during the fabrication (bending)/ processing of the rebar. Our previous study showed that the importance and understanding on the extent of local corrosion of scratched epoxy coated steel in sat. $Ca(OH)_2$ with or without 3% of $Cl^-$ ions (added as NaCl) studied by EIS, SECM [21].

However, there is a limited published literature in higher pH solution by *in-situ* measurements on the understanding of local corrosion mechanism of steels [22-24]. In addition, the level of chloride ion concentration in sat. $Ca(OH)_2$ is one of the critical factors that determine the corrosion behavior of reinforcing steels in contact with concrete structures and depends on the dissolved oxygen concentration, pH and chloride binding, $[Cl^-]/[OH^-]$ ratio and not a unique value due to the differences in the procedures [25,26]. It is important to study the corrosion mechanism of coated steels in sat. $Ca(OH)_2$ with varying concentration of $Cl^-$ ions on the understanding of degradation of coating and local corrosion of steels in high pH solutions. The choice of concentration of NaCl preferred in the present study is (0.0085, 0.085, 0.17 and 0.51M (0.05 to 3% NaCl)) based on our previous study. The concentration of NaCl is limited to 0 - 0.51 M (0-3%) because of change in corrosion behavior was significant at 0.5% NaCl [27]. Moreover, threshold level of chlorides in sat. $Ca(OH)_2$ on the corrosion of steel bars found to be below 3% NaCl [28]. Hence, in the present study, an attempt is made to understand extent of local corrosion process of scratched epoxy coated carbon steel in sat. $Ca(OH)_2$ containing varying concentration of $Cl^-$ ions with different time intervals by LEIS and LEIS mapping.

## Experimental Details

### Preparation of specimens

The chemical composition of the carbon steel specimen was given in a Table 1 as per JIS-SM (Japanese industrial standards-sheet metal). The specimen, with a surface area of $1.7 \times 1.7 \text{ cm}^2$, was polished using silicon carbide (SiC) papers upto 800 grit. After polishing, the samples surface were rinsed with distilled water and dried with compressed air, and cleaned with ethanol before coating.

### Preparation of coatings and electrolyte solution

The organic coating used in this investigation was commercially available fast drying epoxy. The liquid epoxy resin was a blended with multifunctional low molecular weight diluents and the diglycedal ether of bis-phenol-A; the aliphatic amines were used as a curing agent. The weight ratio of the epoxy resin to the curing agent was 2:1. The epoxy resin was coated using a drawdown bar at a constant speed and then kept at room temperature for 24 h. This led to the formation of uniform coating with thickness of about ~40 μm. After coating, an artificial scratch in the epoxy coating was produced by using a driller to produce a scratch of 1000 μm width and length about 10 mm. The samples were

then exposed to aerated aqueous solutions of saturated $Ca(OH)_2$ with varying concentration of $Cl^-$ ions added as NaCl (0.0085, 0.085, 0.17 and 0.51 M). The test electrolyte was prepared using analytical grade chemicals then filtered using Whatman 42 filter paper.

### LEIS measurements

The LEIS measurements were performed using Model 470 scanning electrochemical work station, which comprises a Potentioste/ Galvanostate/FRA (model 3300) with lock-in amplifier, environmental tri-cell system, a video microscope system to position the micro-probe over the working electrode (WE). The schematic of the electrode installation and experimental set-up of LEIS is shown in Figure 1 [27]. Scratched epoxy coated carbon steel was a working electrode (WE) while saturated calomel electrode (SCE) and a graphite rod were used as the reference and auxiliary electrodes, respectively. The relative location of the microprobe to the WE was monitored by the camera system, which can be adjusted by a stepper motor in the x, y, z directions. The scanning microprobe was operated in two modes. The first mode was used for LEIS measurements. The microprobe having a 5-6 μm tip was set directly above the scratched epoxy coated carbon steel to measure the typical impedance response. The distance between the tip of the microprobe and the surface of the WE was ~ 50 μm, which was adjusted and monitored with the help of a video camera TV system, supplied along with the workstation. During LEIS measurements, an AC amplitude signal of 50 mV was applied to the electrode system. The frequency range used for the study was 20,000 to 0.1 Hz. The second mode was used for LEIS mapping at a fixed frequency of 10 Hz. The tip of the microprobe was stepped over a designated area of the electrode surface. The scanning was performed by the moving of the tip of the microprobe in the x-y axis while the x-y scales were set as a $7 \times 1$ mm. The step size ($100 \times 200$ μm in x-y direction) was controlled to obtain a plot of 71 lines × 6 lines. The 3D plots were obtained from 3D Isoplot software. Specimens were mounted horizontally facing upwards. Before all experiments, the scratched epoxy coated carbon steels (WE) was kept at open-circuit potential (OCP) in the test solution for 1800 s and LEIS measurements were made at their respective OCP's for different time intervals for 1-10 h and 1-5 days of immersion.

## Results and Discussion

### LEIS measurements for 1-10 h immersion

LEIS measurements performed on scratched epoxy coated carbon steel in sat. $Ca(OH)_2$ and sat. $Ca(OH)_2$ with 0.0085, 0.085, 0.17 and 0.51 M for 1-10 h. LEIS Nyquist plots of scratched epoxy coated carbon steel in sat. $Ca(OH)_2$ with varying concentration of $Cl^-$ ions for 5 h are shown in Figure 2a. The corresponding Bode impedance and Bode phase angle plots are shown in Figure 2b and 2c. It is evident from Figure 2a that the capacitive part at high frequency range is associated with the pore resistance and its corresponding capacitance at the defect of steel surface. The capacitive part in the low frequency region is related to the double layer capacitance and charge-transfer resistance of corroding steel at the defect of base metal. Since, the LEIS is measured at single point at the defect site, the measured impedance at high frequency corresponds to the pore impedance during 1-10 h immersion [13]. The proposed electrical circuit model is shown in Figure 3. In this model, $R_s$ represents the solution resistance while $R_{pore}$ and $CPE_{pore}$ correspond to the pore resistance and the corresponding pore capacitance, while $R_{ct}$ and $CPE_{dl}$ are the charge transfer resistance and double layer capacitance of the system, respectively. The impedance of CPE is given by $Z_{CPE}=1/Q(j\omega)^n$, Where 'n' is the CPE exponent. The capacitance element $Q$ *(CPE)* will be pure capacitance when n=1 while

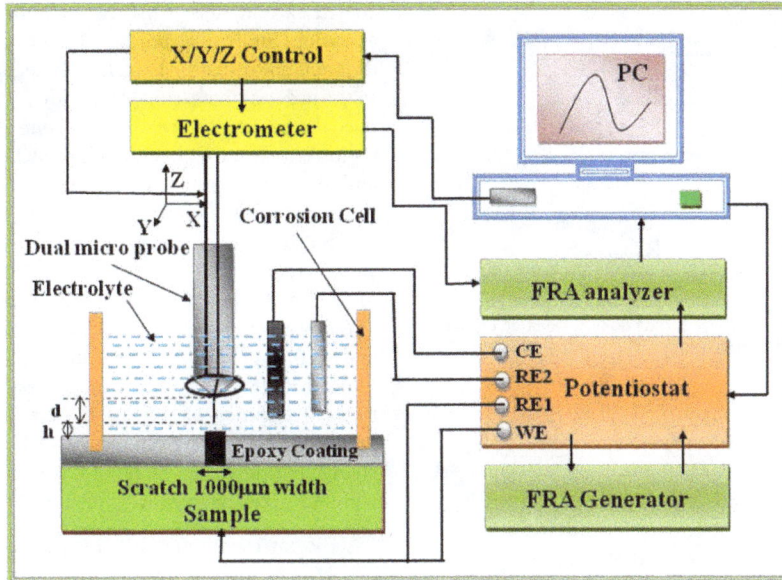

**Figure 1:** The schematic of the electrode installation and experimental set-up of LEIS, from Ref. [27].

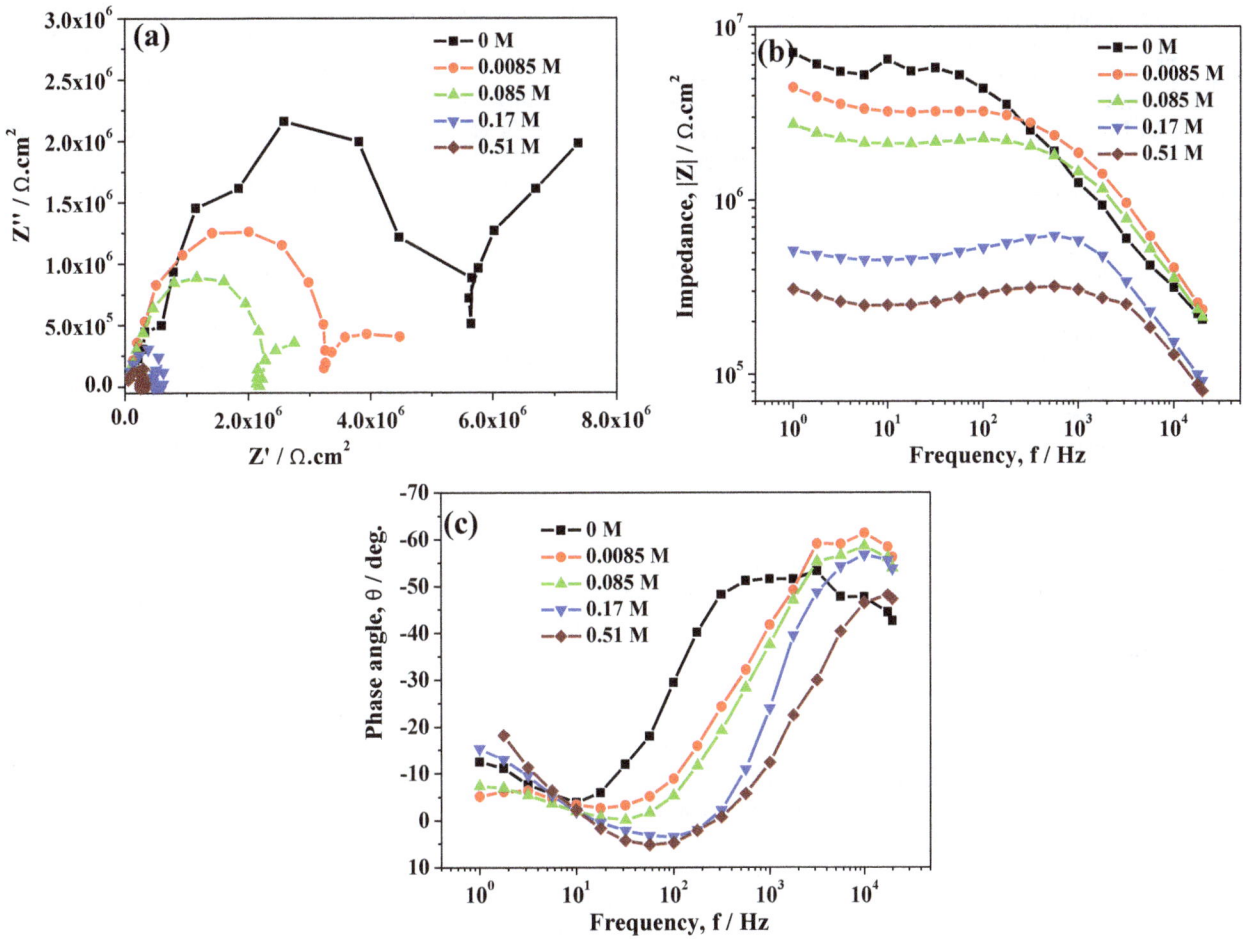

**Figure 2:** LEIS Nyquist (a) and Bode (b, c) plots of scratched epoxy coated carbon steel in sat. Ca(OH)$_2$ with varying concentration of Cl$^-$ ions at their respective open circuit potentials.

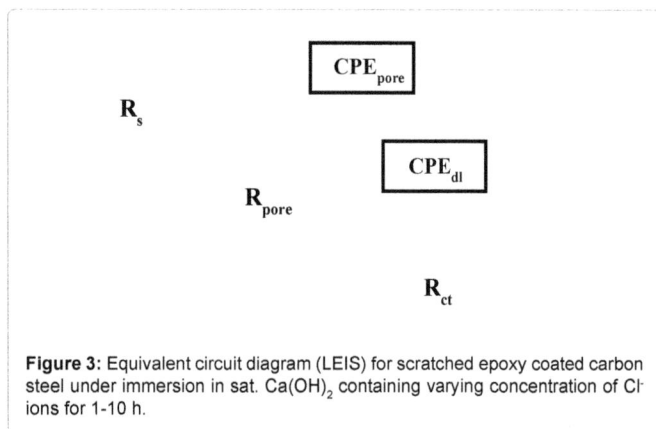

**Figure 3:** Equivalent circuit diagram (LEIS) for scratched epoxy coated carbon steel under immersion in sat. $Ca(OH)_2$ containing varying concentration of $Cl^-$ ions for 1-10 h.

| Sample | Elements (mass %) | | | | | | | | |
|---|---|---|---|---|---|---|---|---|---|
| | C | Si | Mn | P | S | Al | N | O | Fe |
| Carbon steel | 0.10 | 0.30 | 0.70 | 0.01 | 0.003 | 0.03 | 0.003 | 0.002 | Bal. |

**Table 1:** Chemical composition of carbon steel specimen.

it will be pure resistance when n=0. Q is called as *CPE* when 0<n<1 and it prevails under conditions of surface heterogeneity [29]. The local electrochemical parameters derived after fitting the LEIS data are shown in Figure 4. Since, the value of 'n' lies between 0.58 and 0.92 in sat. $Ca(OH)_2$ with varying concentration of $Cl^-$ ions, the choice of CPE as the circuit element seems to be appropriate. The observed impedance for scratched epoxy coated carbon steel is decreased with an increase in concentration of $Cl^-$ ions in sat. $Ca(OH)_2$ (Figure 2a and 2b). This is clearly indicated by the depressed semicircles (Figure 2a) with an increase concentration of $Cl^-$ ions in sat. $Ca(OH)_2$. The pore resistance, $R_{pore}$ of scratched epoxy coated carbon steel is decreased from the range of $3.93 \times 10^6$ to $2.75 \times 10^5$ ohm.cm² with the corresponding increase in pore capacitance, $CPE_{pore}$ from $2.75 \times 10^{-9}$ to $1.48 \times 10^{-7}$ S.sⁿ.cm⁻² with an increase in concentration of $Cl^-$ ions. The charge transfer resistance, $R_{ct}$ decreases from $6.38 \times 10^6$ to $2.84 \times 10^6$ ohm.cm² with corresponding increase in double layer capacitance, $CPE_{dl}$ from $5.53 \times 10^{-8}$ to $6.80 \times 10^{-6}$ S.sⁿ.cm⁻² with an increase in concentration of $Cl^-$ ions in sat. $Ca(OH)_2$ respectively. However, the corrosion resistance values of coated steel in sat. $Ca(OH)_2$ with each concentration $Cl^-$ ions are not changed significantly with increase in immersion time from 1-10 h. In the absence of $Cl^-$ ions, inhibitive ability of the $OH^-$ ions are responsible for the formation of better passive film, while the presence of $Cl^-$ ions in sat. $Ca(OH)_2$, promotes preferential adsorption of $Cl^-$ ions on the electrode surface and enhances the corrosion rate of steels [30,31]. In the presence study, the significant decrease in corrosion resistance is observed for higher concentration of $Cl^-$ ions (0.17 and 0.51M). The decrease in impedance with an increase in concentration of $Cl^-$ ions in in sat. $Ca(OH)_2$ is due to the increasing the local dissolution of Fe.

The corrosion performance of scratched epoxy coated carbon steel is to be observed at low frequency in EIS measurements. Hence, the LEIS are performed at low frequency of 10 Hz over the scratch and coated area. This provides a better understanding of the corrosion mechanism over the designated area. The LEIS maps acquired at different zones (scratch, scratch front and coated area) of the same sample covering a large area is useful to get a better understanding of the rate of corrosion at these zones. The 3D LEIS maps measured over a scanning area of 7 mm × 1 mm (x-y scale) over the scratched epoxy coated carbon steel in sat. $Ca(OH)_2$ with varying concentration of $Cl^-$ ions are shown in Figure 5. In Figure 5, |Z| represents the measured impedance (3-D

impedance distribution along the x-y axis) over the scratched epoxy coated carbon steel, which is an indicator of the electrode stability at various individual points. The magnitude of |Z| is represented by different color shades; blue, light blue, light green, dark green and red, in the order of increasing |Z| (Figure 5). It is evident that the |Z| is high over the coating (represented as zone H), moderate in the interfacial region between the coating and scratched area (represented as zone I) and low in the scratched area (represented as zone L). The average |Z| measured at zone H is above ~ $1 \times 10^7$ ohm.cm² whereas it decreased to ~ $6 \times 10^6$ ohm.cm² in zone L in a sat. $Ca(OH)_2$, while the impedance values are maintained with immersion time for 1-10 h. By the addition of $Cl^-$ ions to the sat. $Ca(OH)_2$, the average |Z| starts to decrease both at the H zone as well as at the L zone from $1 \times 10^7$ to $1 \times 10^6$ ohm.cm² and $4.76 \times 10^6$ to $2.88 \times 10^5$ ohm.cm² respectively. These inferences indicate that the local corrosion process is enhanced by the $Cl^-$ ions due to the dissolution of Fe from the scratch area. However, the addition of 0.0085 M of $Cl^-$ ions in sat. $Ca(OH)_2$ does not show any significant change in impedance values at L zone. The observed results are in line with our previously reported results [27].

The corresponding LEIS line profiles are shown in Figure 6, which clearly indicates the decrease in |Z| in the L-zone due to the higher dissolution of Fe at the scratched area with increase in concentration of Cl ions. The |Z| measured at the L zone as a function of concentration of $Cl^-$ ions is plotted in Figure 7. It is evident from the Figure 7 that an increase in concentration of $Cl^-$ ions has led to an increase in the rate of corrosion of scratched epoxy coated steel at the L zone. The shape of the curves in Figure 6 further confirms the inferences made from Figure 5 that at higher concentration of $Cl^-$ ions, the extent dissolution of Fe at scratched area becomes higher. Further, the continuous decrease in |Z| over the coating and the scratched area indicates that a continuous dissolution of Fe and the easy availability of corrosive intermediate species (containing Fe(II)/Fe(III) compounds) near the scratched area which could be deposited over the coating due to the higher pH from the bulk solution [32]. However, there is no sign of additional defects over the coating that has been observed from the LEIS maps with the addition of $Cl^-$ ions to the sat. $Ca(OH)_2$. On the other hand, the decrease in |Z| over the coating might be due to the water uptake of the epoxy coating [10,33]. Further, the experiments have been conducted by long term immersion for 1-5 days on the better understanding of local corrosion process at scratch and coating area of scratched epoxy coated carbon steel in varying concentration of $Cl^-$ ions.

## LEIS measurements for 1-5 days of immersion

LEIS Nyquist plots of scratched epoxy coated carbon steel in sat. $Ca(OH)_2$ with 0.51 M $Cl^-$ ions for 5 days of immersion are shown in Figure 8a. The corresponding Bode impedance and Bode phase angle plots are shown in Figures 8b and 8c. It is evident from Figure 8a that the capacitive part at high frequency range is associated with the corrosion product layer at scratch, the inductive loop at intermediate frequencies is due to the adsorption of corrosion species involved in the local corrosion process and, the capacitive part in the low frequency region is related to the double layer capacitance and charge transfer resistance of corroding steel [34,35]. The proposed electrical circuit model and fittings values are shown in Figures 9 and 10 for sat. $Ca(OH)_2$ with varying concentration of $Cl^-$ ions respectively. In this model, $R_s$ represents the solution resistance while $R_{corr\ product\ layer}$ and $CPE_{corr\ product\ layer}$ correspond to the resistance of corrosion product layer and the corresponding capacitance of the corrosion product layer, respectively. The L is the inductance of adsorbed intermediate corrosive species involved in the local corrosion process while $R_{ct}$ and

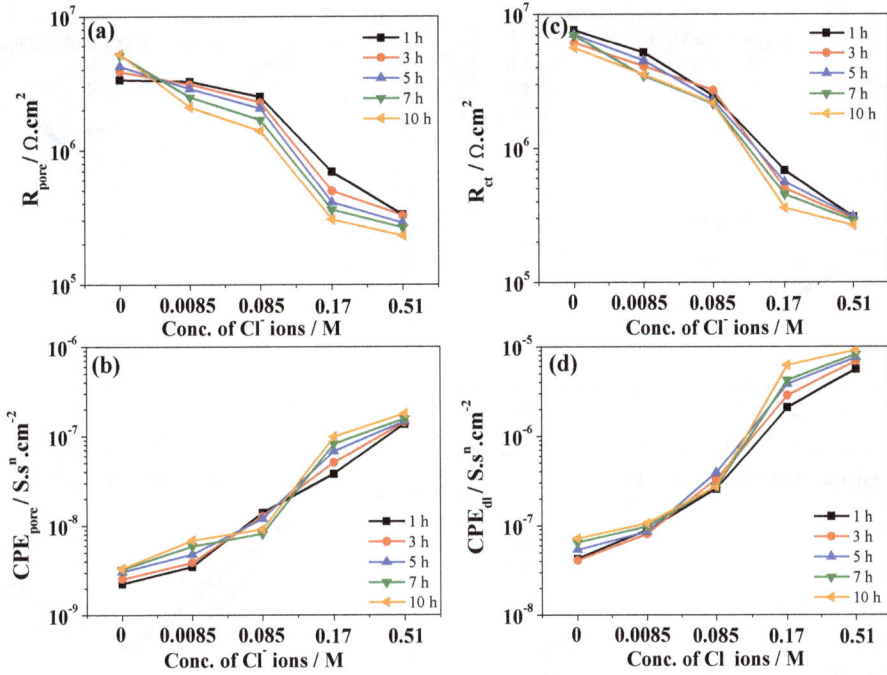

**Figure 4:** Local impedance values of (a) $R_{pore}$, (b) $CPE_{pore}$, (c) $R_{ct}$ and (d) $CPE_{dl}$ for scratched epoxy coated carbon steel after wet/dry cycles test in sat. Ca(OH)$_2$ with varying concentration of Cl$^-$ ions recorded at their respective OCPs for 1-10 h.

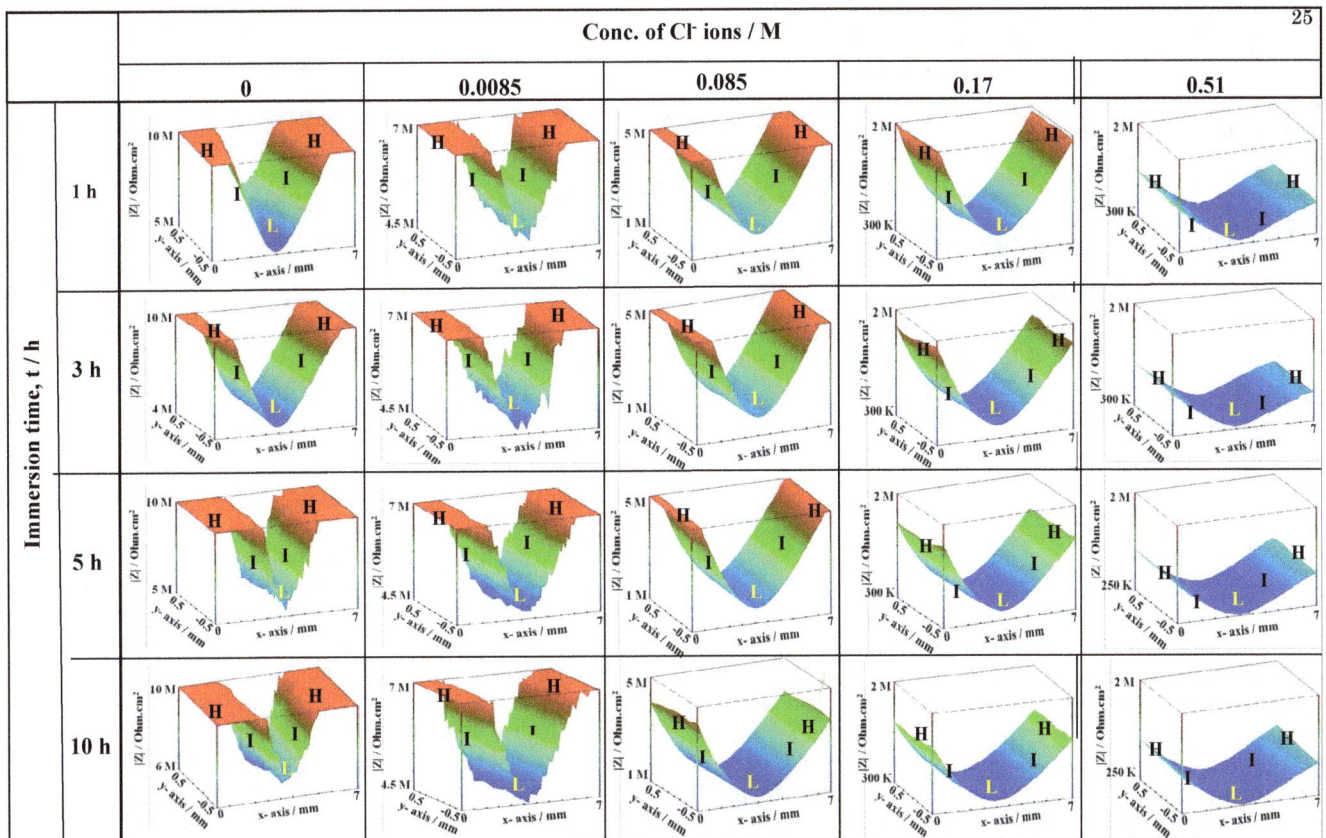

**Figure 5:** The 3D LEIS maps measured over a scanning area of 7 mm × 1 mm (x-y scale) over the scratched epoxy coated carbon steel in sat. Ca(OH)$_2$ with varying concentration of Cl- ions for 1-10h immersion. Scale: 7 mm × 1 mm in x and y directions, vertical direction is |Z|.

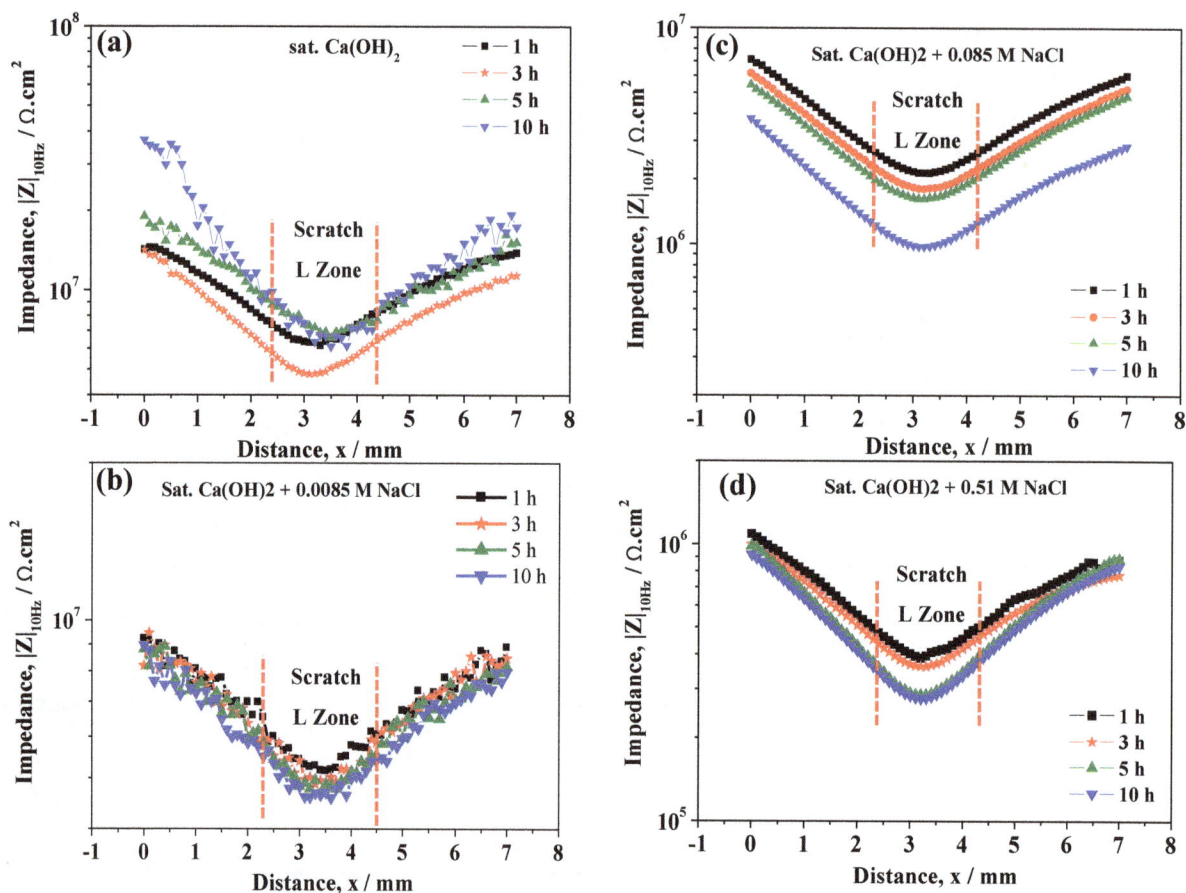

**Figure 6:** LEIS line profile measured at L zone of scratched epoxy coated carbon steel in sat. Ca(OH)$_2$ containing varying concentration of Cl$^-$ ions for 1-10 h: (a) 0 M, (b) 0.0085 M, (c) 0.085 M and (d) 0.51 M.

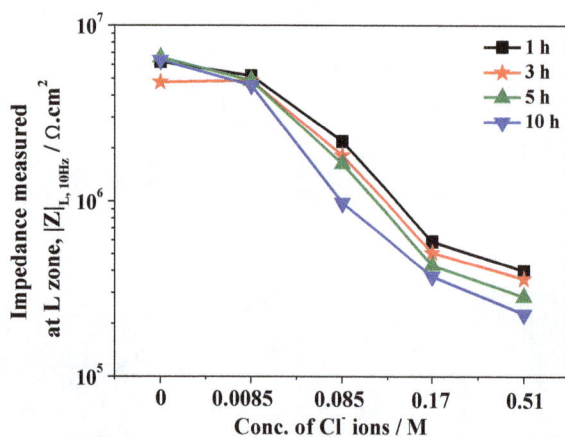

**Figure 7:** Impedance measured at L zone (|Z|$_{10Hz}$) over scratched epoxy coated carbon steel in sat. Ca(OH)$_2$ with varying concentration of Cl$^-$ ions.

CPE$_{dl}$ are the charge transfer resistance and double layer capacitance of the system, respectively. It is evident from Figure 10, the resistance of corrosion product layer is decreased from $1.01 \times 10^5$ to $1.24 \times 10^4$ ohm. cm$^2$ (one fold decrease) by the addition 0.51 M Cl$^-$ ions from 1$^{st}$ to 3$^{rd}$ day of immersion. With increase in immersion time from 3$^{rd}$ to 5$^{th}$ day, resistance of corrosion product layer is slightly increased to $2.66 \times 10^4$

ohm.cm$^2$. This is due to the formation of corrosion products that makes difficulty in charge transfer at the interface of corrosion product layer and bare steel interface. However, the charge transfer is not completely restricted due to the formation of intermediate corrosion products [21]. The resistance of the corrosion product layer formed on scratched area is higher ($9.47 \times 10^6$ to $1.49 \times 10^7$ ohm.cm$^2$) in sat. Ca(OH)$_2$ for 1-5 day of immersion. In the presence of 0.085 M Cl$^-$ ions, the resistance of the corrosion product layer is initially increased from $9.95 \times 10^5$ to $3.57 \times 10^6$ ohm.cm$^2$ then it is decreased from $3.57 \times 10^6$ to $1.81 \times 10^5$ ohm.cm$^2$ with increase in immersion time for 1-5 days. On the other hand, the resistance of R$_{corr\ product\ layer}$ is higher in sat. Ca(OH)$_2$ ($1 \times 10^7$-$1.40 \times 10^7$ ohm.cm$^2$) as compared to 0.0085 M Cl$^-$ added solution ($4.85 \times 10^6$-$7.66 \times 10^6$ ohm.cm$^2$) for 1-5 days of immersion. In this case, the R$_{corr\ poduct\ layer}$ is increased with immersion time for 0 and 0.0085 M Cl$^-$ ions as compared to other concentration of Cl$^-$ ions added solutions. These inferences clearly indicate that R$_{corr\ product\ layer}$ depends on Cl$^-$ ion concentrations and the corresponding chemical compositions of corrosion product layers [23,36]. Moreover, Cl$^-$ to OH$^-$ ion concentration is also important factor that determines the chemical compositions of the corrosion product layers [23]. The charge transfer resistance, R$_{ct}$ of corroding steel is increased from $1.29 \times 10^6$ to $1.28 \times 10^7$ ohm.cm$^2$ 1-2 days of immersion time and it retained at $7.93 \times 10^6$ ohm.cm$^2$ with increase in immersion time in the absence of Cl$^-$ ion solution. Similarly, the R$_{ct}$ values are increased from $1.05 \times 10^6$ to $6.49 \times 10^6$ for 1$^{st}$ day of immersion and then sustained in the resistance values ($2.90 \times 10^6$)

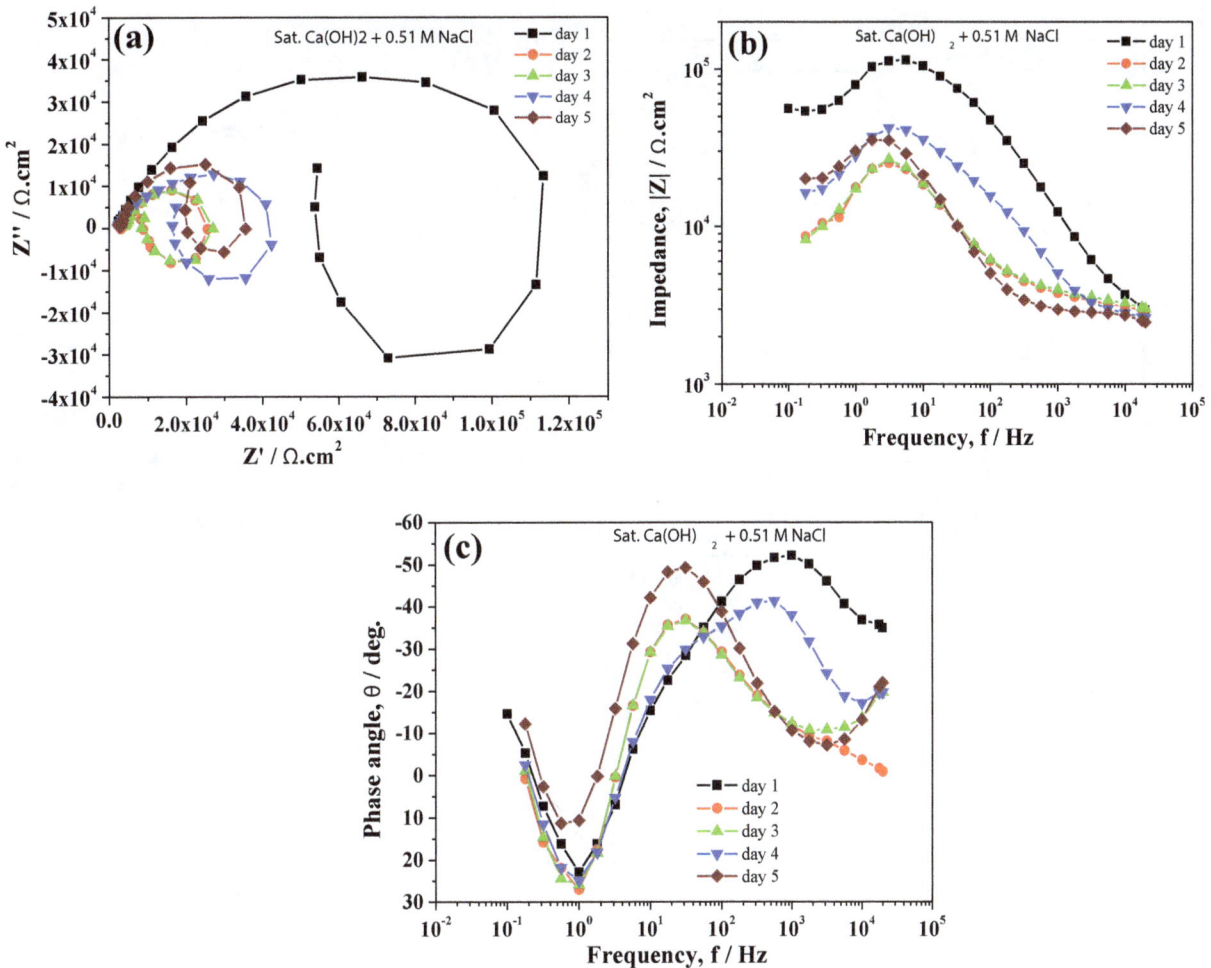

**Figure 8:** LEIS Nyquist (a) and Bode (b, c) plots of scratched epoxy coated carbon steel immersed in sat. Ca(OH)$_2$ with 0.51 M Cl$^-$ ions for 1-5 days at their respective open circuit potentials.

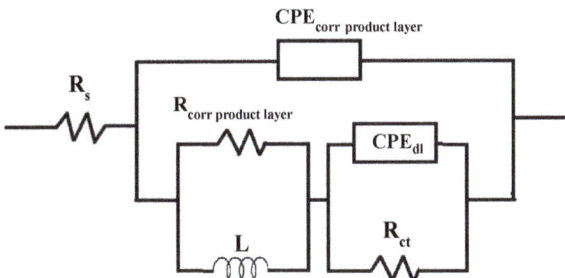

**Figure 9:** Equivalent circuit diagram for scratched epoxy coated carbon steel under immersion in sat. Ca(OH)$_2$ containing varying concentration of Cl$^-$ ions for 1-5 days.

with increase in immersion time in 0.0085 M Cl$^-$ ion solution. The R$_{ct}$ values (absence of Cl$^-$ ions) are comparable with R$_{ct}$ values of 0.0085 M Cl$^-$ ions increase in immersion time. The increase in concentration of Cl$^-$ ions in to 0.085 M, the R$_{ct}$ values are increased initially from 4.20 × 10$^5$ (day 1) to 1.16 × 10$^6$ ohm.cm$^2$ (day 2) and then start to decrease with immersion time from 1.16 × 10$^6$ to 1.01 × 10$^5$ ohm.cm$^2$. However,

in case of sat. Ca(OH)$_2$ with 0.51 M Cl, the R$_{ct}$ values start to decrease from 1.24 × 10$^5$ to 1.12 × 10$^4$ ohm.cm$^2$ for 1$^{st}$ to 2$^{nd}$ day of immersion then starts to increase slightly from 1.12 × 10$^4$ to 3.18 × 10$^4$ ohm.cm$^2$ with increase in immersion time.

The various concentrations of chloride threshold levels on the corrosion resistance of steels have been reported by various researchers [28,37,38]. Li and Sagues [37] studied the chloride threshold level in saturated Ca(OH)$_2$ was found to be 0.01-0.04 M. Gouda [38] has been reported that the 0.007 M could be the chloride threshold level (added as NaCl) in a saturated Ca(OH)$_2$ solution, which did not affect the passivity of the steel. Moreno et al. [28] have been reported that the 0.02% Cl (0.0034 M) addition in to the sat. Ca(OH)$_2$ solution has no appreciable change in anodic behavior of steel and it was observed that Cl$^-$ ion concentration equal to or higher than 0.05% (0.0085 M) induce pitting of the steel. However, the epoxy-coated steel can gave a good protection; long term performance even on severe exposure to chloride conditions have been reported with considering properly coated and handled steels [39,40]. Al-Amoudi et al. [40] found that the chloride threshold level of epoxy-coated steel with various degrees of coating damage (1 and 2%) and the chloride threshold level was about 2 and 0.4% (0.35 and 0.068 M) by weight of cement respectively. In the present

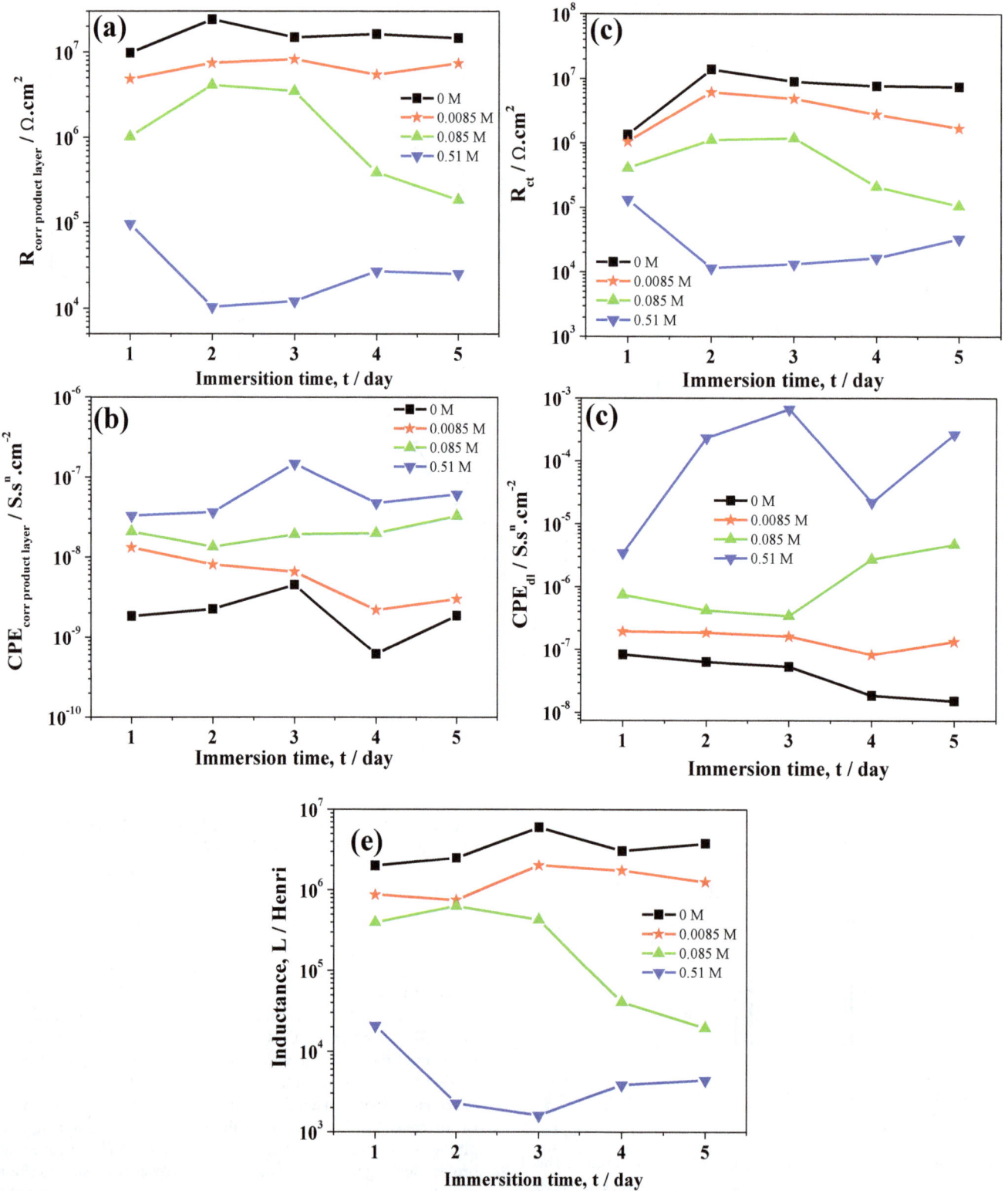

**Figure 10:** Local Impedance values of (a) $R_{corr\ product\ layer}$, (b) $CPE_{corr\ product\ layer}$, (c) $R_{ct}$, (d) $CPE_{ct}$ and (e) L for scratched epoxy coated carbon steel under immersion in sat. $Ca(OH)_2$ with varying concentration of $Cl^-$ ions recorded at their respective OCPs for 1-5 days.

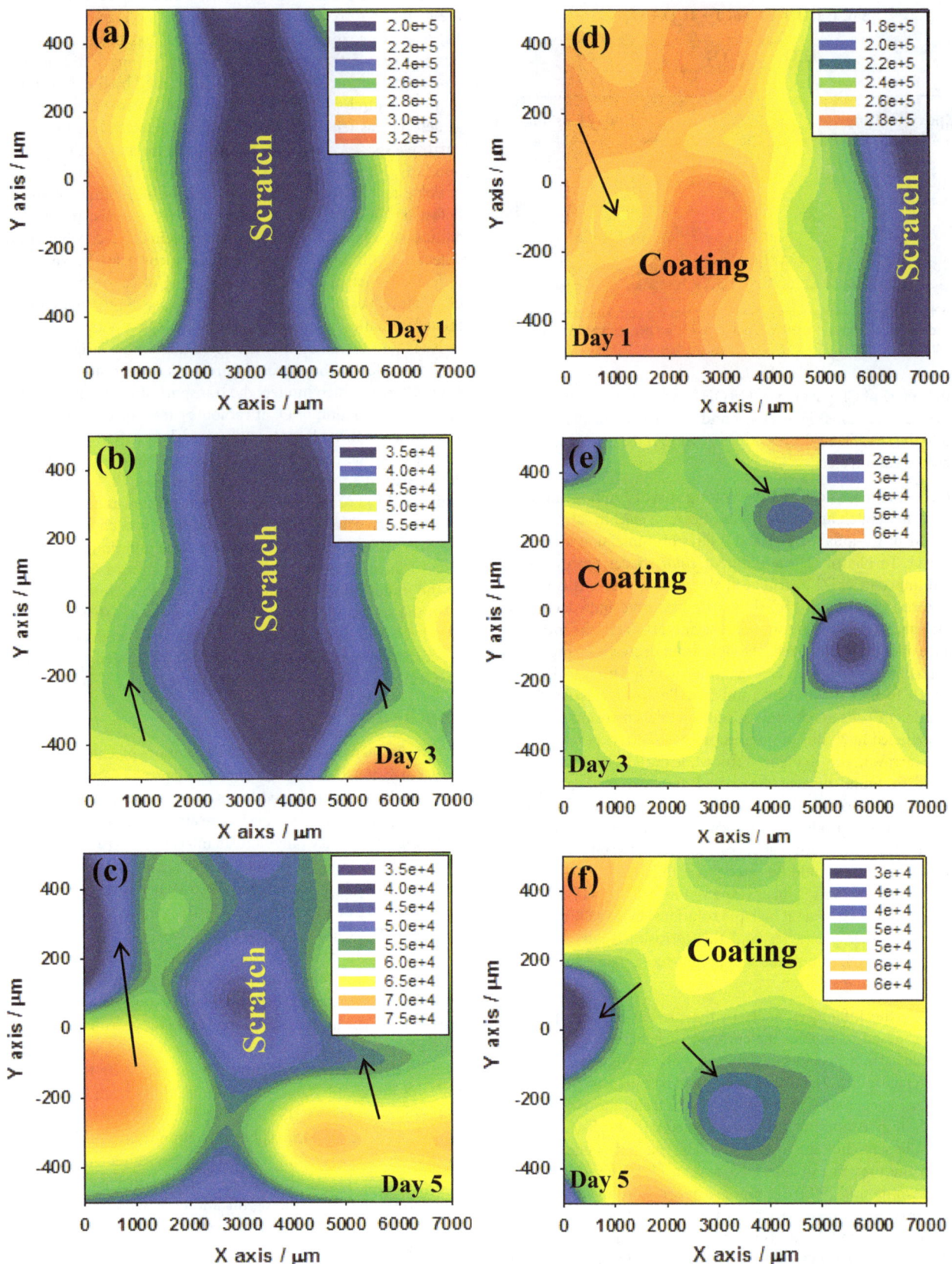

**Figure 11:** 2D LEIS maps measured over a scanning area of 7 mm × 1 mm (x-y scale) over the scratched epoxy coated carbon steel in sat. Ca(OH)$_2$ with 0.51 M of Cl$^-$ ions : (a, b and c) at scratch area, (d, e and f) over coating area for 1, 3 and 5 days of immersion respectively ; different color shads indicates difference in |Z| at different zones.

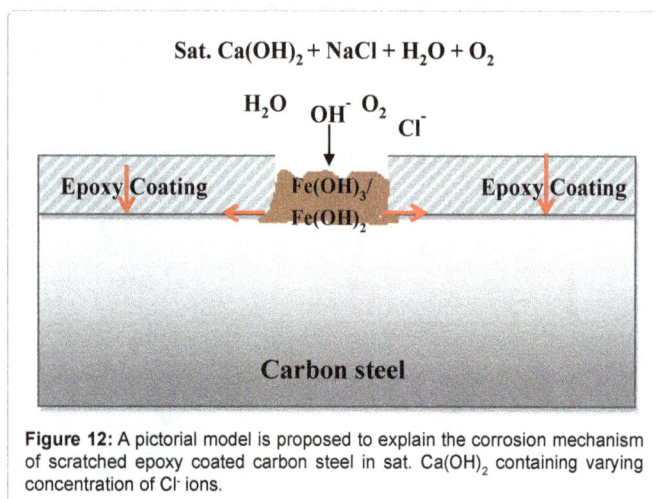

**Figure 12:** A pictorial model is proposed to explain the corrosion mechanism of scratched epoxy coated carbon steel in sat. Ca(OH)$_2$ containing varying concentration of Cl$^-$ ions.

study, 0.0085 M of Cl$^-$ ions in sat. Ca(OH)$_2$ is found to be comparable with absence of Cl- ions (Figures 4, 7 and 10), beyond 0.0085 M of Cl$^-$ ions led to the significant decrease in corrosion resistance of scratched epoxy coated carbon steel.

The 2D LEIS maps acquired at scanning area of 7 mm × 1 mm (x-y scale) over the scratch and coated area of carbon steel in sat. Ca(OH)$_2$ with 0.51 M of Cl$^-$ ions for 1, 3 and 5 days of immersion are shown in Figure 11. In Figure 11, the different color shades represents the measured impedance, |Z| distribution along the x-y axis. It is evident from (Figure 11a, 11b and 11c) shows the measured impedance over the scratch area, while the Figure 11d, 11e and 11f shows the measured impedance over the coated area. The significant decrease in impedance at scratch is observed with increase in immersion time from 1$^{st}$ day to 3$^{rd}$ day (2.5 × 10$^5$ to 3.5 × 10$^4$ ohm.cm$^2$) and then slightly increased at 5$^{th}$ day (5 × 10$^4$ ohm.cm$^2$) at scratch area. In addition, the corrosion at scratch area is broadened in 3$^{rd}$ day of immersion due to direct penetration of electrolyte (arrow marks in Figure 11b), while the additional corrosion spots could be found in over the coated area (arrow marks shown in Figure 11e). The corrosion at scratch area is restricted due to the formation corrosion products with increase in immersion time to 5$^{th}$ day (Figure 11c). The additional corrosion spots at scratch front and away from the scratch could be extended with larger corrosion spots (Figure 11e and 11f). This is mainly due to the formation of corrosion products at the scratch area (at high pH solutions) eventually corrosion starts at some other places (around the corrosion spots) [21]. The observed results are in line with LEIS parameters (Figure 10) for 0.51 M Cl$^-$ ions containing solution.

Based on the inferences made in the present study, a pictorial model is proposed to explain the corrosion mechanism of scratched epoxy coated carbon steel in sat. Ca(OH)$_2$ with varying concentration of Cl$^-$ ions (Figure 12). The dissolution kinetics is increased at scratch area of epoxy coated steel with increase of concentration of Cl$^-$ ions. In case of immersion, the hydroxides (Fe(OH)$_2$ or Fe(OH)$_3$) are formed at the scratched area while at a later stage it is converted to FeOOH in the presence of dissolved oxygen [41]. The formation of green rust compounds, GR(Cl$^-$) are also possible as an intermediate corrosion products when the Cl$^-$ ions are present in sat. Ca(OH)$_2$ [42,43]. However, the GR(Cl$^-$) could be converted easily to (γ, α and β)-FeOOH in the presence oxygen as final corrosion products with immersion time. Once the corrosion products (Fe(OH)$_2$ or Fe(OH)$_3$ or FeOOH) are formed at the scratch area, the oxygen diffusion could be restricted and eventually the cathodic process could occurs

at some other places away from the scratch or scratch front leads to the cathodic disbandment of coating (arrow marks shown in Figure 12) leads to coating degradation [21]. It has been shown that 1000 μm diameter defect with relatively open geometry did not show any blocking effect, which was mainly dominated by the diffusion process in near-neutral pH solution. Moreover, corrosion product formed at defects was left easily from the large defect, while the diffusion of corrosive species could participate on the local corrosion process [13]. Similarly, Philippe et al. [11] have been reported that the degradation of polyester coil coated galvanized steel with 250 μm diameter artificial defect in NaCl by LEIS at different time intervals. It was found the additional corrosion spots away from original defect, while the original defect was still showed the occurrence of corrosion in NaCl due to the diffusion of active species through a plug of corrosion product at the defect. Jorcin et al. [14] have explored the use of LEIS mapping to assess the delamination phenomena at the steel/epoxy-vinyl primer interface in NaCl and identified that the delamination has originated from the artificial defects. The corrosion products formed at the defect acts as anodic zones, which hinders the transport of O$_2$ at this defect site and favors the reduction of O$_2$ at cathodic sites, especially over the coated area, which in turn promotes cathodic delamination at the steel-coating interface. In the present study, the corrosion occurs at initially at the defect area due to the preferential adsorption Cl$^-$ ions and due to the high pH (12.7) of solution it could be deposited as a corrosion product at the scratch area. Hence, the corrosion at scratch is restricted due to the formation of corrosion products. The additional corrosion spots (zones) formed around the original scratch area which is generated by the under-film corrosion beginning at the scratch front and away from the scratch due to the cathodic reactions. This has been clearly visualized by LEIS mapping which provides better understanding on the corrosion mechanism of scratched epoxy coated carbon steel in sat. Ca(OH)$_2$ in Cl$^-$ ions containing solutions.

## Conclusion

The influence of varying concentration of chloride ions on the local corrosion behavior of scratched epoxy coated carbon steel in sat. Ca(OH)$_2$ is investigated by *in-situ* manner using LEIS. The LEIS responses measured at the defect are attributed to the pore impedance in the high-frequency range and an interfacial corrosion reaction in the low-frequency range of corroding steel within 1-10 h immersion. The pore resistance, R$_{pore}$ of scratched epoxy coated carbon steel is decreased from the range of 3.93 × 10$^6$ to 2.75 × 10$^5$ ohm.cm$^2$, the charge transfer resistance; R$_{ct}$ is also decreased from 6.38 × 10$^6$ to 2.84 × 10$^6$ ohm.cm$^2$ with increase in concentration of Cl$^-$ ions from 0-0.51 M. The LEIS maps acquired at 10 Hz provides a better recognition of anodic and cathodic areas on the local corrosion process of scratched epoxy coated carbon steel. The decrease of change in |Z| range from 6.27 × 10$^6$ to 4.85 × 10$^6$ ohm.cm$^2$ is marginal by the addition of 0.0085 M Cl$^-$ ions into the sat. Ca(OH)$_2$, whereas it decreased from 4.85 × 10$^6$ to 2.88 × 10$^5$ ohm. cm$^2$ with increase in concentration of Cl ions from 0.0085 to 0.51 M. The |Z| measured by LEIS reveals a continuous decrease in impedance at the scratch due to the higher dissolution of Fe with an increase in concentration of Cl$^-$ ions, which is further validated by the variation in |Z| by LEIS maps. This clearly indicates that an increase of NaCl concentration leads to the significant increase in the corrosion rate of the scratched epoxy coated carbon steel due to preferential adsorption of Cl$^-$ ions on the electrode.

On the other hand, LEIS Nyquist plots clearly showed that the measured impedance at high frequency is related to resistance of corrosion product layer (R$_{corr\ product\ layer}$) formed at the defect which acts as anodic area and the low frequency part are related to corroding of carbon steel with immersion of 1-5 days. The resistance of corrosion product layer is decreased from $1.01 \times 10^5$ to $1.24 \times 10^4$ ohm.cm$^2$ (one fold decrease) by the addition of 0.51 M Cl$^-$ ions from 1$^{st}$ to 3$^{rd}$ day of immersion. With increase in immersion time from 3$^{rd}$ to 5$^{th}$ day, resistance of corrosion product layer is slightly increased to $2.66 \times 10^4$ ohm.cm$^2$. This is due to the formation of corrosion products that makes difficulty in charge transfer at the interface of corrosion product layer and bare steel in which the charge transfer is not completely restricted due to the formation of intermediate corrosion products. The R$_{corr}$ $_{product\ layer}$ values are much higher in sat. Ca(OH)$_2$ and with 0.0085 M Cl$^-$ ions as compared to 0.085 and 0.51 M Cl$^-$ ions and it varied slightly higher/lower impedance values. These inferences indicate that R$_{corr}$ $_{product\ layer}$ depends on Cl$^-$ ion concentrations and the corresponding chemical compositions of corrosion product layers. This has been clearly indicated in the 2D topographic images that corrosion occurred at scratch along with scratch front as well as away from the scratch with increase in immersion time. The different color zone at scratch area revealed that the formation of corrosion products followed by the localized corrosion due to the porous nature of the passive layer. The 2D topographic image results are well agreement with LEIS parameters for 1-5 days of immersion. No significant reduction in LEIS values at 1-10 h immersion, |Z| values at L zone, LEIS in 1-5 days of immersion in 0 and 0.0085 M Cl$^-$ ions. This is due to the presence of sufficient concentration of inhibitive OH$^-$ ions in solution. The inferences made in the present study points out that beyond 0.0085 M NaCl is likely to increase the corrosion rate of scratched epoxy coated carbon steel.

## Acknowledgements

This work was supported by Council for Science, Technology and Innovation (CSTI), Cross-ministerial Strategic Innovation Promotion Program (SIP), "Infrastructure maintenance, renovation and management" (Funding agency: JST).

## References

1. Mansfeld F (1993) Models for the impedance behavior of protective coatings and cases of localized corrosion. Electrochim Acta 38: 1891-1897.

2. Walter GW (1986) A review of impedance plot methods used for corrosion performance analysis of painted metals. Corros Sci 26: 681-703.

3. He J, Gelling VJ, Tallman DE, Bierwagen GP (2000) A scanning vibrating electrode study of chromated-epoxy primer on steel and aluminum. J Electrochem Soc 147: 3661-3666.

4. Maile FJ, Schauer T, Eisenbach CD (2000) Evaluation of the delamination of coatings with scanning reference electrode technique. Prog Org Coat 38: 117-120.

5. Nazarov A, Le Bozec N, Thierry D, Le Calvé P, Pautasso JP (2012) Scanning kelvin probe investigation of corrosion under thick marine paint systems applied on carbon steel. Corrosion 68: 720-729.

6. Bastos AC, Simo̅es AM, Gonza´lez S, Gonza´lez-Garci´a Y, Souto RM (2004) Imaging concentration profiles of redox-active species in open-circuit corrosion processes with the scanning electrochemical microscope. Electrochem Commun 6: 1212-1215.

7. Lillard RS, Kruger J, Tait WS, Moran PJ (1995) Using local electrochemical impedance spectroscopy to examine coating failure. Corrosion 51: 251-259.

8. Van der Weidje DH, van Westing EPM, der Wit JWH (1996) EIS measurements on artificial blisters in organic coatings. Electrochim Acta 41: 1103-1107.

9. Cambier SM, Verreault D, Frankel GS (2014) Raman investigation of anodic undermining of coated steel during environmental exposure. Corrosion 70: 1219-1229.

10. Nguyen T, Martin JW (2004) Modes and mechanisms for the degradation of fusion bonded epoxy-coated steel in a marine concrete environment. JCT Research 1: 81-92.

11. Philippe LVS, Walter GW, Lyon SB (2003) Investigating localized degradation of organic coatings comparison of electrochemical impedance spectroscopy with local electrochemical impedance spectroscopy. J Electrochem Soc 150: B111-B119.

12. Zou F, Thierry D (1997) Localized electrochemical impedance spectroscopy for studying the degradation of organic coatings. Electrochim Acta 42: 3293-3301.

13. Zhong C, Tang X, Cheng YF (2008) Corrosion of steel under the defected coating studied by localized electrochemical impedance spectroscopy. Electrochim Acta 53 4740-4747.

14. Jorcin JB, Aragon E, Merlatti C, Pébère N (2006) Delaminated areas beneath organic coating: A local electrochemical impedance approach. Corros Sci 48: 1779-1790.

15. Raman V, Nishimura T (2009) Monitoring of Environmental Factors and Corrosion Analysis of Reinforcing Steel in Mortar. Mater Trans 50: 799-805.

16. Dong Y, Zhou Q (2014) Relationship Between Ion Transport and the failure behavior of epoxy resin coatings. corros Sci 78: 22-28.

17. Vakili H, Ramezanzadeh B, Amini R (2015) The corrosion performance and adhesion properties of the epoxy coating applied on the steel substrates treated by cerium-based conversion coatings. Corros Sci 94: 466-475.

18. Nishimura T (2016) Corrosion estimation of epoxy coated high tensile strength steel measured by statistical analysis and TEM - EELS. Mater Trans 57: 52-57.

19. Souto RM, Gonza´lez-Garcia Y, Gonza´lez S, Burstein GT (2004) Damage to paint coatings caused by electrolyte immersion as observed in situ by scanning electrochemical microscopy. Corros Sci 46: 2621- 2628.

20. Souto RM, Gonza´lez-Garcia Y, Gonza´lez S (2005) In situ monitoring of electroactive species by using the scanning electrochemical microscope. Application to the investigation of degradation processes at defective coated metals. Corros Scie 47: 3312-3323.

21. Balusamy T, Nishimura T (2016) In-situ corrosion monitoring of scratched epoxy coated carbon steel in saturated Ca(OH)$_2$ with or without 3% NaCl by scanning electrochemical microscopy and electrochemical impedance spectroscopy. Amer J Anal Chem 7: 533-547.

22. Schaller RF, Thomas S, Birbilis N, Scully JR (2015) Spatially resolved mapping of the relative concentration of dissolved hydrogen using the scanning electrochemical microscope. Electrochem Commun 51: 54-58.

23. Grousset S, Kergourlay F, Neff D, Foy E, Gallias JL, et al. (2015) In situ monitoring of corrosion processes by coupled micro-XRF/micro-XRD mapping to understand the degradation mechanisms of reinforcing bars in hydraulic binders from historic monuments. J Anal At Spectrom 30: 721-729.

24. Lin B, Hu R, Ye C, Li Y, Lin C (2010) A study on the initiation of pitting corrosion in carbon steel in chloride-containing media using scanning electrochemical probes. Electrochim Acta 55: 6542-6545.

25. Glass GK, Buenfeld NR (1997) The presentation of the chloride threshold level for corrosion of steel in concrete. Corros Sci 39: 1001-1013.

26. Shi X, Nguyen TA, Kumar P, Liu Y (2011) A phenomenological model for the chloride threshold of pitting corrosion of steel in simulated concrete pore solutions. Anti-Corros Method Mater 58: 179-189.

27. Balusamy T, Nishimura T (2016) In-situ monitoring of local corrosion process of scratched epoxy coated carbon steel in simulated pore solution containing varying percentage of chloride ions by localized electrochemical impedance spectroscopy. Electrochim Acta 199: 305-313.

28. Moreno M, Morris W, Alvarez MG, Duff GS (2004) Corrosion of reinforcing steel in simulated concrete pore solutions Effect of carbonation and chloride content. Corros Sci 46: 2681-2699.

29. Orazem ME, Tribollet B (2008) Electrochemical Impedance Spectroscopy. John Wiley & Sons, USA.

30. Saremi M, Mahallati E (2002) A study on chloride-induced depassivation of mild steel in simulated concrete pore solution. Cem Concr Res 32: 1915-1921.

31. Burstein GT, Davies DH (1980) The effects of anions on the behaviour of scratched iron electrodes in aqueous solutions. Corros Sci 20: 1143-1155.

32. Sagoe-Crentsil KK, Glasse FP (1993) Green rust, Iron solubility and the role of chloride in the corrosion of steel at high pH. Cem Concr Res 23: 785-791.

33. Park JH, Lee GD, Ooshige H, Nishikata A, Tsuru T (2003) Monitoring of water uptake in organic coatings under cyclic wet-dry condition. Corros Sci 45: 1881-1894.

34. Zhang GA, Cheng YF (2009) Micro-electrochemical characterization of corrosion of welded X70 pipeline steel in near-neutral pH solution. Corros Sci 51: 1714-1724.

35. Meng GZ, Zhang C, Cheng YF (2008) Effects of corrosion product deposit on the subsequent cathodic and anodic reactions of X-70 steel in near-neutral pH solution. Corros Sci 50: 3116-3122.

36. Nishimura T, Katayama H, Noda K, Kodama T (2000) Electrochemical behavior of rust formed on carbon steel in a wet/dry environment containing chloride ions. Corrosion 56: 935-941.

37. Li L, Sagues AA (1999) Effect of chloride concentration on the pitting and repassivation potentials of reinforcing steel in alkaline solutions. Corrosion/99, paper No. 567, NACE, Houston, TX.

38. Gouda VK (1970) Corrosion and corrosion inhibition of reinforcing steel 1: Immersion in alkaline solution Br. Corros J 5: 198-203.

39. Erdogdu S, Bremner TW, Kondratova IL (2001) Accelerated testing of plain and epoxy-coated reinforcement in simulated seawater and chloride solutions. Cem Concr Res 31: 861-867.

40. Al-Amoudi OSB, Maslehuddin M, Ibrahin M,(2004) Long-term performance of fusion-bonded epoxy-coated steel bars in chloride-contaminated concrete. ACI Material Journal 101: 303-309.

41. Volpi E, Olietti A, Stefanoni M, Trasatti SP (2015) Electrochemical characterization of mild steel in alkaline solutions simulating concrete environment. J Electroanaly Chem 736: 38-46.

42. Génin JMR, Dhouibi L, Refait PH, Abdelmoula M, Triki E (2002) Influence of phosphate on the corrosion products of iron in chloride-polluted concrete-simulating solutions: ferrihydrite vs green rust. Corrosion 58: 467-478.

43. Dhouibi L, Refait PH, Triki E, Génin JMR (2006) Interactions between nitrites and Fe (II)-containing phases during corrosion of iron in concrete-simulating electrolytes. J Mater Sci 41: 4928-4936.

# 10

# Determination of Effects of Sample Processing on *Hibiscus sabdariffa* L. Using Tri-step Infrared Spectroscopy

Yew-Keong Choong[1]*, Nor Syaidatul Akmal Mohd Yousof[1], Mohd Isa Wasiman[1], Jamia Azdina Jamal[2] and Zhari Ismail[3]

[1]*Phytochemistry Unit, Herbal Medicine Research Centre, Institute for Medical Research, Jalan Pahang, 50588 Kuala Lumpur, Malaysia*
[2]*Drug and Herbal Research, Faculty of Pharmacy, UKM, Bangi, Selangor, Malaysia*
[3]*School of Pharmaceutical Science, USM, Gelugor, Penang, Malaysia*

## Abstract

*Hibiscus sabdariffa* tea is a widely used medicinal beverage and a treatment for high blood pressure and high blood cholesterol in many parts of the world. Many studies on *H. sabdariffa* have been conducted including extraction and identification of main biocompounds. However, information on the effects of processing the plant is scarce. This is important as sample processing procedure influence the composition of the end product. Hence, the main objective of this present study was to examine the effect of sample processing (non-extracted, ethanol extract and water extract) on *H. sabdariffa* composition. Fourier Transform Infrared (FTIR) was used for the process of identification. The powdered sample of *H. sabdariffa* (FT34) was obtained from a local company in Peninsula Malaysia. A fresh sample obtained from the same company was processed in the Phytochemistry Laboratory, Institute for Medical Research and labelled as FT35. Sample and potassium bromide (KBr) were mixed (1:250) to form a 1-2 mm transparent disk under 9.80 psi in vacuum. The FTIR Spectra were recorded with 32 scans and 0.2 cms$^{-1}$ OPD speed. Spectra of FT34 and FT35 raw samples indicated obvious differences in the range of 1500-1135 cm$^{-1}$. The FT34 ethanol extract using trifluoroacetic acid (TFA) showed that the peak at 1629 cm$^{-1}$ was the highest in the range of 1800-1500 cm$^{-1}$, whereas for FT35, the highest peak was 1739 cm$^{-1}$. The peak at 1071 cm$^{-1}$ of FT35 was the only one compatible to standard dephinidin-3-*O*-sambubioside and cyanidin-3-*O*-sambubioside which are used for qualification of sample content. In fact, both standards showed up as different chromatographs in thin layer chromatography. Water extract of FT35 showed a peak at 1676 cm$^{-1}$ which was not detected in water extract spectrum of FT34, while the pattern of spectrum varied within the range of 1300-400 cm$^{-1}$. Second derivative spectra enhanced the comparable base peaks of both sample and the target standards. There were five matched ethanol extract base peaks, indicating the macro-fingerprint of *H. sabdariffa*. Two dimensional correlation spectrum of FT34 raw powder showed different correlation spot especially in the cluster of 1425 cm$^{-1}$ to 1743 cm$^{-1}$ compared with FT35. The three-stage infrared spectroscopy comprehensively analysed the holographic spectra and hierarchically characterized the integrated constituents involved.

**Keywords:** *Hibiscus sabdariffa*; Sample processing; Tri-step infrared

## Introduction

*Hibiscus sabdariffa* Linn has been well known in Malaysia as roselle with reddish calyces resembles blossom of roses [1]. The taste is sour and known as "Asam paya" locally. This plant is classified under the Family Malvaceae and is an annual dicotyledonous herbaceous shrub [2]. This plant thrives in Malaysia because of its strong adaptive growth in multi-type of soils except clay. Each cycle of planting requires 85 days for first harvest and the same plant could be harvested for more than 5 times. However, the harvest yield is reduced during the raining season.

Phytochemical study on roselle has revealed its main chemical composition [2]. The presence of phenolics content in the plant consists mainly of anthocyanins especially delphinidin-3-*O*-glucoside, delphinidin-3-*O*-sambubioside, and cyanidin-3-*O*-sambubioside which could be the therapeutic marker of roselle [3]. Typical bioactive roles of this natural substances and the synergistic effect with other correlated components provide promising medicinal potential for the therapy of infection and lower the blood pressure [4], lower blood cholesterol level [5] as well as lower overweight [6]. Its broad range of therapeutic effect has been shown in many studies.

Manufactory of roselle products is rightly focussed on instead of the raw materials seems products preformed higher commercial value. However, information on the effects of processing the plant is scarce. This is important as sample processing really influence the composition of the end product(s). There are many factors that interfere with the quality of roselle products in Malaysia. The most common factors affecting sample processing method include temperature control in the oven, deseed procedure, additives added and human error, similar to those in food processing.

Fourier transform infrared and 2DIR correlation spectroscopy [7] has widely been used in quality control of herbal product [8] due to it is rapid and accuracy in identification. It is now recognised as one of the reliable analysis methods in pharmacopeia. Therefore, this technique was used to study roselle samples prepared using different sample processing methods.

Hence, the main objective of this present study was to examine the effect of sample processing on roselle composition (non-extracted, ethanol extract and water extract) using tri-step infrared spectroscopy.

## Materials and Methods

### Sample source

The powdered sample (calyces) of *H. sabdariffa* (FT34) was obtained from a local company in Peninsular Malaysia. The *H. sabdariffa* fruits from the plantation are routinely harvested and sent to the company for processing. After the deseeding and washing by the machine, after which they were immediately dried in the oven at 60°C for 3-4 days. A batch of fresh sample of roselle obtained from the same

---

***Corresponding author:** Yew-Keong Choong, Phytochemistry Unit, Herbal Medicine Research Centre, Institute for Medical Research, Jalan Pahang, 50588 Kuala Lumpur, Malaysia, E-mail: yewkeong11@yahoo.co.uk

company was processed in the Phytochemistry Laboratory in Institute for Medical Research (IMR), Kuala Lumpur and labelled as FT35. This samples were deseeded with hole puncher such as cork borer. Later cleaned with water and further air dried until 80% dryness before oven dried at 40°C for 3-4 days until a constant weight was obtained. The sample was then pulverized with a blender with the finest blades to a coarsely powdered material.

### Authentication

A voucher specimen (PID 050515-05) was submitted to Forest Biodiversity Unit at Forest Research Institute Malaysia (FRIM).

### Ethanol extraction with trifluoroacetic acid

The dried powdered calyx of *H. sabdariffa L.* (250 g) was extracted three times using a mixture of 2000 ml ethanol gradient grade (LiChrosolv® Reag. Ph Eurand) and 0.1% of Trifluoroacetic acid (TFA) and sonicated for 30 minutes at room temperature (25°C). The solution was then filtered using filter paper (Whatman™, Grade 1, circle, diam. 320 mm). The accumulated filtrate was dried by using a Rotavapor™ (brand) and Mivac Concentrator™ (brand) and stored in amber vial at -20°C until used.

### Water extraction with trifluoroacetic acid

The dried and powdered calyx of *H. sabdariffa L.* (250 g) was extracted three times using 2500 ml hot double distilled/MilliQ Water (80°C) mixed with 0.1% of Trifluoroacetic acid (TFA) and sonicated for 30 minutes at room temperature (25°C). The solution was filtered by a few layers of cotton ball (cotton wool) (China National Chemical I/E Corp) into a bottle. The accumulated water filtrate was frozen at -40°C before proceeding to freeze drying. The freeze-dried extract was stored in sample bottle at -20°C.

**Apparatus:** Spectrum GX Fourier-transform infrared (FTIR) spectrometer (Perkin-Elmer) with an attached DTGS detector was used as the main equipment for the whole experiment. This system was set up at a range of 4000-400 cm$^{-1}$ with a resolution of 4 cm$^{-1}$. Spectra were obtained after a total of 32 scans. The dynamic FTIR spectra were recorded with the above mentioned spectrometer combined with a Love Control Corporation's portable programmable temperature Controller (Model 50-886) with a range of 50-120°C.

**Procedures of making 1D and 2D FTIR spectrum:** About 1 mg of roselle samples was mixed with 250 mg dehydrated potassium bromide (KBr) powder. The mixture was further processed until a thin mini disk was formed. The mini disk was positioned on the sample reservoir in the system for the spectrum capture. Spectrum was accepted when 60-80% transmission was achieved; nonetheless the disk may be reformed by adding either more sample or KBr. The second derivative IR spectra were an intermediate 13-points smoothing of the basic IR spectra taken at room temperature.

The 2DIR spectroscopy was carried out using similar disks with thermal perturbation ranging from 30°C to 120°C. The dynamic spectra were analyzed and transformed into 2D and autopeak diagram with TD software developed by the Analysis Center of the Chemistry Department, Tsinghua University.

## Results and Discussion

### IR spectra of roselle samples and theirs extract

**Comparison of IR spectra of roselle raw material:** Owing to the uncertain quantity of sample added to KBr for more than 60%

transmittance, the maximum and minimum absorbance reading of FT35 and FT34 spectrum were quantified and shown to vary within 1.1 and 0.5, respectively. Therefore the comparison was made by focussing on the width and the shape of the peaks. Both raw material extracts have similar pattern of peak from the range of 1800-1600 cm$^{-1}$. The increasing sequence started from peak 1789 cm$^{-1}$, 1741 cm$^{-1}$ and 1630 cm$^{-1}$ respectively. The strong peak at 1741/1743 cm$^{-1}$ was the stretching mode of C=O in esters and the two peaks at 1230/1226 cm$^{-1}$ and 1156/1153 cm$^{-1}$ may be assigned to the C-O bonds. In contrast, peak 1789/1788 of both raw materials were very scarce in general spectra which was assigned to C=O in acid anhydride group. The obvious dissimilarity of both raw material spectra was located at the range of 1500-1150 cm$^{-1}$. Most of the peaks in this range were different in term of their position and shape except two pairs of peaks 1226/1230 of FT 34 and 1153/1156 and FT35, respectively (Figure 1).

Table 1 showed the assignment of each peak in this region. The triple peaks of FT34 in the range of 1150-1000 cm$^{-1}$ position were lower than the peak 1639 cm$^{-1}$ when compared with parallel y, while, this position was opposite in FT35. The peaks below 1150 cm$^{-1}$ indicated content of carbohydrate of the subject [9]. Therefore the content of carbohydrate of FT35 was more intense than FT34. In FT34, when taking 1065 cm$^{-1}$ at the centre, both side peaks (1098 cm$^{-1}$) were slightly higher than peak 1032 cm$^{-1}$, while this condition was also opposite in FT35. However, the peak 1098 cm$^{-1}$ and peak 1065 cm$^{-1}$ of FT34 spectrum were clearer than peak 1032 cm$^{-1}$. The peak 957 cm$^{-1}$ was more clearly shown in FT34 rather than FT35. The peak 715/714 cm$^{-1}$ was the band that is present in mono-, 1:3-, 1:3:5-, 1:2:3-substituted phenyls.

**Comparison of roselle ethanol extract with TFA and standards:** Comparison of the roselle ethanol extract with TFA and the anthocyanin standards was categorised into three sections: 1900-1550 cm$^{-1}$, 1549-1000 cm$^{-1}$, 999-400 cm$^{-1}$. The absence of any peak at the beginning section of 1900-1700 cm$^{-1}$ was observed in both anthocyanin standards. For these standards, the main absorbance peaks were located in the range of 1650-900 cm$^{-1}$, which was also found in the second and third sections of the ethanol extract with TFA. The appearances of peaks from 1645 until 1440 cm$^{-1}$ in both standards were clear footage of their identification in the spectra. In FT34, sequence of 3 peak heights increased gradually from 1788 cm$^{-1}$, while peak 1739 cm$^{-1}$ of FT35 was highest between 1792 and 1632 cm$^{-1}$. Only peak 1071 cm$^{-1}$ of FT35 (1069 cm$^{-1}$ of FT34) was comparable with standard d-3-O-sambubioside and c-3-O-sambubioside. Besides, peak 718 cm$^{-1}$ of both ethanol extract matched with d-3-O-sambubioside but not c-3-O-sambubioside, indicating that most probably the content of ethanol extract with TFA consisted of d-3-O-sambubioside. The peak 1739 cm$^{-1}$ and peak 1194 cm$^{-1}$ of FT35 were slightly more intense than the same position of peaks in FT34 and this could be due to the different process of sample preparation (Figure 2).

The purpose of usage of TFA in the extraction was to stabilise the extracted components. There are minor differences in extraction with or without TFA in term of extracted contents based on spectroscopy analysis.

**Comparison of roselle water extract with TFA and standards:** The roselle water extract was oxidised when exposed to air and extraction had to be replicated until transmission of the spectrum achieved 50%. There were only four peaks in 1D FTIR spectra. The peak 1071 cm$^{-1}$ was the only one appeared in each spectrum, and it was also found in the ethanol extract. Besides, the peaks around 861 and 721 cm$^{-1}$ (718 cm$^{-1}$ in ethanol extract) were maintained in water extract which were also associated well with both ethanol extracts and d-3-O-sambubioside.

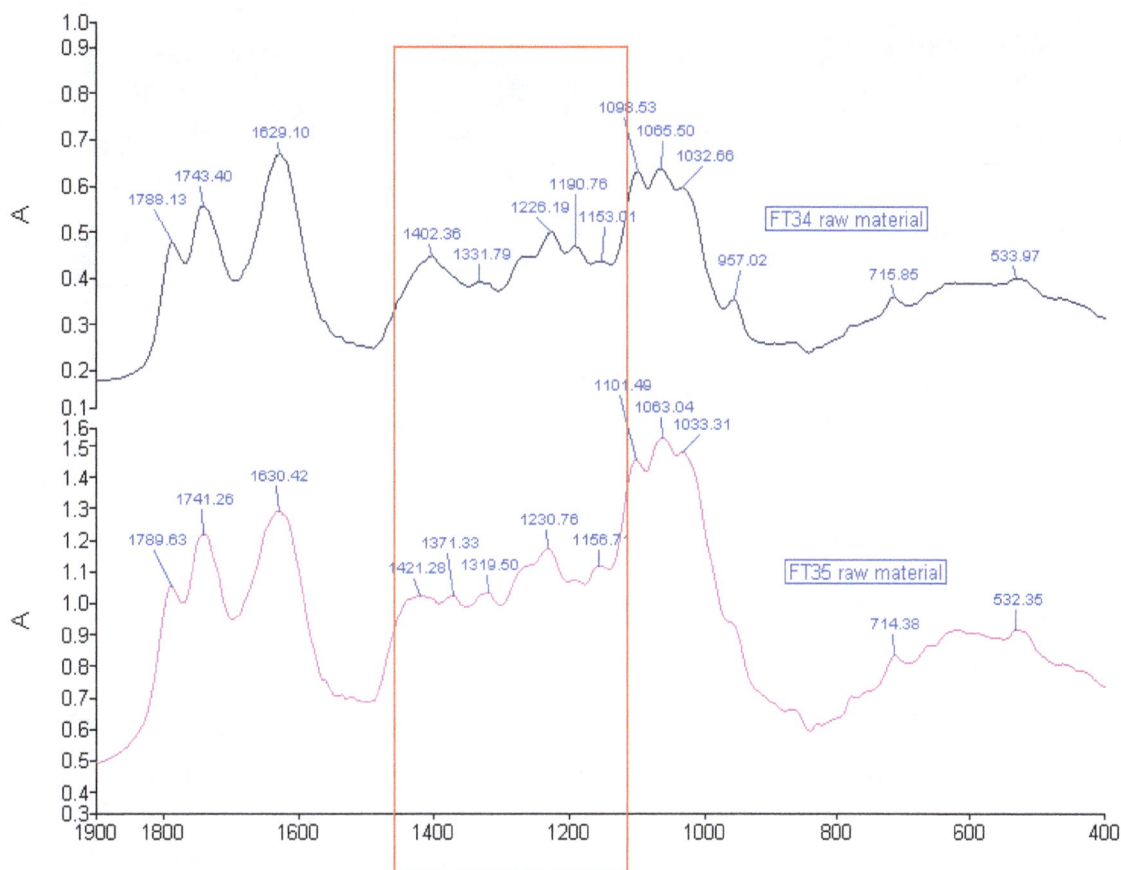

**Figure 1:** FTIR spectra of roselle raw material in the range of 1900-400 cm$^{-1}$. (1) FT34: raw material from company; (2) FT35: raw material processed by phytochemistry laboratory, IMR. The main differences of both spectra were highlighted in box.

| Band(cm$^{-1}$) | Vibration mode | Main attribution | Raw material | |
|---|---|---|---|---|
| | | | FT34 | FT35 |
| 1800-1750 | R<br>    C=CH$_2$<br>R | Methylene, overtone of δ ' CH (out of plane) | 1788 | 1789 |
| 1745 | vC=O | ketone | 1743 | 1741 |
| 1630 | Ortho-CO-C$_6$H$_4$-OH | Influenced by I,M, and steric effects of substituent | 1629 | 1630 |
| 1420 | δ C-H of CH$_2$ | Shifted lower from the usual 1470cm-1 position because it is flanked by an aromatic ring and an acetylenic bond | - | 1421 |
| 1400 | v C=N | coupled with v C=C | 1402 | - |
| 1370-1330 | R-SO$_2$-N | sulfoamide | | 1371 |
| | | sulfoamide | 1331 | |
| 1350-1310 | R-SO$_2$-R' | Sulfone | - | 1319 |
| 1226 | δ C-H | In -plane | 1226 | 1230 |
| 1195 | $v_{as}$ SO$_3^-$ | | 1190 | - |
| 1153 | | Twist-boat form of cyclohexane | 1153 | 1156 |
| | -C-H | In-plane CH pending modes | 1098 | 1101 |
| | -C-H | In-plane CH pending modes | 1065 | 1063 |
| 1034 | $v_s$ SO$_3^-$ | | 1032 | 1033 |
| 964 | C=C | Twist of trans substituted ethylenes | 957 | - |
| 717 | | Band that is present in mono-1:3-, 1:3:5-, and 1:2:3-substituted phenyls | 715 | 714 |
| 529 | v S-S | Two polarized bands generally appear in the narrow range 525-510 cm$^{-1}$ (medium intensity) | 533 | 532 |

**Table 1:** The assignment of each peak in different region.

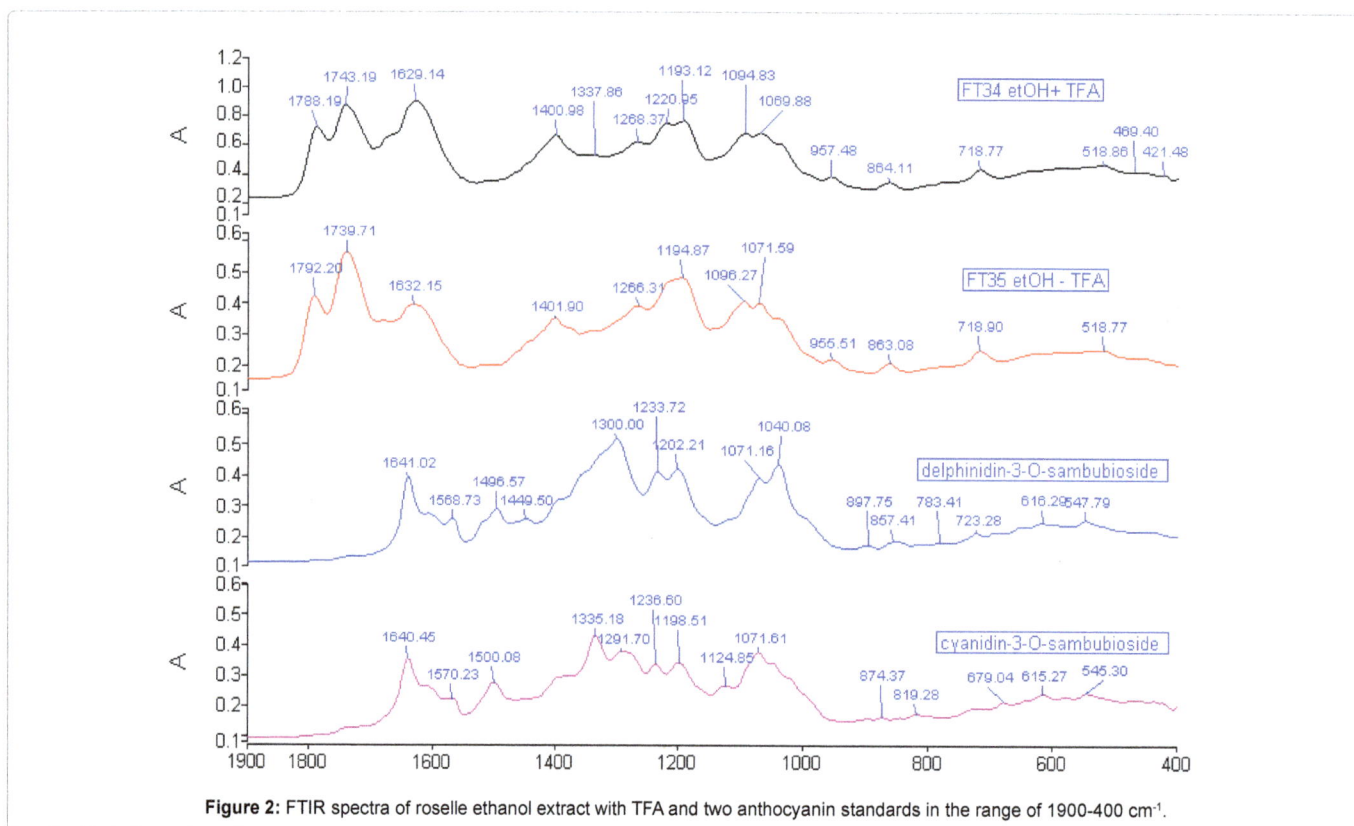

**Figure 2:** FTIR spectra of roselle ethanol extract with TFA and two anthocyanin standards in the range of 1900-400 cm$^{-1}$.

Both water and ethanol roselle extracts showed the existence of anthocyanin content (Figure 3).

### Second derivation spectra of roselle sample and their extracts

**Comparison of 2$^{nd}$ derivative spectra of roselle raw material:** The differences of both raw roselle material second derivative spectra were highlighted in Figure 4. For FT34, the range between 1402 and 1331 cm$^{-1}$ of 1DFTIR has been derived into two small peaks; the peak 1402 cm$^{-1}$ was reserved and peak 1320 cm$^{-1}$ was shown in second derivative spectrum instead of 1331 cm$^{-1}$. In contrast, similar derivative pattern was observed in FT35 for the range between 1421 and 1371 cm$^{-1}$, but the peak 1439 was shown and the peak 1371 cm$^{-1}$ was reserved. Hence the peak 1402 cm$^{-1}$ of FT34 and peak 1371 cm$^{-1}$ of FT35 were the dominant peaks for both differently processed samples. The derived peak 1320 cm$^{-1}$ of FT34 was matched with shape derived peak 1318 cm$^{-1}$ of FT35. Minor changes were observed in peak 1226 and 1190 cm$^{-1}$ of FT34 in 2$^{nd}$ derivative spectra, however, the peak 1230 cm$^{-1}$ of FT35 was derived to two sharp peaks in this region.

**Comparison of 2$^{nd}$ derivative spectra of roselle ethanol extract with TFA:** The similarity of 2$^{nd}$ derivative spectra were detected in at least four regions along the range of 1900-400 cm$^{-1}$ in Figure 5. There were only two pairs of peaks associated with these four spectra, i.e., peaks around 1400 and 1071 cm$^{-1}$ which were the solid evidence of the presence of anthocyanin in the content of roselle ethanol extract. Four regions of 2$^{nd}$ derivative spectra were matched for both roselle ethanol extract and d-3-O-sambubioside. The peak 1522 cm$^{-1}$ of FT35 was matched with standard d-3-O-sambubioside spectrum. In detail, there was 14.3% matched base peak of FT34 with d-3-O-sambubioside compared to 16.7% of FT35 with d-3-O-sambubioside. On the other hand, 7.1% of matched base peak of FT34 with c-3-O-sambubioside

compared to 4.8% of FT35. Even though these were not significant, nevertheless these differences indicated the dissimilarity of the content of FT34 and FT35.

**Comparison of 2$^{nd}$ derivative spectra of roselle water extract with TFA:** More peaks were detected in 2$^{nd}$ derivative spectra. Eight peaks of extract matched with standards. FT34 water extract with TFA have a congested region of base peaks which contained 15 small peaks in the range of 1650-1350 cm$^{-1}$. Five peaks occurred and matched with the carbohydrate region in roselle extract and standards (Figure 6).

### 2DIR synchronous correlation spectral

**Comparison of 2DIR synchronous of FT34 and FT35 raw material**

*FT34 (In the range of 1800-1000 cm$^{-1}$ and 1000-400 cm$^{-1}$):* Raw material represented the actual substance of the sample before the extraction. In this case, FT34 has clean clear-cut edge square created between autopeak 1425 cm$^{-1}$ and 1800 cm$^{-1}$ (Figure 7a). The region was built up by 3 small squares. The centre was the peak with highest intensity from the whole range of this spectrum which showed 2 layers of viewing from top. The background was scattered with small negative blue squares. The important portion was concentrated at the upper right square, as mentioned above. The highest peak of this spectrum was similar to the spectrum of FT34 without TFA (data not shown), i.e., the FT34 ethanol without TFA was not able to extract the assignment compound with the peak 1630 cm$^{-1}$. The positive crosspeaks related in this square also recorded higher intensity. Plenty of compounds were located in the region centered with autopeak 680 cm$^{-1}$ (Figure 7c). The area of square was included with a board range of various colours. The autopeak spectrum could be combination of components of peaks or

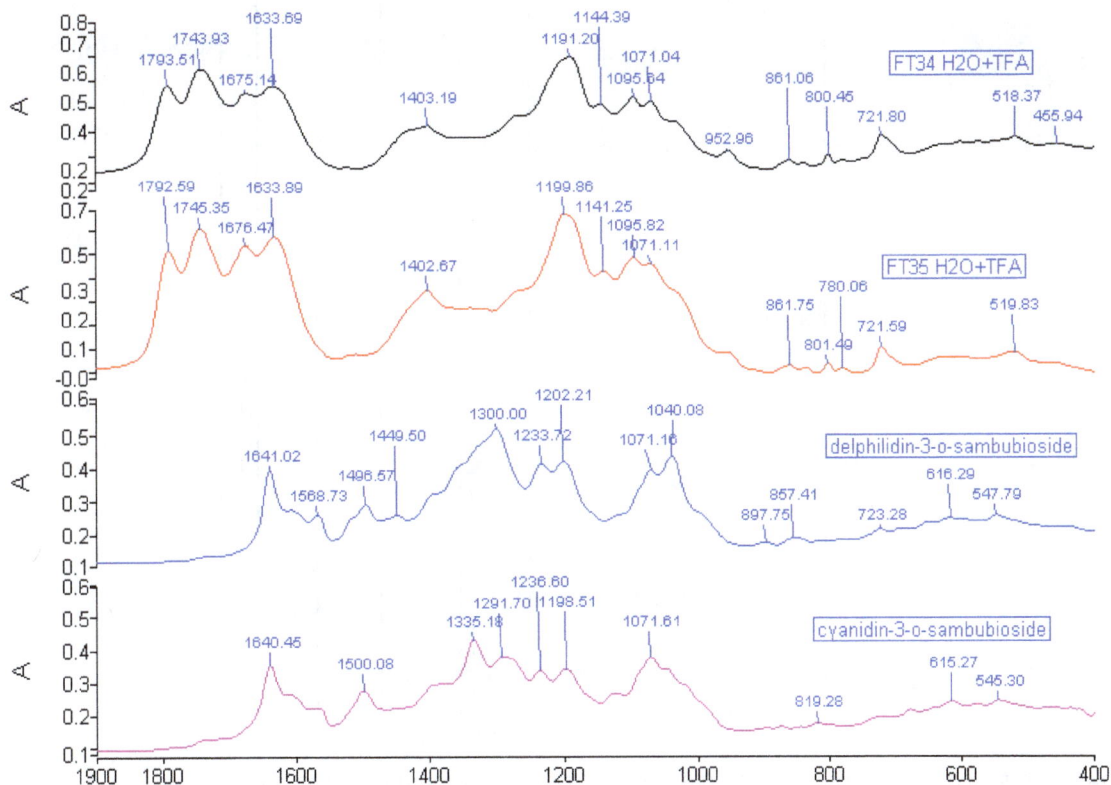

**Figure 3:** FTIR spectra of roselle water extract with TFA and two anthocyanin standards in the range of 1900-400 cm$^{-1}$.

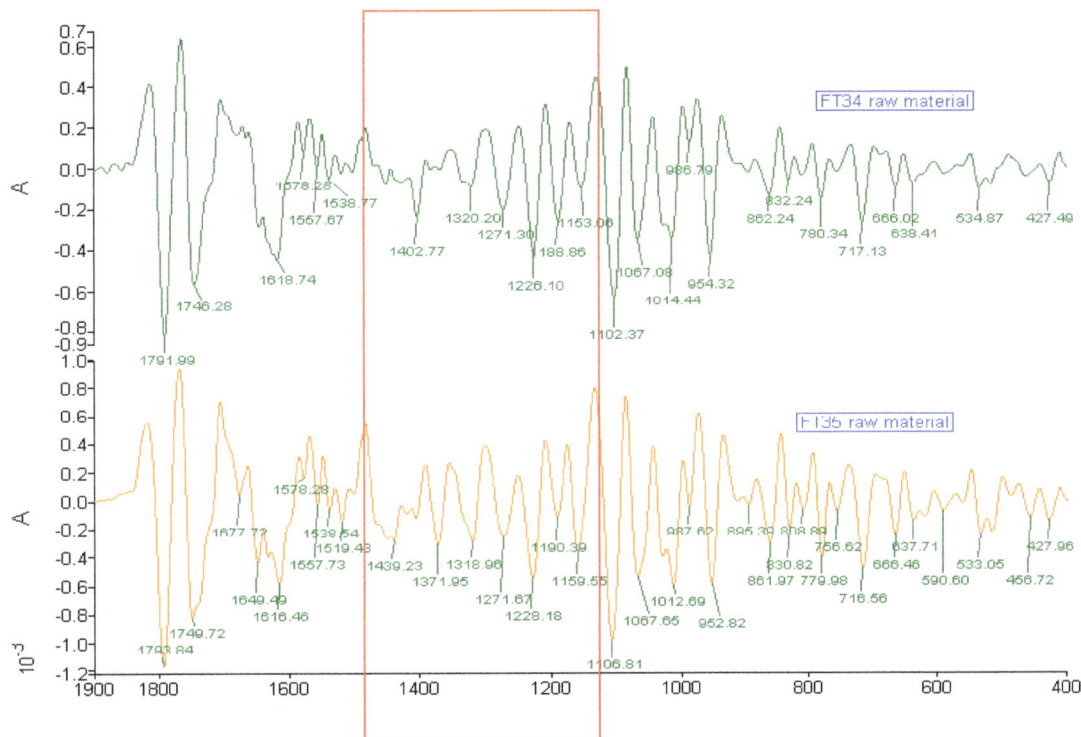

**Figure 4:** Second derivative spectra of roselle raw material in the range of 1900-400 cm$^{-1}$. (1) FT34: raw material from company; (2) FT35: raw material processed by phytochemistry laboratory, IMR. The main differences of both spectra were highlighted in box.

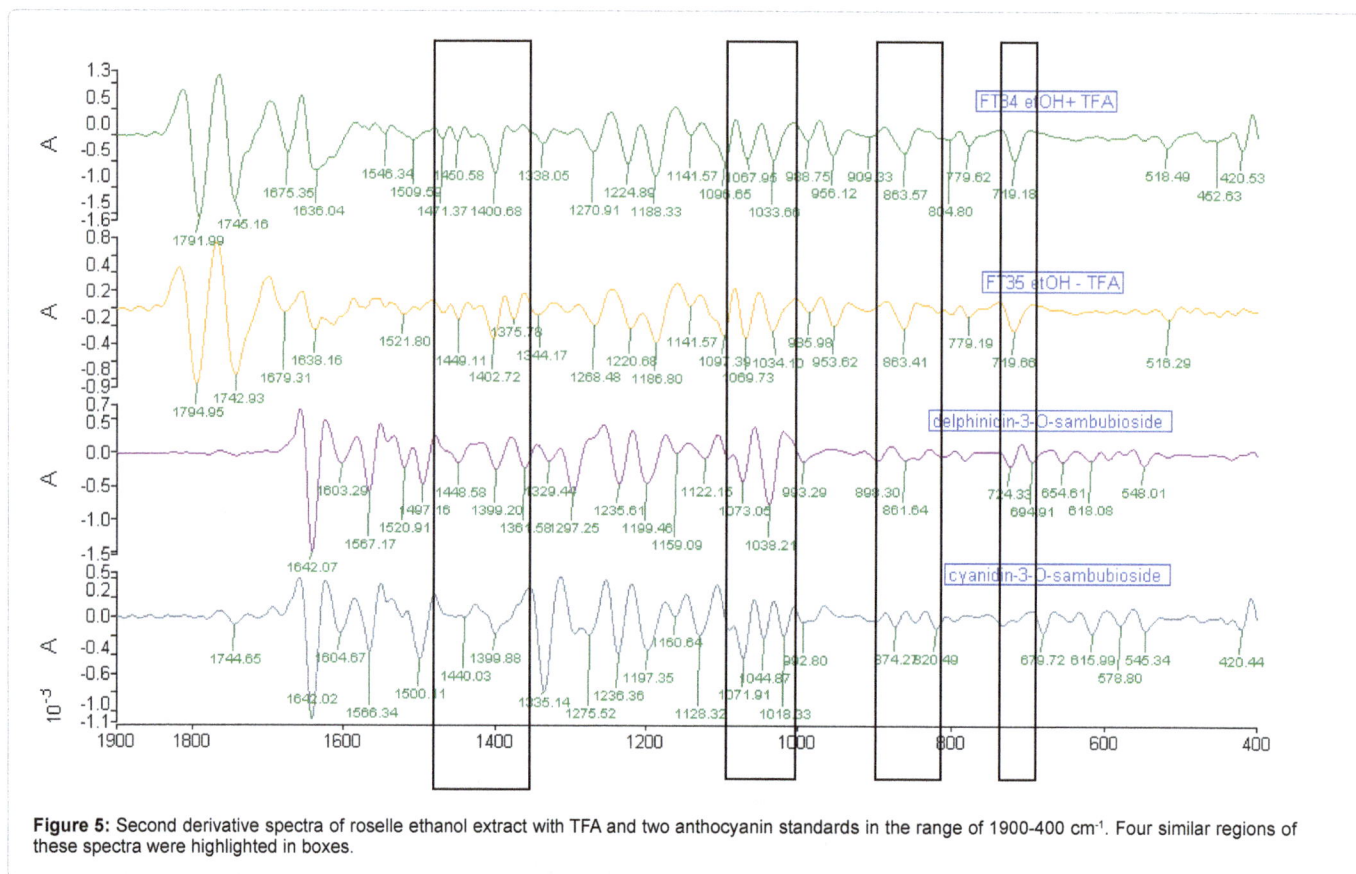

**Figure 5:** Second derivative spectra of roselle ethanol extract with TFA and two anthocyanin standards in the range of 1900-400 cm$^{-1}$. Four similar regions of these spectra were highlighted in boxes.

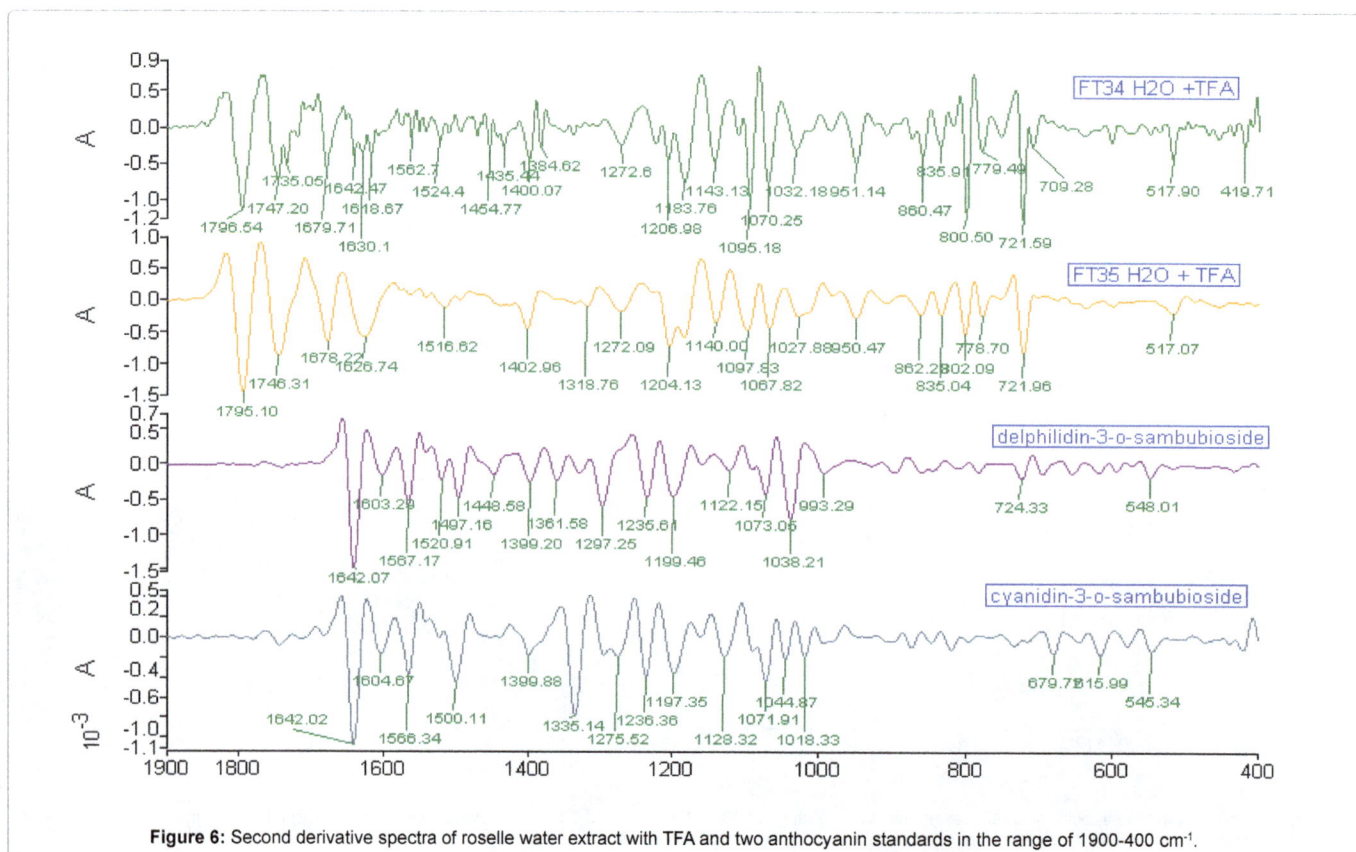

**Figure 6:** Second derivative spectra of roselle water extract with TFA and two anthocyanin standards in the range of 1900-400 cm$^{-1}$.

some smaller hidden ones forming the huge peak. The raw material found in this region was even more complicated than the spectrum which has gone through the extraction process.

**FT 35 (In the range of 1800-1000 cm$^{-1}$ and 1000-400 cm$^{-1}$):** The transmission of FT35 raw material spectrum (Figure 7e) was 40.46%, which was much lower than % transmission of FT34 raw material. However, the spectrum showed higher intensity of more autopeaks in this range, especially the range from 1425-1000 cm$^{-1}$ compared with FT34 raw material. The structure of colours scattered evenly along the diagonal line. The correlation square created by autopeak at 1700 cm$^{-1}$ with 1600 cm$^{-1}$ correlated with positive crosspeak at (1700, 1600) and (1600,1700). Another correlation square created by autopeak 1600 cm$^{-1}$ and 1425 cm$^{-1}$ but correlated negatively with crosspeak at (1600, 1425) and (1425, 1600). The bigger correlation squares created by all the positive peaks included autopeak 1425 cm$^{-1}$ to 1212 cm$^{-1}$ and (1425, 1212) and (1212, 1425). Therefore there was no compression of correlation square in FT35 raw material, with about 2/3 of the spectrum as positive peaks. The broad square fulfilled almost 2/3 of the spectrum (Figure 7g). The highest autopeak at 810 cm$^{-1}$ was similar to FT34, whereby the intensity of highest autopeak was five times higher than FT34. A combination of many component peaks formed the huge twin sharp peaks on top.

**Comparison of 2DIR synchronous of FT34 and FT35 ethanol extract with TFA**

*FT34 (In the range of 1800-1000 cm$^{-1}$ and 1000-400 cm$^{-1}$):* At least 6 autopeaks were formed with very regular correlation squares along the range 1850-1200 cm$^{-1}$ (Figure 8a). The positive cross peaks were scattered with the 50 cm$^{-1}$ interval with another and there were more peaks in this range compared with FT35. A total of 2/3 of the clusters was presented in red, showing that the positive crosspeaks were more than negative crosspeaks. Ethanol and TFA exhibited the strong coefficient of extraction in the range of amide I and amide II bond, including flavonoids in this range. Very clear illustration was found in 2/3 of the cluster in this area and another 3 small clusters divided by almost blank partition (Figure 8c). The centre of the major cluster was located in the highest autopeak at 600 cm$^{-1}$, surrounded by 2 correlation squares with positive crosspeaks, indicating that the centralised peak at 600 cm$^{-1}$ was important and correlated with many compounds crossing with it. The huge peak could be combination or overlapping with other smaller peaks. The intensity of this range was lower than FT35.

*FT35 (In the range of 1800-1000 cm$^{-1}$ and 1000-400 cm$^{-1}$):* The ethanol extract with TFA of FT35 lacked peaks in the range of 1800-1000 cm-1 (Figure 8e-81). The red area of peaks spotted in two straight axes along wavenumber 1700 cm$^{-1}$ perpendicular to each other. The autopeak spectrum showed few sharps autopeaks but not as intense as FT34 with TFA. The square created by the negative crosspeaks indicated the presence of certain substances. Figure 8g showed the 2DIR spectrum of FT35 ethanol extract with TFA in the range of 1000-400 cm$^{-1}$. The higher point of this range was 3.8 times more than FT34. FT35 also showed another sharp autopeak at 480 cm$^{-1}$. The polar compound especially the saccharides are notably more intense than FT34 within this range of wavenumber, showing that ethanol favourably extracted the polar saccharide compounds in FT35 compared to FT34 in this range.

(8a) -2DIR 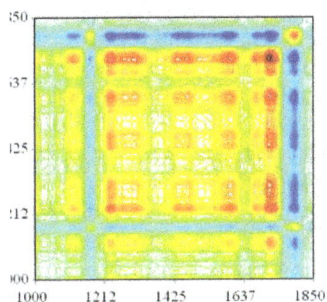 (8b)-autopeak 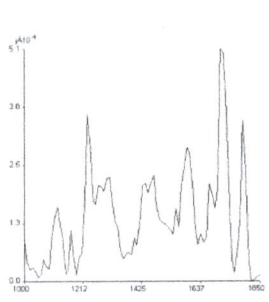 (8c)-2DIR 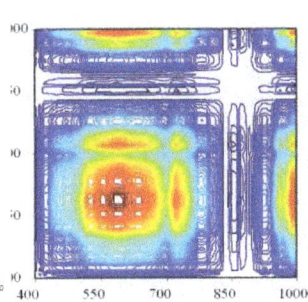 (8d)-autopeak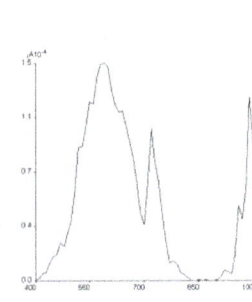

(8e)-2DIR 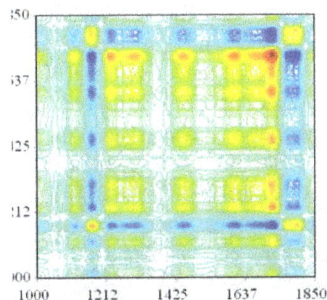 (8f)- autopeak 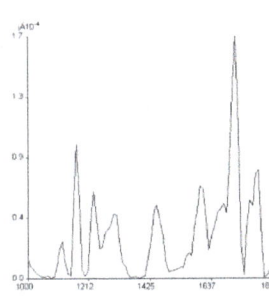 (8g)-2DIR 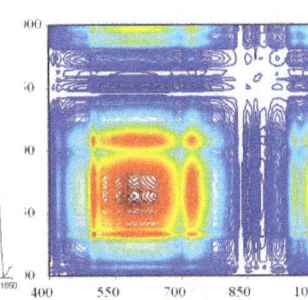 (8h)-autopeak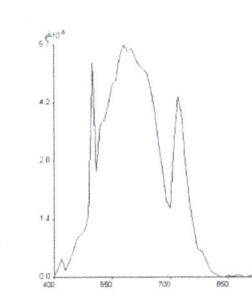

**Figure 7:** 2DIR synchronous and autopeak spectra of roselle raw material. 7a: FT34 2DIR in the range 1850-1000 cm$^{-1}$, 7b: FT34 autopeak in the range 1850-1000 cm$^{-1}$. 7c: FT34 2DIR in the range of 1000-400 cm$^{-1}$; 7d: FT34 autopeak in the range 1000-400 cm$^{-1}$. 7e: FT35 2DIR in the range of 1850-1000 cm$^{-1}$. 7f: FT35 autopeak in the range of 1850-1000 cm$^{-1}$; 7g: FT35 2DIR in the range of 1000-400 cm$^{-1}$; 7h: FT35 autopeak in the range of 1000-400 cm$^{-1}$.

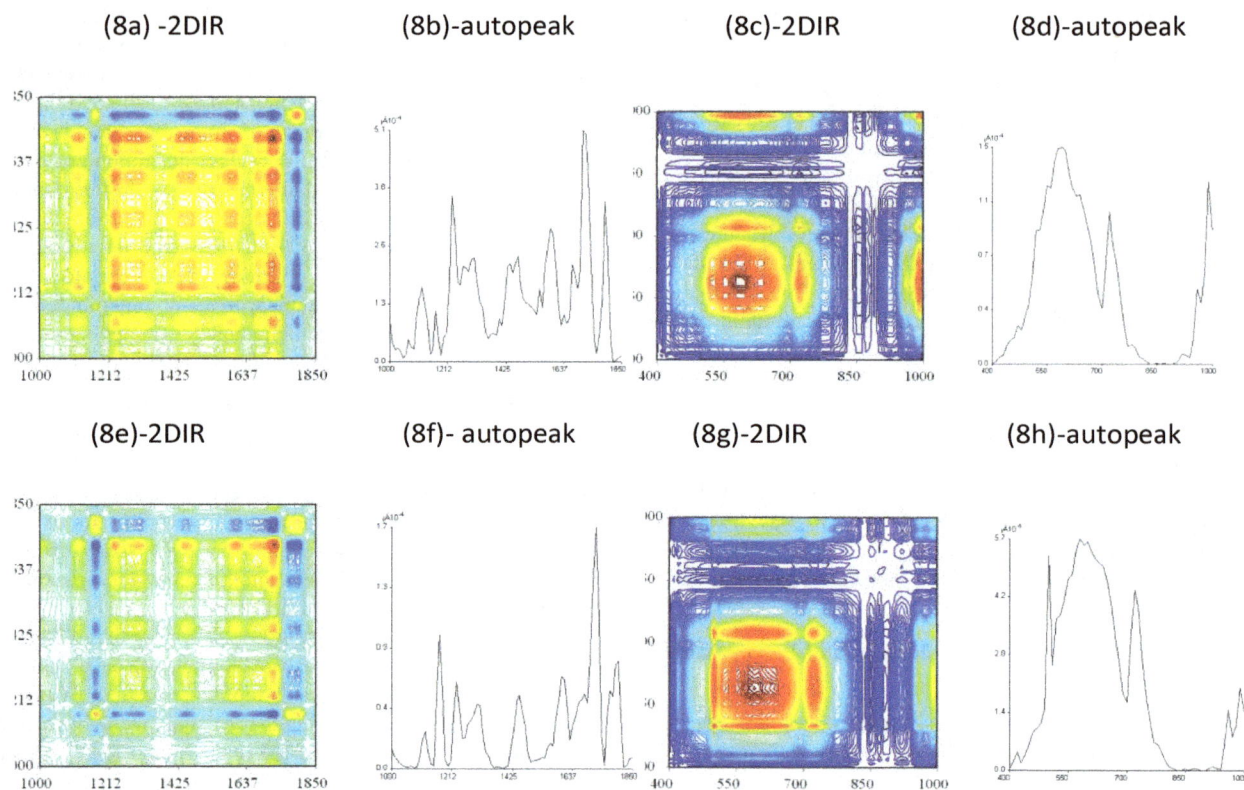

**Figure 8:** 2DIR synchronous and autopeak spectra of roselle ethanol extract with TFA. 8a: FT34 2DIR in the range 1850-1000 cm$^{-1}$, 8b: FT34 autopeak in the range 1850-1000 cm$^{-1}$. 8c: FT34 2DIR in the range of 1000-400 cm$^{-1}$; 8d: FT34 autopeak in the range 1000-400 cm$^{-1}$. 8e: FT35 2DIR in the range of 1850-1000 cm$^{-1}$. 8f: FT35 autopeak in the range of 1850-1000 cm$^{-1}$; 8g: FT35 2DIR in the range of 1000-400 cm$^{-1}$; 8h: FT35 autopeak in the range of 1000-400 cm$^{-1}$.

## Comparison of 2DIR synchronous of FT34 and FT35 water extract with TFA

*FT34 (In the range of 1800-1000 cm$^{-1}$ and 1000-400 cm$^{-1}$):* 2DIR water extract of FT34 presented very different spectra compared to ethanol extract (Figure 9a), indicating that the content of the roselle water extract was dissimilar with ethanol extract in certain aspects. Four obvious points at four corners made up of correlation squares by autopeak 1700, (1700, 1212), 1212, (1212, 1700) were the only focus points in this region. However the autopeak around 1212 cm$^{-1}$ was the highest points compared with others. Most of the peaks were evenly scattered and consisted of mixed positive and negative crosspeaks. Four shape peaks had been found along the diagonal line (Figure 9c). All were separated at least 100 cm$^{-1}$ from each other. The red board peak from 550-650 cm$^{-1}$ was centered at 600 cm$^{-1}$, and the cross view showed the peaks of this 2DIR spectrum had sharper area. When compared with FT34 ethanol extract with TFA, this range of spectrum contained shaper peaks because water is the most polar solvent which is able to extract the polar compounds left after ethanol extraction with TFA.

*FT35 (in the range of 1800-1000 cm$^{-1}$ and 1000-400 cm$^{-1}$):* The autopeak 1212 cm$^{-1}$ was 7 times more intense than FT35 ethanol extract with TFA (Figure 9e). While the autopeak 1212 cm$^{-1}$ was the major autopeak in this spectrum, it is also a major peak in FT34 water extract with TFA, both showed that the differences from ethanol extract e in the range of 1212-1100 cm$^{-1}$ has higher intensity than the range of 1850-1500 cm$^{-1}$. The upper right corner showed 9 points correlation squares in the small region, which was a specific marker for FT35 water extract. It could be explained that the groups of band intensity changes

due to the same component within the 9 assigned peaks. This 2DIR spectrum has fewer positive crosspeaks and negative crosspeaks. When compared with FT34 water extract with TFA, this spectrum has regular peaks assigned in the range of 1850-1500 cm$^{-1}$. Figure 9g showed the spectrum has lower intensity compared with FT34 water extract. The broad red mixed up with orange and yellow at the range 850-550 cm$^{-1}$ indicated there are overlapping of many component of the substances. The autopeak at 830 cm$^{-1}$ was one time higher than broad autopeak with the area 700-500 cm$^{-1}$ which was opposite in FT34.

## Conclusion

The differences of spectrum for both samples with different sample processing method obviously indicated that the composition of the end product relies on the appropriate processing method. The decoilation of seed was initially needed before the washing process since some of the residue or debris need to be totally cleaned. The temparature was considered one of the main factors affecting the end product. FT 35 was air-dried until 80% dry before further drying in the oven at 45°C for 3-4 days. FT35 exhibited more intense anthocyanin content especially in 2DIR spectra compared with FT34. Basically the IR analysis accurately differentiate the contents of both samples from raw material. The second derivative spectrum is another step to enhance the analysis by deriving the original spectrum peak to the diagnostic peak.

### Acknowledgements

The authors thank the Director General of Health for the permission to publish and the Director of Institute for Medical Research (IMR), Kuala Lumpur for the support. This work was financially supported by the NKEA AGRICULTURE

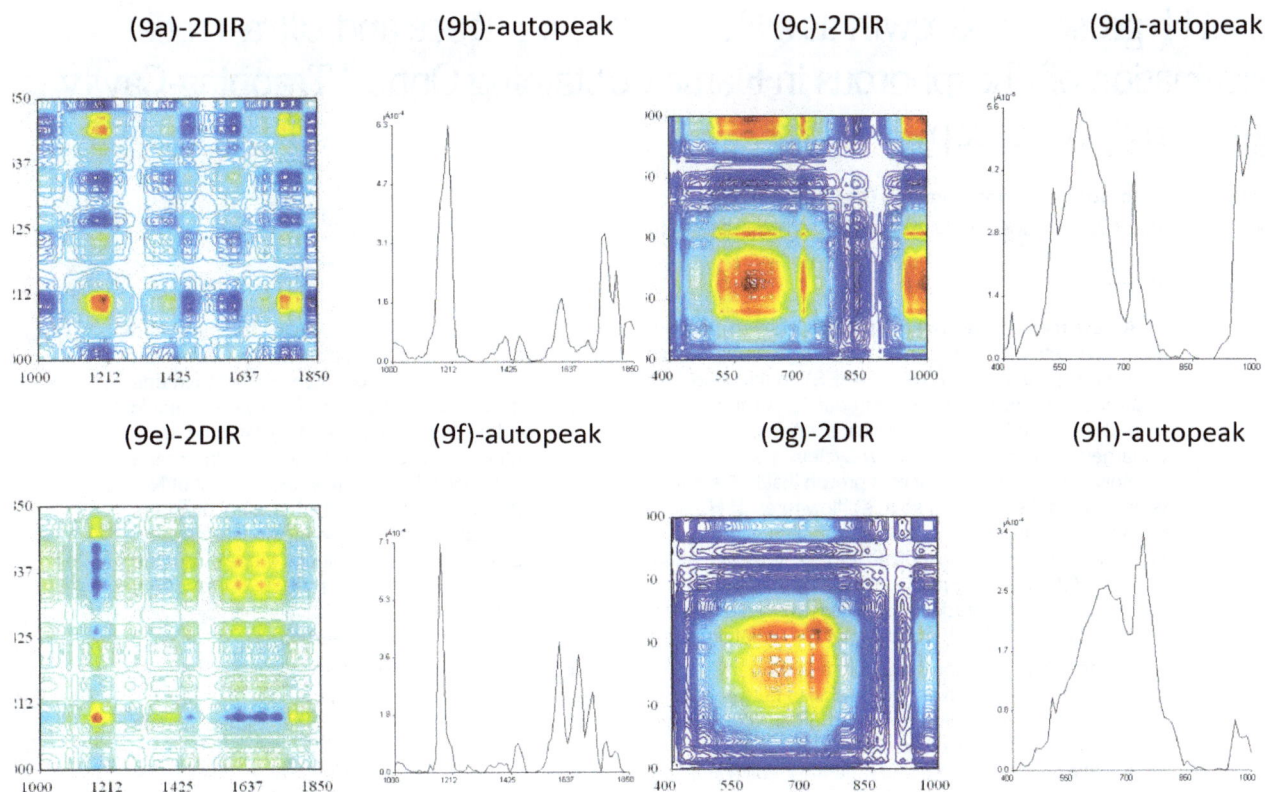

**Figure 9:** 2DIR synchronous and autopeak spectra of roselle water extract with TFA. 9a: FT34 2DIR in the range 1850-1000 cm$^{-1}$, 9b: FT34 autopeak in the range 1850-1000 cm$^{-1}$. 9c: FT34 2DIR in the range of 1000-400 cm$^{-1}$; 9d: FT34 autopeak in the range 1000-400 cm$^{-1}$. 9e: FT35 2DIR in the range of 1850-1000 cm$^{-1}$. 9f: FT35 autopeak in the range of 1850-1000 cm$^{-1}$; 9g: FT35 2DIR in the range of 1000-400 cm$^{-1}$; 9h: FT35 autopeak in the range of 1000-400 cm$^{-1}$.

(EPP#1), NKEA Research Grant Scheme (NRGS) (Grant No: NH1014D060). The authors also thank Dr. Fairulnizal (IMR), Dr. Maizatul Hashim (IMR) and Dr. Lee Han Lim (IMR) for critically reviewing this paper and technical support from Perkin Elmer Sdn Bhd. (Malaysia). The previous support of Islam Development Bank (IDB) for training in FTIR and 2DIR under the Merit Fellowship is gratefully acknowledged.

## References

1. Norhaizan M, Fong SH, Amin I, Chew LY (2010) Antioxidant activity in different parts of roselle (Hibiscus sabdariffa L.) extracts and potential exploitation on the seeds. Food chemistry 122: 1055-1060.

2. Mohamed Salem ZM, Olivares-Perez J, Salem AZM (2014) Studies on biological activities and phytochemicals composition of Hibiscus species- A review. Life Sci J 11: 1-8.

3. Ali BH, Wabel NA, Blunden G (2005) Pharmacological and toxicological aspects of Hibiscus sabdariffa L. Phytotheraphy Research 19: 369-375.

4. Herrera-Arellano A, Flores-Romero S, Chavez-soto S, Tortoriello MAJ (2004) Effectiveness and tolerability of a standardized extract from Hibiscus sabdariffa: a controlled and patients with mild to moderate hypertension: a controlled and randomized a clinical trial. Phytomedicine 11: 375-382.

5. Lin TL, Lin HH, Chen CC, Lin MC, Chen MC, et al. (2007) Hibiscus sabdariffa extract reduces serum cholesterol in men and women. Nutrition Research 27: 140-145.

6. Alarcon-Aguilar FJ, Zamilpa Perez-Garcia A, Almanza-Perez MD, RomeroNurez JCE, Campos-sepulveda EA, et al. (2007) Effect of Hibiscus-sabdariffa on obesity in MSG mice. Journal of Ethnopharmacology 114: 66-71.

7. Noda I, Ozaki Y (2004) Two-Dimensional Correlation Spectroscopy Application in Vibrational Optical Spectroscopy. John-Wiley and Sons Ltd., Chichaster, West Sussex, UK.

8. Sun SQ, Zhou Q (2003) Atlas of Two-dimensional correlational Infrared Spectroscopy for traditional Chinese Medicine Identification. Chemical Industry Press, Beijing, China.

9. Nakanishi K, Solomon PH (1977) Infrared Absorption Spectroscopy. 2nd edn. Holden-day, San Franciso, USA.

# 11

# Aerosol Hygroscopic Growth as a New Factor for Trace and Ultra-Trace Determination of Phosphorous in Flame Containing Optical Trapping-Cavity Ring-Down Spectroscopy

**Mohammad Mahdi Doroodmand\* and Fatemeh Ghasemi**

*Department of Chemistry, College of Sciences, Shiraz University, Shiraz 71454, Iran*

### Abstract

A new method has been introduced based on aerosol hygroscopic growth as a new factor for trace and ultra-trace determination of phosphorous in flame containing optical trapping-cavity ring down aerosol extinction (emission) spectrometer (OT-CRD-AES). In this study, a cavity ring down has been designed using hydrogen and air as fuel and oxidant during introduction of the aerosols containing phosphorous species using an ultrasonic generator (humidifier) from an acidic solution by a flow rate of $N_2$, followed by detection of the Mie scattering using a charged coupling device (CCD) system. Parameters having strong influence during following scattering of the aerosols during their hygroscopic growth inside the humidified flame ($H_2$/air), include: influence and amount of $Na^+$ as radiation buffer (as light source), flow rates of $H_2$, air and $N_2$, kind and concentrations of acid, evaluation of the aerosols inside flame, etc. These parameters were optimized using simplex and one at a time methods. Based on the figures of merit under optimized condition, two linear calibration curves with reverse slopes were evaluated between 10.0 - 250.0 ng mL$^{-1}$ and 1.0 - 20.0 µg mL$^{-1}$ with correlation coefficients ($R^2$) the same as 0.999 and 0.998, respectively. The calibration sensitivities were also estimated to 57.46 and 0.348 (a.u), respectively, with detection limit of 5.0 ng mL$^{-1}$. The mechanism of the radiation (Mie scattering) was also evidenced based on

i) dependency of the scattered radiations to the quantity of an alkali ions such as Na+ as well as the humidity,
ii) presence of acceptable correlation between the response of the cavity with turbidometry,
iii) observation of blue shift from green (color related to the luminescence of HPO\* in $H_2$/air flame) to blue (scattered radiation) and finally
iv) effect of hydration number during stability and growth of the aerosols inside the flames.

No serious interference was evaluated during analysis of at least 500-fold excess of various foreign species. However, the only observed interference was evaluated during introduction of 200-fold excess of $SO_4^{2-}$. Good correlation was also evaluated between the results obtained from this technique and ion exchange chromatography during analysis of wastewater samples that clearly revealed the reliability of this method for phosphorous detection and determination at µg mL$^{-1}$ and ng mL$^{-1}$ levels.

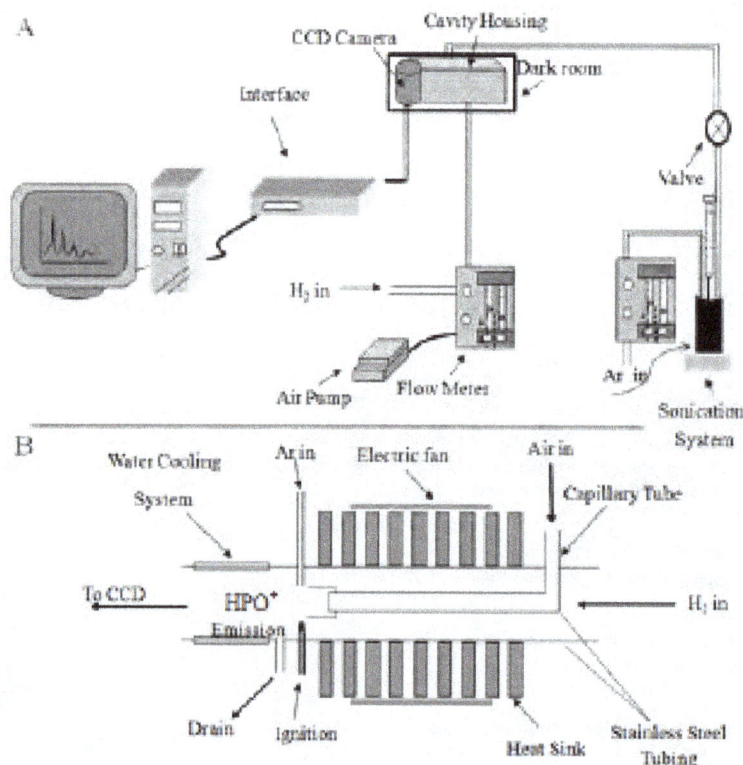

**Keywords:** Aerosol hygroscopic growth; Aerosol; Mie scattering; Reducing flame; Phosphorus determination

## Introduction

Duo to the limitations such as effect of pH, low sensitivity, wide range of geometries, and/or because of significant dependency of the binding affinity of anionic species on solvent, anion recognition has been considered as a challenging task during the last decades [1]. These phenomena point to the importance of the introduction of a selective, sensitive and reliable method for recognition of anionic species [2]. Among various forms of anionic species, detection and determination of phosphorous species in the environmental samples has been important due to the essential information for monitoring the health of ecosystems, investigation of biogeochemical processes and checking compliance with legislation [3]. Also control of the quantity of phosphorous in fertilizers leads to an excessive growth (eutrophication) of aquatic plants and algae that disrupts aquatic life cycles [4]. Based on the perspective of biology, phosphorous species such as phosphate anion is considered as one of the most important electrolytes that plays role as an essential component in all living organisms [5].

Phosphorous also has major role in biological processes such as in the structure of ATP (Adenosine triphosphate) and DNA (Deoxyribonucleic acid) and also during control of pH in blood or lymph fluid [5]. In a clinical setting, phosphate level in serum is determined as part of a routine blood analysis [5]. Also knowledge about phosphorous level in the body fluids can provide useful information about several diseases such as hyperparathyroidism, vitamin D deficiency, and Fanconi syndrome [6]. Concentration fluctuations of salivary phosphate have been investigated as indicators of ovulation of women, uremic state, and risk of development of dental caries and formation of dental calculus [7]. All the mentioned information clearly shows that, analysis of phosphorous is considered as a biomarker for different diagnostic tests [5].

According to the literature review, concentration of phosphorous often varies between 0.2-10.0 mg $L^{-1}$ in natural and wastewater samples and in the range of 0.2 to 50 mg $kg^{-1}$ in soil [5]. In addition, maximum permissible concentration of phosphorous species in river and wastewater samples is estimated to be between $0.32 \times 10^{-6}$ and $0.143 \times 10^{-3}$ mol $L^{-1}$, respectively [5]. As a diagnostic fluid, the concentration of phosphate ions in human saliva is usually found to vary between 5.0-14.0 mmol $L^{-1}$ [8,9]. Whereas the concentration of phosphorous species is estimated between 0.81–1.45 mmol $L^{-1}$ in the adult human serum [10,11]. Precise attention to these, the level of concentration of phosphorous in biological and environmental samples clearly points to the importance of introduction of accurate and sensitive methods for trace and ultra-trace phosphorous detection at both µg $mL^{-1}$ and ng $mL^{-1}$ levels.

During the last decades several analytical methods such as spectroscopy [12], potentiometry [13], voltammetry [14], chemiluminescence (CL) [15], immunoassay [15], and ion exchange chromatographic techniques [15] have been introduced for detection and determination of phosphorous in various real samples such as clinical, environmental, industrial, and biological samples. But most of these analytical techniques are often limited to low sensitivity, less improved detection limits, small sensitivity, negligible interference from real sample matrix, cost and/or fast response time for selective determination of phosphorous species [16-28]. Among these analytical techniques, CL can be considered as a promising detection system [15]. Compared to other analytical techniques [5] CL possesses

significant advantages such as simplicity, low cost and high sensitivity and selectivity [5]. Consequently, CL-based detection has become considered as a quite useful and interesting technique for scientists in a wide variety of disciplines [29-31]. But in spite of great advantages of CL-based detection system, this technique has been still limited because of limitations such as low selectivity; small sensitivity and/or less improved detection limit [5]. These problems have not still completely solved even in the presence of catalytic reagents such as metal nanoparticles [5].

It seems that focusing on sample introduction can solve the current problems such as low selectivity; small sensitivity and less improved detection limit. Based on various types of sample introductions [32], nebulization seems to be considered as an appropriate method for direct introduction of samples by spraying process during the formation of aerosols.

Aerosols are defined as small solid or liquid particles suspended in the atmosphere for a period of a few hours to a few days depending on their size [33]. They have a complex chemical composition that mainly depends on their source, which may be natural and/or anthropogenic [33]. The interest to the molecular aerosols as large molecules with size distribution from sub nanometers to many microns has been increased significantly over the past few years [34]. This interest is due to the critical role in atmospheric/industrial processes, in air pollution where they pose health risks, and in astrophysical and astrochemical processes [34]. To improve the understanding of the impact of aerosols in these various fields, it is crucial to study their physical (size, shape, architecture, phase behavior), and chemical (composition, reactivity) properties, along with their formation processes such as aggregation, chemical reaction, agglomeration, and coagulation [34].

Most researchers efforts in aerosol characterization focus on physical properties such as size distributions [35]. Comparatively study on chemical properties of aerosol like chemical composition, thermodynamics, surface properties, spectral fingerprints, refractive index, etc. has been traditionally left behind [36]. To achieve a better understanding of the above-mentioned chemical properties of aerosol particles, it is necessary to have a fundamental study on single aerosol particles in different sizes, compositions, optical properties (absorbing or nonabsorbent), and surrounding environments using a highly capable technique [36].

During the last decades various researches have been focused on some physical and chemical properties of aerosols using optical trapping-cavity ring down spectroscopy (OT-CRDS) in combination with conventional aerosol characterization methods/techniques during the hygroscopic process [35]. In this work it is aimed to focus on a novel intrinsic behavior of hygroscopic growth factor of aerosols containing phosphorous compounds introduced to a novel OT-CRDS named as "Hygroscopy", for their ultra-trace and trace detection and determination of at both ng $mL^{-1}$ and µg $mL^{-1}$ levels.

*Corresponding author: Mohammad Mahdi Doroodmand, Department of Chemistry, College of Sciences, Shiraz University, Shiraz 71454, Iran
E-mail: oroodmand@shirazu.ac.ir

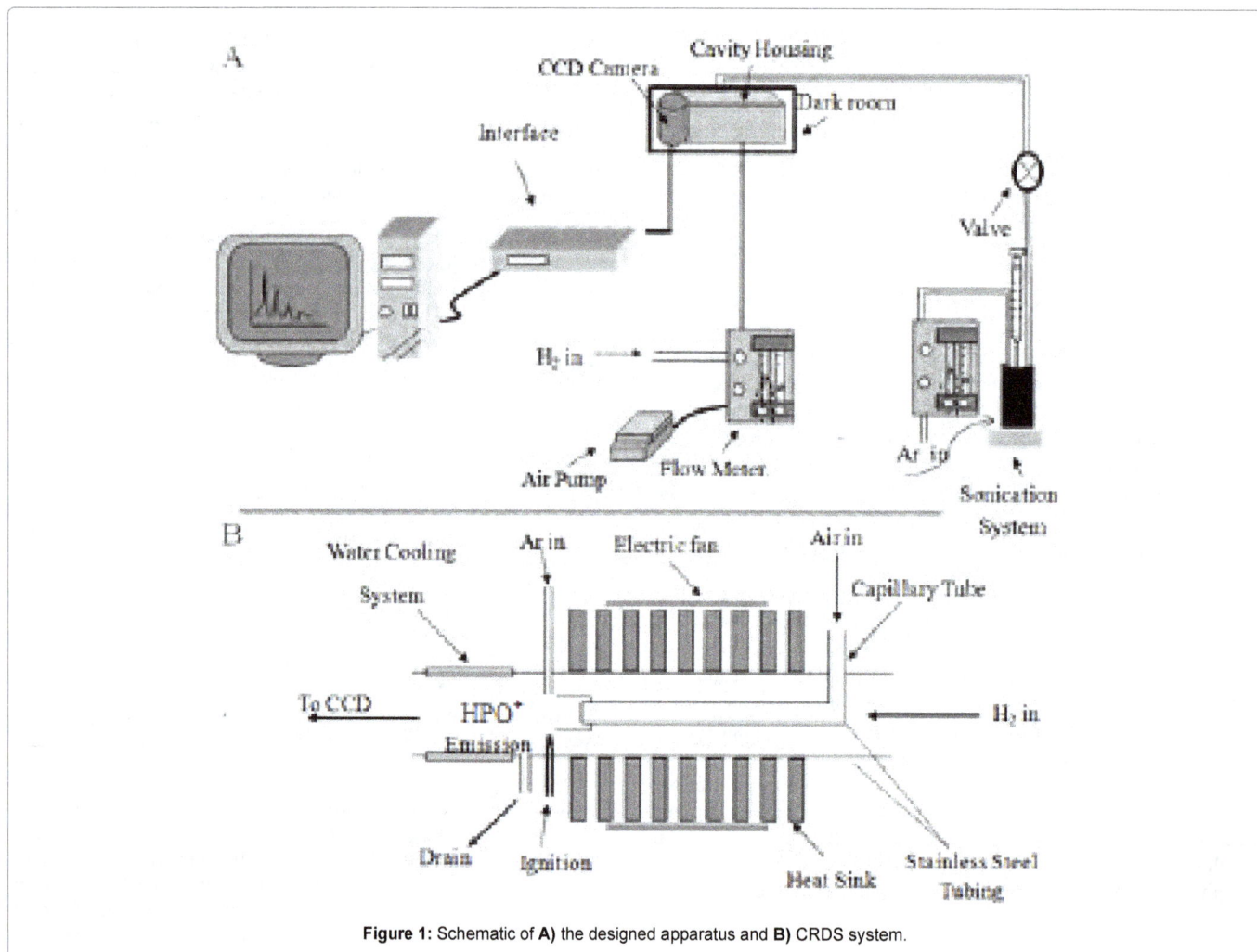

**Figure 1:** Schematic of **A)** the designed apparatus and **B)** CRDS system.

## Experimental

### Reagents

Stock solutions of 1000.0 µg mL$^{-1}$ PO$_4^{3-}$, PO$_2^{3-}$, H$_2$PO$_2^-$ and P$_2$O$_5$ (PO$_4^{3-}$) were individually prepared by dissolving 1.885, 1.506, 0.675, and 0.7474 g of, Na$_2$HPO$_4$.12H$_2$O (Merck, Darmstadt, Germany), Na$_3$PO$_3$.5H$_2$O (Fluka), Ca(H$_2$PO$_2$)$_2$ (Merck, Darmstadt, Germany) and P$_2$O$_5$ (Fluka), respectively, in 500 mL volumetric flask and diluted to the mark with deionized water. Standard solutions were prepared daily by successive dilution of the stock solutions. 500 mL solutions of different acids with pH 3.0 were prepared by dilution of each concentrated HCl (37%, Merck, Darmstadt, Germany), HNO$_3$ (65%, Merck, Darmstadt, Germany), HClO$_4$ (65% Merck, Darmstadt, Germany), acetic acid (100% Merck, Darmstadt, Germany) and H$_3$BO$_3$ (Fluka).

### Instrumentation

Detail of cavity ring down spectroscopy (CRDS) system is almost similar the molecular emission cavity analysis (MECA) system, described in the previous study [37]. In this experiment, some modifications are applied to the cavity in order to be used for phosphorous determination. The schematic of the modified cavity is shown in Figure 1. Briefly it consists of two concentric stainless steel tubes. The outer tube is 12.0 mm OD, 10.0 mm ID and the inner is a capillary tube with 3.0 mm OD

tube. The end of the capillary tube is positioned 4.0 cm shorter from the inside tip of the outer tube. The inner capillary tube transfers air (air pump, model: pyeunicam ltd) and the outer carries hydrogen gas (H$_2$ cylinder, Isfahan Petrochemical Company, purity: 99.996%). The flow of hydrogen and air is controlled using flow controllers. An ignition system is used for automatic ignition of the hydrogen/air flame. A drain is also located at the end part of the outer tube to transfer the generated water. To cool the cavity, small fans are used as shown in Figure 1A. The end part of the outer stainless steel tube is also cooled through water circulation. To introduce the samples into the flame, a reaction tubing cell with 20.0 mL volume is fabricated using glass. A sonicator (model: MIST MAKER, frequency: 500-KHz) is also positioned at the bottom of the reaction cell for the generation of aerosols. The generated aerosols are then carried to the flame through a tygon tubing (internal diameter: 3.0 mm) using N$_2$ gas (N$_2$ cylinder, purity: 99.9, Parsbaloon, Iran, Shiraz).

In this system, the blue emission of HPO* is monitored using a CCD camera. The CCD camera used as a detector is placed in front of the cavity. The cavity system was also placed inside a box (dark room, Figure 1B) to protect the system from any environmental stray lights. The image of the CCD is saved in a computer. The blue component of the color related to the chemiluminescence (CL) of phosphorous is analyzed for further processing using a program written in Visual Basic.

For phosphorous speciation, an injection port is introduced for injection of phosphorous-containing compounds into the glass reaction cell (height: 12.5, diameter: 5.0). For phosphorous determination in the CRDS system using a stainless steel as the CL support. The phosphorous-containing aerosols are directly introduced to the reaction cell during introduction of 3.0 mL of $PO_4^{3-}$ along with setting the flow rate of $N_2$ as carrier gas to the cavity.

The temperature of the chemical reactions cell is controlled using an electric furnace surrounded the cell. The $N_2$-bubbling reaction cell contains 20.0 mL of solution of $HClO_4$ acid (pH=2.6) as well as 3.0 mL of solution unknown phosphorous determination. In this reaction cell, bubbled nitrogen is considered as carrier gas. Phosphorus-containing compound is introduced to this reaction cell through the injection port by a syringe using a septum. For turbidity analysis a turbidimeter (model: DRT-100) is adopted. The humidity of the system is also measured using a humidity sensor (model: GCH-2018).

## Procedure

For phosphorus determination, a 3.0-mL sample container is half-filled with of $HClO_4$ (pH=2.6). The temperature of the solution was set to 33°C using an electric furnace surrounded the reaction cell. The flow rates of hydrogen and air are controlled at 256 and 102 mL min$^{-1}$, respectively. By injection of 3.0 mL of phosphorus solution into the vessel through the injection port, the generated phosphorus-containing aerosols are then swept into the cavity by a stream of $N_2$.

## Real sample analysis

The application of the recommended method is adopted for selective determination of phosphorus in various drinking water samples. For this purpose the sample preparation was achieved using the procedure reported in Ref [38]. Then the samples are individually diluted 10-time

and determined using standard addition method. The reliability of these results is evaluated using ion exchange chromatography (IEC, model: ASI, 310).

## Results and Discussion

"Hygroscopy" is the ability of a substance to attract and hold water molecules from the surrounding environment [34]. This property is achieved through either absorption or adsorption on the absorbing or adsorbing materials such as phosphorous species [39]. This property causes an increase in volume, stickiness, or other physical characteristics of the material, as water molecules become suspended among the material species in the process [39]. The quantity of this parameter is usually evaluated and estimated in a term called "Hygroscopic Growth Factor". Atmospheric aerosols can scatter and absorb the incident light. Therefore the atmospheric visibility is reduced with the increase of their mass [39]. Furthermore, the visibility would sharply decrease when the ambient relative humidity (RH) is high at the same level of aerosol mass [39]. It seems that this factor can significantly promote the luminescent property of species such as phosphorous, even in a cool flame of $H_2$/air, especially when having water molecules as the product of the flame reaction. Therefore, it is expected to achieve major improvement in some figures of merit such as sensitivity and detection limit.

Among the current reported analytical techniques, "Cavity Ring-Down spectroscopy" (CRDS) seemed to be considered as selective and sensitive detection system for determination purposes. This method was also considered as a highly sensitive optical spectroscopic technique that enabled measurement of absolute optical extinction by samples that scatter and absorbed light. The CRDS has been widely used to study gaseous samples which absorb light at specific wavelengths, and in turn to determine mole fractions down to the parts per trillion levels [40-42]. Briefly a typical CRDS setup consisted of a laser that is used to illuminate a high-finesse optical cavity, in which its simplest form consists of two highly reflective mirrors. When the laser was in resonance with a cavity

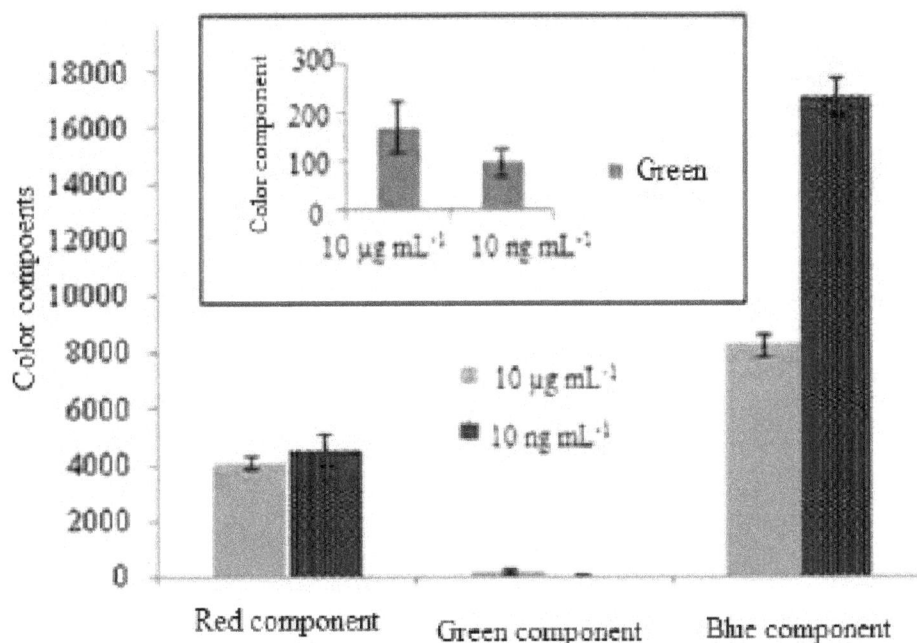

**Figure 2:** Red and blue components vs. different concentrations of $HPO_4^{-2}$ (10.0 ng mL$^{-1}$ and 10.0 μg mL$^{-1}$). Trace: histogram for green component.

mode, intensity built up in the cavity due to constructive interference. The laser was then turned off in order to allow the measurement of the exponentially decaying light intensity leaking from the cavity [40-42].

In the CRDS system, the laser light was reflected back for about thousands of times between the mirrors giving an effective path length for the extinction on the order of a few kilometers. Some characteristics of this analytical technique included the independency of both the exponentially decaying light and the ring-down time from the fluctuations of the light source that caused to get accurate results during the detection purposes without any fluctuations [40-42]. These characteristics of the CRDS led to have significant advantages such as

i)   high sensitivity due to the multi-pass nature (i.e., long path length) of the detection cell,

ii)  possibility to shot variations in laser intensity during the measurement in a fixed rate constant,

iii) wide range use for a given set of mirrors; typically, ± 5% of the center wavelength,

iv)  high throughput because of the individual ring down events, occurred in the millisecond time scale and

v)   finally, no need for a fluorophore.

which made it more attractive for some techniques such as rapidly pre-dissociating. However, this method had disadvantages [40-42]. For instance, the spectra could not be acquired quickly due to the use of a monochromatic laser source. Also, analytes were limited only to the availability of tunable laser light at the appropriate wavelength as well as the availability of high reflectance mirrors at those wavelengths and iii) high expense. The requirement for laser systems and high reflectivity mirrors often made CRDS orders of magnitude more expensive than some alternative spectroscopic techniques [40-42].

To solve these problems, hereby in this study for the first time a new, simple, sensitive and selective methodology has been introduced for trace and ultra-trace detection of phosphorous in various real samples without needing any external laser. In another word, presence of trace amount of $Na^+$ as radiation buffer in the matrix of the sample solution provided the capability for playing role as initial light source during evaluation of the intensity of scattering throughout the determination process via the formation of aerosol. This detection system was based on the observation of blue emission (instead of green phosphorous emission) during the introduction of phosphorous samples to the analyzing volume via formation of aerosols. As shown in Figure 2 the RGB components for 10.0 ng mL$^{-1}$ and 10.0 µg mL$^{-1}$

of $PO_4^{3-}$ clearly revealed the blue component obvious. This detection system was therefore considered as an appropriate detection system for phosphorous detection and determination.

As shown in Figure 2, the green component can be attributed to the molecular emission of phosphorus species. As expected, the green component for 10.0 µg mL$^{-1}$ (green component=171 a.u.) phosphorus species was more intensive than that for 10 ng mL$^{-1}$ (green component=98 a.u.). Also, the red component can be attributed to the emission intensity of sodium introduced to the flame during phosphorous analysis. As evaluated partially, the same red components were detected during introduction of $PO_4^{3-}$ (from $Na_2HPO_4$) with 10.0 ng mL$^{-1}$ and 10.0 µg mL$^{-1}$. Unexpectedly, the intensity of the blue component for 10.0 ng mL$^{-1}$ was significantly higher than that of 10.0 µg mL$^{-1}$.

To interpret this phenomenon the influence of aerosols was evaluated in detail. For this purpose, aerosols containing phosphorous species were often generated via ultrasonically irradiation using a water humidifier (frequency: 500 KHz), followed by transferring into the analyzing system (CRDS) using $N_2$ as carrier gas.

Parameters having strong influence during following luminescence and scattering of the generated aerosols as well as during their hygroscopic growth inside the humidified flame ($H_2$/air) were the amount of $Na^+$ as radiation buffer (initial light source(, flow rates of $H_2$, air and $N_2$ gases, kind and concentrations of acid during formation, introduction and evaluation of the aerosols inside flame, etc. These parameters should be well optimized to reach the highest sensitivity for HPO$^{\cdot}$ emission on the stainless steel as support during introduction of phosphorous species through the port to the reaction cell. In this study, the optimization process was achieved using both one at a time and simplex methods as reported in detail in the supporting information 1.1 (Sections: 1.1. S.P.-1.3. (S.P.)).

## Effect of $Na^+$ as radiation buffer

In this study, it was observed that no significant light emission was detected in the absence of alkali ions such as $Na^+$ during only analysis of ultra-trace phosphorous species. The phenomenon was clearly evidenced when preparing phosphorus solution using phosphorus salts such as $Ca(H_2PO_2)_2$ or $P_2O_5$ according to the histogram shown in Figure 3.

As phosphorus pentoxide was a chemical compound with molecular formula $P_4O_{10}$, this white crystalline solid reagent was the anhydride of phosphoric acid. Phosphorus pentoxide was therefore a

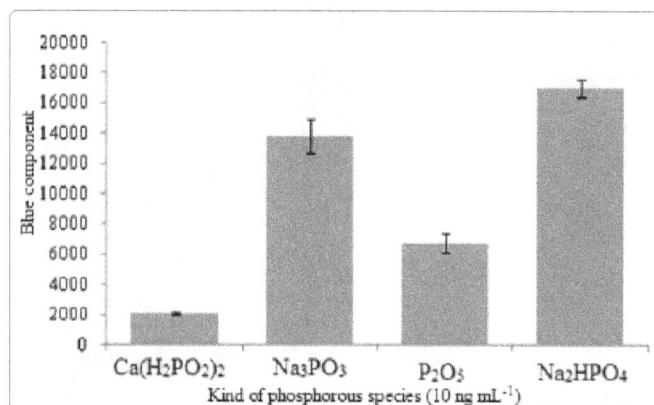

**Figure 3:** Effect of $Na^+$ as radiation buffer during analysis of various types of phosphorous species (10.0 ng mL$^{-1}$) at pH pH~2.6.

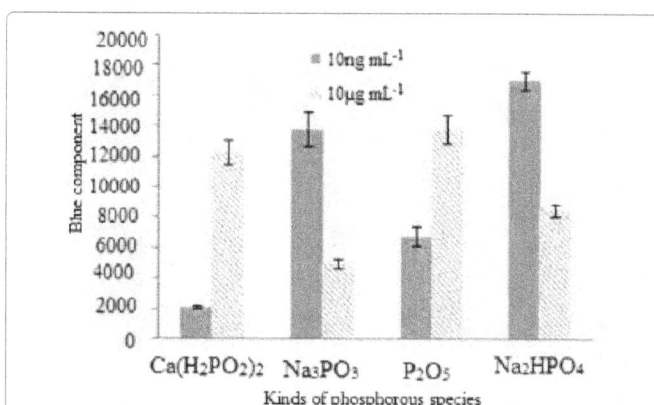

**Figure 4:** Effect of $Na^+$ as radiation buffer trace and ultra-trace analyses of phosphorous species.

potent dehydrating agent as indicated by the exothermic nature of its hydrolysis (Eq. 1.). In fact, both phosphorus salts using each $P_2O_5$ and $NaH_2PO_4$ reagents in aqueous media produced the same anion. This can be considered as suitable probe during using the effective role of $Na^+$ as radiation buffer. It seemed that $Na^+$ radiation with (wavelength=589.0 nm) played role as initial light source (pulsed laser), which made it as low cost and excellent module for phosphorus determination during several reflections between the aerosols inside the flame that acted as mirrors. This effect was attributed to the interaction between hydrogen radicals and $Na^+$ (as radiation buffer) during scattering process according to the following reactions based on the Le Chatelier's principle (Eqs. 2 and 3). Therefore, trace quantity of sodium metals (with zero oxidation state) played role as the source of light at $\lambda_{max}$ of 589.0 nm. It should be noted that this effect was dependent to the $H_2$ flow rates. At high $H_2$ flow rates from one side, high concentrations of radical hydrogen reduced $Na^+$ to the metallic form in the analyzing volume. From the other side, the temperature of the flame became so high that the generated sodium metal was oxidized inside the bulk solution in the flame. Consequently, competition between the other two factors (temperature and reducing behavior of the flame) determined the sensitivity of the system during analysis of ultra-trace of phosphorous species by scattering process.

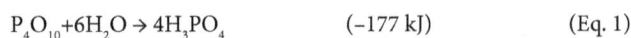

$$P_4O_{10}+6H_2O \rightarrow 4H_3PO_4 \qquad (-177\ kJ) \qquad (Eq.\ 1)$$

It be also noted that due to the necessity of the ring-down spectroscopy to an initial light source during the scattering process, trace amount of $Na^+$ (~ 5 μg) was needed. This low quantity also prevented the light interfering effect of the sodium-containing impurities during the analysis of ultra-trace of phosphorous species using image processing. However this effect was further limited due to the high acidic environment of the solution (pH: 2.6) according to the Le Chatelier's principle based on Eq. 2. In addition, Figure 4 compares the emission and scattering. As clearly shown, during analysis of phosphorous species with 10.0 μg mL$^{-1}$ concentration, the emission was predominant. But during ultra-trace determination of phosphorus species (10.0 ng mL$^{-1}$) reverse sensitivities (blue component intensity) were detected due to the presence of $Na^+$ in some phosphorous species ($Na_3PO_3$, and $Na_2HPO_4$).

$$Na^++H^0 \qquad \rightarrow \qquad Na^*+H^+ \quad in\ the\ bulk\ solution \qquad (Eq.\ 2)$$

$$Na^* \qquad \rightarrow \qquad Na+h\nu \qquad (Eq.\ 3)$$

In this study, selection of $Na^+$ as radiation buffer was attributed to its emission wavelength ($\lambda_{max}$=589.0 nm), which was partially in the middle of the visible light range. For more evaluation of this phenomenon, the effect of $Na^+$ was compared to that of $K^+$ ($\lambda_{max}$=766 nm). Based on result, no significant response was observed when $K^+$ was used instead $Na^+$. This phenomenon was probably due to wavelength emission of potassium that was shifted to red. Therefore $Na^+$ solution was considered as the selective radiation buffer during the detection of the phosphorous emission radiations.

## Chemical stability of the phosphorous solution

For investigating the stability of the phosphorous solutions, effect of some species such as ethylene diamine tetra acetic acid (EDTA) was also investigated. For this purpose, 3.0 mL of EDTA solution (1.0 mM) at pH ~2.6 was added to phosphorous standard solution (10.0 μg mL$^{-1}$). Based on the results no significant change was observed in the results during the analysis in the presence and absence of EDTA. This result clearly revealed the stability of the phosphorous standard solutions during the analyses.

## Proposed mechanism of the detection system

The mechanism of the radiation (Mie scattering) was also evidenced based on briefly

**i)** dependency of the scattered intensity to the presence of $Na^+$,

**ii)** relationship between the scattered and both the humidity of the detection system,

**iii)** good correlation between the response of the cavity with turbidimetry,

**iv)** observation of blue shift from green (emission light related to the luminescence of HPO* in $H_2$/air flame) to blue (scattered radiation) and

**v)** the effect of hydration number during stability and growth of the aerosols [39] inside the flames.

Details of these evidences are as follows:

As among the geometrical shapes sphere has minimum surface-to-volume ratio therefore, maximum surface tension was estimated for sphere in comparison with other geometries. Consequently, aerosols intrinsically tend to have spherical shape. In many optical studies of aerosols, particles can be assumed to be spherical in shape, allowing the application of Mie scattering theory [34]. A Mie resonance often exhibits more than one maximum in the radial distribution of the light intensity within the droplet [34]. If the levitated particle is a homogeneous liquid droplet, then the light-scattering pattern consists of a parallel fringe structure (Mie scattering) [34]. This phenomenon was observed during analysis of phosphorous species at ng mL$^{-1}$ levels.

Good correlation as well as the same behavior was obtained

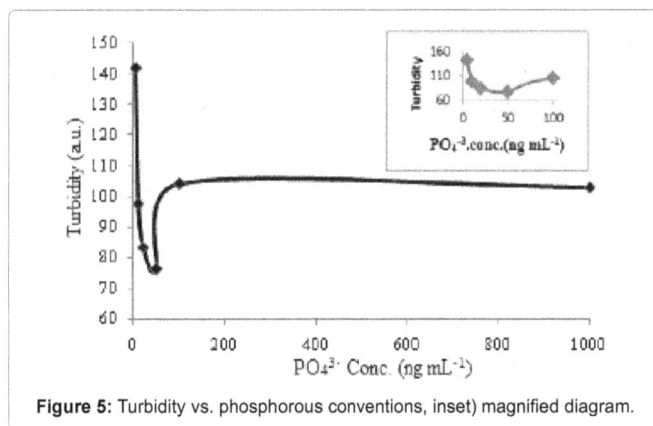

**Figure 5:** Turbidity vs. phosphorous conventions, inset) magnified diagram.

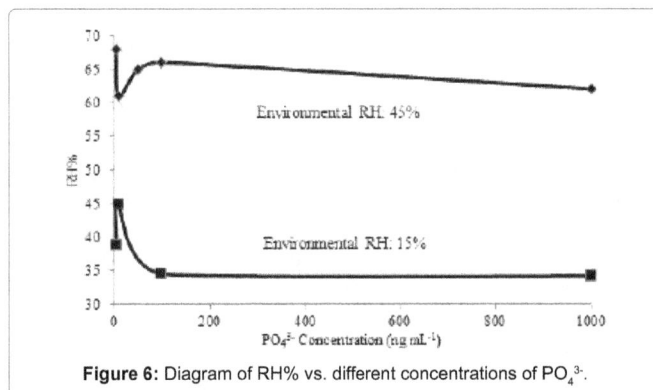

**Figure 6:** Diagram of RH% vs. different concentrations of $PO_4^{3-}$.

between the results of the CRDS and the turbidimetry. The results are shown in Figure 5. It should be noted that due to low precision, small sensitivity, no linearity as well as presence of high interference, turbidimetry cannot be adopted for detection of phosphorous species instead of CRDS. Consequently, turbidimetry as the detection system was only used to approve the scattering phenomenon.

Phosphorus species like any other salt had tendency to hydration. This phenomenon, therefore, changed the size and thickness of the hydrated layer [39]. This effect also converted the hygroscopic growth factor, and thus changing in the scattering and turbidity.

In the $H_2$/air flame, $H_2O$ was considered as a product according to the flowing reactions (Eqs. 4 and 5). Based on these results, aerosols containing compounds such as phosphorus species provide the ability to absorb $H_2O$ molecules from surrounding environment their size is larger [39]. The main reason for this was that the mass scattering efficiency of aerosols would increase when water soluble aerosols grow to become larger in diameter. Consequently, at low concentrations the scattering phenomenon was majorly occurred in comparison with luminescence radiation. This phenomenon can also be interpreted somewhat like the formation of crystal salts in a supper saturated solution. The lower concentration of salts, the larger crystalline size is generated; therefore, larger aerosols (i.e., more hydration no.) are expected from diluted phosphorous species.

$$H_2+O_2 \qquad \rightarrow \qquad 2OH \qquad \text{(Eq. 4)}$$

$$H_2+OH^{\cdot} \qquad \rightarrow \qquad H^{\cdot}+H_2O \qquad \text{(Eq. 5)}$$

A reverse correlation was observed between the relative humidity percentage (RH%) and concentrations of $PO_4^{3-}$ at ng mL$^{-1}$ levels in different environments such as about 15 and 45% RH according to the results shown in Figure 6. This behavior again pointed to the effective role of hygroscopic growth factor during scattering the phosphorous species inside $H_2$/air flame, which generated water as the product of the reaction between hydrogen and oxygen (Eqs. 4 and 5). This result was in good agreement with the results obtained during estimation of the scattering radiation using instruments such as online forward-scatter visibility meter, integrating nephelometer and multi-angle absorption photometer [39].

Hygroscopic growth factor was also defined as the ratio of aerosol scattering coefficient, $f_{(RH)}$ at wet condition to that at dry condition (RH $\leq$ 30%) according to the equation reported in Ref. [39]. In this system, based on the results shown in Figure 6, the RH% of the two different conditions (RH% 15 and 45) was found to be 1.36, which pointed to the effective role of the environmental RH%. This factor was considered as aerosol particle backscatter coefficient, which was dependent on the size and morphology of the aerosols particles [43]. The relationship between the hygroscopic growth effects and the aerosol volume concentration was evaluated by observing a stronger increase in the fine mode volume concentration of the phosphorous-containing aerosols during increase in the quantity of RH% inside the $H_2$/air flame.

The relationship between the size of aerosols and the phosphorous concentration can also be evaluated via following the hydration number of the phosphorous species. In this study, the same behavior can be observed for phosphorous species inside a humid flame generated using $H_2$ and $O_2$. However, hydration number of phosphorus aerosol can be discussed using the extended Debye-Hückel equation using the parameter called "mean distance approach of the hydrated ions". Based on this term, the thicker the hydrated layer of phosphorus species, the more was the activity coefficient and the higher was the activity of phosphorus species.

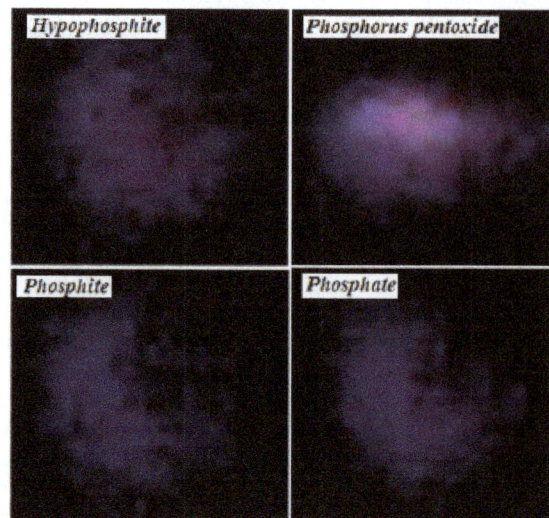

**Figure 7:** CCD images related to the introduction of various phosphorous species to the CRDS system.

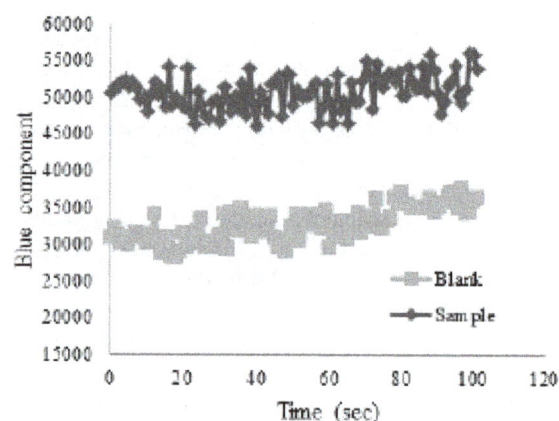

**Figure 8:** Trace diagram during analysis of $PO_4^{3-}$ 10.0 µg mL$^{-1}$.

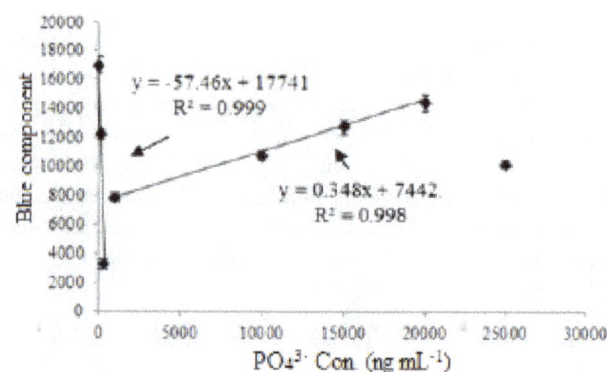

**Figure 9:** Calibration curve of phosphorous compounds at ng mL$^{-1}$ and µg mL$^{-1}$ levels.

The intensity of the scattered light was independent from that proposed by Saha equation [44]. The Saha ionization equation was an expression that relates the ionization state of an element to the temperature and pressure. In this study the Saha equation described the

| Linear dynamic range | Correlation coefficient (R) | Calibration sensitivity | RSD (%) | Detection limit (ng mL$^{-1}$) |
|---|---|---|---|---|
| $PO_4^{3-}$ (1-20 µg mL$^{-1}$) | 0.998 | 0.348 | 11.0 | - |
| $PO_4^{3-}$ (10-250 ng mL$^{-1}$) | 0.999 | -57.46 | 12.0 | 5 |

**Table 1:** Analytical figures of merit during phosphorus detection and determination. Note: The data are the average of 70 sequential analyses.

| Foreign species | Tolerance ratio | Interfering effect (%) | Comments |
|---|---|---|---|
| $CH_3COO^-$, $Cl^-$, $Br^-$, $ClO_4^-$, $I^-$, $CN^-$, $COO^-$, $CO_3^{2-}$ | 1000 | No interference | -No interaction with phosphorus species<br>-Not emission radiation |
| $NO_3^-$ | 500 | ~3% enhancement | -No interaction with phosphorus species<br>-Low emission radiation |
| $K^+$, $Co^{2+}$ | 500 | No interference | -No interaction with phosphorus species<br>-Not emission radiation |
| $NH_4^+$ | 1000 | No interference | -No interaction with phosphorus species<br>-Not emission radiation |
| $Ca^{2+}$ | 500 | No interference | interaction with phosphorus species and formation precipitate Calcium phosphate ($K_{SP}$=2.07 × 10$^{-33}$) |
| $Fe^{3+}$ | 500 | No interference | interaction with phosphorus species and formation precipitate iron phosphate ($K_{SP}$=1.3 × 10$^{-22}$) |
| $Ni^{2+}$ | 500 | No interference | interaction with phosphorus species and formation precipitate nickel phosphate ($K_{SP}$=4.74 × 10$^{-32}$) |
| $SO_4^{2-}$ | 200 | ~20% Enhancement | Molecular emission of sulfur |

**Table 2:** Effect of foreign species on phosphorus determination.

| Methods | Analyzed sample | Linear range | Detection limit | Reference |
|---|---|---|---|---|
| Spectrophotometry | Seawater | 0.034-1.134 µM (3.213-107.163 ng mL$^{-1}$) | 1.4 nM (0.1323 ng mL$^{-1}$) | [48] |
| Fluorimetry | River and marine waters | 0.3-4.0 µM (28.35-378 ng mL$^{-1}$) | 0.3 µM (28.35 ng mL$^{-1}$) | [49] |
| Spectrophotometry | Wastewater | 0.026-0.485 mM (2475-45832 ng mL$^{-1}$) | 7.4 µM (699.3 ng mL$^{-1}$) | [50] |
| Electrochemiluminescence | Water | 2.0 × 10$^{-10}$-1.0 × 10$^{-8}$ gmL$^{-1}$ (0.2-10 ng mL$^{-1}$) | 8.0 × 10$^{-11}$ gmL$^{-1}$ (0.08 ng mL$^{-1}$) | [36] |
| Electrochemistry | Human serum sample | 1.0 × 10$^{-6}$- 100.0 × 10$^{-6}$ mol L$^{-1}$ (94.5-945000 ng mL$^{-1}$) | 3 × 10$^{-6}$ mol L$^{-1}$ (283.5 ng mL$^{-1}$) | [51] |
| Ion exchange chromatography | Drug product | 2-200 µg mL$^{-1}$ (2000-200000 ng mL$^{-1}$) | 1 µg mL$^{-1}$ (1000 ng mL$^{-1}$) | [52] |
| Present study | Drinking water | 10.0-250.0 ng mL$^{-1}$ and 1.0-20.0 µg mL$^{-1}$ or (1000-20000 ng mL$^{-1}$) | 5.0 ng mL$^{-1}$ | This work |
| Real sample | Proposed method (ng mL$^{-1}$) | $^1$Ion exchange chromatography (ng mL$^{-1}$) | Relative error (%) | |
| Drinking water 1 | 19.00 | 19.51 | -2.60 | |
| Drinking water 2 | 17.01 | 17.30 | -1.68 | |
| Drinking water 3 | 10.15 | 10.43 | -2.68 | |
| Well water1 | 25.82 | 26.35 | -2.01 | |
| Well water 2 | 8.71 | 8.43 | +3.32 | |

**Table 3:** Real sample analysis. Where, $^1$Ion exchange chromatography was considered as Ref. method [47].

| Electrochemistry | Human serum sample | 1.0 × 10$^{-6}$- 100.0 × 10$^{-6}$ mol L$^{-1}$ (94.5-945000 ng mL$^{-1}$) | 3 × 10$^{-6}$ mol L$^{-1}$ (283.5 ng mL$^{-1}$) | [51] |
|---|---|---|---|---|
| Ion exchange chromatography | Drug product | 2-200 µg mL$^{-1}$ (2000-200000 ng mL$^{-1}$) | 1 µg mL$^{-1}$ (1000 ng mL$^{-1}$) | [52] |
| Present study | Drinking water | 10.0-250.0 ng mL$^{-1}$ and 1.0-20.0 µg mL$^{-1}$ or (1000-20000 ng mL$^{-1}$) | 5.0 ng mL$^{-1}$ | This work |

**Table 4:** Comparison between the introduced method and the previously reported method.

degree of ionization of the atomizer as a function of the temperature, density, and ionization energies of the species [45]. If Saha ionization occurred, the amount of emission would increase a little. However the Saha equation only held weakly ionized plasmas for in which the Deby length was large [46,47]. The independency of the scattered light with the Saha phenomenon was evidenced via observation of no significant response during independently analysis of inorganic species such as

$Na^+$, $K^+$ or $H_3PO_4$. Consequently, this phenomenon clearly pointed to the presence of strong interaction between $Na^+$ and phosphorous species during introduction of aerosols to the $H_2$/air flame.

Based on this principle, trace quantity of $Na^+$ ions as radiation buffer was stabilized inside the flame and played a role such as the incident laser light for scattering process, resulting to get blue radiation instead of green chemiluminescence of phosphorous species or even

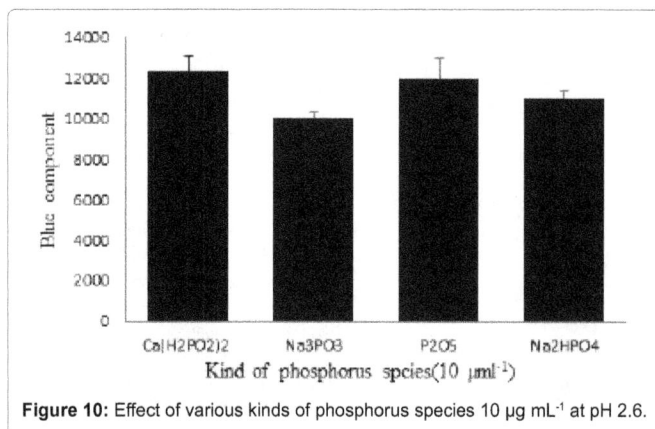

**Figure 10:** Effect of various kinds of phosphorus species 10 μg mL⁻¹ at pH 2.6.

the emission of $Na^+/Na$ inside the flame. All these evidences pointed to the effective role of a parameter called hygroscopic growth factor during formation of aerosols.

## Analytical figures of merit

Figure 7 exhibits the CCD images related to the introduction of various phosphorous species to the CRDS system. Figure 8 shows sample trace diagram during analysis of $po_4^{3-}$ (10.0 μg mL⁻¹) during the image analysis. The calibration curves had also been shown in Figure 9. As shown, the calibration curves had two significant linear ranges. Based on the calibration curve, sensitivity with positive slope was observed in the emission intensity depending on the concentration of phosphorus species at μg mL⁻¹ levels during both scattering and luminescence phenomena. Whereas reverse behavior was exhibited during analysis of phosphorous compounds at ng mL⁻¹ levels by the scattering phenomena. Based on the literature, this was related to the "Aerosol Hygroscopic Growth Factor" [39]. Two linear calibration curves with reverse slope was therefore observed between 10.0 – 250.0 ng mL⁻¹ and 1.0 – 20.0 μg mL⁻¹ with correlation coefficients ($R^2$) of 0.998 and 0.999, respectively (Figure 9). The calibration sensitivity was also estimated to be 0.348 and -57.46 (a.u.), respectively.

The detection limit was defined as the concentration of phosphorous species giving a signal equal to the blank signal plus triple values of the standard deviation of the blank. Based on this definition, the limit of detection was found as 5.0 ng mL⁻¹.

The relative standard deviation (RSD%, reproducibility) for 5 replicate analyses for each 10.0 ng mL⁻¹ and 10.0 μg mL⁻¹ $PO_4^{3-}$ was found to be 12.0 and 11.0%, respectively. Based on the definition of the response time, 90% of maximum response ($t_{90}$), the maximum response time was evaluated to be ~10.0 s. Table 1 shows analytical figures of merit during phosphorus detection and determination.

Figure 10 compares the HPO* emissions measured during the introduction of 3.0 mL of each phosphorus in 10.0 μg mL⁻¹ solution to the reaction cell containing 20.0 mL $HClO_4$ at pH 2.6. As shown partially the same emission intensity was detected during introduction of 10.0 μg mL⁻¹ of each phosphorous species such as $Na_2HPO_4$ (RSD=4.3%, n=28), $Na_3PO_3$ (RSD%=6.1%, n=37), $Ca(H_2PO_2)_2$ (RSD=6.1%, n=38), and $P_2O_5$ (RSD=6.8%, n=40) at pH ~2.6.

No serious interference was evaluated during analysis of at least 500-fold excess of various anions such as $CH_3COO^-$, $Cl^-$, $Br^-$, $ClO_4^-$, $I^-$, $CN^-$, $CO_3^{2-}$, $NO_3^-$, $I_3^-$ and various cations such as, $NH_4^+$, $Na^+$, $Fe^{3+}$, $K^+$, $Ni^{2+}$, $CO^{2+}$, $Ca^{2+}$ to a 10.0 μg mL⁻¹ and 10.0 ng mL⁻¹ of phosphate

standard solution. The results are reported in Table 2. The only observed interference was evaluated during introduction of 200-fold excess of $SO_4^{2-}$. These results clearly pointed to the selectivity of the recommended technique for rapid and sensitive determination of phosphorous species in various real samples without any interfering effects.

## Real sample analyses

The reliability of this method was evaluated via selective determination of phosphorus in various drinking water samples according to the recommended preparation procedure. For this purpose standard addition method was used during the analyses of some drinking water samples. The results are reported in Table 3. As shown good correlation was evaluated between the results obtained from this technique and ion exchange chromatography during analysis of drinking water samples that clearly revealed the reliability of this method for detection and determination of species such as phosphorus compounds.

## Conclusions

In this study, a new method has been introduced based on aerosol hygroscopic growth as a new factor for trace and ultra-trace determination of phosphorus in flame containing OT-CRDS. The advantages and disadvantages of the technique for phosphorous determination have been compared to the articles shown in Table 4. Compared to these reports, this technique has significant characteristics such as high sensitivity, high selectivity, capability to determine phosphorus compounds in a wide range between 10.0 - 250.0 ng mL⁻¹ and 1.0 to 20.0 μg mL⁻¹ with improved detection limit, simplicity, and low cost. To the best of our knowledge this study is the first report that Mie scattering is followed for determination purposes using a simple design of OT-CRDS.

### Acknowledgements

The authors wish to acknowledge the support of this work by the Shiraz University Research Council.

### References

1. Bao XP, Zhou YH, Yu JH (2010) N-Salicyloyltryptamine: An efficient fluorescent turn-on chemosensor for F− and AcO− based on an increase in the rigidity of the receptor. Luminescence 130: 392-398.

2. Kaur N, Kaur S, Kaur A, Saluja P, Sharma H, et al. (2014) Nanoparticle-based, organic receptor coupled fluorescent chemosensors for the determination of phosphate. Luminescence 145: 175-179.

3. Worsfold PJ, Gimbert LJ, Mankasingh U, Omaka ON, Hanrahan G, et al. (2005) Sampling, sample treatment and quality assurance issues for the determination of phosphorus species in natural waters and soils. Talanta 66: 273-293.

4. Mulkerrins D, Dobson A, Colleran E (2004) Parameters affecting biological phosphate removal from wastewaters. Environment International 30: 249-259.

5. Berchmans S, Issa TB, Singh P (2012) Determination of inorganic phosphate by electroanalytical methods: A review. Anal Chim Acta 729: 7-20.

6. Kawasaki H, Sato K, Hasegawa JOY, Yuki H (1989) Determination of inorganic phosphate by flow injection method with immobilized enzymes and chemiluminescence detection. Anal Biochem 812: 366-370.

7. Kwan RCH, Leung HF, Hon PYT, Cheung HCF, Hirota K, et al. (2005) Amperometric biosensor for determining human salivary phosphate. Anal Biochem 343: 263-267.

8. Larsen MJ, Jensen AF, Madsen DM, Pearce EI (1999) Individual variations of pH, buffer capacity, and concentrations of calcium and phosphate in unstimulated whole saliva. Archiv Oral Biolog 44: 111-117.

9. Tobey SL, Anslyn EV (2003) Determination of inorganic phosphate in serum and saliva using a synthetic receptor. Organic Lett 5: 2029-2031.

10. Carey C, Vogel G (2000) Measurement of calcium activity in oral fluids by ion selective electrode: method evaluation and simplified calculation of ion acitivity products. Res Nation Instit Standard Technol 105: 267-274.

11. Gutiérrez OM, Isakova T, Enfield G, Wolf M (2011) Impact of poverty on serum phosphate concentrations in the third national health and nutrition examination survey. Renal Nutrition 21: 140-148.

12. Mesquita RBR, Ferreira MTSOB, Tóth IV, Bordalo AA, McKelvie ID, et al. (2011) Development of a flow method for the determination of phosphate in estuarine and freshwaters-Comparison of flow cells in spectrophotometric sequential injection analysis. Anal Chim Acta 701: 15-22.

13. Davey DE, Mulcahy GR (1990) O'Connell, Flow-injection determination of phosphate with a cadmium ion-selective electrode. Talanta 37: 683-687.

14. Gale PhA, Hursthouse MB, Light ME, Sessler JL, Warriner CN, et al. (2001) Ferrocene-substituted calix 4 pyrrole: a new electrochemical sensor for anions involving CH□ anion hydrogen bonds. Tetrahedron Lett 42: 6759-6762.

15. Lara FJ, García-Campaña AM, Aaron JJ (2010) Analytical applications of photoinduced chemiluminescence in flow systems-A review. Anal Chim Acta 679: 17-30.

16. Grabner EW, Vermes I, König KH (1986) A phosphate-sensitive electrode based on BiPO modified glassy carbon. Electroanal Chem Interf Electrochem 214: 135-140.

17. Quintana JB, Rodil R, Reemtsma (2006) Determination of phosphoric acid mono-and diesters in municipal wastewater by solid-phase extraction and ion-pair liquid chromatography-tandem mass spectrometry. Anal Chem 781: 644-1650.

18. Zyryanov GV, Palacios MA, Anzenbacher P (2007) Rational Design of a Fluorescence-Turn-On Sensor Array for Phosphates in Blood Serum. Angewandt Chem International 46: 7849-7852.

19. Zhang JZ, Chi J (2002) Automated analysis of nanomolar concentrations of phosphate in natural waters with liquid waveguide. Environmen Sci Technol 36: 1048-1053.

20. Gilbert AT Jenkins S, Browning S (2009) Development of an amperometric assay for phosphate ions in urine based on a chemically modified screen-printed carbon electrode. Anal Biochem 393: 242-247.

21. De Marco R, Clarke G, Pejcic B (2007) Ion-Selective Electrode Potentiometry in Environmental Analysis. Electroanal 19: 1987-2001.

22. Parra A, Ramon M, Alonso K, Sh Lemos G, Vieira EC (2005) Nogueira, Flow injection potentiometric system for the simultaneous determination of inositol phosphates and phosphate: phosphorus nutritional evaluation on seeds and grains. Agricultur Food Chem 53: 7644-7648.

23. Ganjali MR, Norouzi P, Qomi M, Salavati-Niasari M (2006) Highly selective and sensitive monohydrogen phosphate membrane sensor based on molybdenum acetylacetonate. Anal Chim Acta 567: 196-201.

24. Akyilmaz E, Yorganci E (2007) Construction of an amperometric pyruvate oxidase enzyme electrode for determination of pyruvate and phosphate. Electrochim Acta 52: 7972-7977.

25. Kwan RCH, Leung HF, Hon PYT, Barford JP, Renneberg R (2005) A screen-printed biosensor using pyruvate oxidase for rapid determination of phosphate in synthetic wastewater. Appl Microbiol Biotechnol 66: 377-383.

26. Villalba MM, McKeegan KJ, Vaughan DH, Cardosi MF, Davis J (2009) Bioelectroanalytical determination of phosphate: A review. Molecul Catal B: Enzym. 59: 1-8.

27. Preechaworapun A, Dai Z, Xiang Y, Chailapakul O, Wang J (2008) Investigation of the enzyme hydrolysis products of the substrates of alkaline phosphatase in electrochemical immunosensing. Talanta 76: 424-431.

28. Mousty Ch, Cosnier S, Shan D, Mu Sh (2001) Trienzymatic biosensor for the determination of inorganic phosphate. Anal Chim Acta 443: 1-8.

29. Gámiz-Gracia L, García-Campaña AM, Huertas-Pérez JF, Lara FJ (2009) Chemiluminescence detection in liquid chromatography: Applications to clinical, pharmaceutical, environmental and food analysis-A review. Anal Chim Acta 640: 7-28.

30. García-Campaña AM, Lara FJ, Gámiz-Gracia L, Huertas-Pérez JF (2009) Chemiluminescence detection coupled to capillary electrophoresis. TrAC Trend. Anal Chem 28: 973-986.

31. Liu M, Lin Z, Lin JM (2010) A review on applications of chemiluminescence detection in food analysis. Anal Chim Acta 670: 1-10.

32. Browner RF, Boorn AW (1984) Sample introduction techniques for atomic spectroscopy. Anal Chem 56: 875A-888A.

33. Signorell R, Reid JP (2010) Fundamentals and Applications in Aerosol Spectroscopy. CRC Press.

34. Kulkarni P, Baron PA, Willeke K (2011) Aerosol Measurement: Principles, Techniques, and Applications. John Wiley and Sons.

35. Tang Y, Huang Y, Li L, Chen H, Chen J, et al. (2014) Characterization of aerosol optical properties, chemical composition and mixing states in the winter season in Shanghai, China. Environment Sci 26: 2412-2422.

36. Ashkin A, Dziedzic J (1975) Optical levitation of liquid drops by radiation pressure. Science 187: 1073-1075.

37. Doroodmand MM, Shafiee Z (2014) Design of a new flame-containing molecular emission cavity for speciation of sulfide, sulfite, sulfate, thiosulfate and thiocyanate in wastewater: catalytic behaviour of hydrogen ion. Internation J Environment Anal Chem 94: 1223-1242.

38. Masson P, Morel Ch, Martin E, Oberson A, Friesen D (2001) Comparison of soluble P in soil water extracts determined by ion chromatography, colorimetric, and inductively coupled plasma techniques in PPB range. Communicati Soil Sci Plant Anal 32: 2241-2253.

39. Liu X, Gu J, Li Y, Cheng Y, Qu Y, et al. (2013) Increase of aerosol scattering by hygroscopic growth: Observation, modeling, and implications on visibility. Atmosphere Res 132: 91-101.

40. Stelmaszczyk K, Fechner M, Fechner M, Rohwetter P, Queiber M, et al. (2009) Towards supercontinuum cavity ring-down spectroscopy. Appl Phys B 94: 369-373.

41. Stelmaszczyk K, Rohwetter Ph, Fechner M, Queiber M, CzySewski A, et al. (2009) Cavity ring-down absorption spectrography based on filament-generated supercontinuum light. Optics express 17: 3673-3678.

42. Nakaema WM, Hao ZQ, Rohwetter Ph, Wöste L, Stelmaszczyk K (2011) PCF-based cavity enhanced spectroscopic sensors for simultaneous multicomponent trace gas analysis. Sensors 11: 1620-1640.

43. Granados-Muñoz MJ, Navas-Guzmán F, Bravo-Aranda JA, Guerrero-Rascado JL, Lyamani H, et al. (2014) Hygroscopic growth of atmospheric aerosol particles based on active remote sensing and radiosounding measurements. Atmosphere Measur Techniqu Discuss 7: 10293-10326.

44. Saha MN (1920) Ionization in the solar chromosphere, London Edinburgh Dublin Philosophical Magazin. J Sci 40: 472-488.

45. Fridman A (2008) Plasma Chemistry. Cambridge University Press.

46. van de Sanden MCM, Schram PPJM, Peeters AG, van der Mullen JJAM, Kroesen GMW (1989) Thermodynamic generalization of the Saha equation for a two-temperature plasma. Phys Rev A 40: 5273.

47. Ruiz-Calero V, Galceran M (2005) Ion chromatographic separations of phosphorus. species: a review. Talanta 66: 376-410.

48. Ma J, Yuan D, Liang Y (2008) Sequential injection analysis of nanomolar soluble reactive phosphorus in seawater with HLB solid phase extraction. Marin Chem 111: 151-159.

49. Frank C, Schroeder F, Ebinghaus R, Ruck W (2006) A fast sequential injection analysis system for the simultaneous determination of ammonia and phosphate. Microchimica Acta 154: 31-38.

50. Mas F, Muñoz A, Estela JM (2000) Estela Simultaneous spectrophotometric determination of phosphate and silicate by a stopped-flow sequential injection method. International Journal of Environmental Analytical Chemistry 77: 185-202.

51. Rahman Md, Park DS, Chang S, McNeil CJ, Shim YB (2006) The biosensor based on the pyruvate oxidase modified conducting polymer for phosphate ions determinations. Biosens Bioelectron 21: 1116-1124.

52. Samy R, Faustino PJ, Adams W, Yu L, Khan MA, et al. (2010) Development and validation of an ion chromatography method for the determination of phosphate-binding of lanthanum carbonate. Pharmaceut Biomed Anal 51: 1108-1112.

# 12

# Mechanistic Evaluation of Matrix Effect on Three Different Model of Mass Spectrometer by Using a Model Drug

Poonam Vats[1]*, S Manaswita Verma[2] and Tausif Monif[1]

[1]*Department of Clinical Pharmacology and Pharmacokinetics, Sun Pharmaceutical Industries Ltd, Gurgaon-122 015, Haryana, India*
[2]*Department of Pharmaceutical Sciences, Birla Institute of Technology, Mesra, Ranchi-835 215, Jharkhand, India*

## Abstract

In this research work we have tried to emphasis that the synergistic approach is required wherein during method development sample processing and instrument i.e., mass spectrometer shall be evaluated in synchronized manner to get the best possible combination of processing technique and instrument model to get a method with no or minimal matrix effect. To perform this research work polar molecule niacin is considered as a model drug to evaluate the synergistic approach. Plasma was treated with three different conventional sample preparation procedures (PPT, LLE and SPE) and samples processed with the three methodology were then analyzed on three different instrument models of AB Sciex i.e., API-3000, API-3200, API-4000. Except for type of sample processing technique and instrument model, rest of the parameters like aliquot volume, internal standard working solution volume and chromatographic conditions were kept constant to avoid contribution of these factors due to these variables. Samples were analyzed using Inertsil® CN-3 as an analytical column and mobile phase consist of acetonitrile - solution-1(0.002% formic acid in water, v/v) in the ratio of 70:30, v/v. Result evidently showed that matrix effect was minimized through SPE technique over LLE technique and behavior of the co-elute matrix also changed significantly with ion-source design of the mass spectrometer. Consequently, the development result clearly showed that matrix effect was nullified by samples prepared by SPE technique and analyzed on API-3000. Results obtained from API-3000 and API-4000 models were also comparable in terms of matrix effect. API-3000 in combination with solid phase extraction procedure was selected to further validate the method as API-4000 showed charge competition of internal standard. Overall, the results indicate that extraction procedure plays a crucial role to control matrix effect but to get the best result LC have to be coupled with mass spectrometer with proper ion source.

**Keywords:** Matrix effect; Process efficiency; Sample extraction technique; LC-MS/MS; Instrument model

## Introduction

Liquid chromatography-tandem mass spectrometer (LC-MS) systems using an electrospray ion source coupled with tandem mass analyzers (LC-ESI-MS/MS) have been applied to a wide variety of studies in pharmaceutical analysis and life sciences. With LC-ESI-MS/MS now considered the benchmark for measurement of drugs and their metabolites in biological matrices, the high selectivity of tandem mass spectrometry, with successive mass filtrations, leads to little or no observed interference even though there may be relatively high concentrations of co-extracted and co-eluted matrix components present. These characteristics have led to a growing trend of high-throughput analysis that incorporates little or no sample preparation and minimal chromatographic retention. Moreover, LC-MS has unprecedented capabilities especially for the molecules which are incompatible with GC-MS due to high polarity, high mass and thermo labile in nature and allowed for the elimination of derivatisation steps prior to injects, saving on time and reagent cost. With LC-ESI-MS/MS having these characteristics of high selectivity, sensitivity and throughput, it is not surprising that this technology is being increasingly used in the clinical laboratory. Though it is considered as a powerful tool for the quantitation of drug at a very low level, it s selectivity was challenged by Tang et al. [1] while studying the effect of conductivity and ion intensity in electro spray. They coined the term matrix effect, wherein along with the analyte of interest other endogenous matter of matrix also ionize and results in suppression or enhancement of response of analyte. Thereby accuracy of the data is compromised. Therefore ion-suppression and/or ion-enhancement due to matrix effect has been one of the major unknown variables of concern that could adversely affect the accuracy and precision of the assay results for biological sample assay in LC-MS/MS methods [2].

Eeckhaut et al. [3], in their review article on matrix effect discussed in detail about the assessment of matrix effect, parameters like sample preparation technique, chromatographic conditions and mass spectrometric conditions that needs to be optimized to the best to eliminate matrix effect. In their review article they emphasised that during validation of any bioanalytical technique, quantitative assessment of matrix effect is must to ensure precision and accuracy of data. Large amount of literature is available on the factors which causes the occurrence of matrix effect in the developed method. These include endogenous substances in the matrix [4,5], exogenous materials such as buffers, ion pairing agents [6], co-medication [7], type of soft ionization technique i.e., atmospheric pressure chemical ionization (APCI) or electro spray ionization (ESI), source design [8]. At the same time, number of approaches was reported to minimize or to eliminate the matrix effect while developing the methods in biological matrix. These include change in chromatographic conditions [9,10], sample preparation techniques [11,12], selection of ionization source and polarity [13-15], stable isotope labeled- internal standard, decrease the amount of sample [16].

Therefore upon literature review, it was evident that to control the matrix effect various parameters which included sample processing

*Corresponding author:* Poonam Vats, Department of Clinical Pharmacology and Pharmacokinetics, Sun Pharmaceutical Industries Ltd, Gurgaon-122 015, Haryana, India, E-mail: poonam.vats@sunpharma.com

techniques, the mode of ionization, and optimization of pretreatment of sample have been investigated, but the role of instrument model based on source design and the sensitivity of instrument have not been reported till date. In this research work, we investigated the role of conventional sample processing techniques and instrument models in controlling the matrix effect. Samples processed by conventional sample processing techniques (i.e., PPT, LLE and SPE) were analyzed on different models of mass spectrometer (i.e., API-3000, API-3200 and API-4000).

Based on the level of sensitivity, different models of instruments have been designed. This research work involves comparison of API-3200, API-3000 and API-4000 models of mass spectrometer from AB Sciex [17]. The orders of sensitivity of the machines are: API-4000>API-3000>API-3200. API-4000 and API-3200 has the same source design and API-3000 has different source design. This difference in source designs also contributes to the matrix effect. In some cases, problem of matrix effects could be resolved by using a MS instrument of other model or manufacturer [18,19]. In today's scenario, many regulatory bodies ask to demonstrate that there will not be any matrix effect if two different models of MS with different source designs are being used like API-3200 and API-3000 of sciex in analysis of single compound or drug.

In some research work, role of ionization source was also studied. A detailed review of literature suggested that the best approach to eliminate or minimize matrix effect in LC-MS/MS include use of APCI ionization source [5,20-22] for analysis. AS per the study done by Liang et al. [23] being a less sensitive source, matrix effect observed on APCI is also less. However, the APCI source has its own boundaries for use as it cannot be used for thermo labile molecules and has a low sensitivity. In case of the model drug selected for research, satisfactory sensitivity could not be attained on the APCI source. ESI source is the workhorse for routine bioanalysis in industries. Therefore research was conducted using the ESI source on which required sensitivity could be attained for the molecule.

In this research work, we emphasized that there is no generic method for detecting and eliminating the matrix effect. For each molecule it needs to be investigated independently. Consequently, while developing bio-analytical methodology for any molecule, optimization and selection of sample processing methodology and instrument model and ionization technique shall be done in combination. Synergistic approach is required to evaluate matrix effect wherein all the parameters need to be optimized in combination and not individually.

### Rational for selection of model drug

It is reported that polar compound exhibits higher degree of matrix effect in reverse phase chromatography [24]. Therefore, niacin is selected as a model drug in this research work, as it is polar in nature. Niacin is easily ionizable because of which its extraction procedure could be easily developed with multiple processing techniques. The required sensitivity for this analyte could also be achieved on all the three instrument models. To develop a bio-analytical method in reverse chromatography for polar molecule with minimal or no matrix effect is a challenging task. Since in the reverse phase chromatography, co-elution of phospholipids may also take place with similar retention time of target analyte. This is probably the cause for higher value of matrix effect for polar molecule in reverse phase chromatography. For development an assay method for niacin, reverse phase chromatography is selected due to shorter analysis time. As matrix effect is mainly due

to endogenous phospholipids, SPE techniques helps in the effective cleaning of the sample when compared to other techniques because it selectively helps to remove the phospholipids from SPE cartridge by washing and extract the analyte of interest during extraction.

## Experimental

### Chemicals and materials

Working standards of Niacin and Nicotinic-d4 acid (internal standard, IS structure shown in Figure 1) were purchased from USP and CDN isotopes respectively. All reagents used were of ACS grade or higher, with solvents of HPLC grade or higher. Acetonitrile were procured from Fluka (Sigma-Aldrich, steinheim, USA). Milli-Q water (Millipore, Moscheim Cedex, France) was used in the preparation of solutions. Blank human plasma containing lithium heparin was obtained from Panexcell clinical lab private limited, Navi Mumbai, India.

### LC-MS/MS instrumentation and chromatographic conditions

**Chromatographic conditions:** LC-MS/MS analysis was performed using Shimadzu Prominence LC system (Shimadzu Corporation, Kyoto, Japan) consisting of two delivery pumps (LC-20AD), an on-line solvent degasser (DGU-20A3), an auto sampler (SIL-HTc) and a column oven (CBM-20A) coupled to AB Sciex mass spectrometer. Chromatographic separation was carried out on a Shimadzu scientific instrument (Shimadzu Corporation; Kyoto, Japan) with Inertsil® CN-3 column (100 × 4.6 mm, 5 μm). A mobile phase consisting of acetonitrile and solution-1 (0.002% Formic acid water, v/v) in the ratio of 70:30, v/v was delivered at a flow rate of 1.0 mL/min with 25% split ratio. The total analysis run time for each sample was 7.0 min. The column oven and auto sampler temperatures were maintained at 45 ± 1°C and 10 ± 1°C, respectively.

**Mass spectrometric conditions:** Samples prepared by different techniques were analyzed with API-3200; API-3000 and API-4000 triple quadrupole mass spectrometer (MDS Sciex, Toronto, Canada) equipped with an electrospray ionization source operating in positive polarity. Ultra high purity nitrogen was used as the nebulizer, auxiliary, collision and curtain gases. Analytes were monitored by MRM transitions of m/z 124.1/ 80.1 for niacin and 128.1/84.1 for nicotinic-d4 acid with dwell time of 400 millisec for both. Data acquisition and processing were performed using Analyst version 1.4.2 software (MDS Sciex, Toronto, Canada). Product ion spectra are shown in Figure 2 for niacin and IS. The parameters optimized on three models of instruments i.e., API-3000; API-3200 and API-4000 by infusing solution of niacin and nicotinic-d4 acid into the mass spectrometer, are as follows: On API-3200/4000 - collision activated dissociation gas (CAD): 10, curtain gas (CUR): 25, nebulizer gas (GS1): 50 and heater gas (GS2): 55, ion spray voltage: 5500 V, source temperature: 500°C; declustering potential (DP): 45 V, entrance potential (EP): 10

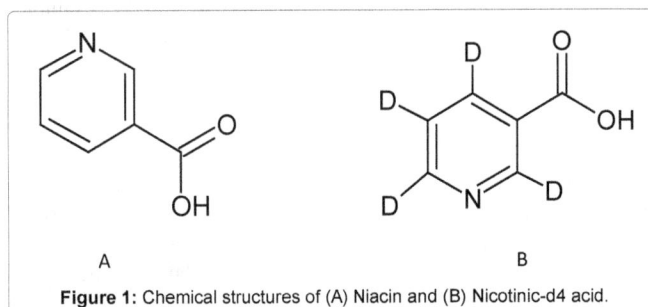

**Figure 1:** Chemical structures of (A) Niacin and (B) Nicotinic-d4 acid.

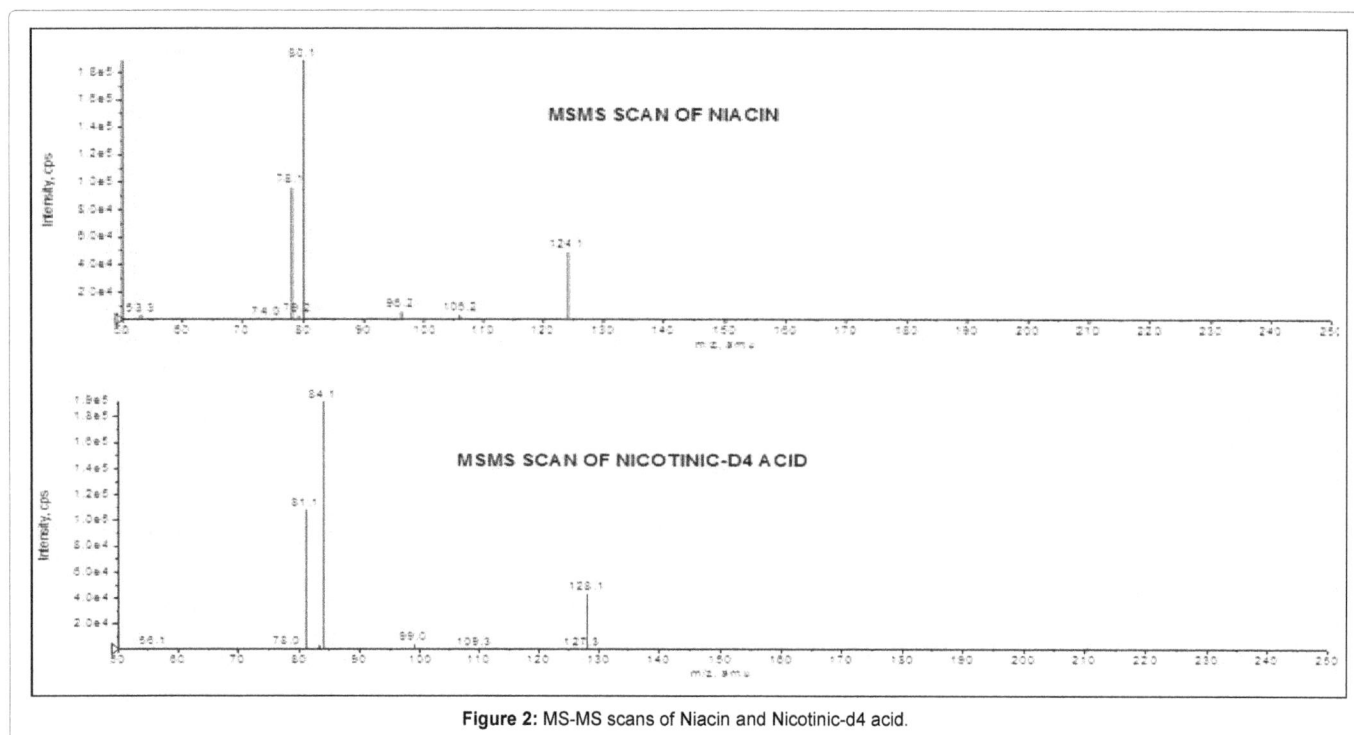

**Figure 2:** MS-MS scans of Niacin and Nicotinic-d4 acid.

V, collision energy (CE): 11 V and collision cell exit potential (CXP): 2 V for both niacin and nicotinic-d4 acid.

On API-3000 - CAD: 8, CUR: 15 and nebulizer gas (NEB): 8, ion spray voltage: 5500 V, source temperature, 475°C, DP: 50 V, EP: 10 V, Focusing Potential (FP): 120 V, CE: 30 V and CXP: 14 V for both niacin and nicotinic-d4 acid.

In API-3200 and API-4000, mass parameters were kept same as both the instruments have identical turbo-V source design. Since API-3000 has different source design, mass parameters were optimized independently to get the desired response.

## Sample preparation

**Preparation of stock solutions, calibration standards and quality control samples:** Stock solutions were prepared in dimethyl sulfoxide - acetonitrile (50: 50, v/v) at concentration of 5 mg/mL and 1 mg/mL for Niacin and Nicotinic-d4 acid, respectively. Working solutions of niacin prepared in methanol-water (50:50; v/v), were used to prepare calibration standards (CS) at eight different concentrations (50.6-25022.7 ng/mL) and at four concentrations (50.7, 129.9, 9992.7 and 19985.4 ng/mL) for quality controls (QC) by 2% spiking in human plasma. Blank human lithium heparin plasma was screened prior to spiking to ensure that it was free from endogenous interference at retention times of niacin and IS. The (bulk) spiked CS and QC samples were stored below -50°C and protected from light until analysis. The IS working solution (500.0 ng/mL) was prepared in methanol-water (50:50, v/v).

## Experimental design

The main aim of this experimental design was to evaluate the role of sample processing and instrument model on Matrix effect (flow diagram, Figure 3) in synergy. So, selected parameters like recovery (process efficiency), absolute matrix effect and relative matrix effect performed according to USFDA [25], EMEA [26] guidelines with described sample processing techniques and analyzed on different models of mass spectrometry.

## Matrix effect

Matrix effects are of two types, absolute and relative. The two main techniques used to determine the degree of matrix effects (or absolute matrix effect) qualitatively as well as quantitatively on an LC-ESI-MS/MS method are post-column infusion and post-extraction addition method respectively.

**Post-column infusion method:** Matrix ion suppression effects on the MRM LC-MS/MS sensitivity were evaluated by the post-column analyte infusion experiment. A reference solution containing 500.0 ng/mL of niacin and IS in mobile phase was infused post-column via a "T" connector into the mobile phase at 10 μL/min, employing an infusion pump. Aliquots of 10 μL of extracted blank plasma were then injected into the column by the auto sampler, and MRM LC-MS/MS chromatograms were acquired for niacin and IS. Any dip in the baseline upon injection of blank plasma would indicate ion suppression.

**Post extraction addition technique:** As extraction protocol involves a terminal drying step, hence spiking (addition of reference sample) was carried out in post-extracted blank plasma (normal, hemolyzed plasma and lipemic plasma; 6 lots of each type of plasma) sample to perform matrix effect. The concentration of both niacin and IS in reference sample representing the final extracted concentrations in QC samples (at LQC, MQC and HQC level). The control sample (neat sample) was a reference solution prepared at an appropriate concentration in mobile phase.

Absolute matrix effect (ion suppression / ion-enhancement) was determined by the following equation:

The positive value represents the % of ion suppression and the negative value represents the % of ion enhancement.

For calculating the IS-normalized matrix effect of the method, peak area ratio (analyte peak area/ IS peak area) was considered instead of

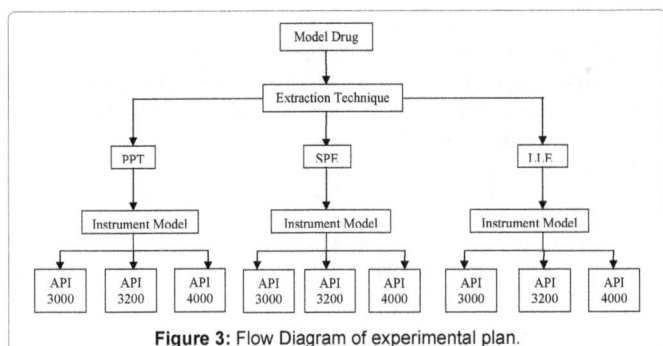

**Figure 3:** Flow Diagram of experimental plan.

peak area of the analyte and by following equation it was calculated. This can be a good indicator of variability contributed by the internal standard.

$$\text{IS – normalized matrix effect} = \left[ \frac{\text{Mean peak area ratio in post extracted sample}}{\text{Mean peak area ratio in neat solution}} \right]$$

This approach can be useful for evaluating matrix effect during clinical sample analysis, as variation in area response of internal standard can be used as a parameter to assess matrix effect in biostudy.

**Relative matrix effect:** The more important parameter in the evaluation of a bio-analytical method in biological matrix is the demonstration of the absence of "relative matrix effect". The word relative referring to the comparison of matrix effect values between different lots (source) of biological matrices. Relative matrix effect was evaluated using eighteen different lots of human lithium heparin plasma including six different lots of hemolyzed and six different lots of lipemic plasma, processed in duplicate samples at limit of quantification quality control (LOQQC) and HQC levels and the area ratio (i.e., peak area response of analyte / peak area response of IS) was used to check the acceptability of the result. The standard deviation for each lot was calculated along with % CV and % accuracy at each level. The precision (%CV) at HQC level and LOQQC level from the nominal concentration was expected to be <15.0 and <20.0 respectively. Similarly the mean accuracy at HQC level should be within ± 15.0% and for LOQQC it should be ± 20.0% of the nominal concentration.

**Process efficiency:** Process efficiency (PE) or recovery was determined by measuring the mean peak area response of six replicates of extracted QC samples (at LQC, MQC, and HQC level) against the mean peak area response of neat aqueous solution. PE of niacin was estimated by using the following equation:

$$\% \text{ PE} = \frac{\text{Mean peak area response of analyte in extracted samples}}{\text{Mean peak area response of analyte in neat sample}} \times 100$$

**Acetonitrile extraction:** For sample preparation, 100 µL of plasma and 50 µL of IS dilution (500.0 ng/mL), 750 µL of 1% formic acid in Acetonitrile was added. This mixture was then vortexed for 30 s and centrifuged (5 min at 4000 rpm) for precipitation of proteins. The clear supernatant was separated and then dried at 50°C under a stream of nitrogen at 20 psi. The dried residue was reconstituted individually with 500 µL of mobile phase and 10 µL of the reconstituted sample was injected into LC-MS/MS system for analysis. Since samples were precipitated by using acetonitrile, term acetonitrile extraction has been used.

**Liquid-liquid extraction:** Plasma sample (100 µL) was aliquoted in polypropylene tube (16 × 125 mm) and 50 µL of the IS working solution (500.0 ng/mL) was added and 750 µL of 1% formic acid in water was added and vortexed (approximately for 30 s), followed by

the addition of 3 mL of an extraction solvent (ethyl acetate). The sample was extracted on a reciprocating shaker at 100 rpm for 30 min. After centrifugation at 4000 rpm for 5 min, the aqueous layer was frozen in a dry ice-methanol bath. The organic layer was decanted into a glass tube (13 × 100 mm) and evaporated to dryness under nitrogen stream at 50°C. The dried residue was reconstituted individually with 500 µL of mobile phase and 10 µL of the reconstituted sample was injected into LC-MS/MS system for analysis.

**Solid phase extraction:** 100 µL of plasma sample was pipetted into polypropylene tubes (12 × 75 mm) and 50 µL of ISTD working solution (500.0 ng/mL of IS) was added with the use of multistepper. Samples were vortexed approximately for 30 s. Samples were pretreated with 750 µL of 1% formic acid solution and vortexed again (approximately for 30 s). Before extraction, the sorbent of the extraction cartridge (Oasis⁺ MCX, 30 mg/1 cc) was conditioned with 1 mL of 10% liquor ammonia in acetonitrile followed by 1 mL of solution-1. Then, the pretreated samples were loaded onto the cartridge (Oasis⁺ MCX, 30 mg/1 cc) and spun in centrifugation at 1500 rpm (or 453 g) for 1 min at 2-10°C. The cartridges were washed with 1 mL of 1% formic acid followed by 1 mL of methanol. Compounds were then eluted with 1 mL of 10% liquor ammonia in acetonitrile solution. Extraction was performed on refrigerated centrifuge. After each addition, samples were centrifuged at 1500 rpm for 1 minute at 2-10°C. The extracted samples were evaporated to dryness at 20 psi and 50°C under a stream of dry nitrogen using a Zymark TurboVap LV evaporator (Caliper, Hopkinton, MA, USA). The dried residue was reconstituted individually with 500 µL of mobile phase and 10 µL of the reconstituted sample was injected into LC-MS/MS system for analysis.

## Results and Discussion

In this experimental work our aim was systematic assessment of matrix effect by evaluating conventional sample processing techniques and different API instrument models in combination and strategic steps to be followed to nullify the matrix effect. During conduct of the experiment, samples processed with all the three processing techniques were analyzed on all the three instrument models. Results were then compared across the models and sample processing techniques to mark out the best combination with respect to process efficiency and matrix effect. Using a Stable isotope labeled internal standard is the best way to control or eliminate matrix effect. So, Nicotinic-d4 acid was selected as an internal standard.

### Optimization of mass parameters

Initially mass parameters were tuned in APCI and ESI ion sources, but inadequate response was observed in APCI ion source and desired sensitivity could not be achieved for niacin in all the API models under the developed chromatographic conditions. In ESI source, niacin and IS formed protonated molecules [M+H]⁺. Several fragment ions were observed in the product ion spectra of both niacin as well as IS. Fragment ion 80.1 was selected for niacin and 84.1 for nicotinic-d4 acid as these ions are in abundant, selective and produced stable response (Figure 2). During mass parameter optimization it was observed that mass parameters CE and CAD played vital role in achieving highest sensitivity with stable response for niacin.

### Optimization of chromatographic conditions

Chromatographic analyses of niacin and IS was carried out under isocratic conditions to obtain adequate response, sharp peak shape, and a shorter analysis time. The use of volatile buffers like ammonium formate and ammonium acetate (in combination of methanol-acetonitrile) for the separation of niacin had been also evaluated. It was observed that the pH of mobile phase and selection of column were critical parameters. Chromatographic separation was tried using various combinations of methanol-acetonitrile, acidic buffers and additives (like formic acid, glacial acetic acid and liquor ammonia solution) on different C18, C8, cyano and phenyl reversed phase columns with 5 μm particle size (viz., several like Xterra C18, Chromolith RP-18, Atlantis HILLIC, Ascentis C8 Poroshell 120 EC-C18, Ascentis express, Sunshell C18, Kinetex C18, Hypurity advance, Zorbax SB-C18, Zorbax SB-CN, Discovery C18, Unisol C18, Luna C18(2), kinetex PFP and ACE C18 PFP). The analytes showed nonlinear behavior on Chromolith RP-18 column while HILLIC column was marked unsuitable due to co-eluting matrix compounds especially with haemolysed plasma samples. In most of the columns, the endogenous peak (adjacent to the peak of interest) merged with the target analyte i.e., niacin. Column with fused core technology was also evaluated to attain resolution between niacin and the endogenous peak, but due to high back pressure these column could not be used.

Inertsil˙ CN-3 column provided good peak shape for niacin with acetonitrile as organic component in combination with formic acid solution in mobile phase. Niacin and the endogenous compound were chromatographically well separated on Inertsil˙ CN-3 (150 × 4.6 mm, 5 μm) column with high S/N ratio for niacin. This could be due to lower carbon loading of the column, enabling the selectivity by base material of the column and lower carbon load reduces RT of niacin and increased high throughput. Additionally, column oven temperature was optimized to 45°C. With the increase in temperature to 45°C, resolution of peak of analyte from the other endogenous substances improved further.

## Control of variables

In order to draw the accurate results, except for the extraction techniques and instrument models, other parameters which includes volume of plasma (aliquot volume); volume of addition for IS working solution, the concentration of IS working solution, pretreatment of the samples prior to extraction; final solvent composition of reconstituted extract, final reconstitution volume of the extracted samples, model of HPLC (Shimadzu) and LC parameters (like mobile phase, flow rate, analytical column, injection volume and rinsing solution) to be used on three different API models i.e., API-3200, API-3000, API-4000 for analysis were kept constant across analysis. Here, various variable factors had to be controlled to ensure precise comparison of results. Therefore for each extraction procedure, the equal volume of plasma (100 μL) was aliquoted, so that the starting endogenous matrix levels are identical in all methods; pretreatment was kept exactly same so that sample constitution is same prior to extraction; reconstitution composition is chosen same for all methods, to solubilize both highly polar and non-polar compounds, dried extract reconstituted with same volume (500 μL) so that the final concentration of endogenous matter is same across different processing techniques, injection volume was kept constant so that equal volume of sample is loaded onto the analytical column. Same analytical column was used on three instruments so that chromatographic separation and retention time is also same across instrument models. Chromatographic conditions like rinsing solution (acetonitrile-methanol-water, 60:20:20, v/v/v) composition and mobile phase composition was also kept alike on all three models for identical chromatographic separation.

## Optimization of sample preparation

As the purpose of this research is to evaluate the role of sample processing technique on matrix effect, initially individual method was optimized for each extraction technique which gives better results in terms of higher process efficiency and less matrix effect. In general, high throughput sample analysis has led the common practice of preparing samples by the simplest, fastest method possible, which often means using PPT. Although the PPT is quick and easy step to extract the target analyte from the biological matrix but fails to effectively remove enough of the endogenous plasma components (i.e., phospholipids).

During extraction step optimization, different protein precipitating agents like ACN, MeOH, acidic MeOH, basic MeOH, acidic ACN, and basic ACN were evaluated to increase the process efficiency. It was observed that in acidic condition, the process efficiency of niacin is increased but high value of % ion suppression (Table 1) was also noted. High value of % ion suppression leads to significant decrease in the S/N ratio of niacin and observed S/N value was less than 3 which is not an acceptable value as per current regulatory guidelines. Hence, further matrix evaluation exercises were not carried out with PPT technique as it was evident from the obtained results that PPT will lead to massive matrix effect. Besides this, samples processed by PPT technique were not clean and injection of these samples onto analytical column resulted in fluctuation of area response and retention time of analytes. This fluctuation might be due to the deposition of phospholipids on the analytical column and its impact on the subsequent analysis.

Niacin being a polar compound and zwitterionic in nature, liquid-liquid extraction was a method of choice as it can be easily be extracted out in any polar solvent [27,28]. Initially for LLE, ethyl acetate, tertiary butyl methyl ether (TBME), diethyl ether was tried to check the extraction efficiency for higher recovery. From the Table 2 data, we concluded that due to polar nature of niacin, the process efficiency is maximum in ethyl acetate which is also polar in nature. To further enhance the process efficiency and minimize the matrix

| Liquid-liquid extraction[a] | | | Protein precipitation | |
|---|---|---|---|---|
| Extraction condition | PE (%) | % Ion Suppression | Precipitating agent | % Ion Suppression |
| 500 μL of water | 48.9 | 32.2 | Acetonitrile | 62.6 |
| 500 μL of 1% formic acid in water (v/v) | 67.2 | 18.6 | 1% formic acid in acetonitrile (v/v) | 56.4 |
| 500 μL of 1% Liq. Ammonia in water (v/v) | 57.6 | 24.1 | 1% Liq. Ammonia in acetonitrile (v/v) | 71.3 |
| 750 μL of 1% formic acid in water (v/v) | 74.7 | 12.1 | Methanol | 52.7 |
| | | | 1% formic acid in methanol (v/v) | 76.1 |
| | | | 1% Liq. Ammonia in methanol (v/v) | 76.1 |

[a]By using ethyl acetate based on results from Table 1.

**Table 1:** Process efficiency and matrix effect of niacin using LLE and PPT under different extraction conditions at LQC level analyzed on model API-3000.

| Extracting solvent | Process Efficiency (%) |
|---|---|
| Ethyl acetate | 65.4 |
| Di ethyl ether | 48.6 |
| Tertiary butyl methyl ether | 41.3 |

Preparation of sample: 500 μL of 1% formic acid in water solution as pretreatment for 100 μL of plasma.

**Table 2:** Optimization of extracting solvent for liquid-liquid extraction based on process efficiency at LQC level analyzed on model API-3000.

effect composition of pretreatment solution was optimized. From various trials it was observed that sample pretreatment should be done with 750 µL of 1% formic acid in water (v/v) to get higher recovery and less matrix effect. Addition of acidic solution during sample preparation step caused the ionization of phospholipids and the ionized phospholipids (hydrophilic nature) remains in the aqueous layer which leads to minimized matrix effect. Finally LLE technique were developed using ethyl acetate as extracting solvent and formic acid solution as a pretreatment solution.

During optimization of SPE technique different cartridges like reverse phase and polymer based were evaluated to get the higher recovery. Since Niacin is polar compound with acidic nature, ion exchange cartridges were preferred over normal polymer or silica based cartridges. Oasis MCX cartridge particularly has given better results as the target analyte is polar in nature and easily ionizable with acidified pretreatment solution, matrix effect was nullified but required S/N ratio at limit of quantification (LOQ) level could not be achieved. So, several trials were taken for optimization of conditioning, washing and elution steps (Table 3). Finally, acidified wash solution and basic elution solution is found suitable for higher recovery and almost zero matrix effect.

All the three processing techniques were optimized on API-3000 and then bioanalytical methodologies were extended to other two models of mass spectrometer i.e., API-3200 and API-4000.

### Matrix effect and process efficiency

***Post-column infusion method:*** In post column infusion method, first we analyzed a reference solution (aqueous sample) to locate the retention time (RT) of target analyte, which is the most important factor for qualitative assessment of matrix effect by 'T' joint experiment, where any dip in the baseline upon injection of blank plasma would indicate ion suppression. Niacin and nicotinic-d4 acid were eluted from the column at the retention time (RT) region of 2.8 to 2.9 min with set chromatographic conditions. In PPT technique, severe perturbations in the response were seen with all the precipitating agents as evident by post column infusion experiment. A massive ion suppression (~75%) was observed at the retention time of niacin and nicotinic-d4 acid and also between 4.5-5.0 min using acidic acetonitrile as precipitating agent as shown in Figure 4A. Additionally, significant enhancement was also observed in the region of 2.5-3.0 min. Replacing acetonitrile with methanol resulted in the decreased response. Though considerable improvement of ion suppression was observed, when 'T' joint experiment was performed using the blank sample prepared with LLE technique (Figure 4B). The suppression zone is slightly separated from the RT of the analyte (2.83 min) and hence the absolute matrix effect is improved as compare to PPT technique. In same chromatographic conditions, when 'T' joint experiment was performed using the blank sample prepared with SPE technique (Figure 4C), no suppression zone

was observed at the RT of analyte (2.83 min). Blank plasma sample prepared by SPE technique was also analyzed on API-4000 and API-3200 instrument to perform 'T' joint experiment as shown in Figure 4D and 4E respectively. On API-3200, zone of ion-enhancement was observed at the retention time of analyte and IS (Figure 4D). Similar result was obtained for matrix suppression zone on API-3000 and API-4000 for SPE.

***Post-extraction method and absolute matrix effect:*** During initial stage of method development, significant matrix related problem were observed in post column evaluation with PPT technique and process efficiency was quite low for niacin, hence further research work was not continued using PPT technique. Absolute Matrix effect was evaluated using eighteen lots of human plasma, with the samples prepared by the extraction methodologies of all the three types and then analyzed on different instrument models by post extraction addition method. The matrix ionization suppression or enhancement of the analyte was assessed at three QC concentration levels. Upon comparison of results tabulated in Table 4 following observations were made-

a) Ion-suppression and ion-enhancement across different QC levels in each technique on each model were consistent.

b) For LLE methods ion-suppression was observed on API-3000 and API-4000, whereas on API-3200, ion-enhancement was observed.

c) With SPE method, significant decrease in matrix effect was observed on API-3000 and API-4000 when compared with the LLE method. However on API-3200, ion enhancement was still observed as it was observed with the LLE method.

d) Minimum matrix effect was observed in sample prepared by SPE technique when compared between SPE and LLE techniques. Similar trend was observed on all the models of instrument that is API-3200, API-3000 and API-4000.

e) Best combination wherein % ion-suppression within 2 was observed was when sample processed by SPE method were analyzed on API-3000.

f) Another interesting observation made was that charge competition between analyte to IS was much more on API-4000 when compared to API-3000.

Global matrix effect data for SPE and LLE method are shown in Figure 5, from where we concluded that ion suppression is minimal for developed SPE method over the LLE method and it is also concluded that API-3000 data is more promising over API-4000.

Upon comparison of results across instrument model, on API-3000 and API-4000, comparable results in terms of % ion suppression were observed. However on API-3200, substantial ion enhancement was

| Conditioning Optimization | Washing Optimization | Elution Optimization | S/N ratio | PE (%) | %Ion suppression |
|---|---|---|---|---|---|
| 0.5 mL of MeOH followed by 0.5 mL of water | 1 mL of MeOH followed by 1 mL of water | 1 mL of 10% of Liq.NH$_3$ in MeOH | Fail | 81.4 | 8.1 |
| 0.5 mL of MeOH followed by 0.5 mL of water | 1 mL of MeOH followed by 1 mL of water | 1 mL of 10% of Liq.NH$_3$ in ACN | Fail | 89.5 | 4.9 |
| 0.5 mL of MeOH followed by 0.5 mL of water | 1 mL of 1% formic acid in water (v/v) followed by 1 mL of MeOH | 1 mL of 10% of Liq.NH$_3$ in ACN | Fail | 95.5 | 3.6 |
| 1 mL of 10% of Liq.NH$_3$ in ACN and 1 mL of 1% Formic acid in water | 1 mL of 1% formic acid in water (v/v) followed by 1 mL of MeOH | 1 mL of 10% of Liq.NH$_3$ in ACN | Fail | 99.5 | 0.5 |

Preparation of sample: 750 µL of 1% formic acid in water solution as pretreatment for 100 µL of plasma based on results from Table 2.

**Table 3:** Process efficiency and matrix effect of Niacin after SPE with Oasis MCX cartridge at LQC level analyzed on model API-3000.

4A: PPT(API-3000)

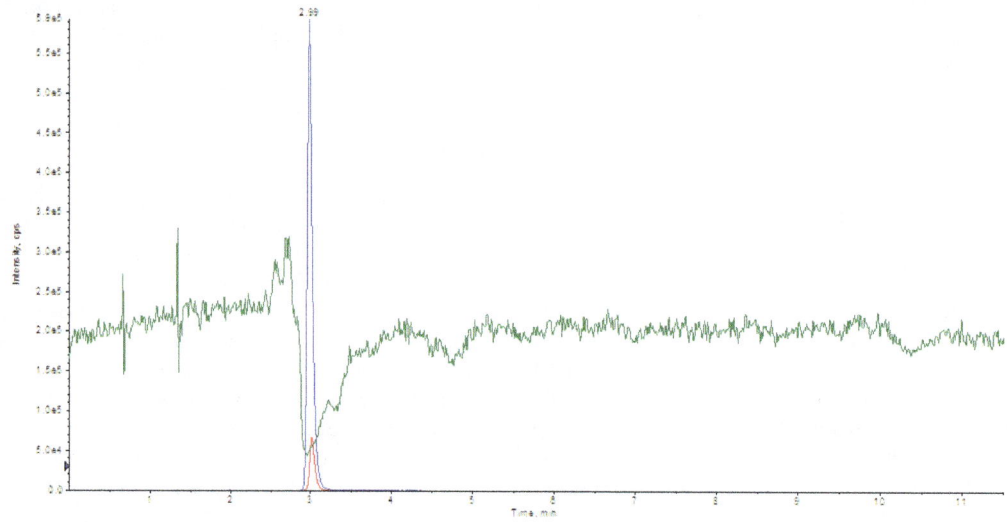

4B: LLE(API-3000)

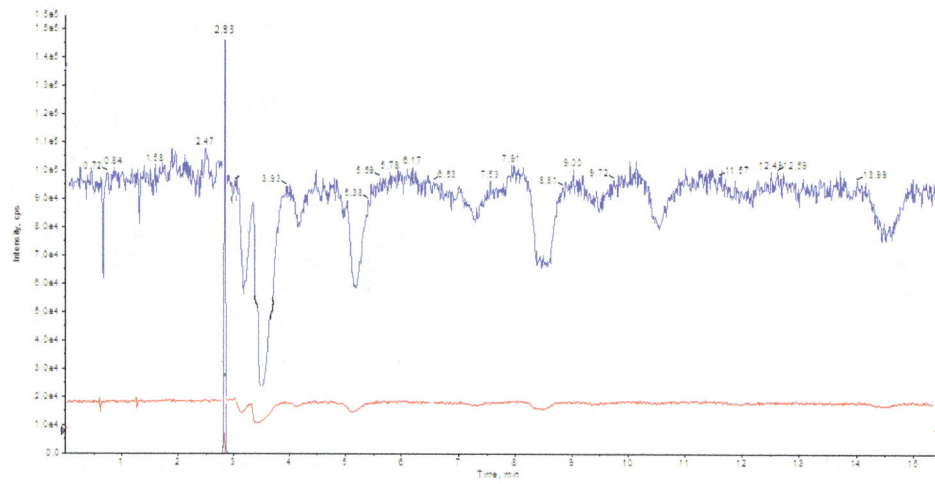

4C: SPE(API-3000)

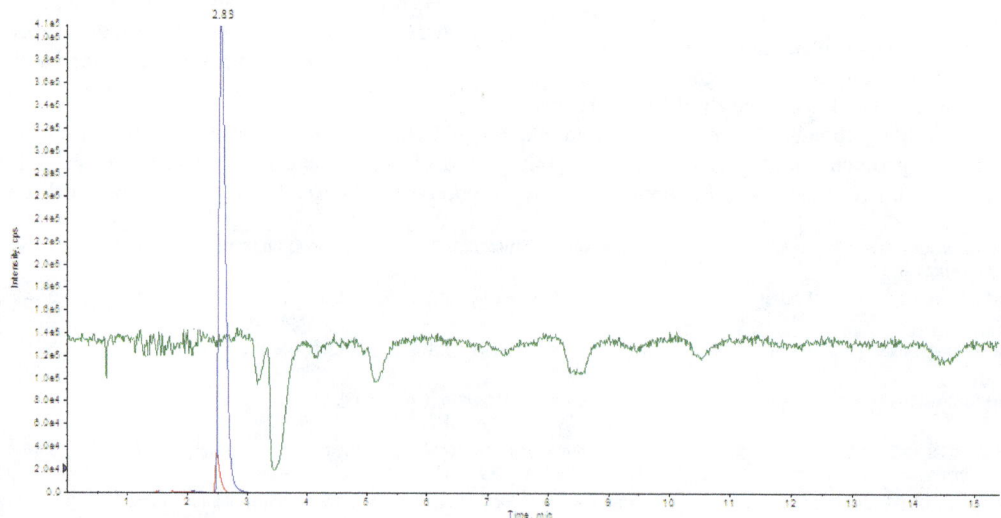

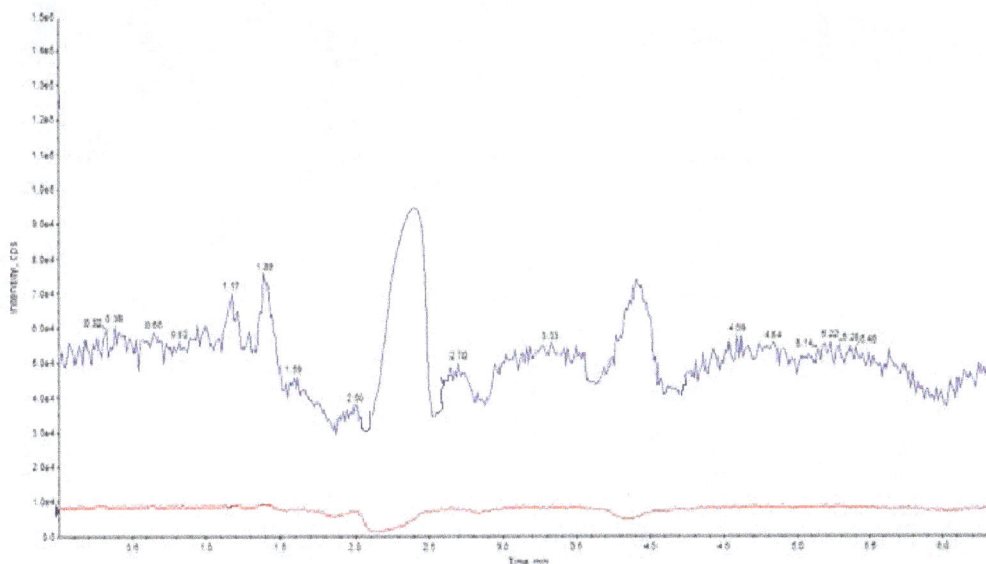

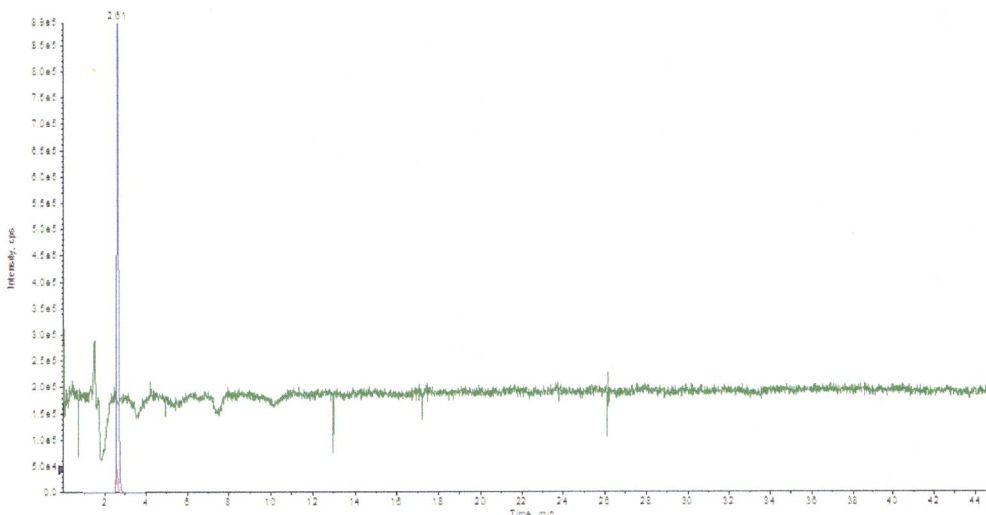

**Figure 4:** Qualitative assessment of matrix effect through 'T'-joint experiment (API-3000).

observed with both the processing methods (Table 4). The mechanism of matrix induced ion suppression or ion enhancement is still not fully understood, this is in part due to the fact that the mechanism of electrospray ionization has proven to be very difficult to establish. Since exact mechanism of ion suppression is still not very clear [22], the ion enhancement observed on API-3200 can be attributed to the competition between a ions of niacin and the co eluting, undetected matrix components.

***Process efficiency:*** Process efficiency (PE) for niacin was estimated at three QC concentration levels prepared by different extraction techniques and analyzed on different models of mass spectrometer. The results shown in Table 5 indicates that PE was less in LLE when compared to SPE technique, almost 10-25% of analyte signal has been suppressed by co-eluting substances using LLE samples, although it was consistent across QC levels independent of instrument model. It indicates that co-eluted of endogenous substances has caused grater

ion-suppression of the analyte and resulted in less recovery.

***Relative matrix effect:*** Results of relative matrix effect were found to be acceptable as accuracy and precision of QC samples were within the limits. Variation caused by different type of plasma (i.e., normal plasma, haemolyzed plasma and lipemic plasma) are presented in Table 6, which indicates that the use of different plasma types, including hemolyzed and lipemic did not cause much variation in the estimation of concentration at LOQQC and HQC levels. Results on instrument models API-4000 and API-3000 were better compared to API-3200. The relative matrix effect results indicate that source design has no impact on matrix effect for niacin under the optimized conditions. The results of API-3000 and API-4000 were almost comparable although the sources of API-3000 and API-4000 were completely different. But, API-3000 was found to be most effective as API-4000 produced charge competition on the peak area response of IS as is evident from Figure 6. To rule out the sensitivity factor of the instrument on charge

| Matrix type | Ion enhancement/Ion suppression of Niacin in API-3000 ion source across the QC levels | | | | | |
|---|---|---|---|---|---|---|
| | LQC | | MQC | | HQC | |
| | SPE | LLE | SPE | LLE | SPE | LLE |
| NP | 2.55 | 9.20 | -0.31 | 7.35 | -0.91 | 4.26 |
| HP | 1.47 | 12.65 | 1.99 | 7.76 | 0.29 | 10.59 |
| LP | 1.53 | 11.12 | 0.81 | 8.30 | -0.97 | 10.89 |
| Matrix type | Ion enhancement/Ion suppression of Niacin in API-3200 ion source across the QC levels | | | | | |
| | LQC | | MQC | | HQC | |
| | SPE | LLE | SPE | LLE | SPE | LLE |
| NP | -19.38 | -41.45 | -31.45 | -65.28 | -26.53 | -52.94 |
| HP | -23.71 | -78.92 | -31.59 | -21.43 | -46.42 | -67.52 |
| LP | -17.78 | -72.03 | -32.27 | -49.74 | -34.69 | -58.99 |
| Matrix type | Ion enhancement/Ion suppression of Niacin in API-4000 ion source across the QC levels | | | | | |
| | LQC | | MQC | | HQC | |
| | SPE | LLE | SPE | LLE | SPE | LLE |
| NP | 4.58 | 10.97 | -0.87 | 11.81 | 1.38 | 12.15 |
| HP | 0.91 | 10.85 | 0.73 | 19.02 | -0.37 | 10.85 |
| LP | -0.04 | 12.65 | 1.78 | 16.42 | 0.07 | 11.61 |

NP=Normal Plasma; HP=Hemolyzed Plasma; LP=Lipemic Plasma

Positive value indicates ion-suppression and negative value indicate ion-enhancement

**Table 4:** Ion-enhancement/ion-suppression for niacin across QC level analyzed on different API models prepared with different extraction techniques (n=6).

| QC level | API-4000 | | API-3200 | | API-3000 | |
|---|---|---|---|---|---|---|
| | SPE | LLE | SPE | LLE | SPE | LLE |
| LQC | 94.5 | 78.2 | 83.5 | 69.2 | 99.6 | 75.1 |
| MQC | 91.5 | 74.4 | 85.1 | 61.9 | 99.5 | 72.6 |
| HQC | 91.0 | 77.1 | 87.9 | 67.2 | 101.1 | 69.3 |

**Table 5:** Process efficiency of niacin across QC level (sample prepared by SPE and LLE and analyzed on different API model; n=6).

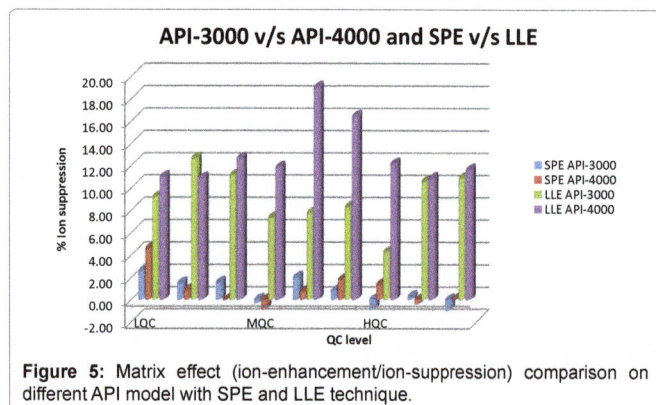

**Figure 5:** Matrix effect (ion-enhancement/ion-suppression) comparison on different API model with SPE and LLE technique.

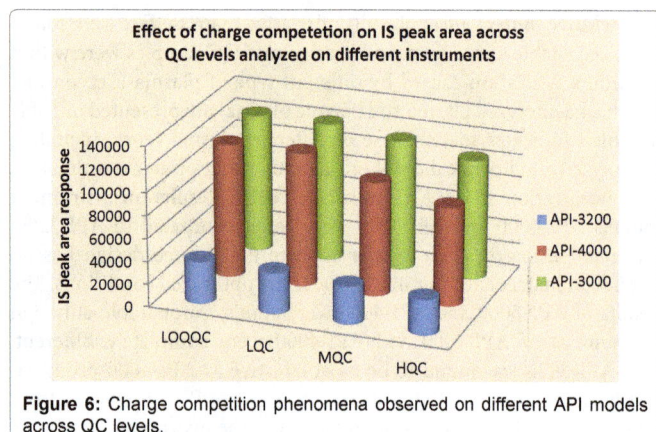

**Figure 6:** Charge competition phenomena observed on different API models across QC levels.

competition, mass parameters are optimized in such way to produce the same LOQ area response on both instruments, but still charge competition phenomena was observed on API-4000.

Based on the above stated observations it was concluded that, the order of choice of instrument shall be API-3000>API-4000>API-3200. This research work clearly indicates that during method development of any compound two important parameters that is sample preparartion technique and mass spectrometer based on the sensitivity of the instrument shall be selected in combination to nullify matrix effect.

**Selection of internal standard:** In LC-MS/MS analysis, selection of IS with similar chromatographic and mass spectrometric behavior to that of analyte is of utmost priority. It is usually assumed that an analyte and stable isotopic labeled internal standard have identical physicochemical properties and thus, isotopic labeled internal standard are chosen for a correct quantification of analyte in LC-MS/MS assay.

Isotopic labeled internal standard (nicotinic-d4 acid) was selected as an internal standard for assay method of niacin due to the following reasons: a) In reverse phase liquid chromatography, similar retention time was observed for nicotinic-d4 acid and niacin, b) In mass spectroscopy similar fragmentation was observed for both niacin and nicotinic-d4 acid, c) Equivalent extraction recovery (process efficiency, PE) was observed that was observed for niacin. Niacin and nicotinic-d4 acid are eluted from the analytical column at same retention time, which causes the similar degree of ion-suppression are imposed on the two compounds. Similar type of ion-suppression for both two compounds uphold the area ratio in post-extracted sample as well as neat solution and the value of IS-normalized matrix effect falls within 0.8-1.2, which is as per the current regulatory guidelines.

**Method validation:** Based on the method development data, even though % ion-suppression on API-3000 and API-4000 varied only by 0.2%, API-3000 was chosen for method validation as on API-4000 charge competition was observed between analyte and IS i.e., with increase in the concentration of analyte, IS peak area decrease. A complete method validation was performed on API-3000 using SPE extraction technique as per the current USFDA, EMEA guidelines. The calibration curve was shown to be linear from 50.6 ng/mL to 25022.7

ng/mL for niacin in plasma. Calibration curve was constructed using peak area ratio of analyte to internal standard and by applying linear, weighted least squares regression analysis with weighting factor of $1/(concentration)^2$. The correlation coefficient (r) was greater than 0.99 during the course of precision and accuracy batches. The results of three precision and accuracy batches are summarized in Table 7. The intraday precision and inter-day precision (%CV) ranged from 2.0-14.2 and the intra-day and inter-day accuracy ranged from 88.6-109.2%. Stability of stock solutions of niacin and IS were established for 16 days at 1-10°C and % stability of niacin and IS were 96.2 and 100.3 respectively. Niacin was proved to be stable in plasma for three freeze-thaw cycles. Bench top stability of niacin was established for 6.78 h in human plasma in ice cold water bath and under low light conditions. Auto sampler stability was assessed for ~77.0 h and long term stability was established at -50°C for 133 days. The observed mean nominal concentration of niacin was found to be within ± 15% of their respective nominal concentration and % CV was less than 15 at LQC and HQC levels (Table 8).

## Conclusion

Much of the literatures propose that the best approach to eliminate or minimize matrix effect is to use APCI source and solid phase extraction technique for sample preparation. Based on our results it was concluded that there is no generic method to eliminate matrix effect in bioanalysis. Each molecule needs to be investigated independently for matrix effect. Like in case of model drug-niacin, SPE is the technique of choice to minimize matrix effect, on turbo ion source on instrument model API-3000 without compromising on the recovery and sensitivity of the method. On API-4000, matrix effect observed was comparable to API-3000 but due to charge competition observed between analyte and IS, API-3000 was instrument of choice for further consideration. In charge competition as the concentration of analyte increased, area response of IS (Nicotinic-d4 acid) decreased. Therefore, in the presence of charge competition, internal standard could not be used as an indicator of matrix effect. In BE study, it will not be possible to identify that variability observed in area response of internal standard is due to change in concentration of analyte or it is due to change in the matrix of the subject.

In this research work, we emphasized that while developing bio-analytical methodology for any molecule, optimization and selection of sample processing methodology and instrument model and ionization technique shall be done in combination. Additionally, selection of instrument model shall be done based on the requirement of sensitivity of the bioanalytical method for an analyte, as selection of highly sensitive instrument may lead to charge competition or enhance the matrix effect as sensitivity for endogenous phospholipids may increase proportionally.

### Acknowledgements

This paper is part of PhD Thesis of Poonam Vats. The author would like to thanks Sun pharmaceutical Industries Limited, Gurgaon, India, for carrying out this work.

| Plasma type | Calculated concentration (ng/mL) | | | | | | | | | | | |
| --- | --- | --- | --- | --- | --- | --- | --- | --- | --- | --- | --- | --- |
| | API-3000 | | | | API-3200 | | | | API-4000 | | | |
| | LOQQC | | HQC | | LOQQC | | HQC | | LOQQC | | HQC | |
| | SPE | LLE | SPE | LLE | SPE | LLE | SPE | LLE | SPE | LLE | SPE | LLE |
| Normal | 49.3 | 49.1 | 20386.4 | 19978.7 | 55.0 | 48.8 | 19655.2 | 19458.6 | 57.8 | 53.9 | 21148.8 | 20725.9 |
| Heamolyzed | 49.3 | 52.0 | 20638.6 | 21257.8 | 52.0 | 48.3 | 19776.9 | 19381.4 | 53.0 | 53.0 | 19983.1 | 19183.8 |
| Lipemic | 50.3 | 56.5 | 20712.0 | 20090.7 | 56.5 | 51.8 | 19725.0 | 20316.8 | 58.8 | 54.2 | 20668.1 | 21288.2 |

Nominal value at LOQQC level 50.7 ng/mL and at HQC level 19985.4 ng/mL

**Table 6:** Relative matrix effect (n=6).

| QC level | %intra-run accuracy[a] | %Inter-run accuracy[b] | %intra-run precision[c] | %Inter-run precision[b] |
| --- | --- | --- | --- | --- |
| LOQQC (50.7 ng/mL) | 109.2 | 104.3 | 14.2 | 13.8 |
| LQC (129.9ng/mL) | 100.3 | 99.4 | 2.0 | 2.3 |
| MQC (9992.7 ng/mL) | 95.0 | 94.8 | 2.0 | 2.1 |
| HQC (19985.4 ng/mL) | 88.7 | 88.6 | 2.4 | 2.1 |

[a]n=6, expressed as 100 × (mean calculated concentration)/(nominal concentration)
[b]Values obtained from all three runs (n=18)
[c]n=6

**Table 7:** Intra-and inter-run.

| Stability | Level | A | %CV | B | %CV | % Change |
| --- | --- | --- | --- | --- | --- | --- |
| Auto sampler stability (~77 h, 10°C | LQC | 125.2 | 3.8 | 124.5 | 2.7 | -0.56 |
| | HQC | 19379.1 | 1.3 | 19557.5 | 2.0 | 2.02 |
| Bench top stability (~6.78 h, in ice cold water bath) | LQC | 123.0 | 5.4 | 125.5 | 3.1 | 2.03 |
| | HQC | 19828.5 | 1.0 | 19370.7 | 1.0 | -2.31 |
| Freeze-thaw stability (Thee freeze-thaw cycle) | LQC | 123.0 | 5.4 | 118.9 | 1.9 | -3.37 |
| | HQC | 19828.5 | 1.0 | 19733.9 | 1.0 | -0.48 |
| Long term stability (133 days, below -50°C | LQC | 126.3 | 4.1 | 123.1 | 2.2 | -2.53 |
| | HQC | 19633.2 | 1.3 | 19841. | 1.1 | 1.06 |

A=comparison sample concentration (ng/mL); B=stability sample concentration (ng/mL);
CV=coefficient of variation;

**Table 8:** Stability of niacin under different storage conditions (n=4).

## References

1. Tang L, Kebarle P (1993) Dependence of ion intensity in electrospray mass spectrometry on the concentration of the analytes in the electrosprayed solution. Analytical Chemistry 65: 3654-3668.

2. Chow F, Ocampo A, Vogel P, Lum S, Tran N (2009) Current challenges for FDA-regulated bioanalytical laboratories performing human BA/BE studies; Part III: selected discussion topics in bioanalytical LC/MS/MS method validation. The Quality Assurance Journal 12: 22-30.

3. Van Eeckhaut A, Lanckmans K, Sarre S, Smolders I, Michotte Y (2009) Validation of bioanalytical LC-MS/MS assays: evaluation of matrix effects. J Chromatogr B Analyt Technol Biomed Life Sci 877: 2198-2207.

4. Little JL, Wempe MF, Buchanan CM (2006) Liquid chromatography-mass spectrometry/mass spectrometry method development for drug metabolism studies: Examining lipid matrix ionization effects in plasma. J Chromatogr B Analyt Technol Biomed Life Sci 833: 219-230.

5. Ismaiel OA, Halquist MS, Elmamly MY, Shalaby A, Karnes HT (2007) Monitoring phospholipids for assessment of matrix effect in a liquid chromatography-tandem mass spectrometry method for hydrocodone and pseudoephedrine in human plasma. J Chromatogr B Analyt Technol Biomed Life Sci 859: 84-93.

6. Gustavsson SA, Samskog J, Markides KE, Långström B (2001) Studies of signal suppression in liquid chromatography-electrospray ionization mass spectrometry using volatile ion-pairing reagents. J Chromatogr A 937: 41-47.

7. Leverence R, Avery MJ, Kavetskaia O, Bi H, Hop CE, et al. (2007) Signal suppression/enhancement in HPLC-ESI-MS/MS from concomitant medications. Biomed Chromatogr 21: 1143-1150.

8. Mei H, Hsieh Y, Nardo C, Xu X, Wang S, et al. (2003) Investigation of matrix effects in bioanalytical high-performance liquid chromatography/tandem mass spectrometric assays: application to drug discovery. Rapid Commun Mass Spectrom 17: 97-103.

9. Chambers E, Wagrowski-Diehl DM, Lu Z, Mazzeo JR (2007) Systematic and comprehensive strategy for reducing matrix effects in LC/MS/MS analyses. J Chromatogr B Analyt Technol Biomed Life Sci 852: 22-34.

10. Trufelli H, Palma P, Famiglini G, Cappiello A (2011) An overview of matrix effects in liquid chromatography-mass spectrometry. Mass Spectrom Rev 30: 491-509.

11. Li KM, Rivory LP, Clarke SJ (2006) Solid-phase (SPE) techniques for sample preparation in clinical and pharmaceutical analysis: a brief overview. Current Pharmaceutical Analysis 2: 95-102.

12. Bylda C, Thiele R, Kobold U, Volmer DA (2014) Recent advances in sample preparation techniques to overcome difficulties encountered during quantitative analysis of small molecules from biofluids using LC-MS/MS. Analyst 139: 2265-2276.

13. Dams R, Huestis MA, Lambert WE, Murphy CM (2003) Matrix effect in bio-analysis of illicit drugs with LC-MS/MS: influence of ionization type, sample preparation, and biofluid. J Am Soc Mass Spectrom 14: 1290-1294.

14. Remane D, Meyer MR, Wissenbach DK, Maurer HH (2010) Ion suppression and enhancement effects of co-eluting analytes in multi analyte approaches: systematic investigation using ultra-high performance liquid chromatography/mass spectrometry with atmospheric-pressure chemical ionization or electrospray ionization. Rapid Commun Mass Spectrom 24: 3103-3108.

15. Ghosh C, Shinde CP, Chakraborthy BS (2012) Influence of ionization source design on matrix effects during LC-ESI-MS/MS analysis. J Chromatogr B Analyt Technol Biomed Life Sci 893-894: 193-200.

16. Schuhmacher J, Zimmer D, Tesche F, Pickard V (2003) Matrix effects during analysis of plasma samples by electrospray and atmospheric pressure chemical ionization mass spectrometry: practical approaches to their elimination. Rapid Commun Mass Spectrom 17: 1950-1957.

17. AB Sciex (2008) Hardware manual of API-4000.

18. Mei H (2005) Matrix effects: causes and solutions. In: Using mass spectrometry for drug metabolism studies. CRC Press pp: 103-149.

19. Niessen WM, Manini P, Andreoli R (2006) Matrix effects in quantitative pesticide analysis using liquid chromatography-mass spectrometry. Mass Spectrom Rev 25: 881-899.

20. Heller DN (2007) Ruggedness testing of quantitative atmospheric pressure ionization mass spectrometry methods: the effect of co-injected matrix on matrix effects. Rapid Commun Mass Spectrom 21: 644-652.

21. Shen JX, Motyka RJ, Roach JP, Hayes RN (2005) Minimization of ion suppression in LC-MS/MS analysis through the application of strong cation exchange solid-phase extraction (SCX-SPE). J Pharm Biomed Anal 37: 359-367.

22. Jessome LL, Volmer DA (2006) Ion Suppression: A Major Concern in Mass Spectrometry. LCGC solutions for separation scientists 24: 498-510.

23. Liang HR, Takagak T, Foltz RL, Bennett P (2005) Quantitative determination of endogenous sorbitol and fructose in human nerve tissues by atmospheric-pressure chemical ionization liquid chromatography/tandem mass spectrometry. Rapid Commun Mass Spectrom 19: 2284-2294.

24. Bonfiglio R, King RC, Olah TV, Merkle K (1999) The effects of sample preparation methods on the variability of the electrospray ionization response for model drug compounds. Rapid Commun Mass Spectrom 13: 1175-1185.

25. USFDA (2001) Guidance for Industry: Bioanalytical Method Validation.

26. European Medical Agency (2011) Guideline on bioanalytical method validation.

27. Peoples MC, Halquist MS, Ismaiel O, El-Mammli MY, Shalaby A, et al. (2008) Assessment of matrix effect and determination of niacin in human plasma using liquid-liquid extraction and liquid chromatography-tandem mass spectrometry. Biomed Chromatogr 22: 1272-1278.

28. Tang L, Kebarle P (1991) Effect of conductivity of the electrospray solution on the electrospray current. Factors determining Analyte sensitivity in electrospray mass spectrometry. Analytical Chemistry 63: 2709-2715.

# 13

# Determination of Intramolecular $^{13}$C Isotope Distribution of Pyruvate by Headspace Solid Phase Microextraction-Gas Chromatography-Pyrolysis-Gas Chromatography-Combustion-Isotope Ratio Mass Spectrometry (HS-SPME-GC-Py-GC-C-IRMS) Method

Tarin Nimmanwudipong[1], Naizhong Zhang[3], Alexis Gilbert[2], Keita Yamada[3] and Naohiro Yoshida[1,2,3]

[1]Department of Environmental Science and Technology, Tokyo Institute of Technology, 4259 Nagatsuta, Midori-ku, Yokohama 226-8502, Japan
[2]Earth-Life Science Institute (WPI-ELSI), Tokyo Institute of Technology, Meguro, Tokyo 152-8551, Japan
[3]Department of Environmental Chemistry and Engineering, Tokyo Institute of Technology, 4259 Nagatsuta, Midori-ku, Yokohama 226-8502, Japan

## Abstract

This paper presents the improvement of analytical methods for intramolecular carbon isotope compositions ($\delta^{13}$C) of pyruvate. Decarboxylated by $H_2O_2$, pyruvate yields acetic acid and $CO_2$. Headspace solid phase micro-extraction-gas chromatography-pyrolysis-gas chromatography-combustion-isotope ratio mass spectrometry (HS-SPME-GC-Py-GC-C-IRMS) was used to measure the intramolecular $\delta^{13}$C values of acetic acid. $\delta^{13}$C value of $CO_2$ can be later calculated using mass balance equation. The method's consistency was confirmed by comparison of the $\delta^{13}$C value of $CO_2$ from calculation to its direct measurement. Results of this study confirmed the method improvement because pyruvate $^{13}$C intramolecular distribution patterns were obtained. Two intramolecular $^{13}$C distribution patterns for commercial chemical reagents were found using this developed method. Intramolecular $^{13}$C distribution patterns for pyruvate were found for application in dietary supplements. Its origin was inferred. The method presented herein is expected to be a useful tool for categorization of pyruvate into different intramolecular $^{13}$C distribution patterns, which might indicate different production processes or raw materials.

**Keywords:** Pyruvate; Dietary supplement; Intramolecular isotope distribution; Food authenticity; Quality control

## Introduction

Isotope analysis has been applied to facilitate identification of the origin, metabolic pathways, and biosphere-atmosphere interaction of organic materials [1-4]. Compound Specific Isotope Analysis (CSIA) is typically used to ascertain the isotopic composition of a target compound. However, in some cases, CSIA data alone are not sufficient for identification of the compound's origin. Position-Specific Isotope Analysis (PSIA) has provided information related to the heterogeneous isotope distribution of organic compounds including amino acids, acetic acid, fatty acids, sugars, ethanol, and hydrocarbons [5-9]. This information is crucially important for the investigation of synthetic processes and metabolic pathways of the target compound. These isotope analysis techniques are also applied in the food industry for quality control [1,2].

Pyruvate, a key metabolite for carbohydrate metabolism, is necessary to trigger a plant's citric acid cycle, fat, and protein metabolism. It can also be useful as a dietary supplement to increase the metabolic rate [10]. Pyruvate influences the isotopic contents of respired $CO_2$ and its related metabolites. Therefore, its isotope signature would be beneficial for studying the authenticity and metabolic pathways in plants.

The main objective of this study is to improve the analytical method for intramolecular $^{13}$C distribution of pyruvate, which can be degraded into acetic acid and $CO_2$ using $H_2O_2$ [11,12]. Results confirm the success of this technique: we obtained a pyruvate $^{13}$C intramolecular distribution pattern.

Additionally, we applied this method to ascertain the intramolecular $^{13}$C distribution in a pyruvate sample from pyruvate supplement pills. Dietary supplements are convenient choices that provide essential nutrients. Because of their high demand, the number of manufacturers has increased rapidly in the past few years. One way to remain competitive in the market is to decrease manufacturing costs to the greatest extent possible. Commonly, people prefer authentic products from natural sources. However, the same commercial synthetic substance, which is obtainable by the derivatization of petroleum or coal, offers a rapidly producible and cheaper alternative than the natural extract from biogenic sources [13-15]. According to the Food and Drug Administration (FDA) in some countries, synthetic substances are illegal: their manufacture is prohibited [16]. To minimize the risk of illegality, a method is needed to help differentiate between synthetic and natural substances [17-19]. This study considered the potential origin of pyruvate in dietary supplement based on the intramolecular $^{13}$C distribution.

## Materials and Methods

### Annotations

The carbon isotope composition in per mil (‰) concentrations is expressed as the $\delta^{13}$C value, the carbon isotope ratio ($^{13}$C/$^{12}$C) of the sample against an international standard (VPDB).

---

**Corresponding author:** Tarin Nimmanwudipong, Department of Environmental Science and Technology, Tokyo Institute of Technology, 4259 Nagatsuta, Midori-ku, Yokohama 226-8502, Japan, E-mail: Nimmanwudipong.t.aa@m.titech.ac.jp

$$\delta^{13}C = \left[\left(^{13}C/^{12}C\right)_{sample} / \left(^{13}C/^{12}C\right)_{standard}\right) - 1] \tag{1}$$

For this study, pyruvate samples were measured after decarboxylation, which yields acetic acid and carbon dioxide. Bulk and intramolecular $\delta^{13}C$ values of pyruvate are definable in a mass balance equation as

$$\delta^{13}C_{Pyruvate} = [2(\delta^{13}C_{AcOH}) + \delta^{13}C_{CO_2}]/3 \tag{2}$$

Therein, $\delta^{13}C_{AcOH}$ value is the bulk carbon isotope composition of acetic acid and $\delta^{13}C_{CO2}$ value is the carbon isotope composition of $CO_2$ from pyruvate decarboxylation in this study. Actually, bulk and intramolecular $\delta^{13}C$ values of acetic acid are definable in a mass balance equation as

$$\delta^{13}C_{AcOH} = (\delta^{13}C_{CH_3} + \delta^{13}C_{COOH})/2 \tag{3}$$

Where $\delta^{13}C_{CH3}$ value and $\delta^{13}C_{COOH}$ value respectively represents the carbon isotope composition of methyl and carboxyl carbon atom of acetic acid. Each carbon position of pyruvate was designated as C-1 (carboxyl part), C-2 (carbonyl part), and C-3 (methyl part).

## Chemicals

Four sodium pyruvates designated as A (Tokyo Chemical Industry Co. Ltd., Tokyo, Japan), B (MP Biomedicals, LLC, CA, USA), C (Sigma-Aldrich Corp., St. Louis, MO, USA), and D (Wako Pure Chemical Industries Ltd., Osaka, Japan) were used for this study. Four pyruvate supplement samples, designated as DP1 (Earth Natural Supplements, Florida, USA), DP2 (Best Naturals, NJ, USA), DP3 (Source Naturals, Inc., CA, USA) in capsule pills, and DP4 (Now Foods, IL, USA) in tablet pills were purchased. Hydrogen peroxide (30%) and hydrochloric acid (0.1 mM) (Wako Pure Chemical Industries Ltd.) were used respectively for pyruvate decarboxylation and pH adjustment. A tin capsule (0.15 ml; Ø 5/19 mm; LÜDI Swiss AG, Switzerland) was used to contain sodium pyruvate samples for laser spectroscopy analysis.

## Degradation of pyruvate

$H_2O_2$-catalyzed decarboxylation of pyruvate is described according to the following scheme:

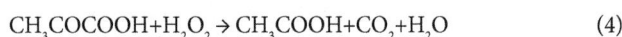

$$CH_3COCOOH + H_2O_2 \rightarrow CH_3COOH + CO_2 + H_2O \tag{4}$$

In this study, sodium pyruvate samples (A, B, C, and D) were degraded using 30% hydrogen peroxide, then yielding acetic acid, $CO_2$ and $H_2O$ as products, as described below. The yields of acetic acid at 10, 30, 60 and 120 min of degradation time were determined using ion chromatography (IC-20 Dionex™; Thermo Fisher Scientific Inc., Bremen, Germany).

For the degradation of sodium pyruvate, a pyruvate aqueous solution was prepared at 85 mM diluting with distilled water. For pyruvate supplement samples, pyruvate was separated from other ingredients before dilution with distilled water. One pill of 750 mg (DP2, DP3) and 1000 mg (DP1, DP4) was used for pyruvate extraction. Regarding to packages' label, pyruvate supplementary samples have main ingredients consists of pyruvate salt, gelatin (contained capsule), stearate, and cellulose. The gelatin container was taken off and discarded (DP1, DP2, DP3); the powder sample was kept. The DP4 tablet was crushed to powder in ceramic mortar. Remaining powder of samples was diluted in 100 mL milliQ water. Because of the lack of water solubility, stearate and cellulose were separated using microfiltration three times using a 20 μm filter. Then the pyruvate aqueous solution was obtained from supplement samples.

In a 20 mL gas-tight vial, 1 mL of each pyruvate aqueous solution sample was put in and topped with rubber cap for analysis. To samples for pyruvate degradation, 0.2 mL of $H_2O_2$ was added. After complete degradation, the samples' pH was adjusted to 1.0-2.0 pH by adding 0.2 mL of 0.1 mol/L HCl. The $CO_2$ derived from the degradation was collected and purified by repeated cryogenic method and trapped in the Pyrex® sealed tube for $\delta^{13}C$ analysis.

## Carbon isotopic analysis

Intramolecular $\delta^{13}C$ value of acetic acid derived from pyruvate degradation was measured using HS-SPME-GC-Py-GC-C-IRMS [20]. The system consists of a first gas chromatograph, (Trace™ GC Ultra; Thermo Fisher Scientific Inc.) equipped with a capillary column (Nukol™, 30 m × 0.32 mm i.d., 1 μm film thickness; Supelco, PA, USA), connected to a second gas chromatograph (HP 6890 series; Hewlett-Packard Co., PA, USA) equipped with a second capillary column (HP-Plot Q 30 m × 0.32 mm i.d., 20 μm film thickness; Agilent Technologies Inc., CA, USA). Two gas chromatographs were connected through a pyrolysis furnace part (ceramic tube, 25 cm × 0.5 mm i.d.) operated at 1000°C for pyrolysis of acetic acid. The pyrolytic products were separated using a second capillary column and were introduced into a combustion furnace (ceramic tube, 25 cm × 0.5 mm i.d., packed with CuO, NiO, and Pt wires) operated at 960°C. The second chromatograph was connected via Thermo GC Isolink™ and Conflo-IV™ interfaces (both from Thermo Fisher Scientific Inc.) to a mass spectrometer (Finnigan Delta V™; Thermo Fisher Scientific Inc.). A transfer line between chromatographs was made using deactivated fused silica capillary (0.32 mm i.d.; GL Sciences Inc., Japan).

Acetic acid from pyruvate degradation was extracted using an SPME device, equipped with 85 μm SPME fiber coated with carboxen/polydimethylsiloxane (Carboxen/PDMS stableflex™; Supelco, PA, USA). Extraction was conducted in a thermostatic chamber controlled to 25°C: the non-stirred samples condition. The extraction time was 60 min. After HS-SPME extraction, the fiber was inserted into the injection port of the first gas chromatograph at 250°C. Helium was used as a carrier gas for all experiments. Chromatographic conditions were the following: 2.0 mL/min flow rate of carrier gas and 10:1 split ratio. The first oven temperature program was the following: 100°C (5 min), then rising to 190°C (10 min) at the rate of 15°C/min, and finally at 200°C (2 min) at the rate of 15°C/min. The second gas chromatograph was kept constantly at 40°C.

The dual-inlet system of an isotope ratio mass spectrometer (MAT 253™, Thermo Fisher Scientific Inc.) was used for the measurement of $\delta^{13}C$ value of $CO_2$ derived from pyruvate degradation. Bulk $\delta^{13}C$ values of sodium pyruvate (A, B, C, and D) were measured using cavity ring-down laser spectroscopy (Picarro G1121i; Picarro Inc., CA, USA).

## Results and Discussion

### Completeness of reaction and consistency of method

The experiment of pyruvate decarboxylation by $H_2O_2$ was conducted respectively in ranges of 10, 30, 60 and 120 min. The acetic acid yield was measured using ion chromatography and was calculated using a calibration curve of the acetic acid standard. As shown in Table 1, the yield of acetic acid reaches 99% at 60 min reaction time. At the 120 min range, it also had the same number around 99%, which implies that the reaction is completed at 60 min. For subsequent experiments, we used 60 min as the decarboxylation time.

The consistency of $\delta^{13}C_{C-1}$ ($\delta^{13}C$ value of C-1 of pyruvate) was confirmed by comparison of $\delta^{13}C_{CO2}$ between the value calculated using the mass balance equation of pyruvate (equation 2) and from direct measurement, which are expected to be the same. Table 2 shows that the differences of $\delta^{13}C_{CO2}$ values from the two methods were approximately 0.6‰, which is an acceptable range, showing that usage of the mass balance equation can obtain $\delta^{13}C_{CO2}$ value (equation 2). This consistency of method has also confirmed the acceptable use of HS-SPME-GC-Py-GC-C-IRMS, for which $\delta^{13}C$ of C-2 and C-3 ($\delta^{13}C$ value of C-2 and C-3 of pyruvate) are obtainable in a single step. Without the measurement of bulk $\delta^{13}C_{AcOH}$ value, the $\delta^{13}C$ measurement can reduce the unexpected errors occurs by duplicate sample preparation or switching between configuration systems [20]. $\delta^{13}C_{C-1}$ value can be calculated later using the mass balance equation of pyruvate.

## Bulk and intramolecular $\delta^{13}C$ isotope distribution of sodium pyruvate

Details of $\delta^{13}C$ values of sodium pyruvate samples are presented in Table 3. First, we obtained bulk $\delta^{13}C$ of sodium pyruvate, which are -22.6‰ (A), -22.6‰ (B), -21.3‰ (C), and -23.1‰ (D). For intramolecular $\delta^{13}C$ values, samples A, C, and D have $\delta^{13}C$ values in the pattern of C-3>C-1>C-2, whereas sample B has the $\delta^{13}C$ pattern of C-2>C-3>C-1. Figure 1 clarifies that we obtained intramolecular $\delta^{13}C$ distribution of pyruvate of two kinds. Moreover, same bulk $\delta^{13}C$ value of A and B samples have different patterns of intramolecular $\delta^{13}C$ values. These indicate that these pyruvates are potentially derived from different production processes or raw materials.

Pyruvate can be synthesized using chemical production, with

| Degradation time (min.) | Yield of acetic acid (%) |
|---|---|
| 10 | 76.4 |
| 30 | 86.2 |
| 60 | 99.1 |
| 120 | 99.9 |

Table 1: Yield percentage of acetic acid by degradation time.

| Sample (n=3) | Calculation (‰) | Measurement (‰) | Difference (‰) |
|---|---|---|---|
| A | -20.0 ± 0.7 | -19.4 ± 0.1 | 0.6 |
| B | -25.3 ± 0.6 | -24.8 ± 0.3 | 0.5 |
| C | -15.2 ± 0.7 | -15.5 ± 0.0 | 0.3 |
| D | -20.9 ± 0.1 | -20.3 ± 0.3 | 0.6 |

Table 2: Mass balance calculation and direct measurement of $\delta^{13}C_{CO2}$ (C-1).

| Sample | $\delta^{13}C_{C-1}$ (‰) | $\delta^{13}C_{C-2}$ (‰) | $\delta^{13}C_{C-3}$ (‰) | $\delta^{13}C_{AcOH}$ (‰) | Bulk $\delta^{13}C$ (‰) |
|---|---|---|---|---|---|
| A | -20.0 ± 0.7[a] | -36.5 ± 0.3 | -11.3 ± 0.4 | -23.89 ± 0.5 | -22.6 ± 0.2 |
| B | -25.3 ± 0.6 | -19.9 ± 0.2 | -22.7 ± 0.3 | -21.26 ± 104 | -22.6 ± 0.2 |
| C | -15.2 ± 0.7 | -36.6 ± 0.4 | -12.0 ± 0.5 | -24.27 ± 0.6 | -21.3 ± 0.2 |
| D | -20.9 ± 0.1 | -35.6 ± 0.5 | -12.7 ± 0.5 | -24.14 ± 0.7 | -23.1 ± 0.0 |
| DP1 | -36.6 ± 0.8 | -21.2 ± 0.3 | -29.3 ± 0.1 | -25.23 ± 0.3 | -29.0 ± 1.5 |
| DP2 | -34.8 ± 1.5 | -16.5 ± 0.3 | -23.3 ± 0.3 | -19.92 ± 0.4 | -24.9 ± 2.7 |
| DP3 | -39.0 ± 1.5 | -17.6 ± 0.0 | -23.3 ± 0.3 | -20.46 ± 0.3 | -26.7 ± 2.6 |
| DP4 | -34.9 ± 0.3 | -18.4 ± 0.2 | -24.6 ± 0.2 | -21.51 ± 0.3 | -26.0 ± 0.7 |
| Sodium pyruvate using $H_2O_2$ degradation [11] | -22.3 | -19.6 | -21.5 | -20.57 | -21.2 |

[a]Standard deviation from the mean (n=3)

Table 3: $\delta^{13}C$ measurement of sodium pyruvate samples and degraded fragments with $\delta^{13}C$ calculated from the mass balance equation and measurement values of $\delta^{13}C$.

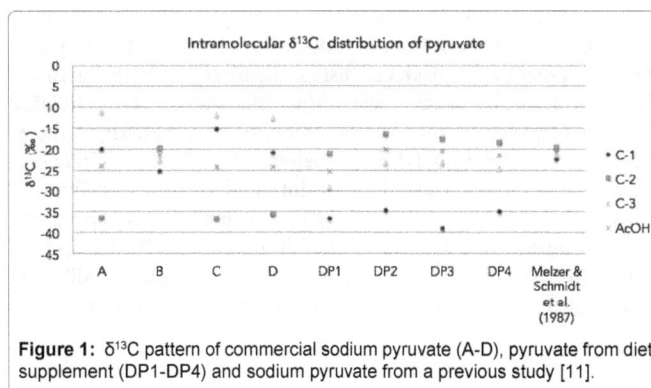

Figure 1: $\delta^{13}C$ pattern of commercial sodium pyruvate (A-D), pyruvate from diet supplement (DP1-DP4) and sodium pyruvate from a previous study [11].

tartaric acid and $KHSO_4$ as substrates [21]. Tartaric acid has two main pathways for chemical synthesis [22,23]. First is tartaric acid obtained from petroleum by-products, which inherit the $\delta^{13}C$ value from hydrocarbon substrate. Second is tartaric acid from cyanohydrin synthesis, which inherits the $\delta^{13}C$ value from the initial substrate (3 carbons from glyceraldehyde and 1 carbon from the cyano group). Recently, Zyakun et al. determined the intramolecular $\delta^{13}C$ value in synthetic tartaric acid from chemical synthesis, finding $^{13}C$ depletion in its carboxyl carbon [23]. In general, without isotope fractionation, pyruvate is expected to inherit $\delta^{13}C$ value of the beginning tartaric acid. However, isotope fractionation can occur during actual production processes. Our hypothesis according to previous works is tartaric acid from chemical synthesis, which also has a similar trend of $\delta^{13}C$ values to acetic acid from chemical synthesis, which has depleted carboxyl carbon [8,23]. For intramolecular $\delta^{13}C$ value of pyruvate, the depletion of $\delta^{13}C$ value in carboxyl carbon (C-1) from C-2 and C-3 has been found in sample B. This trend is similar to the trend of intramolecular $\delta^{13}C$ values of acetic acid [8,24] and tartaric acid [23] from chemical synthesis. We might infer that sample B had high potential to be produced by a chemical synthesis method, along with a good agreement to the $\delta^{13}C$ values of previous studies. However, without details of proprietary synthetic process of sample, the discussion about the exact pattern remains unclear. Further details related to isotope fractionation, which possibly occurred in production process, must be clarified in future works' discussion for concrete references.

Another pattern of intramolecular $\delta^{13}C$ values might have a different mode of production or substrate. Pyruvate can also be produced using biotechnological methods. Biotechnological methods have at least three methods: direct fermentation method, the resting cell method, and the enzymatic method [25]. The enzymatic method is simple, with a high conversion rate of the substrate. For example, lactate can be the substrate for pyruvate production using L-lactate catalyzed by glycolate oxidase in Hansenula polymorpha [26]. However, the high price of raw materials and some complicated processes for removal of by-products of production are shortcomings related to industrialize enzymatic methods for pyruvate production. Consequently, direct fermentation and the resting cell methods have higher potential for mass production of pyruvate. Samples A, C, and D have found enrichment in $\delta^{13}C$ value in C-1 than C-2 and C-3, which also have the same trend of $\delta^{13}C$ values as those of biological products reported from previous studies [8,23]. In this case, samples A, C, and D should have been regarded as products from biotechnological methods. We also considered the pattern of $\delta^{13}C$ values of the sodium pyruvate sample that used $H_2O_2$ degradation in a previous study [11], which has a similar pattern to that of sample B and which should fall into the category of chemical synthesis production, from the $^{13}C$ depletion in its carboxyl carbon than its C-2 and C-3.

Considering pyruvate supplement samples, we found their intramolecular $^{13}C$ distribution patterns to be similar to sample B, which is potentially, produced using chemical synthesis methods. However, if natural tartaric acid is the initial substance in chemical synthesis of pyruvate, then the intramolecular $^{13}C$ pattern might be different, according to the different pattern of $\delta^{13}C$ values of biogenic and abiogenic tartaric acid [23]. These intramolecular $^{13}C$ distributions of pyruvate can help us categorize the production process of pyruvate, although further investigation of the intramolecular $^{13}C$ distribution pattern from plenty of natural samples and samples that are different from known processes must be done for additional explanations.

## Conclusion

Adoption of HS-SPME-GC-Py-GC-C-IRMS produces a more convenient analytical method for the determination of intramolecular $\delta^{13}C$ values in pyruvate. Pyruvate samples in this study have two patterns that are useful for categorizing samples into different production processes. Further studies of the natural pattern of the pyruvate from plants can be a good first step, followed by studies of pyruvate production by different known processes. These will help to distinguish the pyruvate samples into the correct categories of origin processes.

### Acknowledgements

This study was supported by a Grant in-Aid for Scientific Research (S) (23224013). A. Gilbert appreciates a Grant-in-Aid for Young Scientists (B) (15K17774), and thanks MEXT, Japan, for financial support.

### References

1. Rossmann (2001) Determination of stable isotope ratios in food analysis. Food Rev Int 17: 347-381.

2. Förstel H (2007) The natural fingerprint of stable isotopes--use of IRMS to test food authenticity. Anal Bioanal Chem 388: 541-544.

3. Brenna JT (2001) Natural intramolecular isotope measurements in physiology: elements of the case for an effort toward high-precision position-specific isotope analysis. Rapid Commun Mass Spectrom 15: 1252-1262.

4. Gilbert A, Robins RJ, Remaud GS, Tcherkez GG (2012) Intramolecular $^{13}C$ pattern in hexoses from autotrophic and heterotrophic $C_3$ plant tissues. Proc Natl Acad Sci USA 109: 18204-18209.

5. Gilbert A, Silvestre V, Robins RJ, Remaud GS (2009) Accurate quantitative isotopic $^{13}C$ NMR spectroscopy for the determination of the intramolecular distribution of $^{13}C$ in glucose at natural abundance. Anal Chem 81: 8978-8985.

6. Gilbert A, Silvestre V, Robins RJ, Remaud GS, Tcherkez G (2012) Biochemical and physiological determinants of intramolecular isotope patterns in sucrose from $C_3$, $C_4$ and CAM plants accessed by isotopic $^{13}C$ NMR spectrometry: a viewpoint. Nat Prod Rep 29: 476-486.

7. Hayes JM (2001) Fractionation of Carbon and Hydrogen Isotopes in Biosynthetic Processes. Rev Mineral Geochem 43: 225-277.

8. Meinschein WG, Rinaldi GG, Hayes JM, Schoeller DA (1974) Intramolecular isotopic order in biologically produced acetic acid. Biomed Mass Spectrom 1: 172-174.

9. Rinaldi G, Meinschein WG, Hayes JM (1974) Intramolecular carbon isotopic distribution in biologically produced acetoin. Biomed Mass Spectrom 1: 415-417.

10. Stanko RT, Tietze DL, Arch JE (1992) Body composition, energy utilization, and nitrogen metabolism with a severely restricted diet supplemented with dihydroxyacetone and pyruvate. Am J Clin Nutr 55: 771-776.

11. Melzer E, Schmidt HL (1987) Carbon isotope effects on the pyruvate dehydrogenase reaction and their importance for relative carbon-13 depletion in lipids. J Biol Chem 262: 8159-8164.

12. Melzer E, Schmidt HL (1988) Carbon isotope effects on the decarboxylation of carboxylic acids, comparison of the lactate oxidase reaction and the degradation of pyruvate by $H_2O_2$. Biochem J 252: 913-915.

13. Bahl A, Bahl BS (2010) Advance organic chemistry. S Chand and Company Ltd.

14. Chenier PJ (1992) Survey of Industrial Chemistry. 2nd revised edn. Wiley-VCH Publishers, New York, USA.

15. Harold WA, Bryan GR (1996) Industrial Organic Chemicals. Wiley-Interscience, New York, USA.

16. Japan External Trade Organization (2011) Guildbook for export to Japan (Food Articles).

17. Zhang L, Kujawinski DM, Federherr E, Schmidt TC, Jochmann MA (2012) Caffeine in your drink: natural or synthetic? Anal Chem 84: 2805-2810.

18. Suzuki Y, Akamatsu F, Nakashita R, Korenaga T (2010) A Novel Method to Discriminate between Plant- and Petroleum-derived Plastics by Stable Carbon Isotope Analysis. Chemistry Letters 39: 998-999.

19. Calderone G, Guillou C (2008) Analysis of isotopic ratios for the detection of illegal watering of beverages. Food Chem 106: 1399-1405.

20. Nimmanwudipong T, Gilbert A, Yamada K, Yoshida N (2015) Analytical method for simultaneous determination of bulk and intramolecular $^{13}C$-isotope compositions of acetic acid. Rapid Commun Mass Spectrom 29: 2337-2340.

21. Howard JW, Fraser WA (1932) Preparation of pyruvic acid. Org Synth Coll 1: 475-480.

22. Serra F, Reniero F, Guillou CG, Moreno JM, Marinas JM, et al. (2005) $^{13}C$ and $^{18}O$ isotopic analysis to determine the origin of L-tartaric acid. Rapid Commun Mass Spectrom 19: 1227-1230.

23. Zyakun AM, Oganesyants LA, Panasyuk AL, Kuz'mina EI, Shilkin AA, et al. (2015) Site-specific $^{13}C/^{12}C$ isotope abundance ratios in dicarboxylic oxyacids as characteristics of their origin. Rapid Commun Mass Spectrom 29: 2026-2030.

24. Rinaldi G, Meinschein WG, Hayes JM (1974) Carbon isotopic fractionations associated with acetic acid production by Acetobacter suboxydans. Biomed Mass Spectrom 1: 412-414.

25. Li Y, Chen J, Lun SY (2001) Biotechnological production of pyruvic acid. Appl Microbiol Biotechnol 57: 451-459.

26. Anton DL, Dicosmo R (1995) Witterholt Proccess for the preparation of pyruvic acid. WO patent, 95000656.

# 14

# Application of CPE-FAAS Methodology for the Analysis of Trace Heavy Metals in Real Samples using Phenanthraquinone Monophenyl Thiosemicarbazone and Triton X-114

**Magda Akl A\*, Magdy Bekheit M and Ibraheim Helmy**
*Faculty of Science, Mansoura University, Mansoura, Egypt*

### Abstract

In the present study, a cloud point extraction method was used for the preconcentration and extraction of cadmium (II) and zinc (II) ions in different environmental samples. The zinc and cadmium ions formed hydrophobic complexes with phenanthraquinone monophenyl thiosemicarbazone (PPT). These complexes were extracted using Triton X-114 nonionic surfactant. The surfactant-rich phase was diluted with acidified methanol. Then, the concentrations of the metal ions were determined by FAAS. The experimental factors controlling the process of separation are investigated e.g., pH, complexing agent concentration, surfactant's concentration, temperature, and incubation time. The present CPE-FAAS procedure has been used to preconcentrate and determine Cd(II) and Zn(II) metal ions in natural water samples, drug samples and certified reference materials. The LODs for cadmium(II) and zinc(II) ions were 0.38 and 1.85 µg/L, respectively with a preconcentration factor of 100. The recovery % of the extracted Cd(II) and Zn(II), is greater than 90% and the relative standard deviation(RSD,%) is less than 5%.

**Keywords:** CPE; Zinc(II); Cadmium(II); Surfactants; FAAS

## Introduction

Heavy metals in the environment are really increasing due to human activity, particularly mining process and heavy industry in the developing world. This has caused great accumulation of high concentrations of toxic heavy metals in natural waters [1].

Monitoring the presence of toxic trace metals in different matrices is of great importance in order to evaluate the exposure to these heavy metals in the environment. In this sense, cadmium is considered as one of the most toxic elements; it accumulates in humans body mostly in the kidneys and liver and is classified as a rampant toxic element with biological half-life time in the range of 10-30 years [2]. The maximum allowable concentration of cadmium permitted by the American Environmental Protection Agency (US EPA) in standard drinking water is 10 µg $L^{-1}$.

Zinc is considered as an essential micro-nutrient that has many biochemical functions in all living organisms [3]. Deficiency of zinc can lead to many disorders such as growth retardation, diarrhea, inefficient immunological defense, eye and skin lesions, delaying of wound healing and other skin diseases [4]. Excess amounts of zinc are harmful and can be toxic when exposures exceed the physiological needs.

Monitoring trace element levels in environmental samples might be considered a difficult analytical task, mostly because of the complexity of the matrix and the very low concentrations of these elements, which requires a prior preconcentration step followed by sensitive instrumental techniques [5]. Many separation/preconcentration techniques have been developed for trace metal analysis such as solid phase extraction [6], liquid- liquid extraction [7], co-precipitation [8], column extraction [9], ion-selective electrode [10] and cloud point extraction (CPE) [11-16].

Cloud point extraction (CPE) has attracted a great attention because it complies with the "Green Chemistry" principle [17]. CPE is simple, highly efficient, cheap, rapid and of lower toxicity than those procedures using organic solvents.

Cloud point extraction (CPE) is based on the phase behavior of non-ionic surfactants in aqueous solutions. Non-ionic surfactants undergo phase separation upon raising the temperature or the addition of a salting-out agent [18,19]. This procedure has been successfully employed to extract and preconcentrate different trace metals from different samples [20-23].

In recent years, considerable attention has been paid to the use of chelating agents, containing sulphur and nitrogen in analytical chemistry, in separation, purification and estimation of metal ions [24]. The extensive application and rapid growth in the popularity of sulphur ligands is because of their outstanding property as potential donors to form stable and characterized complexes. In addition, the presence of nitrogen along with sulphur tends to reduce the solubility of the complexes, making the isolation of these complexes easier.

Among various organic chelating reagents containing S and N, thiosemicarbazones and their aromatic derivatives occupy a unique place. Thiosemicarbazones are a group of organic compounds prepared by condensing thiosemicarbazide with carbonyl compounds in the presence of few drops of glacial acetic acid. These organic reagents function as good complexing agents and form complexes with various metal ions through thionate Sulphur atom and hydrazine nitrogen atom binding sites.

Phenanthraquinone monophenyl thiosemicarbazone (PPT) reacts with different metal ions such as Ni(II), Zn(II), Cu(II), Cd(II) and Pb(II) to form hydrophobic colored complexes. These metal ions were further determined in media of diverse origin by FAAS and/or ultraviolet-visible (UV-Vis) spectrophotometry [25-28].

*\*Corresponding author:* Magda Akl A, Faculty of Science, Mansoura University, Mansoura, Egypt, E-mail: magdaakl@yahoo.com

In the present work, CPE method has been used for the preconcentration of zinc and cadmium after the formation of hydrophobic complexes with phenanthraquinone monophenyl thiosemicarbazone (PPT). These complexes were simply extracted using Triton-X114 nonionic surfactant followed by FAAS determination. The analytical conditions for the preconcentration, extraction and FAAS determination of the analytes were investigated. The proposed combined CPE-FAAS methodology was successfully used for the analysis of zinc(II) and cadmium(II) in water, drug and certified samples.

## Experimental

### Instruments

A Perkin-Elmer 2380 air-acetylene atomic absorption spectrometer has been used for the determination of Cd(II) and Zn(II) concentrations. To measure the pH of sample solutions, a digital pH meter was used

### Chemicals

All the chemicals used in the presented study are of analytical-reagent grade. Double distilled water (DDW) was used through. Zinc stock solution (1000 mgL$^{-1}$) was prepared by dissolving 2.089 g ZnCl$_2$ in 100 ml of water with the addition of 2 ml of concentrated HCl and dilution to 1 L with bidistilled water. Cadmium stock solution (1000 mgL$^{-1}$) was prepared by 2.03 g of CdCl$_2$.2.5 H$_2$O in 1 L DDW. Triton X-144 was obtained from Sigma. To prepare a 1% (v/v) Triton X-114 stock solution, 1 mL of Triton X-114 is dispersed in 5 mL ethyl alcohol and completed with 100 ml DDW.

Phenanthraquinone monophonyl thiosemicarbazone, PPT was synthesized as previously reported [26]. A 10$^{-2}$ mol. L$^{-1}$ solution of PPT was prepared by dissolving 0.357 gm of phenanthraquinone monophenyl thiosemicarbazone (PPT) in 100 ml of acetone.

### Analytical procedure

An aliquot of 20 ml of aqueous solution that contains 3 μg mL$^{-1}$ of metal ions, 2 × 10$^{-4}$ mol. L$^{-1}$ of PPT and 0.05% (v/v) of Triton X-114 at pH 6 was prepared (Scheme 1). This mixture was shaken for 1 minute and left to stand for 20 minutes in water bath at 60°C. Centrifugation at 3500 rpm for 10 minutes was performed to achieve separation of the two phases. Upon cooling in an ice bath for 15 minutes, the surfactant-rich phase became viscous and the mother aqueous phase was smoothly separated. The micellar phase is dissolved in 0.2 mL of acidified methanol and aspirated directly to the flame to determine the concentration of the investigated metal ions.

Calibration curves were constructed in the concentration ranges 0.40-2.0 mg/l and the concentrations of the investigated metal ions were determined by FAAS at specified wave lengths for each analyte.

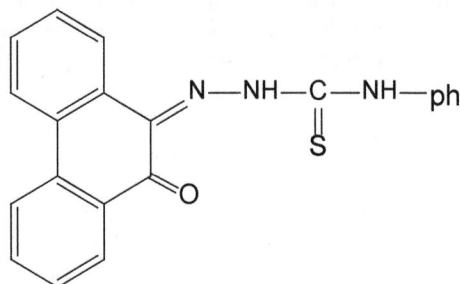

**Scheme 1:** Phenanthraquinone monophenyl thiosemicarbazone (PPT).

For determination of zinc(II) in pharmaceutical samples, two commercial drug samples were selected viz., Vitamax plus and Totavit tablets. The samples were brought into solution using the procedure previously described in the literature [28].

## Results and Discussion

### Cloud point extraction (CPE)

**Effect of pH:** The pH is a very important parameter for both the coacervation of the micelles and the complexation of the ligand with the metal ions. The influence of pH on analytes extraction was investigated by performing a number of CPE experiments and varying the pH of sample solution over a wide range (2 -9) using 3 μg mL$^{-1}$ of Zn(II) and 3 μg mL$^{-1}$ of Cd(II) in the presence of 2 × 10$^{-4}$ molL$^{-1}$ PPT and 0.05% (v/v) Triton X-114. As it can be seen from Figure 1, the maximum recovery was achieved at pH 6.

**Effect of PPT concentration:** The CPE efficiency depends on the hydrophobicity of the ligand and the complex formed. So, in order to evaluate the role of PPT in CPE of Zn(II) and Cd(II), similar CPE experiments were performed by adding different concentrations of PPT to a suitable concentration, 3 μg mL$^{-1}$ of Zn(II), 3 μg mL$^{-1}$ of Cd(II) and 0.05% (v/v) Triton X-114 at pH 6. Figure 2 shows that, at low PPT concentration, the recovery of the analytes is low because the amount of ligand required for complexation of all the amount of the analytes is insufficient. Then, the recovery significantly increases by increasing PPT concentration till maximum recovery was attained at 1.5 × 10$^{-4}$ mol. L$^{-1}$ PPT. The recovery of the analytes remains constant upon the use of higher concentrations of the chelating agent. So, 2 × 10$^{-4}$ molL$^{-1}$ PPT was selected for the subsequent work.

**Effect of triton X-114 concentration:** The effect of Triton X-114 concentration on the extraction efficiency of 3 μg mL$^{-1}$ of Zn(II) and 3 μg mL$^{-1}$ of Cd(II) in the presence of 2 × 10$^{-4}$ mol L$^{-1}$ PPT at pH 6 was studied in the concentration range of 0.02-0.25% (v/v). In Figure 3, the results showed that, at low surfactant concentration the recovery of the analytes is poor because there are few surfactant molecules to entrap the ligand-metal complexes quantitatively [29]. Then, the recovery sharply increases by increasing the concentration of Triton X-114. The highest recoveries were obtained with 0.05% (v/v) Triton X-114. By increasing the surfactant concentration above 0.1% the recovery starts to decrease gradually.

**Effect of temperature and incubation time:** Like the other parameters, temperature and duration of the CPE procedure seem to play significant roles. Raising the incubation temperature above the cloud point temperature is an important factor for obtaining satisfactory extraction and enhancing CPE efficiency in a reduced incubation time. A series of experiments was performed to investigate the effect of temperature on the CPE method under the previous optimum conditions of pH, PPT concentration and Triton X-114 concentration over a wide range of temperatures (20-80°C) through 20 minutes' incubation time. The data in Figure 4 shows that maximum recovery was obtained at range (50-80°C). The enhanced recovery at high temperature is related to the dehydration between hydrogen bonds that occur by increasing temperature and decrease amount of water in surfactant rich phase hence the volume of the phase decrease leading to enhancing the recovery. The effect of the incubation time, on the other hand, was investigated within the ranges 5-60 min. The results, Figure 5, illustrate that an equilibration time of 20 min was quite adequate to achieve quantitative extraction of cadmium(II) and zinc(II) ions [30].

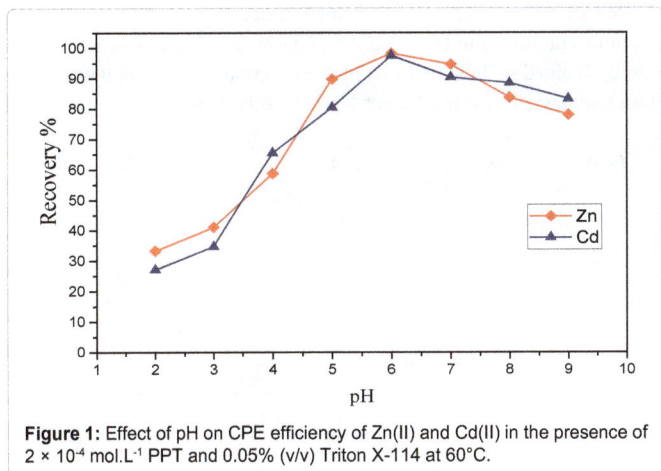

**Figure 1:** Effect of pH on CPE efficiency of Zn(II) and Cd(II) in the presence of $2 \times 10^{-4}$ mol.L$^{-1}$ PPT and 0.05% (v/v) Triton X-114 at 60°C.

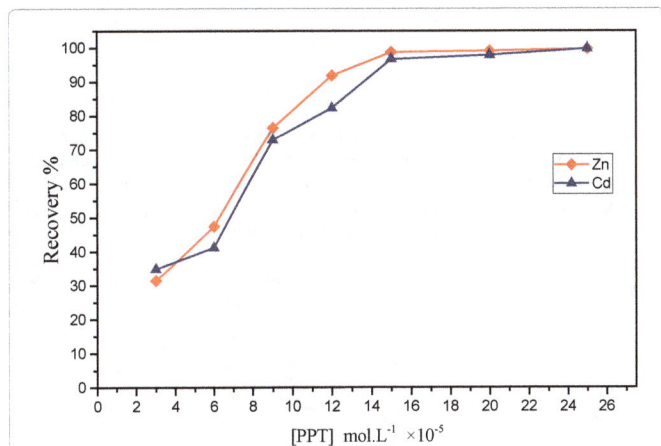

**Figure 2:** Effect of the PPT concentration on the recovery of Zn(II) and Cd(II) in the presence of 0.05% (v/v) Triton X-114 at pH 6, at 60°C.

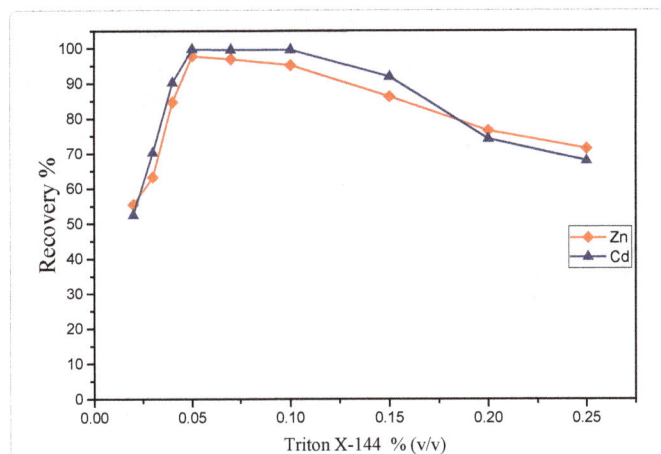

**Figure 3:** Effect of Triton X-114 concentration on the recovery % of Zn(II) and Cd(II) in the presence of $2 \times 10^{-4}$ mol. L$^{-1}$ PPT at pH 6 at 60°C.

**Effect of centrifugation time and rate:** For the best efficiency of the method, it is required to preconcentrate trace amount of metal ions with high sensitivity and in a short time. To test the influence of centrifugation time on the method sensitivity, CPE has been carried out for a series of experiments, in which an aliquot of 10 ml of aqueous solution containing 3 µg mL$^{-1}$ of Zn(II) and 3 µg mL$^{-1}$ of Cd(II) in the

presence of $2 \times 10^{-4}$ mol. L$^{-1}$ PPT and 0.05% (v/v) of Triton X-114 at pH 6 has been incubated at 60°C for 20 minutes and centrifuged at various rates (1500-4500 rpm) and for different time periods (5-20 min). The best conditions for sample centrifugation were 10 min at 3500 rpm.

**Effect of foreign ions:** Because of the high selectivity attributed to flame atomic absorption spectrometry, the only source of low sample recovery must be the preconcentration step. This problem may be attributed to the fact that cations may react with ligand and anions may form stable complex with metal ions and resulting in a decrease in extraction efficiency (Table 1). Therefore studying the effect of foreign ions was carried out using 3 µg mL$^{-1}$ of Zn(II) and 3 µg mL$^{-1}$ of Cd(II) in the presence of an excess amount of PPT ($5 \times 10^{-4}$ mol.L$^{-1}$). The effect of almost foreign ions, even those which formed colored complexes with PPT such as Cu$^{+2}$, Ni$^{+2}$ and Pb$^{+2}$, interference could be eliminated.

## Analytical characteristics

To evaluate the reproducibility of the method, 20 mL of model solution containing metal ions ($n=10$) was used. The relative standard deviation (RSD, %) as precision of the method (RSD, %) was 1.86 and 3.06%, for Cd(II) and Zn(II), respectively (Table 2). The LODs based on three times the SD of the blank for cadmium(II) and zinc(II) ions were 0.38

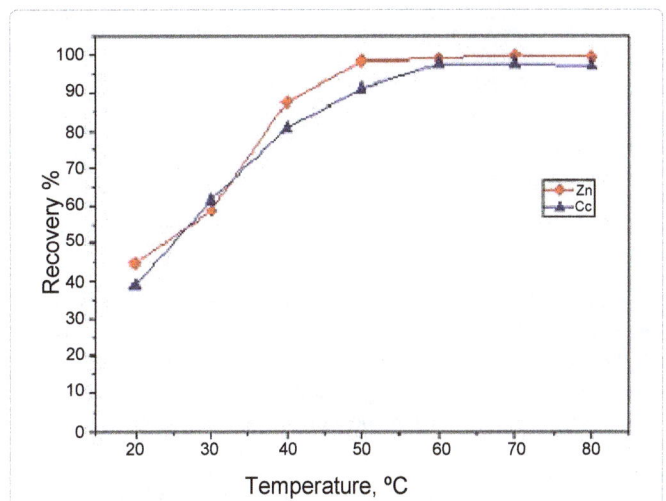

**Figure 4:** Effect of temperature on CPE Recovery of Zn(II) and Cd (II) in the presence of $2 \times 10^{-4}$ mol.L$^{-1}$ PPT and 0.05% (v/v) Triton X-114 at pH 6.

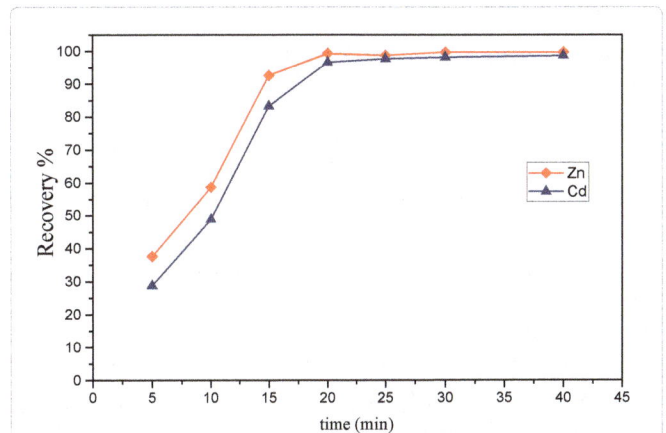

**Figure 5:** Effect of incubation time on CPE Recovery of Zn(II) and Cd(II) in the presence of $2 \times 10^{-4}$ mol.L$^{-1}$ PPT and 0.05% (v/v) Triton X-114 at pH 6 at 60°C.

| Foreign ion | Concentration (mg. L⁻¹) | Recovery % | |
|---|---|---|---|
| | | Zn(II) | Cd(II) |
| $Hg^{+2}$ | 20 | 100 | 96.9 |
| $Ni^{+2}$ | 10 | 100 | 97.7 |
| $Co^{+2}$ | 20 | 97.9 | 98.4 |
| $Cr^{+3}$ | 5 | 98.9 | 99.3 |
| $Fe^{+3}$ | 10 | 99.8 | 98.5 |
| $Cu^{+2}$ | 10 | 98.2 | 97.6 |
| $Pb^{+2}$ | 10 | 99.7 | 99.2 |
| $Ba^{+2}$ | 20 | 99.8 | 98.4 |
| $Mn^{+2}$ | 20 | 100 | 96.8 |
| $Bi^{+3}$ | 20 | 100 | 97.3 |
| $Al^{+3}$ | 20 | 100 | 98.1 |
| $Na^+$ | 230 | 98.8 | 97.9 |
| $SO_4^-$ | 480 | 99.5 | 99.8 |
| $Cl^-$ | 177.5 | 97.9 | 99.6 |
| $NO_2^-$ | 230 | 98.7 | 97.8 |
| $NO_3^-$ | 310 | 98.1 | 100 |
| $CO_3^-$ | 300 | 97.2 | 96.8 |
| $C_2O_4^-$ | 440 | 97.9 | 100 |
| $CH_3COO^-$ | 295 | 98.8 | 98.5 |

**Table 1:** Effect of interfering ion on the recovery of Zn(II) and Cd(II) in the presence of $2 \times 10^{-4}$ PPT and 0.05 % (v/v) Triton X-114 (n=5).

| Analyte | Correlation coefficient | Linear range, mg/L | Regression equation[a] | RSD, % |
|---|---|---|---|---|
| Cd | 0.9997 | 0.40–1.50 | A=0.0471C | 1.86 |
| Zn | 0.9999 | 0.50–2.00 | A=0.0367C–0.00049 | 3.06 |

[a]A=Absorbance, C=concentration

**Table 2:** Analytical characteristics of the calibration curves of the analytes.

and 1.85 μg/L, respectively (n=10). The regression equations and optimum concentration ranges for the metal ions are shown in Table 2.

To verify the accuracy and applicability of proposed CPE procedure, two reference standard materials (lead-zinc sulphide ore-OCrO (COD 161-96) and stream sediment SARM 52) were analyzed. The results are given in Table 3. There is no significant difference between the results obtained by the proposed method and the certified results. The relative standard deviation (as a precision) is less than 5%.

Statistical analysis of the results in Table 3 indicate that the preconcentrated samples are not subject to any systematic error i.e., accurate.

## Applications

**Water analysis:** Various water samples from different origins (Tap water, Nile river and sea water samples) were spiked with different amounts of cadmium and zinc and CPE procedures were employed for determination of the recovery of the analytes in these water samples, Table 4. The results of the present CPE method for Cd(II) and Zn(II) are in good comparison with those gained upon using the well-known APDC/MIBK solvent extraction method (SE), Table 5.

**Analysis of pharmaceutical samples:** Cloud point extraction followed by FAAS determination was applied to determine zinc in some commercial zinc containing pharmaceutical samples. The experimental results agreed well with the given reported values, Table 6.

**Application to synthetic mixtures:** An aliquot of 10 ml of aqueous sample solution containing different compositions of foreign metal ions (Co(II), Ni(II), Hg(II) and Bi(III), Cr(III) and Pb(II), different amounts of the analytes, $2 \times 10^{-4}$ mol L⁻¹ of PPT and 0.05% (v/v) of Triton X-114 at pH

6. The CPE procedures were performed under the previously mentioned optimum conditions and the recovery of the analytes were determined. The results obtained in Table 7 shows high extraction recovery of the analytes from samples containing different synthetic mixtures.

| Element Ore No. | $Cd^{2+}$ | $Zn^{2+}$ |
|---|---|---|
| No. 1[a] | | |
| $(\bar{x})$, ppm | 0.950 | 120 |
| | ± 0.001 | ± 1.70 |
| μ, ppm | 0.950 | 122.0 |
| S | 0.02 | 2.4 |
| $|t|_1$ | 0.00 | 1.86 |
| RSD | 2.1 | 1.75 |
| No. 2[b] | | |
| $(\bar{x})$, ppm | ND | 1.25 |
| | | ± 0.05 |
| μ, ppm | - | 1.320 |
| S | - | 0.06 |
| $|t|_1$ | - | 2.50 |
| RSD | - | 4.30 |

[a] Lead-zinc sulfide ore-OCrO (COD 161-96); [b] Stream sediment SARM 52. ($\bar{x}$): experimental value, (μ) true value; $|t|_1$: for P=0. 05 and n=5 (4 degree of freedom)=2.78, from Ref. [29]. RSD; %: Relative standard deviation.

**Table 3:** Statistical evaluation for analysis of some certified reference samples after CPE (n=5). Comparison of experimental mean ($\bar{x}$) with true value (μ) by $|t|_1$ test

| Water sample | Cd(II) added (μg/ml) | Zn(II) added (μg/ml) | Recovery % | |
|---|---|---|---|---|
| | | | Zn(II) | Cd(II) |
| Tap water (Our laboratory) | 5.00 | 5.00 | 98.9 | 99.1 |
| | 10.00 | 10.00 | 98.8 | 99.7 |
| Nile water (Mansoura city) | 5.00 | 5.00 | 99.3 | 98.5 |
| | 10.00 | 10.00 | 99.7 | 99.3 |
| Sea water (Ras El-Barr city) | 5.00 | 5.00 | 98.7 | 99.5 |
| | 10.00 | 10.00 | 98.6 | 98.6 |

**Table 4:** Recovery of Cd(II) and Zn(II) in different water samples (n=5).

| Sample | Cloud point extraction (PE) | | | Solvent extraction (SE) | | |
|---|---|---|---|---|---|---|
| | Cd(II), μgl⁻¹ | Zn(II), μgl⁻¹ | RSDs, % | Cd(II), μgl⁻¹ | Zn(II), μgl⁻¹ | RSDs,% |
| Tap water (Our Lab Mansoura City) | 0.30 | 185 | 1.0-5.0 | 0.32 | 186.0 | 2.0-3.9 |
| Waste water (Meat Anter, Talkha) | 0.6 | 80 | 2.0-4.0 | 0.62 | 82.0 | 1.2-4.0 |
| Nile River water (Sherbin City) | 0.13 | 26 | 1.5-4.8 | 0.14 | 25.5 | 1.5-4.5 |
| Lake water (Manzalah) | 0.15 | 158 | 1.0-5.0 | 0.16 | 160 | 2.0-4.5 |
| Sea water (Ras Elbar city) | 0.12 | 30.0 | 2.0-4.0 | 0.11 | 31.5 | 1.0-5.0 |

**Table 5:** Determination of Cd(II) and Zn(II) in μgl⁻¹ in natural water samples by the present cloud point extraction (CPE) and the standard APDC/MIBK solvent extraction methods (SE) (n=5 for both methods).

| Sample | $(\bar{x})$ | μ | S | $|t|_1$ | RSD, % |
|---|---|---|---|---|---|
| No. 1[a] | 14.95 | 15.0 | 0.05 | 1.5 | 4.2 |
| No. 2[b] | 15.02 | 15.0 | 0.02 | 1.95 | 2.8. |

[a]Vitamax plus (Galxo Welcome Egypt) capsule; [b]Totavit (Egyphar) tablet. ($\bar{x}$): experimental value, (μ) true value. For P=0.05 and n=5 (4 degree of freedom)=2.78, from Ref. [29]. RSD, %: Relative standard deviation.

**Table 6:** Statistical evaluation for analysis of some pharmaceutical vitamin samples after CPE (n=5). Comparison of experimental mean ($\bar{x}$), mg/capsule with true value (), by $|t|_1$ test.

| Synthetic mixtures composition (µg/ml) | Zn(II) added (µg/ml) | Zn(II) found (µg/ml) | RSD % | Cd(II) added (µg/ml) | Cd(II) found (µg/ml) | RSD, % |
|---|---|---|---|---|---|---|
| Co⁺² (5 µg/ml)+Ni⁺² (5 µg/ml) | 2.00 | 2.97 | 0.81 | 2.00 | 2.94 | 0.88 |
| | 5.00 | 4.96 | 1.25 | 5.00 | 4.89 | 1.11 |
| Hg⁺² (5 µg/ml)+Pb⁺² (5 µg/ml) | 3.00 | 2.87 | 0.48 | 3.00 | 2.93 | 0.37 |
| | 5.00 | 4.94 | 1.03 | 5.00 | 4.87 | 0.67 |
| Co⁺² (3 µg/ml)+Ni⁺² (3 µg/ml)+Cr⁺³ (3 µg/ml) | 3.00 | 2.91 | 0.41 | 3.00 | 2.97 | 1.34 |
| | 5.00 | 4.95 | 0.70 | 5.00 | 4.90 | 1.08 |
| Hg⁺² (3 µg/ml)+Pb⁺² (3 µg/ml)+Bi⁺³ (3 µg/ml) | 3.00 | 2.92 | 1.05 | 3.00 | 2.87 | 0.93 |
| | 5.00 | 4.89 | 1.27 | 5.00 | 4.90 | 0.58 |

**Table 7:** Cloud point extraction of Cd(II) and Zn(II) from synthetic mixtures ($n$=3).

## Conclusion

In the present study, a simple, low cost and reliable combined cloud point extraction-FAAS methodology using the Triton-X114 nonionic surfactant has been used successfully for the determination of cadmium and zinc in water, vitamin and certified samples. The analytical characteristics of the proposed procedure (LOD, Recovery, % and RSD, %) are obtained. The limit of detection of cadmium and zinc gained is in good comparison or superior to the cloud point procedures reported in literature. Two certified reference materials were used to verify the accuracy of the proposed method.

## References

1. Jaric Z, Visnjic-Jeftic G, Cvijanovic Z, Gacic L, Jovanovic S, et al. (2011) Determination of differential heavy metal and trace element accumulation in liver, gills, intestine and muscle of sterlet (Acipenser ruthenus) from the Danube River in Serbia by ICP-OES. Microchemical Journal 98: 77-81.

2. Davis C, Wu P, Zhang X, Hou X, Jones BT (2006) Determination of cadmium in biological samples. Applied Spectroscopy Reviews 41: 35-75.

3. Ferreira HS, dos Santos WN, Fiuza RP, Nóbrega JA, Ferreira SL (2007) Determination of zinc and copper in human hair by slurry sampling employing sequential multi-element flame atomic absorption spectrometry. Microchemical Journal 87: 128-131.

4. Scherz H, Kirchhoff E (2006) Trace elements in foods: zinc contents of raw foods-a comparison of data originating from different geographical regions of the world. Journal of Food Composition and Analysis 19: 420-433.

5. Gundogdu C, Duran HB, Senturk Elci L, Soylak M (2007) Simultaneous preconcentration of trace metals in environmental samples using amberlite XAD-2010/8-hydroxyquinoline system. Acta Chimica Slovenica 54: 308.

6. Ansari R, Kazi TG, Jamali MK, Arain MB, Sherazi ST, et al. (2008) Improved extraction method for the determination of iron, copper, and nickel in new varieties of sunflower oil by atomic absorption spectroscopy. J AOAC Int 91: 400-407.

7. Chand R, Watari T, Inoue K, Kawakita H, Luitel HN, et al. (2009) Selective adsorption of precious metals from hydrochloric acid solutions using porous carbon prepared from barley straw and rice husk. Minerals Engineering 22: 1277-1282.

8. Kagaya S, Takata D, Yoshimori T, Kanbara T, Tohda K (2010) A sensitive and selective method for determination of gold(III) based on electrothermal atomic absorption spectrometry in combination with dispersive liquid-liquid microextraction using dicyclohexylamine. Talanta 80: 1364-1370.

9. Soylak M, Tuzen M (2008) Coprecipitation of gold(III), palladium(II) and lead(II) for their flame atomic absorption spectrometric determinations. J Hazard Mater 152: 656-661.

10. Hu Q, Yang X, Huang Z, Chen J, Yang G (2005) Simultaneous determination of palladium, platinum, rhodium and gold by on-line solid phase extraction and high performance liquid chromatography with 5-(2-hydroxy-5-nitrophenylazo) thiorhodanine as pre-column derivatization regents. J Chromatogr A 1094: 77-82.

11. Gupta VK, Goyal RN, Sharma RA (2009) Comparative studies of ONNO-based ligands as ionophores for palladium ion-selective membrane sensors. Talanta 78: 484-490.

12. da Silva MAM, Frescura VLA, Curtius AJ (2001) Determination of noble metals in biological samples by electrothermal vaporization inductively coupled plasma mass spectrometry, following cloud point extraction. Spectrochimica Acta Part B: Atomic Spectroscopy 56: 1941-1949.

13. Dalali N, Javadi N, Agawal YK (2008) On-line incorporation of cloud point extraction in flame atomic absorption spectrometric determination of silver. Turkish Journal of Chemistry 32: 561-570.

14. Fan Z, Bai F (2007) Determination of trace amounts of silver in various samples by electrothermal atomic absorption spectrometry after sample preparation using cloud point extraction. Atomic Spectroscopy-Norwalk Connecticut 28: 30.

15. Ghaedi M, Shokrollahi A, Niknam K, Niknam E, Najibi A, et al. (2009) Cloud point extraction and flame atomic absorption spectrometric determination of cadmium(II), lead(II), palladium(II) and silver(I) in environmental samples. J Hazard Mater 168: 1022-1027.

16. Manzoori JL, Abdolmohammad-Zadeh H, Amjadi M (2007) Ultra-trace determination of silver in water samples by electrothermal atomic absorption spectrometry after preconcentration with a ligand-less cloud point extraction methodology. J Hazard Mater 144: 458-463.

17. Mortada WI, Ali AAZ, Hassanien MM (2013) Mixed micelle-mediated extraction of alizarin red S complexes of Zr (IV) and Hf (IV) ions prior to their determination by inductively coupled plasma-optical emission spectrometry. Analytical Methods 5: 5234-5240.

18. Anastas P, Eghbali N (2010) Green chemistry: principles and practice. Chem Soc Rev 39: 301-312.

19. Lindqvist O (1995) Environmental impact of mercury and other heavy metals. Journal of power sources 57: 3-7.

20. Magos L, Carson BL, Ellis HV, McCann JL (198) Toxicology and Biological Monitoring of Metals in Humans Including Feasibility and Need. Lewis Publishers, Chelsea, Michigan. Wiley Online Library 7: 328.

21. Ferrer R, Beltran J, Guiteras J (1996) Use of cloud point extraction methodology for the determination of PAHs priority pollutants in water samples by high-performance liquid chromatography with fluorescence detection and wavelength programming. Analytica chimica acta 330: 199-206.

22. Garcia Pinto C, Perez Pavon JL, Moreno Cordero B (1995) Cloud point preconcentration and high-performance liquid chromatographic determination of organophosphorus pesticides with dual electrochemical detection. Analytical Chemistry 67: 2606-2612.

23. Frankewich RP, Hinze WL (1994) Evaluation and optimization of the factors affecting nonionic surfactant-mediated phase separations. Analytical chemistry 66: 944-954.

24. Ali MA, Livingstone S (1974) Metal complexes of sulphur-nitrogen chelating agents. Coordination Chemistry Reviews 13: 101-132.

25. Saxena R, Ahmad S (2010) Synthesis and characterisation of transition metal complexes of 2, 6-diacetylpyridine bis (S-methyl isothiosemicarbazone). Oriental Journal of Chemistry 26: 1507.

26. Akl MA (2001) Spectrophotometric and AAS determinations of trace zinc(II) in natural waters and human blood after preconcentration with phenanthraquinone monophenylthiosemicarbazone. Anal Sci 17: 561-564.

27. Akl MA, Khalifa ME, Ghazy SE, Hassanien MM (2002) Selective flotation-separation and spectrophotometric determination of cadmium using phenanthraquinone monophenythiosemicarbazone. Anal Sci 18: 1235-1240.

28. Akl MA (2006) An improved colorimetric determination of lead(II) in the presence of nonionic surfactant. Anal Sci 22: 1227-1231.

29. Khalifa ME, Akl MA, Ghazy SES (2001) Selective flotation-spectrophotometric determination of trace copper (II) in natural waters, human blood and drug samples using phenanthraquinone monophenylthiosemicarbazone. Chemical and pharmaceutical bulletin 49: 664-668.

30. Miller JC, Miller JN (1986) Statistics for analytical chemistry. Ellis Horwood Limited, England, pp: 43-192.

# 15

# Fabrication of a Dual-Enzyme Sensing System Based on Surface Plasmon Resonance Imager (SPRi) for Simultaneous Determination of Bio-Markers

**Yuhei Hida[1] and Hiroaki Shinohara[1,2]***

[1]*Graduate School of Innovative Life Science for Education, University of Toyama, 3190 Gofuku, Toyama, Japan*
[2]*Graduate School of Science and Engineering for Research, University of Toyama, 3190 Gofuku, Toyama, Japan*

## Abstract

Saccharides such as glucose, galactose and amino acids like lysine, phenylalanine are known as bio-markers of metabolic disorders, and several analytical methods have been developed to determine these bio-markers in blood and urine by enzymatic method, HPLC analysis and so on. Multi-biosensing systems for simultaneous determination of these bio-markers are presently expected to be necessary for troublesome diagnosis. We have previously proposed enzyme-sensors based on a two-dimensional surface plasmon resonance imager (2D-SPRi), which can sensitively monitor refractive index change of local region in the evanescent field on an Au chip. L-lysine sensor, L-glutamate sensor and galactose sensor has been developed each by combining oxidase reactions with 2D-SPR imaging of Os polymer-redox state. In this study, we aimed to develop a dual-enzyme sensor capable of the simultaneous detection of lysine and galactose by means of co-immobilizing LysOx/Os-HRP polymer spot and GalOx/Os-HRP polymer spot adjacently on the Au chip, which is divided into two sections.

**Keywords:** Surface plasmon resonance imager; Bio-markers; Saccharides; HPLC analysis

## Introduction

Saccharides and amino acids are essential metabolites, playing an important role in energy production, because ATP, NADH, and FADH$_2$ as energy carriers are made with saccharides and amino acids in cells through glycolysis and TCA cycle. Therefore, various kinds of sensors, which target some saccharides and amino acids have been developed for diagnoses of metabolic diseases such as diabetes, galactosemia, phenylketonuria, maple syrup urine symptom and high lysine blood symptom, etc. [1-11]. Furthermore, recent statistics data of medical treatments demonstrated that abnormality of the amino acids balance in blood was correlated well with some diseases, and a novel method to diagnose by using the data about free-amino acid concentrations with multi-variable analysis has been proposed recently [12-17]. For example, the Fischer ratio defined as the concentration of branched-chain amino acids (BCAA)/the concentration of aromatic amino acids (AAA), shows less than 2.5 for the blood of cirrhosis patients, while that of the blood of healthy person shows 3.5~4.0 [16]. It is also reported that metabolic syndrome causes variation of some amino acids level in the blood [15].

High performance liquid chromatography (HPLC) and the enzyme method are well known as conventional methods for the amino acids measurement [18-21]. HPLC analysis is good at sensitivity, reproducibility and quantitativity, however, it also has the shortcomings such as time-consuming, high costs and need of some complex treatments before the measurement. On the other hand, enzyme method is simple and easy but not good at multi-determination. Therefore, it is required to develop multi-enzyme sensors, which can determine some saccharides and amino acids simultaneously in easy, quick and sensitive way.

Several types of bio-sensors by means of surface plasmon resonance imager (SPRi) have been already reported. Slight change of refractive index in the evanescent field on the sensor chip enables us to image the surface state. Hence, this technology has been extensively used for observing molecular adsorption / desorption reactions, antigen-antibody binding reactions and DNA hybridization reactions [22-28]. In recent years, Os-HRP complex polymer featuring the ability to change the refractive index in accordance with its redox state was utilized along with glucose oxidase (GOD) to develop a SPR-enzyme sensor for glucose [29-32]. From this basic idea, we started the design and the measurements of multi-enzyme sensor using SPRi.

In this study, we aimed to develop a dual-enzyme sensor capable of the simultaneous detection of L-lysine and galactose by means of co-immobilizing LysOx/Os-HRP polymer spot and GalOx/Os-HRP polymer spot adjacently on the Au chip, which was divided into two sections.

## Materials and Methods

### Reagents and chemicals

All chemicals as enzyme substrates were purchased from Wako (Japan). L-lysine oxidase (from *Trichoderma viride*) and galactose oxidase (from *Dactylium dendroides*) were purchased from Sigma. Os complex polymer containing horseradish peroxidase (Os-HRP poymer) was purchased from BAS (Japan). Glutaraldehyde and the reduced form of nicotinamide adenine dinucleotide (NADH) were purchased from TCI (Japan) and Oriental Yeast Co. (Japan), respectively.

A chip of 50 nm Au thin film-coated high refractive index glass (SF6) (18 × 17 mm) was obtained from BAS (Japan) and flexiPERM® (11 × 7 × 10 mm) was obtained from Greiner Bio One (Germany).

**\*Corresponding author:** Hiroaki Shinohara, Graduate School of Innovative Life Science for Education, University of Toyama, 3190 Gofuku, Toyama 930-8555, Japan, E-mail: hshinoha@eng.u-toyama.ac.jp

## Fabrication of 2D-SPR sensor chip

The multi-sensor chip was prepared as follows. First, the Au thin film of a SPR sensor chip was separated into two sections by cutting the chip surface with a cutter knife. Then Os-HRP polymer solution was dropped on both sides of the cutted Au chip with 2 mm interval and dried at 4°C. Next, mixture of L-lysine oxidase (LysOx) solution or galactose oxidase (GalOx) solution and 1% glutaraldehyde solution was dropped onto each Os polymer spot and dried at 4°C. A silicon well (flexiPERM®) was attached onto the dual-enzyme chip to make measurement chamber, and the chamber was filled with a PB solution (0.1 M, pH 6.8) before measurements. The sensor chip was then set on the 2D-SPR measurement prism.

## 2D-SPR experimental setup and measurement

All sensing experiments were performed with 2D-SPR sensor (SPR imager 04A, NTT Advanced Technology [NTT-AT], Japan), which was equipped with a collimator lens, P and S-polarizer, four kinds of magnification lenses (× 1, 2, 4 and 7) and a CCD camera. A 770 nm LED was used as an incident light source to arouse SPR after passing through a collimator lens and P-polarizer.

SPR curves were obtained by changing measurement angle to determine the resonanse angle (θr) at each enzyme spot, and optimum angle for time-course measurement of reflection intensity before and after injection of the sample solution containing lysine and galactose. Thereafter, time-course measurement of reflection intensity (RI) at the fixed angle was performed. At this time, several sample solutions were injected into the sensor chamber at 1 minute after record starting. RI change was observed at each spot, and all the data were recorded and analyzed by analysis software.

After these experiments, oxidized-state of Os polymer at each enzyme spot was re-reduced by NADH solution to use this sensor repeatedly.

## Results and Discussion

Detection principle of this sensing system for lysine and galactose is described as follows. First, lysine or galactose was oxidized to produce $H_2O_2$ at each enzyme spot when sample solutions containing each substrate are injected into the sensor chamber. Os complex polymer was transformed from reduced form ($Os^{2+}$) to oxidized form ($Os^{3+}$) by $H_2O_2$ with HRP catalysis [29]. Refractive index change caused by the oxidation of Os complex polymer was observed as RI change [30,31], and simultaneous detection of lysine and galactose at each spot was possible.

Figure 1 shows SPR curves at LysOx/Os polymer spot, GalOx/Os polymer spot and bare Au region in the angle scan mode. Resonance angles (θr) were 55.6°, 56.4° and 51.6°, respectively. In case of Os complex polymer and enzymes, whose refraction index are higher than that of water on Au chip, θr showed higher, and it indicates that each enzyme/Os polymer spot was certainly immobilized. It was inferred that higher θr of GalOx/Os polymer as compared with that of LysOx/Os polymer spot may be due to the large extent of cross-linkng between GalOx and glutaraldehyde, and the large mass of GalOx existed on the Os polymer layer.

Time-course measurement was performed at a fixed angle (55.5°) determined by the angle scan measurement. We confirmed previously that the enzyme reaction resulted in negative shift of the θr because oxidized form of Os complex polymer has lower refractive index as compared with reduced form. Consequently, it is expected that the

RI at LysOx/Os polymer spot should increase and the RI at GalOx/Os poymer spot should decrease at θm = 55.5° by addition of lysine and galactose. Figure 2 shows the RI change at the LysOx/Os polymer spot and the GalOx/Os poymer spot and SPR images of those regions. RI increase was observed at LysOx/Os polymer spot upon lysine injection, whereas no significant change of RI at GalOx/Os poymer spot was observed. In contrast, RI decrease clearly occurred only at GalOx/Os poymer spot after galactose addition as we expected.

The results of L-lysine measurement performed in a mixed sample solution containing various concentration of lysine and constant concentration of galactose (1 mM) are shown in Figure 3. RI increased in proportion to the lysine concentration at LysOx/Os polymer spot. On the other hand, no change of the RI was observed at GalOx/Os poymer spot. RI decrease according to the galactose concentration was also observed at GalOx/Os poymer spot, whereas RI change at LysOx/Os polymer spot was nothing (Data not shown).

Figure 4 shows the dependence of reaction rate (response) of each enzyme spot on the lysine and galactose concentration. It was demonstrated that lysine can be determined from 20 to 200 μM without influence of galactose, and galactose also can be sensed in the rage from 50 to 1000 μM without influence of lysine. It was confirmed that calibration curves of dual-sensing were almost same as compared with that of each single sensing. The mean and SD values at LysOx and GalOx spot were calculated at 1.26 ± 0.141%/min and 4.55 ± 0.392%/min in a concentration (100 μM Lys at LysOx spot, 1 mM Galactose at GalOx spot, n=7) on a chip, respectively. And CV values were also calculated at 11.2% and 8.6%, respectively. These results demonstrated that developed dual-enzyme sensing system has capability of measuring these two analytes simultaneously at each enzyme spot without any

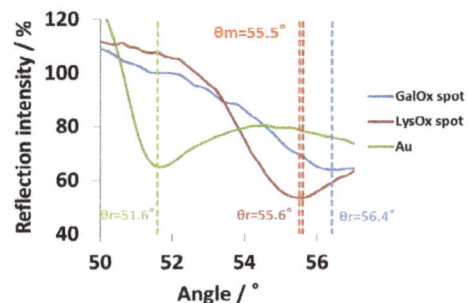

**Figure 1:** SPR curves at LysOx/Os-HRP polymer spot, GalOx/Os-HRP polymer spot and bare Au region.

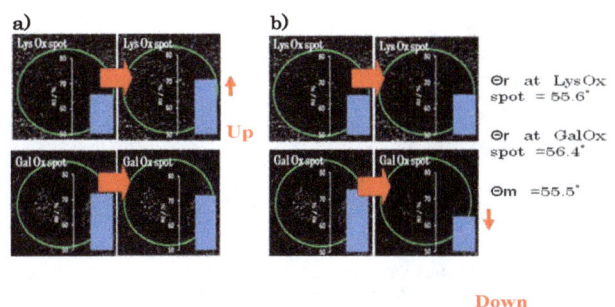

**Figure 2:** SPR images and reflection intensity at each enzyme spot before and after injection of a) 1 mM lysine or b) 1 mM galactose.

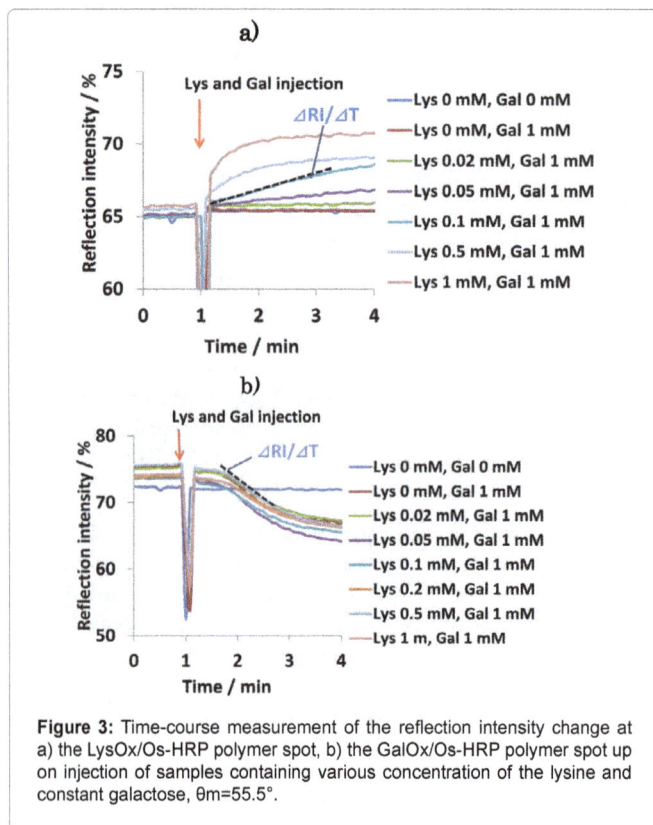

**Figure 3:** Time-course measurement of the reflection intensity change at a) the LysOx/Os-HRP polymer spot, b) the GalOx/Os-HRP polymer spot up on injection of samples containing various concentration of the lysine and constant galactose, θm=55.5°.

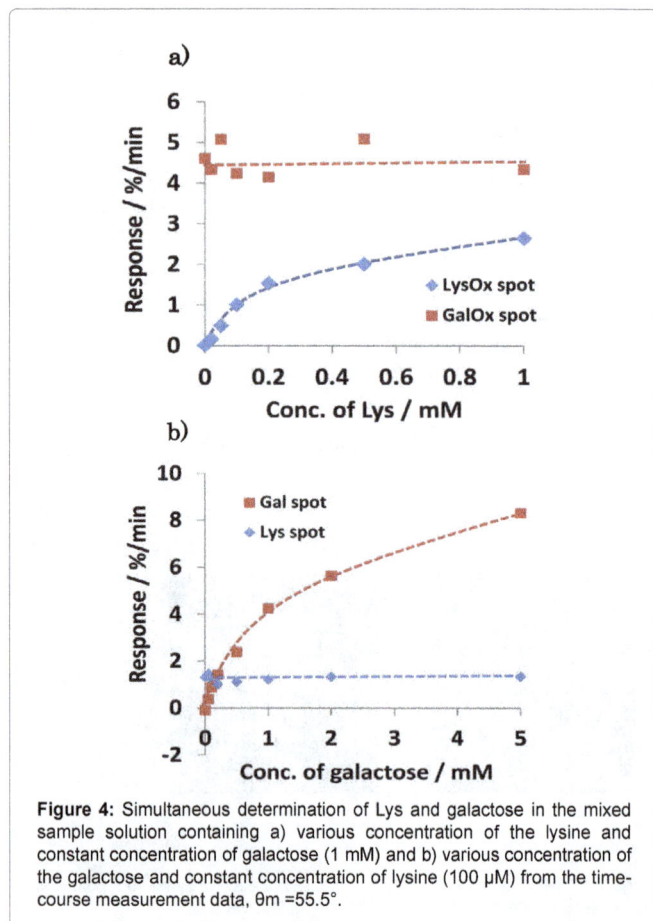

**Figure 4:** Simultaneous determination of Lys and galactose in the mixed sample solution containing a) various concentration of the lysine and constant concentration of galactose (1 mM) and b) various concentration of the galactose and constant concentration of lysine (100 μM) from the time-course measurement data, θm =55.5°.

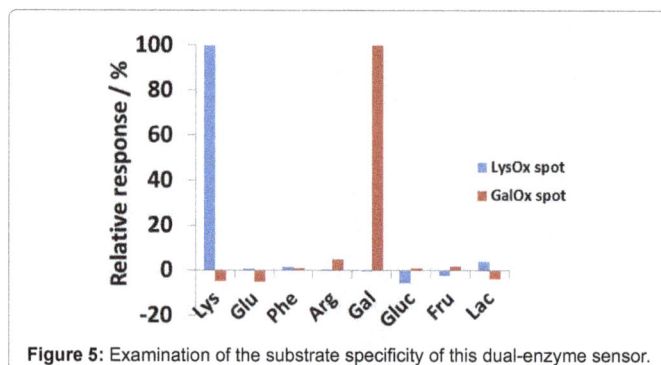

**Figure 5:** Examination of the substrate specificity of this dual-enzyme sensor.

cross-talk. Substrate specificity was also investigated with several kinds of saccharides and amino acids in Figure 5 (It was tested for separate amino acids and saccharides samples). As we expected, each enzyme spot showed high specificity for lysine and galactose respectively comparing other saccharides and amino acids.

## Conclusion

In this paper, we proposed 2D-SPR based dual-enzyme sensing system by which simultaneous determination of each substrate was feasible without cross-talk by using an Au chip separated into two sections. L-lysine oxidase and galactose oxidase were chosen for sensing oxidase in this time and combined with Os-HRP complex polymer to prepare two adjacent enzyme spots (LysOx/HRP-Os spot and GalOx/HRP-Os spot) on an Au sensor chip. L-lysine and galactose were determined specifically at each spot from 20 to 200 μM and from 50 to 1000 μM, respectively. This sensor system also has great advantage of simple and quick measurement (within 2 min without pretreatment). Then, we will try to demonstrate accuracy of this method for plasma sample, in which lysine and galactose were determined with other methods like HPLC as future prospects. These results suggested that our sensing strategy based on a 2D-SPRi was capable to measure many kinds of enzyme substrates simultaneously and applicable to medical screening.

## References

1. Wang TJ, Larson MG, Vasan RS, Cheng S, Rhee EP, et al. (2011) Metabolite profiles and the risk of developing diabetes. Nat Med 17: 448-453.

2. Neville S, O'Sullivan S, Sweeney B, Lynch B, Hanrahan D, et al. (2015) Friedreich Ataxia in Classical Galactosaemia. JIMD Rep. pp: 1-5.

3. Timson DJ (2015) The molecular basis of galactosemia-Past, present and future. Gene (In Press).

4. Pyhtila BM, Shaw KA, Neumann SE, Fridovich-Keil JL (2015) Newborn screening for galactosemia in the United States: looking back, looking around, and looking ahead. JIMD Rep 15: 79-93.

5. Seki M, Takizawa T, Suzuki S, Shimizu T, Shibata H, et al. (2015) Adult phenylketonuria presenting with subacute severe neurologic symptoms. J Clin Neurosci 22: 1361-1363.

6. Lee JY, Chiong MA, Estrada SC, Cutiongco-De la Paz EM, Silao CLT, et al. (2008) Maple syrup urine disease (MSUD)-Clinical profile of 47 Filipino patients. J Inherit Metab Dis 31: S281-S285.

7. Wang XL, Li CJ, Xing Y, Yang YH, Jia JP (2015) Hypervalinemia and hyperleucine-isoleucinemia caused by mutations in the branched-chain-amino-acid aminotransferase gene. J Inherit Metab Dis 38: 855-861.

8. Stabler S, Koyama T, Zhao Z, Martinez-Ferrer M, Allen RH, et al. (2011) Serum Methionine Metabolites Are Risk Factors for Metastatic Prostate Cancer Progression. PLoS ONE 6: e22486.

9. Saudubray JM, Rabier D (2007) Biomarkers identified in inborn errors for lysine, arginine, and ornithine. J Nutr 137: 1669S-1672S.

10. Houten SM, Te Brinke H, Denis S, Ruiter JP, Knegt AC, et al. (2013) Genetic basis of hyperlysinemia. Orphanet J Rare Dis 8: 57.

11. Sacksteder KA, Biery BJ, Morrell JC, Goodman BK, Geisbrecht BV, et al. (2000) Identification of the alpha-aminoadipic semialdehyde synthase gene, which is defective in familial hyperlysinemia. Am J Hum Genet 66: 1736-1743.

12. Noguchi Y, Zhang QW, Sugimoto T, Furuhata Y, Sakai R, et al. (2006) Network analysis of plasma andtissue amino acids and thegeneration of an amino index for potential diagnostic use. Am J Clin Nutr 83: 513S-519S.

13. Miyagi Y, Higashiyama M, Gochi A, Akaike M, Ishikawa T, et al. (2011) Plasma Free Amino Acid Profiling of Five Types of Cancer Patients and Its Application for Early Detection. PLOS ONE 6: e24143.

14. Fukutake N, Ueno M, Hiraoka N, Shimada K, Shiraishi K, et al. (2015) A Novel Multivariate Index for Pancreatic Cancer Detection Based On the Plasma Free Amino Acid Profile. PLOS ONE 10: e0132223.

15. Yamakado M, Nagao K, Imaizumi A, Tani M, Toda A, et al. (2015) Plasma Free Amino Acid Profiles Predict Four-Year Risk of Developing Diabetes, Metabolic Syndrome, Dyslipidemia, and Hypertension in Japanese Population. Scientific Reports 5: 11918.

16. Yanagisawa R, Kataoka M, Inami T, Momose Y, Kawakami T, et al. (2015) Usefulness of circulating amino acid profile and Fischer ratio to predict severity of pulmonary hypertension. Am J Cardiol 115: 831-836.

17. Lake AD, Novak P, Shipkova P, Aranibar N, Robertson DG, et al. (2015) Branched chain amino acid metabolism profiles in progressive human nonalcoholic fatty liver disease. Amino Acids 47: 603-615.

18. Cheng F, Wang Z, Huang Y, Duan Y, Wang X (2015) Investigation of salivary free amino acid profile for early diagnosis of breast cancer with ultra performance liquid chromatography-mass spectrometry. Clin Chim Acta 447: 23-31.

19. Shimbo K, Oonuki T, Yahashi A, Hirayama K, Miyano H (2009) Precolumn derivatization reagents for high-speed analysis of amines and amino acids in biological fluid using liquid chromatography/electrospray ionization tandem mass spectrometry. Rapid Commun Mass Spectrom 23: 1483-1492.

20. Waterval WA, Scheijen JL, Ortmans-Ploemen MM, Habets-van der Poel CD, Bierau J (2009) Quantitative UPLC-MS/MS analysis of underivatised amino acids in body fluids is a reliable tool for the diagnosis and follow-up of patients with inborn errors of metabolism. Clin Chim Acta 407: 36-42.

21. Yoshida H, Kondo K, Yamamoto H, Kageyama N, Ozawa S, et al. (2015) Validation of an analytical method for human plasma free amino acids by high-performance liquid chromatography ionization mass spectrometry using automated precolumn derivatization. J Chromatogr B Analyt Technol Biomed Life Sci 998-999: 88-96.

22. Uludag Y, Tothill IE (2012) Cancer biomarker detection in serum samples using surface plasmon resonance and quartz crystal microbalance sensors with nanoparticle signal amplification. Anal Chem 84: 5898-5904.

23. Wegner GJ, Lee H, Marriott G, Corn RM (2003) Fabrication of Histidine-Tagged Fusion Protein Arrays for Surface Plasmon Resonance Imaging Studies of Protein-Protein and Protein-DNA Interactions. Anal Chem 75: 4740-4746.

24. Sakai T, Shinahara K, Torimaru A, Tanaka H, Shoyama Y, et al. (2004) Sensitive detection of glycyrrhizin and evaluation of the affinity constants by a surface Plasmon resonance-based immunosensor. Anal Sciences 20: 279-283.

25. Altintas Z, Uludaga Y, Gurbuzb Y, Tothill IE (2011) Surface Plasmon resonance based immunosensor for the detection of the cancer biomarker carcinoembryonic antigen. Talanta 86: 377-383.

26. Hemmi A, Mizumura R, Kawanishi R, Nakajima H, Zeng H, et al. (2013) Development of a novel two dimensional surface plasmon resonance sensor using multiplied beam splitting optics. Sensors (Basel) 13: 801-812.

27. Wegner GJ, Wark AW, Lee HJ, Codner E, Saeki T, et al. (2004) Real-Time Surface Plasmon Resonance Imaging Measurements for the Multiplexed Determination of Protein Adsorption/Desorption Kinetics and Surface Enzymatic Reactions on Peptide Microarrays. Anal Chem 76: 5677-5684.

28. Nelson BP, Grimsrud TE, Liles MR, Goodman RM, Corn RM (2001) Surface plasmon resonance imaging measurements of DNA and RNA hybridization adsorption onto DNA microarrays. Anal Chem 73: 1-7.

29. Vreeke M, Maidan R, Heller A (1992) Hydrogen Peroxide and beta-Nicotinamide Adenine Dinucleotide Sensing Amperometric Electrodes Based on Electrical Connection of Horseradish Peroxidase Redox Centers to Electrodes through a Three-Dimensional Electron Relaying Polymer Network. Anal Chem 64: 3084-3090.

30. Koide S, Iwasaki Y, Horiuchi T, Niwa O, Tamiya E, et al. (2000) A novel biosensor using electrochemical surface plasmon resonance measurements. Chem Commun 2000: 741-742.

31. Iwasaki Y, Horiuchi T, Niwa O (2001) Detection of electrochemical enzymatic reactions by surface plasmon resonance measurement. Anal Chem 73: 1595-1598.

32. Iwasaki Y, Tobita T, Kurihara K, Horiuchi T, Suzuki K, et al. (2002) Imaging of electrochemical enzyme sensor on gold electrode using surface plasmon resonance. Biosens Bioelectron 17: 783-788.

# 16

# Assessing Stability, Durability, and Protein Adsorption Behavior of Hydrophilic Silane Coatings in Glass Microchannels

Sean Williams, Neeraja Venkateswaran, Travis Del Bonis O'Donnell, Pete Crisalli, Sameh Helmy, Maria Teresa Napoli* and Sumita Pennathur

*Mechanical Engineering Department, University of California Santa Barbara, Santa Barbara, CA 93106, USA*

## Abstract

Microfluidics-based separation of biomolecules has numerous applications, including fundamental characterization of biomolecules, sequencing of genomes for biological functions, biometric fingerprinting, and identification of pathogens and genetic diseases. One of the main drawbacks, however, for making microfluidic based separations more commercially viable is the non-specific adsorption of biomolecules at the channel walls during separations. Herein, we compare five commonly employed surface coatings, and evaluate their performance in terms of successful silanization of channel surface walls, long term stability, and antifouling performance, using BSA or IgG as model proteins. We compare adsorption of fluorescently-tagged proteins on glass slides with those confined within channels, showing similar behavior with static measurements, but differences when incorporating electrokinetic flow. Based on these data, we find that MPEG is an effective surface coating for applications where long term stability is critical. However, for separation experiments, where the channel is used shortly after coating, a silanized zwitterionic sultone has superior anti-fouling characteristics.

**Keywords:** Non-specific adsorption; Microfluidic separations; Protein-surface interactions; Surface coatings; Silane coatings

## Introduction

Advances in microfabrication technology, and the consequent ability to precisely control geometrical features down to the sub-micron scale, have enabled the development of novel micro- and nanofluidic platforms. This new generation of miniaturized devices allows for faster analysis on smaller sample volumes, integration of multiple functionalities on the same platform, and better sensitivity at a lower cost. Furthermore, such devices have also enabled novel applications and methods that exploit the physics of fluids in micro and nanoconfinements, from novel separation modalities [1,2], to fluid pumping by diffusio-osmosis [3], to energy conversion [4].

As more research focuses on medical and diagnostic applications involving micro- and nanofluidic devices, the issue of non-specific adsorption of biomolecules at the channel surface is gaining increasing attention [5-7]. In the case of micro-scale electrokinetic separations, for example, unwanted wall-sample interactions lead to non-uniform charge density on the inner capillary wall and, thus, to non-uniform electroosmotic flow profiles [6,8], subsequently causing sample dispersion and degradation of separation resolution Furthermore, such interactions affect peak shape and electromigration times, potentially tainting analysis results [7,9,10]. Finally, loss of sample and degradation of device functionality are additional pressing concerns.

In general, the main mechanisms of nonspecific interactions between biomolecules and solid surfaces are electrostatic (ionic), hydrophobic and hydrogen bond interactions [10,11]. Electrostatic interactions arise from Coulombic attraction or repulsion between charged groups on a biomolecule, and are highly affected by pH. Hydrophobic interactions can cause biomolecules to aggregate, and also play a major role in biomolecular adsorption from water onto hydrophobic surfaces [11,12]. Among biomolecules, protein adsorption to surfaces poses a unique challenge, apparent through the plethora of research available [13-18]. Proteins often denature upon adsorbing and become very difficult to remove from the walls, limiting both device reliability and reusability [11,19,20].

A straightforward way to control protein surface adsorption is to use buffers at pH extremes, to either operate above the isoelectric point of proteins, or suppress ionization of silanol groups at the channel walls [10,21]. Under these conditions, electrostatic repulsion forces prevent proteins from adsorbing to the wall. However, the abundance of OH⁻ ions at pH>11, and H⁺ at pH<2 can result in the generation of large currents and Joule heating. In addition, the buffering power of most electrolytes is very poor at these pH extremes [21]. More importantly, these conditions are not suitable to study the activity of proteins in native environments. More sophisticated methods to eliminate or reduce adsorption include studies to control the wall potential by means of an external electric field [22,23], or using additives in the buffer solution to compete for cation-exchange sites on the wall [24-27]. However, by far the most popular method to mitigate adsorption is the use of surface coatings [27-30]. Not only is their use widespread in microscale devices [27,31-33]: research to develop new types of chemical coatings is also flourishing [34,35].

Ideally, surface coatings should possess a variety of properties including: biocompatibility, hydrophilicity, electrical neutrality, hydrogen bond acceptance, self-repulsion, and conformational flexibility [36,37]. In recent years, a number of researchers have synthesized and characterized many surface coatings [9,35,38-40] intended to prevent analyte adsorption within micro- and nanofluidic devices [27,31,32]. However, there is no comparative quantitative data between promising coatings, and meaningful comparisons of data between different experimental studies are not always possible [5,39]. This paper aims to fill this gap by providing a quantitative comparative analysis of surface coatings for the purpose of assessing their properties

*Corresponding author: Maria Teresa Napoli, Mechanical Engineering Department, University of California Santa Barbara, Santa Barbara, CA 93106, USA, E-mail: napoli@engineering.ucsb.edu

and behaviors in controlled conditions relevant to microfluidics, specifically for microfluidic based protein separations. We chose to use bovine serum albumin (BSA) and immunoglobulin G (IgG) as model proteins, given their propensity to adsorb to a variety of different surfaces, and the fact that they are the two most abundant proteins in blood. In addition, the behavior of BSA and IgG may be used as an indication for respectively, other serum or globular proteins (BSA) and antibodies (IgG).

We chose to assess the performance of some of the most common surface coatings: two hydrophilic (2-[methoxy(polyethyleneoxy)] propyl trimethoxysilane (MPEG) and 3-mercaptopropyl (3-MPS) silane), two ionic (a zwitterionic sultone derived silane (ZS), and a zwitterionic phosphate derived silane (ZP)), and 3-aminopropyl (3-APS) silane. We compare these coatings with regards to: a) the ability to successfully silanize planar microfluidic channels (since the procedures we propose are adapted from existing recipes used to coat various silica substrates such as glass slides and beads), b) stability and durability over time via zeta potential ($\zeta$) measurements, and c) antifouling performance through protein interactions.

## Materials and Methods

### Materials

We used two types of fused silica, isotropically-etched microfluidic devices for this study: simple cross (N, W, S: 5 mm long, E: 30 mm long), 20 μm deep × 50 μm wide channels (Dolomite Ltd, UK) for competitive electrophoresis injection experiments, and straight custom-fabricated 20 μm deep × 25 mm long channels for continuous flow experiments (Figure 1a and b respectively). We also used plain Borosilicate coverslips (Corning, Catalog Number 2865-22) as a control surface with no continuous flow or confinement.

### Chemicals and reagents for surface coating

**Synthesis:** All chemicals and reagents were purchased from Sigma-Aldrich with the exception of MPEG-silane which was purchased from Gelest. 2-(dimethyl(3-(trimethoxysilyl)propyl)ammonio)ethane-1-sulfonate (zwitterionic sultone silane, ZS-silane) was prepared by the reaction of 1,3-propanesultone with *N,N*-dimethyl-3-(trimethoxysilyl) propan-1-amine in acetone according to the procedure outlined in Ref. [41]. 2-(dimethyl(3-(trimethoxysilyl)propyl)ammonio)ethyl ethyl phosphate (zwitterionic phosphate silane, ZP-silane) was prepared in two steps according to the procedure outlined in Ref. [42].

**Silanization on glass slides:** Borosilicate coverslips (Corning, Catalog Number 2865-22) were modified by immersion in a solution containing 3-4% w/v silane (in ethanol, water or toluene) and incubated at 40°C for 18 hours. The coverslips were then removed from the incubation solution, washed with ethanol and water, dried, and stored at ambient temperature [43]. MPEG-silane and zwitterionic silanes were incorporated using methods reported in Ref. [42,44].

**Silanization in channels:** MPEG-silane and ZS-silane were prepared as 20 mL of a 4% w/v solution in deionized water. 10 μL of the desired silane solution was added to a single well of the microchannel. Upon full channel wetting (as confirmed by light microscopy), the remaining wells were filled with 10 μL of the silane solution and the entire microchannel chip was immersed in the silane solution and incubated at 80°C for 18 hours. The microchannel chip was then removed from the silane solution and the exterior washed extensively with DI water. The microchannel was rinsed by inducing an electrokinetic flow (applied voltage V=100 V) using DI water for thirty minutes. 3-APS, 3-MPS and ZP-silane were prepared as 20 mL of

a 2% to 4% w/v solution of the silane in EtOH (3-APS and ZP-silane) or toluene (3-MPS) and 10 μL of the silane solution was added to a single well of the microchannel. Upon full channel wetting (as confirmed by light microscopy), the remaining wells were filled with 10 μL of the silane solution and the entire microchannel chip was immersed in the silane solution and incubated at room temperature for 3-APS and ZP-silane or 45°C for 3-MPS, overnight. After modification was complete, the microchannel chip was removed from the silane solution and the exterior of the chip was washed with EtOH. The microchannel was then flushed by inducing an electrokinetic flow of EtOH, 1:1 EtOH:DI water, and DI water in successive 30 minute periods (applied voltage V=100 V). All microchannels were stored in DI water at ambient temperature until use.

**Protein solutions and slide incubation:** We incubated modified coverslips in 10 mM sodium borate buffer (pH 9) with either: a) FITC-labelled bovine serum albumin (BSA), or b) immunoglobulin G (IgG). Figure 2 depicts electrostatic potential maps for both proteins. Protein concentrations for BSA and IgG were prepared in concentrations of 1 mg/mL and 2 mg/mL, respectively. We used all solutions immediately after preparation to prevent protein degradation. For adsorption experiments, we immersed glass slides in the appropriate protein solution and allowed it to incubate for 24 hours in the dark, to prevent bleaching. Before performing fluorescence imaging measurements, all glass slides were soaked twice in DI water for 2 hours with agitation, to remove excess protein.

### Data collection methods

**Fluorescence imaging:** We measured all fluorescence data by recording images with an inverted epifluorescence microscope (Olympus IX70, Olympus, Inc.) fitted with a 20X water immersion objective lens (0.45 NA, Olympus, Inc.). Illumination from a 200 W Hg-arc lamp was filtered with a FITC fluorescence filter cube (Omega, Inc.) containing excitation and emission filters and a dichroic mirror matched to the fluorescence spectrum of FITC. We recorded images using a back illuminated EMCCD camera (Ixon Ultra897, Andor Technology) with a 512 × 512 pixel array and 16-bit digitization to a PC. Frame rate (10 to 20 Hz) and exposure time (0.05 to 0.1 s) varied, depending on the channel depth and analyte, to maximize signal-

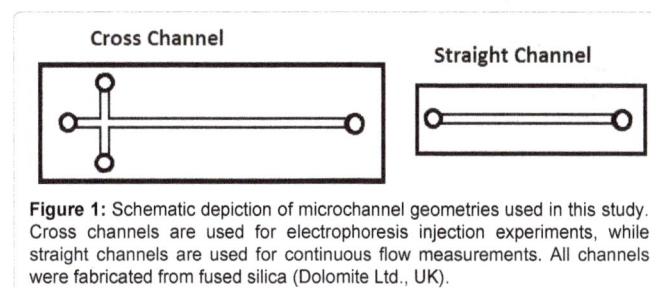

**Figure 1:** Schematic depiction of microchannel geometries used in this study. Cross channels are used for electrophoresis injection experiments, while straight channels are used for continuous flow measurements. All channels were fabricated from fused silica (Dolomite Ltd., UK).

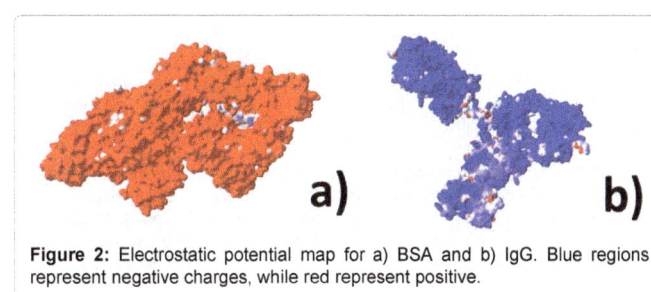

**Figure 2:** Electrostatic potential map for a) BSA and b) IgG. Blue regions represent negative charges, while red represent positive.

to-noise ratio. We performed background subtraction and flatfield correction on all images using custom Matlab (The MathWorks) programs, to further enhance signal to noise ratio [45].

To measure protein adsorption on slides, after silanization and protein incubation as described above, we collected one image at three separate locations. For each location, we computed the average fluorescence value by randomly selecting ten pixels. For each combination of surface coating and protein, we performed experiments on two slides. Results, in terms of average absolute fluorescence and pooled variance, are shown in Figure 3.

For fluorescence data in straight channels, after surface coatings were applied to the channel as described above, the channels were rinsed with DI water, and then filled with 10 mM sodium borate buffer solution (pH=9). We applied an external voltage, V=300 V (Keithley 2410), through platinum electrodes (Omega Eng. Inc., Stamford, CT) to establish an electroosmotic flow. Next, protein solution was added to the inlet ports and fluorescence values were monitored over time, while the protein solution was electrokinetically driven in the channels. After fluorescence reached its maximum value (different for each coating and protein combination), channels were rinsed with a continuous flow of 10 mM sodium borate buffer. We also monitored fluorescence during this phase. Fluorescence values displayed in Figure 4 show the average fluorescence within the channel after background and flat field correction.

**Current monitoring:** Prior to all measurements, we electrokinetically drove deionized water into the straight channels (Figure 1b) by applying a voltage V=100 V (Keithley 2410) until the

current, monitored using a second electrometer (Keithley 2410) stabilized. Next, the channels were filled with a 9 mM sodium borate buffer solution (SBX). After current stabilization, we replaced the 9 mM SBX solution in reservoir 1 by 10 mM SBX, and recorded the increase in current, as the channel filled with the higher concentration buffer. We repeated this process 3-4 times for each coating. To derive the zeta-potential, we solve for bulk average electroosmotic velocity, $v_{EOF}$, from current monitoring using a custom Matlab script in which the length of the channel is divided by the time to fill the channel. From this value, the zeta-potential is found using the classic Helmholtz-Smoluchowski equation:

$$v_{eof} = \frac{\varepsilon E \zeta}{\mu}$$

where $\mu$ is viscosity ($8.90 \times 10^{-4}$ Pa for water), E is the applied electric field, and $\zeta$ is the zeta-potential [46]. For the dielectric constant $\varepsilon$, we use the value of water, given the very low molarity of the buffers used.

**Microfluidic injections:** Microfluidic injections are typically used to assess the electroosmotic mobility of fluorescently tagged species. This technique can also be used to study adsorption of fluorescent species by, for example, examining the fluorescent residue that remains on the channel wall once a fluorescently-tagged protein travels down the channel, or in our case, measuring the peak intensity over time after successive injections. Figure 1a shows a schematic of the channel geometry and our naming convention for each well: N (northern), S (southern), W (western) and E (eastern). Electrical potentials were applied at these wells using platinum electrodes (Omega Eng. Inc., Stamford, CT), connected to a high voltage power supply (LabSmith HVS448). The pre-programmed voltage scheme for sample loading and injection was designed following the recommendations of [47]. Briefly, during the loading phase, the sample solution is placed in the N well and electrodes in the N, W, and E wells are set to positive voltages while the S well is grounded, resulting in electrokinetic flow from all wells towards S. For the injection step, the applied voltages are then switched (W, N, S at high, E at ground), and the sample is injected along the E channel. The electric field during injections is 78.8 V/m. Injection data is recorded 10 mm downstream from the injection point with a high sensitivity EMCCD camera (Andor iXon), fitted to a 20X objective. The resulting electropherograms display fluorescence intensity over time, as plugs corresponding to different analytes pass through the detection point (For example: Figure 5).

For each experiment (Figures 5 and 6), the same sample is injected 20-30 times, by applying the voltage sequence inject - load - inject, with the load step long enough to allow for the previously injected plug to flow towards the S well. In these electropherograms, a net reduction in peak intensity and area indicates protein adsorption.

## Results

### Adsorption measurements on glass slides

We first measured the degree of protein absorption of the five chosen surface coatings (ZS, PEG, ZP, 3-MPS and 3-APS) on glass slides, using FITC-labeled BSA and IgG as model proteins. These surface coatings were chosen from an array of 7 previously analyzed [43]. Specifically, we chose MPEG because it is the proven standard hydrophilic silane coating, and is universally well-known to prevent surface adsorption. ZS and ZP were chosen as possible zwitterionic candidates (hydrophilic with neutral opposing ionic charges/forces), given that zwitterions are known to help prevent protein adsorption. 3-MPS and 3-APS were both chosen as positive controls, but through different mechanisms of

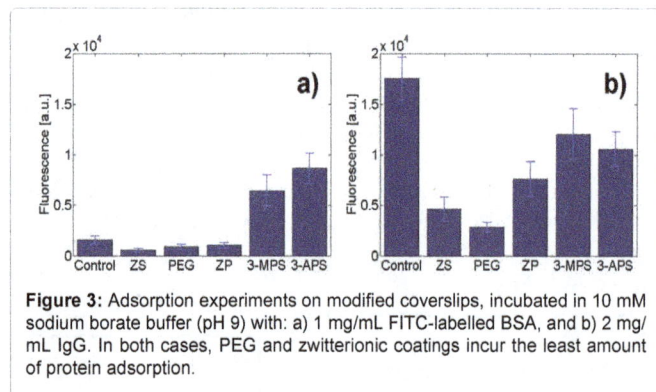

**Figure 3:** Adsorption experiments on modified coverslips, incubated in 10 mM sodium borate buffer (pH 9) with: a) 1 mg/mL FITC-labelled BSA, and b) 2 mg/mL IgG. In both cases, PEG and zwitterionic coatings incur the least amount of protein adsorption.

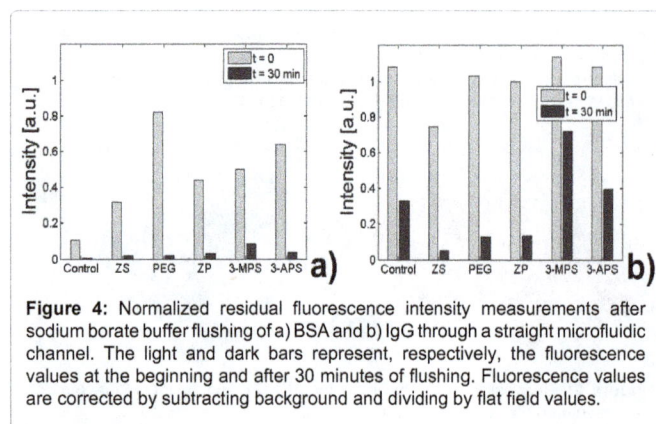

**Figure 4:** Normalized residual fluorescence intensity measurements after sodium borate buffer flushing of a) BSA and b) IgG through a straight microfluidic channel. The light and dark bars represent, respectively, the fluorescence values at the beginning and after 30 minutes of flushing. Fluorescence values are corrected by subtracting background and dividing by flat field values.

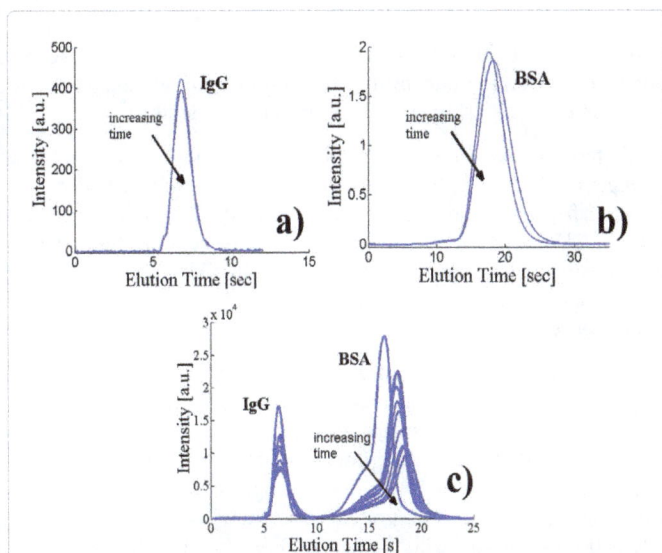

**Figure 5:** Electropherogram of FITC-labeled injections in a ZS-coated microfluidic chip (20 µm deep glass cross channel). Trials chosen to be representative of the trend. Protein sample is suspended in 10 mM sodium borate buffer (pH 9) and composed of: a) IgG, b) BSA, c) IgG and BSA.

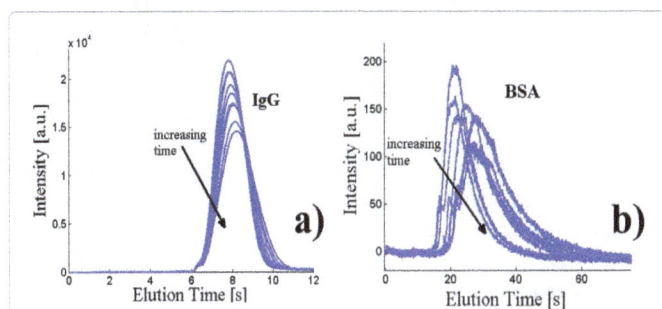

**Figure 6:** Electropherogram of IgG-FITC in a PEG coated (20 µm glass cross channel). Intensity (AU) over Elution time as shown. Trials chosen to be representative of the trend. Area under the curve was calculated for each trial and then percent area is displayed in reference to the initial trial (area max).

plausible adsorption [48,49]. 3-MPS was selected as a positive control for protein binding via disulfide linkages over 3-APS, which only provides charge based interactions.

Based on the data shown in Figure 3, which shows the residual fluorescence on the glass slide after protein adsorbed for 24 hours and was subsequently washed away with DI water. As expected, simple hydrophilic (MPEG) and zwitterionic coatings repelled proteins best, whereas hydrophilic coatings (3-MPS, 3-APS) were not effective at preventing adsorption, potentially due to the aforementioned disulfide linkages with 3-MPS or carboxylate-salt interactions between the protein c-terminus and 3-APS. The MPEG coating exhibited very low adsorption for both BSA and IgG, most likely because the formation of a denser hydration layer at the surface [50]. Zwitterions have the advantage of containing both positive and negative functional groups [51], which are responsible for ionic repulsive forces. In agreement with previous work [41], the heat-treated ZS coating resulted in the least residual fluorescence for BSA, as well as a very low value for IgG. Heat treatment is hypothesized to generate a more stable coating matrix and side chain connections between silanol groups, thus resulting in higher stability. ZP performed better than the control sample for both BSA and IgG, also in agreement with previous work [42].

## Stability of surface coatings

To assess the stability over time of surface coatings we performed current monitoring experiments over a period of 6 weeks (Figure 7). In microfluidics, current monitoring is a procedure that is typically used to measure the channel zeta-potential. The zeta-potential, $\zeta$ is a parameter that characterizes the surface potential at the shear plane of the channel, and thus can be used to characterize electroosmotic flow. It is further useful in studying adsorption, because analyte adsorption at the walls will change the initial surface, and consequently the value of the zeta-potential. Thus, $\zeta$ can be used as an effective indicator of surface properties. Each $\zeta$-potential value presented in Figure 7 represents the average of at least 3 measurements performed at each time point. We also calculated the average zeta potential over the 6 weeks measurements, $\zeta_{avg}$, shown in the table to the right of Figure 7. Assuming the same statistical distribution of the error for each measurement, we compute the variance by the pooled variance method [52]. Finally, we estimate the variability in zeta potential defining $\Delta\zeta_{avg}=(\zeta_{max} - \zeta_{min})/\zeta_{max}$. We note that potential complications for all coatings is that the electrokinetic flow may be causing shearing of the surface coatings, but current monitoring measurements are only performed for a small percentage of the time as compared to the 6 weeks that they were subjected to standard temperature and pressure (STP) conditions.

As expected, the (unmodified) control chip has the highest negative zeta potential [53]. The exposed silanol groups ionize when in contact with an aqueous solution at high pH, resulting in a high negative net surface charge. The control chip was the most stable over time, however with a surprisingly larger than expected variance $\Delta\zeta_{avg}$=9%. We attribute this non-zero variance over time to dissolution of the surface from OH⁻ groups [54] or from chemicals from the ambient atmosphere that attach and change the makeup of the surface [55].

The MPEG coating exhibited the lowest value of $\zeta_{avg}$ over the duration of the six-week experiment. Because the zeta potential was not consistently increasing over time, we attribute the large variation of zeta potential ($\Delta\zeta_{avg}$=54%) to the low absolute $\zeta$-potential value and thus relatively large contribution of experimental errors that are inherent with such a low measurement value. Changes in $\zeta$-potential could be due to oxidative effects on MPEG [37,44,56] induced by electrolysis. However, no electrolytic bubbles were observed, and the external field was a constant 100 V, so we assume that the MPEG coating did not degrade, and instead the variability is due to experimental error.

3-MPS is shown to be very stable, rivaling the stability of the uncoated channel with a $\Delta\zeta_{avg}$=12% (Figure 3). One possible explanation of this stability is disulfide formation between MPS moieties in the

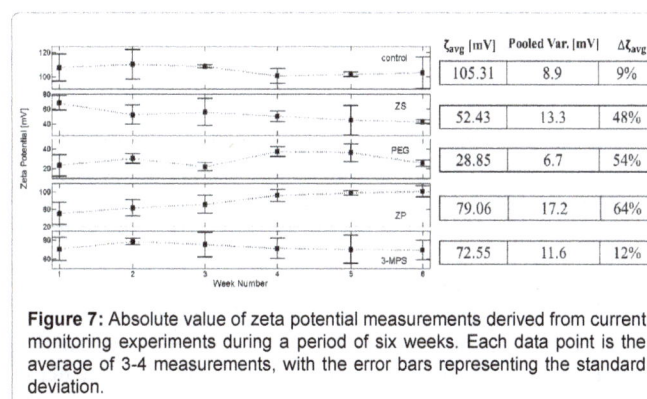

| | $\zeta_{avg}$ [mV] | Pooled Var. [mV] | $\Delta\zeta_{avg}$ |
|---|---|---|---|
| control | 105.31 | 8.9 | 9% |
| ZS | 52.43 | 13.3 | 48% |
| PEG | 28.85 | 6.7 | 54% |
| ZP | 79.06 | 17.2 | 64% |
| 3-MPS | 72.55 | 11.6 | 12% |

**Figure 7:** Absolute value of zeta potential measurements derived from current monitoring experiments during a period of six weeks. Each data point is the average of 3-4 measurements, with the error bars representing the standard deviation.

coating. If this is occurring, then silane hydrolysis would not result in full delamination, and thus the channel would retain its properties over an extended period of time.

Our results for channels coated with ZS do not correlate well with literature findings. Our data shows unstable behavior, with large variation in zeta-potential values over the 6 weeks, $\Delta\zeta_{avg}$=54%, and a large spread in each single measurement, with each measurement decreasing in value. We hypothesize that in our case, full polymerization and formation of a stable monolayer was not achieved, and instead single point silanization may have occurred, with concomitant oligomerization resulting in island morphologies and multi-layer formation. Note that because the zeta potential absolute value was between 40-70 mV, and it was monotonically decreasing over time, we do not believe that user and experimental-error contributed to the large variance.

ZP was the most unstable, with the $\zeta$-potential at the end of 6 weeks approaching that of an uncoated channel, suggesting that the coating completely degraded. Authors in Ref. [51] noted that ZP-like structures form densely packed hydration layers, binding water molecules more strongly than PEG. However, past work has dealt with the case of "static" surfaces that is in the absence of flow. Work by Ref. [42], using the exact same phosphate coating as the one in this study, found similar positive changes in contact angle over time in phosphate buffer, indicating degradation. The changes were attributed to weakly bound silanes diffusing out from the system. In light of these findings, we hypothesize that our observed behavior of ZP coating is the result of interactions in the presence of sodium borate buffer that, in the course of our measurements, remove weakly bound silanes via ionic interactions or hydrolyze the phosphodiester bond of the functional chain, thus shifting charge neutrality and density.

### Adsorption measurements in continuous flow

We next assessed the adsorption of proteins within channels with flow, since fluid flow may change the effects of adsorption as compared to static glass slide measurements. To do so, we monitored adsorption during protein flow and subsequent aqueous-based flushing of both BSA and IgG solutions on various coated microchannels (Figure 4).

As mentioned in the introduction, proteins may be reversibly attracted to hydrophilic surfaces through electrostatic forces. These interactions may lead to irreversible conformational changes and stretching of the proteins across the surface, which results in more electrostatic interactions [57]. Factors affecting such adsorption include: 1) surface coating density, which may generate an entropic and steric environment which is favorable/unfavorable for further attachment; 2) time dependence of the process of permanent adsorption, which is directly dependent on both flow velocity and the total time proteins are flushed through the channel [58]; 3) protein concentration (we used 1 mg/ml, based on results reported in Ref. [36,59]); and 4) washing procedures, which may remove protein not permanently bound, again dependent on both flow velocity and the total time of flushing.

The time t=0 in Figure 4 corresponds to the point where we started flushing the channel with buffer solution, after exposing the channels to a continuous solution of proteins for 20-30 min. During rinsing, fluorescence was measured at regular intervals, and the process was stopped when its value came as close as possible to the initial intensity. For each coating, this time was chosen to allow for the fluorescence signal to reach its maximum value and stabilize (Table 1). Since the coating affects zeta potential and thus electroosmotic flow mobility, for different coatings the tagged proteins take a different amount of time to enter and fill the channel. It is interesting to note, how the

different coatings perform at t=0, with e.g., MPEG showing the largest fluorescence value at the beginning of flushing in case of BSA, but going down to almost zero values by the end of flushing, something that could not be assessed with simple glass coverslip experiments.

In terms of experiments with IgG (Figure 4b and Table 1), we find stronger fouling characteristics than BSA for all coatings considered, as expected from glass slide experiments (Figure 3). Like for the coverslips, MPEG, ZP and ZS have the lowest values of residual fluorescence during rinsing. However, contrary to the glass slide experiments, it is 3-MPS, and not the bare channel, that has the largest value of residual fluorescence, similar to the BSA case.

We posit that the difference we observe in glass coverslips with respect to microfluidic channels stems from time-dependent and flow-dependent behavior of the adsorption process, which becomes apparent only in the latter experiments. In uncoated channels, entropic interactions with the ionized silanols on glass are expected to be a contributing mechanism for protein adsorption [44], with electrostatic interactions associated with the positive BSA sites and the hydrophilic glass surface [60]. The relative ease with which protein was removed from bare glass microchannels may have resulted from the large EOF velocity, due to the large zeta-potential (Figure 7). The EOF velocity may also have hindered attachment of BSA, compared to what is expected with glass slides [61]. In the case of IgG, the uncoated channel showed a 33% residual fluorescence with respect to peak value after 30 minutes. According to Ref. [60], IgG adsorption on borosilicate vials is driven mainly by electrostatic forces, which in turn are the product of pH and ionic forces. Thus, the combination of sodium borate buffer (pH 9) and exposed glass silanols is expected to promote IgG fouling in uncoated channels, in spite of the large EOF (Figure 7). The difference between the adsorption profiles of BSA and IgG from coverslips to channels illustrates the importance of experimental conditions (Figure 3). Specifically, flow velocity can affect BSA adsorption, while the choice of buffer clearly affects IgG adsorption.

Because of the low zeta-potential, MPEG results in a slower EOF velocity for the same applied voltage. As a consequence, the behavior we observe is diffusion limited and matches static coverslip results well. ZS and ZP coatings show excellent anti-fouling properties in all cases. These results are not surprising considering the strong hydration layer formed and described also in Ref. [51], along with potentially additional electrostatic and steric mechanisms described by Ref. [44], and results from other studies [62,42]. However, changes in zeta potential over time (Figure 7) suggest that the quality of the coating may deteriorate with time. Both 3-MPS and 3-APS as expected had the worst performance, potentially due to disulfide linkages with 3-MPS or carboxylate-salt interactions between the protein c-terminus and 3-APS.

### Adsorption measurements in electrokinetic flow

In consideration of data from all these experiments, we chose to further analyze the behavior of ZS and MPEG during electrokinetic separation injections of proteins in cross channel glass chips. The purpose behind evaluating competitive injections was to assess coating performance in discrete trials and with multiple proteins, as opposed to continuous exposure to a single protein. Furthermore, such injections are a cornerstone of microfluidic capillary electrophoresis, which has, through the years, been highlighted as a possible cheaper, faster, alternative method for performing separations of a variety of complex mixtures [63-70].

Figure 5 shows a series of injections through a ZS coated cross channel using sodium borate buffer solution containing FITC-labeled

| | Peak FL BSA | Residual FL BSA | Peak FL IgG | Residual FL IgG | Type | Structure |
|---|---|---|---|---|---|---|
| **Control** | 0.727 (7.5 min) | 0.0006 (0%) | 1.14 (20 min) | 0.329 (29%) | Reference | |
| **MPEG** | 0.96 (30 min) | 0.02 (2.1%) | 1.04 (15 min) | 0.13 (12.5%) | Hydrophilic Experimental | |
| **ZS** | 0.7 (20 min) | 0.02 (2.8%) | 1.01 (30 min) | 0.05 (4.9%) | Zwitterionic Experimental | |
| **3-APS** | 0.77 (15 min) | 0.036 (4.6%) | 1.1 (20 min) | 0.397 (36.1%) | Cationic Control | |
| **ZP** | 0.96 (20 min) | 0.05 (5.2%) | 1.03 (15 min) | 0.132 (12.8%) | Zwitterionic Experimental | |
| **3-MPS** | 0.87 (7.5 min) | 0.084 (9.65%) | 1.14 (20 min) | 0.72 (63.2%) | Fouling Control | |

**Table 1:** Displays relevant numerical data from antifouling experiments with IgG-FITC and BSA-FITC. Peak FL is the peak fluorescence for the trial, residual FL is remainder after 30 min. "Type" corresponds to role in the study.

IgG and BSA. The first two panels, Figure 5a and b, show that when BSA or IgG are injected alone, changes between serial injections are negligible. The arrival of the BSA peak in Figure 5b has a very small delay, which could indicate that over one injection the BSA absorbed to the channel, slightly changing the zeta potential, and thus arriving at a slightly later time. However, the results seem to be well within experimental error.

Figure 5c shows a series of electropherograms with both IgG and BSA injected at the same time. Here, the BSA peak changes drastically between the first two injections, with a major decrease in overall intensity, as well as a major shift in arrival time. The IgG peak intensity is also reduced over time, more significantly than shown in Figure 5a. The marked decrease in fluorescence as the measurements progress in time is an indication of sample adsorption at the channel walls: a process that seems to continue throughout the experiment. The absence of tailing suggests a practically irreversible process [10], which, contrary to experiments in glass slides (Figure 3), mostly affects BSA. It is known that, in the presence of multiple analytes, there is a competitive behavior for available binding sites, leading to one analyte prevailing, based on the mechanism of adsorption for the specific environment [59]. Further, competitive protein experiments are known to show protein dominance based on diffusivity [58] and it has been established that on hydrophilic media such as glass, albumin tends to adsorb first, which is then followed by IgG, based on size differences [71]. Our results also clearly indicate these phenomena at play, which result in drastic differences between single vs. multiple analyte electropherograms, and ultimately separation efficiency. Estephan et al. [44] outlined a mechanism for adsorption via ionic forces where zwitterionic sultone

(and like) do not allow the removal of counterion base pairs from the protein and surface respectively. This therefore inhibits attachment at the active site because of charge neutrality. Another explanation is that within this experimental design, BSA and IgG may be attracted via different mechanisms to ZS, not the glass.

Figure 6a and b show the same experiments, performed in MPEG coated channels. Since the BSA arrived much later than the IgG in a typical injection, and the intensity was very low, we were not able to perform the injection with both proteins in the same channel. However, even with the individual proteins, we can observe a similar behavior as in the ZS coated channel. As with ZS, the coating seems to be particularly effective in preventing IgG adsorption, while the BSA is barely discernible. The very low fluorescence value of the BSA peak suggests that most of the sample is irreversibly adsorbed to the wall already during the first injection. The shape of the BSA peak over time, with a pronounced tail, suggests that adsorption continue also during subsequent injections, but that this phenomenon is at least partially reversible. Similarly, the IgG peak shows a slight shift and broadening, which indicates that there are small amounts of absorption here as well.

## Conclusion

Surface coatings are increasingly common in microfluidics-based applications that focus on biomolecules and protein separations. In this paper, we have compared five commonly used coatings, and compared their performance in terms of successful silanization of channel surfaces, stability over time, and antifouling performance. BSA and IgG have been used as model proteins, given their abundance and adsorption behavior.

Fluorescence measurements of adsorption on glass slides and straight channels confirm that simple hydrophilic (MPEG) and zwitterionic coatings (ZP, ZS) are very effective at preventing adsorption of proteins, whereas hydrophilic coatings (3-MPS, 3-APS) are not. Similarly, measurements of residual fluorescence in straight channels exposed to electrokinetic buffer rinse confirm the superior performance of MPEG, ZP and ZS, as well as the stronger fouling characteristics of IgG compared to BSA. We hypothesize that the poor performance of 3-MPS and 3-APS is due to disulfide linkages with 3-MPS or carboxylate-salt interactions between the protein c-terminus and 3. In addition, flow within microchannels affects adsorption most likely due to a shearing effect at the wall, which can affect MPEG channels adversely.

Finally, given the widespread use of microfluidic separation techniques for identification and characterization of complex samples in a variety of applications, we have compared the behavior of MPEG and ZS during competitive injections of protein mixtures. Although it is not surprising that ZS shows a superior performance compared to MPEG, the very large adsorption of BSA in the MPEG-coated channel is unexpected from the glass slide and channel measurements.

Although ZS showed superior performance in regards to microfluidic separations, it is also important to consider the coating stability. Measurements of zeta-potential over a period of six weeks revealed that zwitterionic coatings tend to degrade over time. In case of ZS, we hypothesize that the behavior observed is the result of an island morphology and multi-layer formation in the coating, due to single point silanization, with concomitant oligomerization. For ZP, we hypothesize that interactions in the presence of sodium borate buffer removed weakly bound silanes via ionic interactions or hydrolyzed the phosphodiester bond of the functional chain, thus shifting charge neutrality and density.

Based on these findings, we conclude that MPEG is the most effective surface coating for applications where stability of the coating over time is critical. However, for separation experiments of BSA- or IgG-like proteins, where the channel is used shortly after coating, silanized ZS has superior anti-fouling characteristics, and should be used for the most accurate measurements. Importantly, our findings also highlight that, when optimizing microfluidic-/surface- based protein assays, sample composition needs to be taken into account, as different proteins may behave differently based on their structure and properties such as molecular weight (BSA: 66.5 kDa, Igg: 150-170 kDa), isoelectric point, flexibility, as well as maybe distribution of hydrophobic groups.

## Acknowledgement

This work was supported in part by the Institute for Collaborative Biotechnologies through grant W911NF-09-0001, and by grant W911NF-12-1-0031, both from the U.S. Army Research Office. The content of the information does not necessarily reflect the position or the policy of the Government, and no official endorsement should be inferred.

## References

1. Pennathur S, Santiago JG (2006) Electrokinetic Transport in Nanochannels. 1. Theory. Anal Chem 78: 972-973.

2. Huber DE, Markel ML, Pennathur S, Patel KD (2009) Oligonucleotide Hybridization and Free-solution Electrokinetic Separation in a Nanofluidic Device. Lab Chip 9: 2933-2940.

3. Ajdari A, Bocquet L (2006) Giant amplification of interfacially driven transport by hydrodynamic slip: diffusio-osmosis and beyond. Phys Rev Lett 96: 186102.

4. van der Heyden FH, Bonthuis DJ, Stein D, Meyer C, Dekker C (2007) Power generation by pressure-driven transport of ions in nanofluidic channels. Nano Lett 7: 1022-1025.

5. Pallandre A, de Lambert B, Attia R, Jonas AM, Viovy JL (2006) Surface treatment and characterization: Perspectives to electrophoresis and lab-on-chips. Electrophoresis 27: 584-610.

6. Taylor RB (1993) Capillary electrophoresis-priciples, practice and applications: S. F. Y. Li, Elsevier, Amsterdam, 1992. Pages: xxvi + 582. Dfl 395.00. ISBN 0-444-89433-0. Talanta 40: 770.

7. Bello MS, Zhukov MY, Righetti PG (1995) Combined effects of non-linear electrophoresis and non-linear chromatography on concentration profiles in capillary electrophoresis. J Chromatogr A 693: 113-130.

8. Dolník V (2006) Capillary electrophoresis of proteins 2003-2005. Electrophoresis 27: 126-141.

9. Gubala V, Siegrist J, Monaghan R, O'Reilly B, Gandhiraman RP, et al. (2013) Simple approach to study biomolecule adsorption in polymeric microfluidic channels. Anal Chim Acta 760: 75-82.

10. Ermakov SV, Zhukov MY, Capelli L, Righetti PG (1995) Wall adsorption in capillary electrophoresis. Experimental study and computer simulation. J Chromatogr A 699: 297-313.

11. Yoon J-Y, Kim C-J, Garrell RL (2003) Preventing biomolecular adsorption in electrowetting-based biofluidic chips. Anal Chem 75: 5097-5102.

12. Kim J (2002) Protein adsorption on polymer particles. J Biomed Mater Res 2: 4373-438.

13. Horbett TA, Brash JL (1995) Proteins at Interfaces II Fundamentals and Applications. ACS Symposium series.

14. Malmsten M (1998) Formation of Adsorbed Protein Layers. J Colloid Interface Sci 207: 186-199.

15. Mrksich M, Whitesides GM (1996) Using self-assembled monolayers to understand the interactions of man-made surfaces with proteins and cells. Annu Rev Biophys Biomol Struct 25: 55-78.

16. Ostuni E, Chapman RG, Liang MN, Meluleni G, Pier G, et al. (2001) Self-Assembled Monolayers That Resist the Adsorption of Cells. Langmuir 17: 6336-6343.

17. Szleifer I (1997) Polymers and proteins: interactions at interfaces. Curr Opin Solid State Mater Sci 2: 337-344.

18. Yuan Y, Oberholzer MR, Lenhoff AM (2000) Size does matter: Electrostatically determined surface coverage trends in protein and colloid adsorption. Colloids Surfaces A Physicochem Eng Asp 165: 125-141.

19. Yoon J-Y, Garrell RL (2008) Biomolecular Adsorption in Microfluidics. In: Li D (Ed.) Encyclopedia of Microfluidics and Nanofluidics, Springer pp. 68-76.

20. Norde W (1986) Adsorption of proteins from solution at the solid-liquid interface. Adv Colloid Interface Sci 25: 267-340.

21. Chen AB, Nashabeh W, Wehr T (2001) CE in Biotechnology: Practical Applications for Protein and Peptide Analyses. Chromatographia CE Series, Springer.

22. Lee C, Blanchard W, Wu C (1990) Direct control of the electroosmosis in capillary zone electrophoresis by using an external electric field. Anal Chem 62: 1550-1552.

23. Hayes MA, Kheterpal I, Ewing AG (1993) Electroosmotic flow control and surface conductance in capillary zone electrophoresis. Anal Chem 65: 2010-2013.

24. Green JS, Jorgenson W (1989) Minimizing Adsorption Of Proteins On Fused Silica In Capillary Zone Electrophoresis By The Addition Of Alkali Metal Salts To The Buffers. J Chromatogr 478: 63-70.

25. Bushey MM, Jorgenson JW (1989) Capillary electrophoresis of proteins in buffers containing high concentrations of zwitterionic salts. J Chromatogr 480: 301-310.

26. Lauer HH, McManigill D (1986) Capillary zone electrophoresis of proteins in untreated fused silica tubing. Anal Chem 58: 166-170.

27. Belder D, Ludwig M (2003) Surface modification in microchip electrophoresis. Electrophoresis 24: 3595-3606.

28. Hjertén S (1985) High-performance electrophoresis Elimination of electroendosmosis and solute adsorption. J Chromatogr A 347: 191-198.

29. Nashabeh W, El Rassi Z (1991) Capillary zone electrophoresis of proteins with hydrophilic fused-silica capillaries. J Chromatogr 559: 367-383.

30. Bruin GJM, Kuhlman RH, Zegers K, Kraak JC, Poppe H (1989) Capillary Zone Electrophoretic Separations Of Proteins In Polyethylene Glycol-Modified Capillaries. J Chromatogr 47: 429-436.

31. van Reenen A, de Jong AM, den Toonder JMJ, Prins MWJ (2014) Integrated lab-on-chip biosensing systems based on magnetic particle actuation - a comprehensive review. Lab Chip 14: 1966-1986.

32. Milanova D, Chambers RD, Bahga SS, Santiago JG (2012) Effect of PVP on the electroosmotic mobility of wet-etched glass microchannels. Electrophoresis 33: 3259-3262.

33. Jiang Y, Wang H, Li S, Wen W (2014) Applications of micro/nanoparticles in microfluidic sensors: A review. Sensors 14: 6952-6964.

34. Banerjee I, Pangule RC, Kane RS (2011) Antifouling coatings: Recent developments in the design of surfaces that prevent fouling by proteins, bacteria, and marine organisms. Adv Mater 23: 690-718.

35. Riche CT, Zhang C, Gupta M, Malmstadt N (2014) Fluoropolymer surface coatings to control droplets in microfluidic devices. Lab Chip 14: 1834-1841.

36. Sapsford KE, Ligler FS (2004) Real-time analysis of protein adsorption to a variety of thin films. Biosens Bioelectron 19: 1045-1055.

37. Luk YY, Kato M, Mrksich M (2000) Self-assembled monolayers of alkanethiolates presenting mannitol groups are inert to protein adsorption and cell attachment. Langmuir 16: 9604-9608.

38. Horvath J, Dolník V (2001) Polymer wall coatings for capillary electrophoresis. Electrophoresis 22: 644-655.

39. Doherty EA, Meagher RJ, Albarghouthi MN, Barron AE (2003) Microchannel wall coatings for protein separations by capillary and chip electrophoresis. Electrophoresis 24: 34-54.

40. Wong I, Ho CM (2009) Surface molecular property modifications for poly(dimethylsiloxane) (PDMS) based microfluidic devices. Microfluid Nanofluidics 7: 291-306.

41. Estephan ZG, Jaber JA, Schlenoff JB (2010) Zwitterion-stabilized silica nanoparticles: toward nonstick nano. Langmuir 26: 16884-16889.

42. Wu L, Guo Z, Meng S, Zhong W, Du Q, et al. (2010) Synthesis of a zwitterionic silane and its application in the surface modification of silicon-based material surfaces for improved hemocompatibility. ACS Appl Mater Interfaces 2: 2781-2788.

43. Pennathur S, Crisalli P (2013) Low Temperature Fabrication and Surface Modification Methods for Fused Silica Micro- and Nanochannels. MRS Proc 1659: 15-26.

44. Estephan ZG, Schlenoff PS, Schlenoff JB (2011) Zwitteration as an alternative to PEGylation. Langmuir 27: 6794-6800.

45. Pennathur S, Santiago JG (2005) Electrokinetic transport in nanochannels. 2. Experiments. Anal Chem 77: 6782-6789.

46. Sze A, Erickson D, Ren L, Li D (2003) Zeta-potential measurement using the Smoluchowski equation and the slope of the current-time relationship in electroosmotic flow. J Colloid Interface Sci 261: 402-410.

47. Bharadwaj R, Santiago JG, Mohammadi B (2002) Design and optimization of on-chip capillary electrophoresis. Electrophoresis 23: 2729-2744.

48. Liu H, May K (2012) Disulfide bond structures of IgG molecules: structural variations, chemical modifications and possible impacts to stability and biological function. MAbs 4: 17-23.

49. Smith EA, Chen W (2008) How to prevent the loss of surface functionality derived from aminosilanes. Langmuir 24: 12405-12409.

50. Li L, Chen S, Zheng J, Ratner BD, Jiang S (2005) Protein Adsorption on Oligo (ethylene glycol) -Terminated Alkanethiolate Self-Assembled Monolayers?: The Molecular Basis for Nonfouling Behavior. J Phys Chem B 109: 2934-2941.

51. Chen S, Li L, Zhao C, Zheng J (2010) Surface hydration: Principles and applications toward low-fouling/nonfouling biomaterials. Polymer 5: 5283-5293.

52. McNaught AD, Wilkinson A (1997) Compendium of Chemical Terminology. Oxford: Blackwell Scientific Publications.

53. Kirby BJ (2010) Micro- and Nanoscale Fluid Mechanics: Transport in Microfluidic Devices. Cambridge University Press.

54. Andersen MB, Bruus H, Bardhan JP, Pennathur S (2011) Streaming current and wall dissolution over 48 h in silica nanochannels. J Colloid Interface Sci 360: 262-271.

55. Jensen KL, Kristensen JT, Crumrine AM, Andersen MB, Bruus H, et al. (2011) Hydronium-dominated ion transport in carbon-dioxide-saturated electrolytes at low salt concentrations in nanochannels. Phys Rev E 83: 056307.

56. Yang W, Zhang L, Wang S, White AD, Jiang S (2009) Functionalizable and ultra-stable nanoparticles coated with zwitterionic poly(carboxybetaine) in undiluted blood serum. Biomaterials 30: 5617-5621.

57. Mathé C, Devineau S, Aude JC, Lagniel G, Chédin S, et al. (2013) Structural determinants for protein adsorption/non-adsorption to silica surface. PLoS One 8: e81346.

58. Vogler EA (2012) Protein adsorption in three dimensions. Biomaterials 33: 1201-1237.

59. Hlady V, Buijs J, Jennissen HP (1999) Methods for studying protein adsorption. Methods Enzymol 309: 402-429.

60. Mathes J, Friess W (2011) Influence of pH and ionic strength on IgG adsorption to vials. Eur J Pharm Biopharm 78: 239-247.

61. Lionello A, Josserand J, Jensen H, Girault HH (2005) Protein adsorption in static microsystems: effect of the surface to volume ratio. Lab Chip 5: 254-260.

62. Chen S, Zheng J, Li L, Jiang S (2005) Strong resistance of phosphorylcholine self-assembled monolayers to protein adsorption: Insights into nonfouling properties of zwitterionic materials. J Am Chem Soc 127: 14473-14478.

63. Schoch RB, Renaud P (2008) Transport phenomena in nanofluidics. Rev Mod Phys 80: 839-883.

64. Karnik R, Castelino K, Majumdar A (2006) Field-effect control of protein transport in a nanofluidic transistor circuit. Appl Phys Lett 88: 1-3.

65. Volkmuth WD, Duke T, Wu MC, Austin RH, Szabo A (1994) DNA electrodiffusion in a 2D array of posts. Phys Rev Lett 72: 2117-2120.

66. Lee JH, Chung S, Kim SJ, Han J (2007) Poly(dimethylsiloxane)-based protein preconcentration using a nanogap generated by junction gap breakdown. Anal Chem 79: 6868-6873.

67. Wang Y-C, Choi MH, Han J (2004) Two-dimensional protein separation with advanced sample and buffer isolation using microfluidic valves. Anal Chem 76: 4426-4431.

68. Kaji N, Tezuka Y, Takamura Y, Ueda M, Nishimoto T, et al. (2004) Separation of long DNA molecules by quartz nanopillar chips under a direct current electric field. Anal Chem 76: 15-22.

69. Doyle PS, Bibette J, Bancaud A, Viovy JL (2002) Self-assembled magnetic matrices for DNA separation chips. Science 295: 2237.

70. Tegenfeldt JO, Prinz C, Cao H, Huang RL, Austin RH, et al. (2004) Micro- and nanofluidics for DNA analysis. Anal Bioanal Chem 378: 1678-1692.

71. Lutanie E, Voegel JC, Schaaf P, Freund M, Cazenave JP, et al. (1992) Competitive adsorption of human immunoglobulin G and albumin: consequences for structure and reactivity of the adsorbed layer. Proc Natl. Acad Sci USA 89: 9890-9894.

# 17

# Nano Scale Potentiometric and Spectrophotometric Assays for 2,4-Dichlorophenoxyacetic Acid

**El-Beshlawy MM***

*Department of Chemistry, Collage of Girls, Ain Shams University, Egypt*

## Abstract

The aim of this work was to investigate the ability of silicate wire ion selective electrode to measure 2,4 dichlorophenoxyacetic acid (2,4-D). The slope of the electrode was 56.2 mV decade -1 in the range from $10^{-6}$ to $10^{-2}$ M with lower limit of detection $8.3 \times 10^{-7}$ M. The electrodes were successfully applied to determination of 2,4-D by separate and standard addition potentiometry. The electrode was used to detect the effect of pH and temperature on 2,4-D placed on baby watercress plant grown in the clay soil and the residue after five day from addition. Leaf chlorophylls quantity measured utilize spectrophotometric or colorimetric assay following extraction of pigments of watercress by ethanol. This work was carried at 286, 440, 662 nm. Beer's law was obeyed in the concentration range of 2.21 mg/ml:2.21 µg/ml. The limits of detection were 2.21 µg/ml.

**Keywords:** 2,4-Dichlorophenoxyacetic acid; Potentiometric; Spectrophotometric; Wire ion-selective electrodes, Watercress; Chlorophyll

## Introduction

2,4-D is one of the most widely used herbicides in the world and is characterized as low-cost, quite efficient even at low concentrations, and moderately hazardous (class II) according to the World Health Organization [1] Although this herbicide easily moves though the soil, its leaching into the groundwater can be minimized due to the degradation by microorganisms and also the absorption by plants. Its primary route of degradation in the environment is by microorganisms, which increases with temperature, humidity, pH, and organic matter content [2,3]. Microbial degradation is a possible route for the breakdown of 2,4-D, but it is very dependent on the characteristics of the water. Laboratory studies have shown that in warm, nutrient rich water that has been previously treated with 2,4-D microbial degradation can be a major factor for dissipation [4]. Residues of 2,4-D and its salts or esters in water are commonly measure by extraction, chemical derivatization, separation by gas-liquid chromatography and electron capture detection [4]. Other methods used in the determination of 2,4-D residues include high-performance liquid and thin-layer chromatography [5-7].

Several methods have been reported for the analysis of 2,4-D in food and environmental samples. These methods include gas chromatography (GC) [8-10]. High-Performance Liquid Chromatography (HPLC) and UV-spectrophotometry [11,12]. However the chromatographic techniques are expensive and not available in many quality laboratories worldwide. Spectrophotometry is probably the most convenient analytical technique for routine analysis because of its inherent simplicity, low cost and wide availability in quality control laboratories [13].

## Experimental Section

### Materials and methods chemicals

All chemicals used were of analytical-reagent grade. 2,4 Dichlorophenoxyacetic acid (Fluka AG) 95%, Sodium silicate (99%), Potassium chloride (99%), Sodium hydroxide (98%), Barium nitrate (98.5%), Ferrous sulphate (98%), L-Glutamic acid (99%), L-Ascorbic acid (99%), Maltose, D-Glucose (extra pure), Acrylamide (99%) and 1,2 Dichloroethane (98%) were provided by (LOBA Chemie) Fluka,

Switzerland, Hydrochloric acid (37%) from (Scharau). Ethanol (96%), from HAYMAN and Methanol (99.8%) from AnalaR Distilled water was used to prepare all solutions.

### Instrumentation

The electrochemical measurements were carried out with HANNA instrument 211 (EZODO) for measuring pH. Sensitive balance model AG204 (METTLER TOLEDO) was used for measurements. Spectral measurements were carried out by using a single beam UV (OPTIMA) SP-SPECTROPHOTOMETER, with quartz cells of 1 cm optical path length.

### Stock standard solution of 2,4-D

An accurately 0.01 M solution was prepared by dissolving 0.212 g of 2,4-D standard in 25% ethanol, transferred into a 100 mL volumetric flask and diluted to the mark with 25% ethanol and mixed well. This stock solution was further diluted with 25% ethanol to obtain working solutions in the ranges of $10^{-3}$:$10^{-11}$ M, reservation solutions in brown bottles inside the refrigerator.

### Construction of wire ion selective electrode

Membrane was prepared by mixing well of 96% sodium silicate, 1% active coal, 3% dioctylephethalate with 1 cm of fusible polyethylene tube. The mixture was poured into 3 cm diameter petri dish with 10 ml tetrahydrofuran (THF). A pure silver wire electrode was dipped in the coating membrane solution several time, allow the film to dry for about five min repeat the process until 2.0 mm diameter. Wire electrode was soaked about 3 h in $1 \times 10^{-3}$ M of 2,4-D. All potentiometric determination were performed using the analytical cell, Ag /membrane/2,4-D test solution//Reference electrode.

***Corresponding authors:** El-Beshlawy MM, Department of Chemistry, Collage of Girls, Ain Shams University, Egypt
E-mail: munmun1231@yahoo.com

## Electrode calibration

Connect the wire electrode to the pH meter in the presence of reference electrode. The calibration of the electrode was preceded using standard solutions of 2,4-D ranging from $10^{-7}$:$10^{-2}$ M.

Separate and spiking addition method was used by the sequence from low to a higher concentration. The measured potential was plotted against logarithm of 2,4-D concentration after one min (Figure 1). The electrode was washed with distilled water and dried with tissue paper and back to base line by immersing in water between measurements, stored in a dark place immersed in $10^{-5}$ M.

## Response times

Calibration curve was measured from the lower ($10^{-7}$ M 2,4-D) to the higher concentrated ($10^{-2}$ M of 2,4-D) solutions the time traces of the calibration curves of silicate electrode was represented in Figure 2.

## Selectivity

Selectivity coefficients $K^{pot}$ 2,4-D of the electrodes towards different species were determined by the standard addition technique method [14] in which the following equation was applied

$$K^{pot}=(a`A - aA)/aB$$

where a`A known activity of primary ion, aA fixed activity of primary ion and aB activity of interfering ions. The results are appear by potting mV in the presence of different interference (Figure 3).

## Effect of pH

The effect of pH of the potential of electrode was investigated. The potential was measured at a 2,4-D concentration solution ($10^{-5}$ M) from the pH value of 3 up to 9. The solution was acidified by the addition of very small volumes of 0.2 N HCl then the pH value was increased gradually using 0.2 N NaOH for each pH value, two pH/mV meters were used to measure the potential. The potential was recorded and plotted against different pH values (Figure 4).

## Analytical application

**Growing watercress:** Watercress seeds were germinated in 250 gm clay dry soil with about 3-10 seeds per pot, after two weak. The temperature of the plants was maintained between 20-25°C.

**Watercress treatment:** A total of 20 pots were used to grow the watercress plants, two pots didn't contain 2,4-D. Stock solution of 2.21 µg /ml of 2,4-D was prepared in 25% ethanol as solvent eighteen pots were treated by 50 ml of the previous stock solution for each pot after adjust number of watercress plants (1, 2, 3, 5, 8,..) and, the pH (2.5-8) and temperature of solutions (-5, 15, 27, 43), plants are irrigated daily with distilled water.

**Residual of 2,4-D:** After two, three and five days from adding pesticides on the pot of watercress plants. 40 ml distilled water are added to each pot. The calibration curve was used to determine the concentration of 2,4-D in each pot plant solution above, all measurements was recorded after one minute of electrode immersing. Effect of pH and temperature on the residual was drawing in Figures 5 and 6.

## Spectrophotometric determination of 2,4-D

**Selection of solvent:** It was important to see the effects of various solvents on the photodegradation of 2,4-D. After assessing the solubility of 2,4-D in different solvent as methanol, ethanol, 25% ethanol and 2,4 dichloroethane. 25% ethanol has been selected as common solvent for developing spectral characteristics, 0.00221 ng /ml solution of 2,4-D present in each solution. The absorbencies are plotted against the different solvent at 286 nm (Figure 7). It has been observed the absorption within half an hour of the beginning of the preparations in each solvent. Blank solution was the solvent.

**Beer's Law concentration range:** The stock solution were suitable diluted with 25% ethanol to get concentration range from 0.00221 g ml$^{-1}$-0.0221 ng/ml for 2,4-D. The absorbance was measured at 286 nm. Standards of the compound of interest are made, and their absorbance and concentration values are used to create a standardization graph (Figure 8). From the calibration curve 2,4-D obey Beer's law between 2.21 mg/ml to 2.21 µg/ml.

**Photo degradation of 2,4-D by UV Light:** The cuvet was filled within about 1.5 ml of the top with the blank and $10^{-3}$ M of 2,4-D solution. Wipe the outside of each cuvet gently with a lint-free tissue to remove any fingerprints or solution that may be on the surface. Adjust the wavelength to 286 nm, the absorbance of $10^{-3}$ M of 2,4-D was determined and plotted against time (Figure 9).

**Effect of 2,4-D on watercress leaf chlorophyll Extract:** Young baby-leaf watercress were acquired directly from organic local farmers in region of South Saudi Arabia. Cut of 7 g of watercress leaf, rinse the leaves with distilled water and pat dry. Tear up the leaves and place

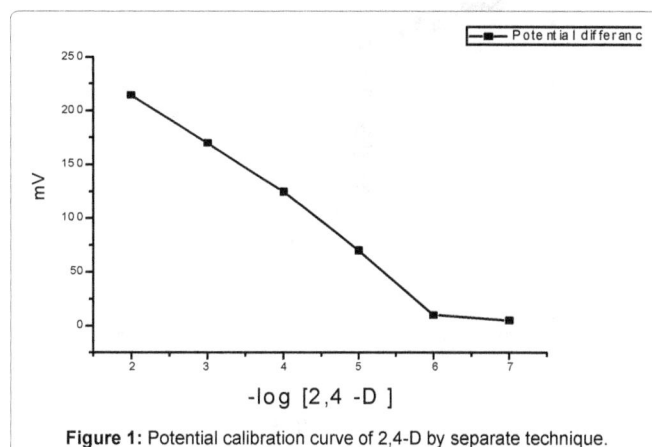

**Figure 1:** Potential calibration curve of 2,4-D by separate technique.

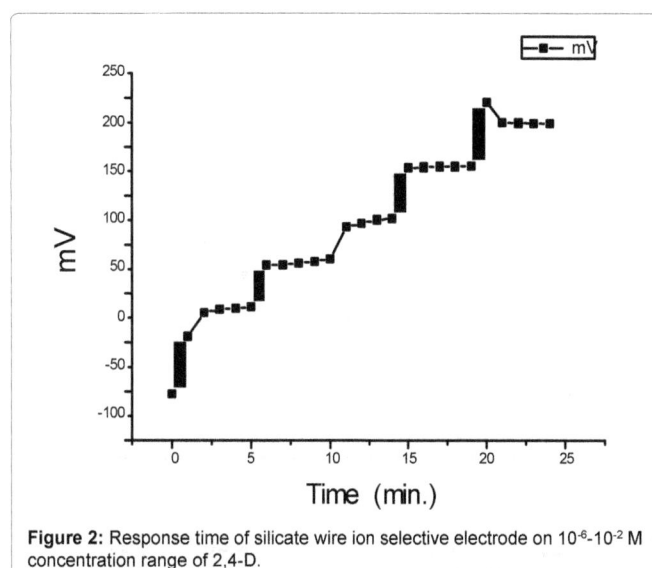

**Figure 2:** Response time of silicate wire ion selective electrode on $10^{-6}$-$10^{-2}$ M concentration range of 2,4-D.

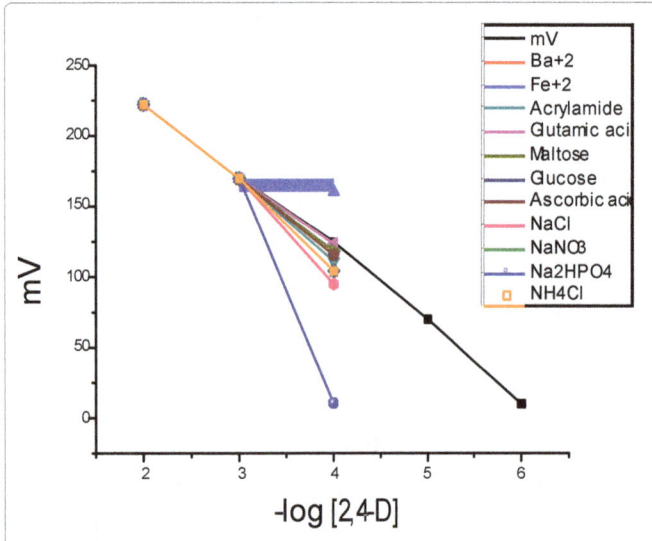

**Figure 3:** Effect of 1ml $10^{-3}$ M interference on 9 ml $10^{-4}$ M of 2,4-D.

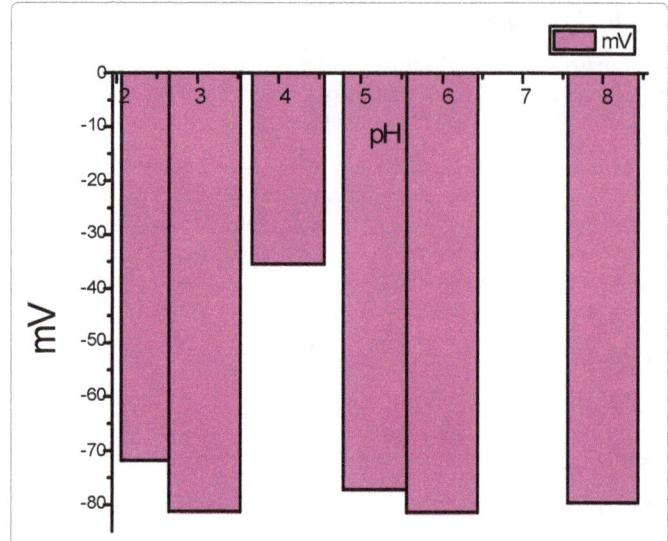

**Figure 6:** Effect of pH on the residual of 2,4-D.

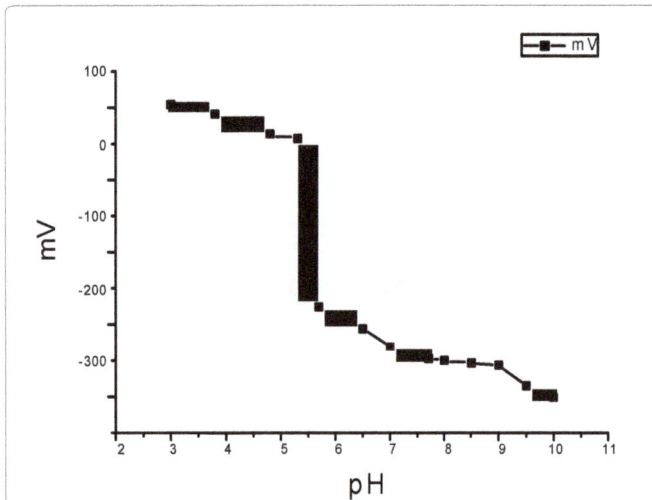

**Figure 4:** Effect of pH on $10^{-5}$ M Solution of 2,4-D.

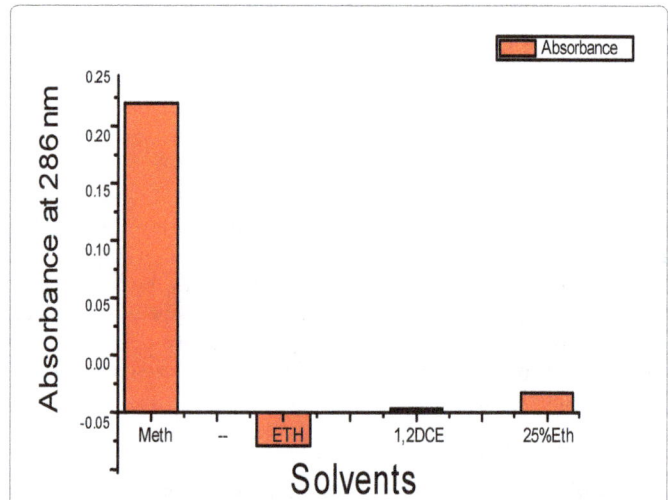

**Figure 7:** Effect of different solvent on the absorbance of $10^{-11}$ M 2,4-D.

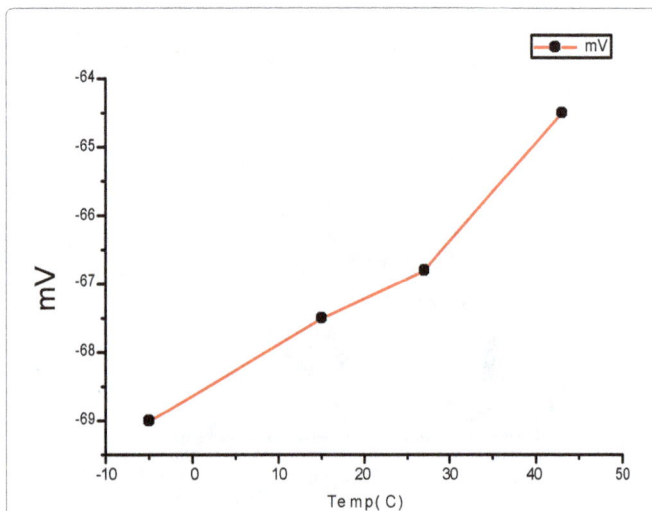

**Figure 5:** Effect of temperature (°C) on the residual of 2,4-D.

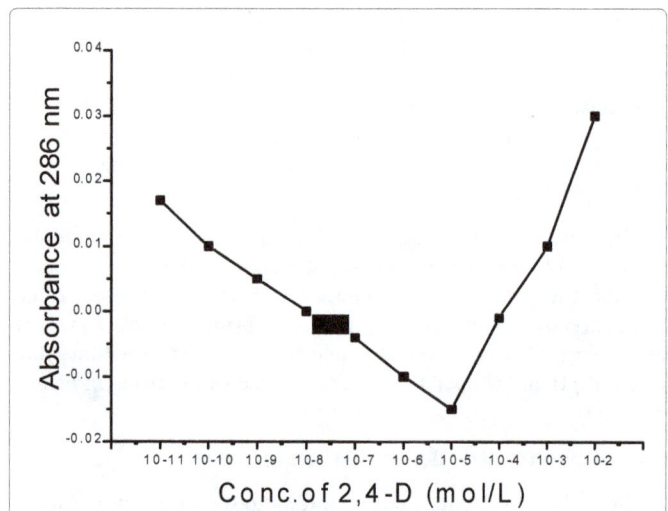

**Figure 8:** Calibration curve of 2,4-D.

them in a 50 ml ethanol half an hour. Ethanol is a water-miscible solvent, removes all pigments from soft-leafed. The blank solution was prepared by 1 ml of chlorophyll extract and 11 ml distilled water and sample stock solutions were prepared by mixing aliquots of different concentration ($10^{-11}$, $10^{-5}$ and $10^{-3}$) of 2,4-D were precisely pipette into 10 ml distelled water and 1 ml chlorophyll extract. The absorbance of solutions was measured at 286, 440, 662 nm and plotted against different concentration of 2,4-D in sample solutions (Figure 10).

**Time considerations:** Effect of time on the absorbance of watercress leaf chlorophyll extract with 1 ml different concentration of ($10^{-11}$, $10^{-5}$ and $10^{-3}$ M) of 2,4-D solutions was determined and plotted (Figure 11).

## Result and Discussion

2,4-D was classified among the phenoxy acid herbicides MCPA as a class-2B carcinogen possibly carcinogenic to humans. The movement of auxin, (2,4-D) is a common herbicide,. Denmark, Norway, Kuwait and the Canadian provinces of Québec and Ontario, 2,4-D use is severely restricted in the country of Belize [15].

2,4-D concentration is one of the affecting factor to the rate of reaction, so that affects to the performance of silicate ion selective electrode. It has been found that separate technique is the best way to apply Nernst equation. Response characteristics of the silicate ion selective electrode appear in Table 1, the results showed that the highest sensitivity is 42.1 ng/ml.

The purpose of this study was to investigate the working electrode conditions to obtain the maximum performance. The research studied of the electrode character was pH and response time. The pH optimum of 2,4-D at a range from 2.5 to 5.5. Response time is a measure of the rate of degradation reaction of 2,4-D. The response time with concentration higher than $10^{-4}$ M were of the order of 15 seconds. The silicate electrode has good selectivity to 2,4-D, in the presence of double-charged metal ions, as $Fe^{+2}$ and Glutamic acid, and has poor sensitivity in the presence of $Na_2HPO_4$.

The residual pesticide was decreases to 1.005, 0.663, 0.221 µg $ml^{-1}$, after two, three, and five days respectively from the original solution (2.21 µg $ml^{-1}$) added on the pots of watercress. The number of plants did not effect on the residual of the pesticide concentration. The residual of 2,4-D degradation of (10.6 µg $ml^{-1}$) by the temperature effect was varied from -5 to 43°C (Figure 5), the large amount of residual of 2,4-D was at 43°C, for a period of time following treatment. It is therefore to be expected that temperature may influence the rate of degradation response and govern to some extent, Plants that put out the pesticide at low temperature longer than the age of the plants that put them at a temperature of exterminator other heat. This research study reports the concentrations of residual 2,4-D (pH 2.5-9) taken up by watercress grown. This study showed that the pH 4 of 2,4-D solution is the lower pesticide absorption by soil and watercress plants [16,17].

It is very important to see the effect of different solvent as methanol, ethanol, 1,2-dichloroethane and 25% ethanol on the photodegradation of $10^{-11}$ M of 2,4-D (Figure 1). It is observed that the degradation of 2,4-D in the ethanol was the maximum and the degradation in methanol was the minimum. The velocity of photodegradation of $10^{-3}$ M of 2,4-D in 25% ethanol is very high (Figure 3).

2,4-D was studied in the concentration range $10^{-11}$:$10^{-2}$ M . The absorbance was decreased from $10^{-11}$:$10^{-5}$ M the reason is the photodegradation of 2,4-D in this region. Microbial degradation is considered and breakdown of 2,4-D . The most important mechanism of microbial degradation involves the removal of the acetic acid side

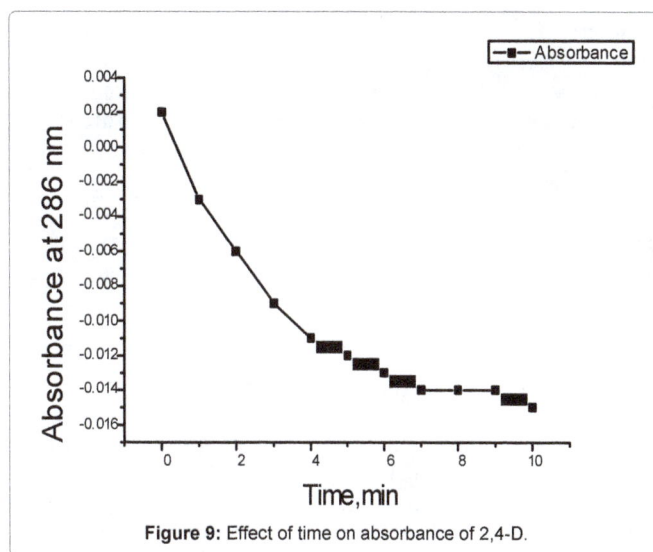

**Figure 9:** Effect of time on absorbance of 2,4-D.

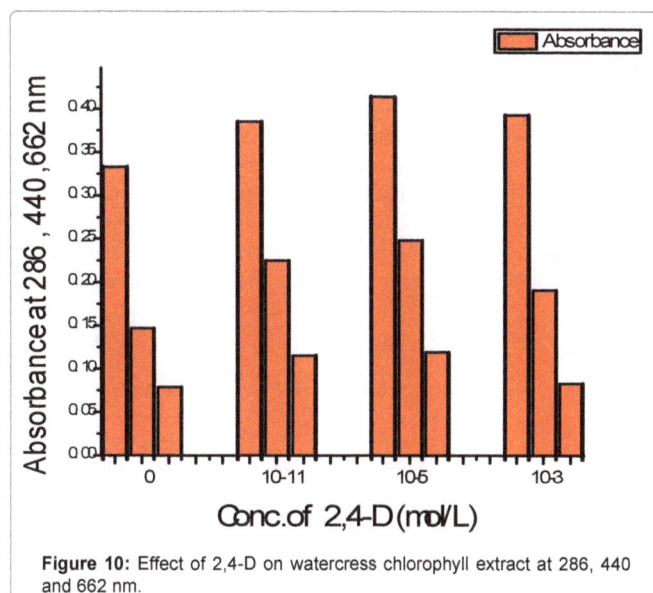

**Figure 10:** Effect of 2,4-D on watercress chlorophyll extract at 286, 440 and 662 nm.

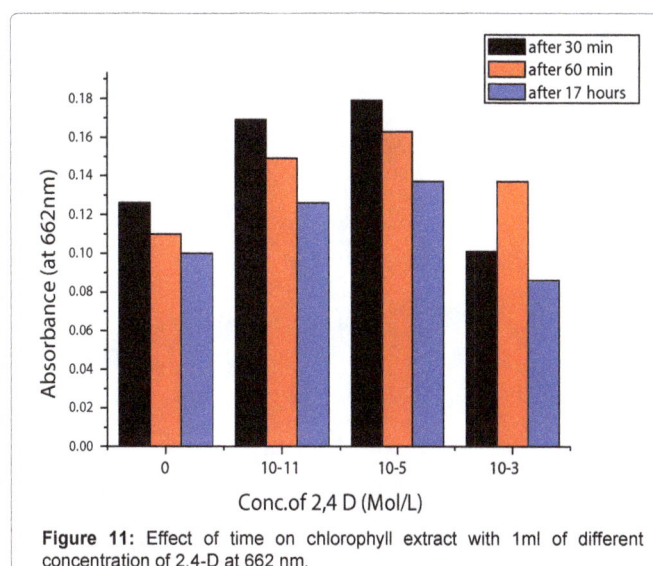

**Figure 11:** Effect of time on chlorophyll extract with 1ml of different concentration of 2,4-D at 662 nm.

| Parameter | Value* |
|---|---|
| Slope,( mV/decade ) | 56.2 ± 2 |
| Intercep , mV | 260 ± 5 |
| Correlation coefficient between mV and -log[H$^+$ ], (r) | -0.014 |
| Detection limit , M | 8.3 × 10$^{-7}$ |
| Response time for 10$^{-4}$ M solution,s | 15 |
| Working pH range | 2.5-5.5 |
| Standard Deviation ( SD) | 2.61 |
| Relative Standard Deviation( RSD) | 4.48 |

*Average of five replicate

**Table 1:** Response characteristics of the silicate wire ion selective electrode.

chain. This is followed by ring cleavage and degradation to produce aliphatic acids such as succinic acid. The rate of microbial degradation is dependent upon the water potential 17, the absorbance increased and obeyed Beer's law from $10^{-5}$:$10^{-2}$ M. The molar absorptivity coefficient equal to 1000 l mol$^{-1}$ cm$^{-1}$. Pigment content of chlorophyll were increased after adding , the reason is photodegradation of 2,4-D in this region.$10^{-3}$ M < $10^{-5}$ M < $10^{-11}$ M of 2,4-D (Figure 10) at 286,440 and 662 nm. The absorption decreasing with the effect of time (Figure 11).

## Conclusion

Silicate wire electrodes was suitable for the determination of 2,4-D with regard to working concentration range, slope, pH range, response time, and selectivity . The electrodes exhibited good reproducibility over a useful lifetime of 2 months. Spectrophotometric study allows understand 2,4-D toxicity and it's excellent deleterious effect. The reproducibility, recovery, and operational simplicity of the methods make them suitable to determined 2,4-D. The detection limit was 2.21 µg ml$^{-1}$.

### Acknowledgements

The author would like to thank the publishers for their interest in reviewing find.

## References

1. WHO (2013) The WHO recommended classification of pesticides by hazard.

2. Maleki N, Safavi A, Shahbaazi HR (2005) Electrochemical determination of 2,4-D at a mercury electrode. Analytica Chimica Acta 530: 69-74.

3. Gaultier J, Farenhorst A, Cathcart J, Goddard T (2008) Degradation of [carboxyl-14C] 2,4-D and ring-U-14C] 2,4-D in 114 agricultural soils as affected by soil organic carbon content. Soil Biol Biochem 40: 217-227.

4. Halter MT (1980) 2,4-D in the Aquatic Environment: Section II in Literature Reviews of Four Selected Herbicides: 2,4-D, Dichlobenil, Diquat, and Endothall. Municipality of Metropolitan Seattle

5. World Health Organization (1984) 2,4-Dichlorophenoxyacetic acid (2,4-D). Environmental Health Criteria 29. International Programme on Chemical Safety, Geneva.

6. Frank R, Logan L, Clegg BS (1991) Pesticide and polychlorinated biphenyl residues in waters at the mouth of the Grand, Saugeen, and Thames Rivers, Ontario, Canada, 1986-1990. Arch Environ Contam Toxicol 21: 585-595.

7. Shane S, Que Hee, Ronald G, Sutherland (1981) The phenoxyalkanoic herbicides: Chemistry, analysis, and environmental pollution. Chemical Rubber Company Series in Pesticide Chemistry. CRC Press, Boca Raton, USA.

8. Tucker SP, Reynolds JM, Wickman DC, Hines CJ, Perkins JB (2001) Development of sampling and analytical methods for concerted determination of commonly used chloroacetanilide, chlorotriazine, and 2,4-D herbicides in hand-wash, dermal-patch, and air samples. Appl Occup Environ Hyg 16: 698-707.

9. Cook LW, Zach FW, Klosterman HJ, Bristol DW (1983) Comparison of free and total residues of (2,4-dichlorophenoxy)acetic acid and 2,4-dichlorophenol in millet resulting from postemergence and preharvest treatment. J Agric Food Chem 31: 268-271.

10. Williams KJ, James CR, Thorpe SA, Reynolds SL (1997) Two analytical methods for the measurement of 2,4-D in oranges: An ELISA screening procedure and a GC-MS confirmatory procedure. Pesticide Science 50: 135-140.

11. Kashyap SM, Pandya GH, Kondawar VK, Gabhane SS (2005) Rapid analysis of 2,4-D in soil samples by modified Soxhlet apparatus using HPLC with UV detection. J Chromatogr Sci 43: 81-86.

12. Park JY, Choi JH, Abd El-Aty AM, Kim BM, Park JH, et al. (2011) Development and validation of an analytical method for determination of endocrine disruptor, 2,4-D, in paddy field water. Biomed Chromatogr 25: 1018-1024.

13. Jan MR, Shah J, Bashir N (2006) Flow injection spectrophotometric determination of 2,4-D herbicide Journal of the Chinese Chemical Society 53: 845-850.

14. Mostafa GA, Hefnawy MM, Al-Majed A (2007) Sensors 7: 3272-3286.

15. IRAC (1987) Overall evaluations of carcinogenicity: an updating of IARC Monographs volumes 1 to 42. IARC Monogr Eval Carcinog Risks Hum Suppl 7: 1-440.

16. Zetterberg G, Busk L, Elovson R, Starec-Nordenhammar I, Ryttman H (1977) The influence of pH on the effects of 2,4-D (2,4-dichlorophenoxyacetic acid, Na salt) on Saccharomyces cerevisiae and Salmonella typhimurium. Mutat Res 42: 3-17.

17. Ghassemi M, Fargo L, Painter P, Quinlivan S, Scofield R, et al. (1981) Environmental Fates and Impacts of Major Forest Use Pesticides, Office of Pesticides and Toxic Substances, Washington DC, USA.

# 18

# Species Selective Measurement of 10 *B. anthracis*-Sterne Spores within 10 Minutes by Surface-Enhanced Raman Spectroscopy

**Chetan Shende[1], Hermes Huang[2], Jay Sperry[3] and Stuart Farquharson[1*]**

[1]*Real-Time Analyzers Inc., 362 Industrial Park Road, Unit 8, Middletown, CT-06457, USA*
[2]*Smiths Detection, 14 Commerce Drive, Danbury, CT 06810, USA*
[3]*University of Rhode Island, 45 Lower College Road, Kingston, RI 02881, USA*

## Abstract

The use of biological warfare agents by terrorists remains a global concern. While there has been substantial effort since the 2001 distribution of *Bacillus anthracis* spores through the US Postal System to develop analyzers to detect this and other biological agents, the analyzers lack sensitivity, lack specificity (produce high false-positive rates), are too slow, or cannot be fielded. For the past decade we have been investigating the ability of surface-enhanced Raman spectroscopy (SERS) to overcome these limitations. Recently, we developed an assay by functionalizing silver nanoparticles with various peptides to selectively bind *B. anthracis*, and then adding acetic acid and silver colloids to release and detect, respectively, dipicolinic acid as a biomarker by SERS. Here we describe the successful measurement of *B. anthracis*-Sterne spores with a 10- to 20-fold selectivity over other *Bacillus* species at $10^5$ spores/mL, using the peptide functionalized SERS assay with a sensitivity capable of detecting 10 spores in a $10^3$ spores/mL sample in 6.5 minutes. This measurement represents 6 orders-of-magnitude improvement over our previous peptide based SERS assay measurements.

**Keywords:** *Bacillus anthracis*; Anthrax; Biological warfare agents; Surface-enhanced Raman spectroscopy

## Introduction

During the most recent 2011 Biological Weapons Convention held in Geneva, US Secretary of State Hillary Rodham Clinton stated "Unfortunately, the ability of terrorists to develop and use these weapons is growing. Terrorist groups have made it known they want to acquire these weapons [1]." Consequently the need for a portable technology that can rapidly identify biological warfare agents with high specificity (no false-positive responses) and sensitivity (e.g. $10^4$ spores for *B. anthracis* [1]) in the field remains. During the past decade, various techniques have been developed to detect and identify biological agents. Notable techniques include mass spectrometry [2], fluorescence [3], luminescence [4], infrared [5,6], and Raman spectroscopy [7], as well as the polymerase chain reaction (PCR) [8]. While all of these techniques have been employed for biological agent detection to some extent (e.g. identification of anthrax on mail sorting equipment by PCR), none of them satisfy all of the requirements of speed, sensitivity, selectivity, and field ruggedness, especially the latter as required by military personnel.

The most successful field techniques are based on immunoassays, such as enzyme-linked immunosorbent assays (ELISA) [9,10]. In an effort to improve upon this technique, a number of other antigen-antibody based binding event reporter methods have been investigated. These techniques include cantilever [11,12], electrochemical [13], magnetic [14], piezoelectric [15], and surface plasmon resonance (SPR) based devices [16]. While these techniques are relatively fast, inexpensive, and easy to use, they suffer from high false-positive and false-negative rates due to a lack of specificity and sensitivity [17]. In the specific case of *B. anthracis*, other *Bacillus* spores, such as *B. cereus*, are common in the environment, and they share the same surface antigens used for antibody binding [18], increasing the false-positive rate.

Recognizing the limitations of all the techniques mentioned above, we have been developing a surface-enhanced Raman spectroscopy (SERS) based assay for the detection of trace quantities of biological

agents, since it is capable of amplifying Raman signal intensities by 6 orders of magnitude or more [19]. We demonstrated the potential of SERS to detect *Bacillus* spores by performing the first measurements of dipicolinic acid (DPA) as a biomarker more than a decade ago [20]. DPA is a suitable biomarker, as it represents approximately 10% of the spore mass in the form of calcium dipicolinate (CaDPA) [21]. In the past few years, we developed and patented a method to extract DPA using acetic acid [22]. This method allowed detecting ~200 spores within 3 minutes [23,24], which is well below the estimated infectious dose of $10^4$ spores for *B. anthracis* [2]. Although the use of DPA as a biomarker is a logical approach, the chances of false-positive identification are significant, since both *Bacillus* and *Clostridium* bacteria contain CaDPA in their protective outer layers. To overcome this limitation, we have been investigating the possibility of functionalizing silver nanoparticles with peptides that specifically bind *B. anthracis*, so that these spores can be selectively measured by SERS. Recently, several peptides have been identified that demonstrate high selectivity towards *B. anthracis* versus other *Bacillus* [25,26]. Peptides offer several advantages over the traditional antibody-antigen binding assays, in that they can be easily synthesized, functionalized, and covalently attached to sensor surfaces. Furthermore, they are more tolerant to heat and moisture, which extends their usable lifetime making them suitable for field use.

Previously, we reported the ability of this approach to detect $10^9$ *B. anthracis*-Sterne spores per mL in less than 20 minutes [27].

*****Corresponding author:** Stuart Farquharson, Real-Time Analyzers Inc., 362 Industrial Park Road, Unit 8, Middletown, CT-06457, USA, E-mail: stu@rta.biz

Unfortunately, the addition of the peptide likely dampened the plasmon field responsible for the SER effect. Here we describe the addition of a second SER-active material to the assay in the form of a silver colloid to measure 10 *B. anthracis*-Sterne spores in a $10^3$ spores/mL sample within 10 minutes, representing an improvement in sensitivity of 6 orders-of-magnitude!

## Materials and Methods

### Materials

All chemicals, reagents, and solvents, including those used to prepare the SER-active sol-gels, were purchased and used as received from Sigma-Aldrich (Milwaukee WI). All *Bacillus* samples were obtained from the American Type Culture Collection (Manassas, VA) and prepared by Professor Jay Sperry (University of Rhode Island) [28]. Stock solutions were serially diluted to produce the measured concentrations. Concentrations were verified by direct count of spores in $4 \times 10^{-3}$ µL triplicate samples using a light microscope [25]. The concentrations of these samples were determined to be $5.7 \times 10^9$, $1.7 \times 10^{10}$, $5.5 \times 10^9$, and $6.1 \times 10^9$ spores/mL for *B. anthracis*-Sterne, *B. cereus*, *B. megaterium* and *B. subtilis*, respectively. The ATYPLPIR peptide [26] used in this study was custom synthesized by New England Peptide (Gardner, MA). Glass capillaries, tubing, syringes and syringe ports were obtained from VWR Scientific (Arlington Heights, IL).

### SER-active sol-gel capillaries

SER-active capillaries (*Simple SERS Sample Capillaries*, RTA) were prepared according to published procedures [28,29] by mixing a silver amine precursor and an alkoxide precursor at 1:1 v/v. The silver amine precursor consisted of a 1:1:2 v/v/v ratio of 1N $AgNO_3/28\%$ $NH_4OH/CH_3OH$, while the alkoxide precursor consisted of methyltrimethoxysilane. The SERS capillaries were prepared by drawing 20 µL of the silver-doped sol-gels into 10 cm long, 0.8 mm inner diameter glass capillaries to produce ~1 cm long sol-gel segments. The segments were allowed to gel and cure for 12 hours, after which the incorporated silver ions were reduced with dilute $NaBH_4$. Silver colloids were prepared from $AgNO_3$ and $NaBH_4$ according to literature [30] with modifications.

### Peptide functionalized SERS capillaries

A cysteine residue was attached to the C-terminus of the peptide sequence, and served to link the peptide to the SER-active metal surface. Cysteine forms a strong covalent bond with silver via the sulfur of its thiol side chain. The peptide functionalization of silver nanoparticles was carried out by adding 10 µL of peptide solution to the SER-active sol-gel segments immobilized within the glass capillaries. After the peptide was allowed to react with the SER-active metal, water was passed through the capillary to remove any unbound peptide.

### SERS analysis

For measurements the SERS capillaries were mounted on an XY stage (Conix Research, Springfield, OR) such that the focal point of an f/0.7 aspheric lens of a fiber optic probe was just inside the capillary glass wall. A software program developed in-house was used to measure 1 min spectra at 9 points spaced 1 mm apart along the length of the metal-doped sol-gel segment. A Fourier transform Raman spectrometer (Real-Time Analyzers, Middletown, CT), equipped with a 785 nm diode laser (Innovative Photonic Solutions, Monmouth Junction, NJ) and a Si-photo-avalanche detector (Perkin Elmer, Stamford, CT) was used to deliver 80 mW of power and collect spectra at 8 cm⁻¹ resolution.

## Results and Discussion

Functionalization of silver nanoparticles with the *B. anthracis* specific peptide was investigated in terms of reaction times and peptide concentrations. In each case, 10 µL of peptide was added to a previously prepared glass capillary, which was sufficient to saturate the 1 cm segment of the SER-active sol-gel. After each experiment, the SER-active sol-gel was flushed with 10 mL of HPLC grade water to remove any unbound peptide. Peptide functionalization to the silver nanoparticles was verified by observing the SERS of the peptide. Initially, 1 mg/mL peptide was allowed to react with the silver for 8 hours to ensure success. This time was decreased by factors of 2 until the SERS intensity of the peptide decreased. This occurred at binding times below 0.5 hours. As shown in Figure 1A, the SER spectrum of the peptide is very similar to cysteine, both dominated by the peak at 657 cm⁻¹ due to the sulfur-silver vibrational mode, as well as cysteine peaks at 863, 1067, and 1394 cm⁻¹ due to a CC stretch, CN stretch, and $CH_2$ wag, respectively [31]. The other amino acids of the peptide do not contribute significantly to the spectrum.

In addition to the reaction time it was important to determine the peptide concentration that would provide monolayer coverage of the silver, so that chemicals or biochemicals introduced with a sample would not produce interfering SER spectral features. To establish monolayer coverage, a high concentration of DPA, 100 µg/mL, was added to the silver nanoparticles functionalized with various amounts of peptide. Since DPA is a highly SER-active molecule, it would produce a spectrum through interaction with any silver surface uncovered by the peptide. DPA produces several characteristic SERS peaks at 659 (coincident with the dominant peptide peak), 812, 1007, and 1381 cm⁻¹, which have been previously assigned to a CC ring bend, a CH out-of-plane bend, the symmetric pyridine ring stretch, and OCO symmetric stretch, respectively [8]. It was found that the DPA spectrum was relatively intense, barely discernible, and absent at 5, 10 and 50 µg/mL peptide, respectively (Figure 1B). Based on these measurements, peptide functionalized capillaries were prepared by reacting 100 µg/mL peptide for 0.5 hours to ensure monolayer coverage.

Measurements of *B. anthracis*-Sterne spores, an avirulent strain of *B. anthracis*, were performed at the University of Rhode Island to determine optimum sample and reagent volumes, and reaction (binding) time for detecting *B. anthracis* by SERS. Once determined, these conditions were used to demonstrate selective binding of the *B. anthracis* peptide versus *B. cereus*, *B. megaterium*, and *B. subtilis*, as well as to determine the limit of detection for *B. anthracis*-Sterne. For each of the *Bacillus* samples, initial 1 mL wet suspensions of the spores were diluted by a factor of 300. These samples were then diluted to produce serial concentrations of $10^5$, $10^4$, $10^3$ and $10^2$ spores/mL. A starting concentration of $10^5$ spores/mL was selected as based on 1) a previous measurement of 10 ng/mL chemical DPA using unfunctionalized SER-active capillaries [32], and 2) the fact that 10 ng DPA corresponds to ~$10^5$ spores. The latter assumes that 10% of the spore mass is DPA [22], and that 1 spore has a mass of ~1 pg [33,34].

The following steps were found to repeatedly produce quality spectra for *B. anthracis*-Sterne at $10^5$ spores/mL: 1) draw 10 µL of spores into the peptide functionalized, SER-active capillaries; 2) allow the spores to bind to the peptides; 3) draw 50 µL of water through the capillary to remove unbound spores; 4) draw 10 µL of silver colloid as prepared [31] into the capillary to coat the spores; 5) draw 10 µL acetic acid solution into the capillary to cause DPA to be released from the spores; 6) mount the capillary on the XY stage of the Raman spectrometer; and 7) measure the SERS at 9 points along the capillary. Measurements of $10^5$ *B. anthracis*-Sterne spores/mL using this procedure produced intense DPA spectra for binding times of 30, 15, and 5 minutes (Figure 2). Since one of the goals was to establish the minimum time to perform the measurement, the 5 min binding time was used for the *B. anthracis*-Ames samples, as it produced a sufficiently intense DPA spectrum.

Next, the other bacilli at $10^5$ spores/mL were measured on the *B. anthracis* specific peptide functionalized SERS capillaries. Again, measurements followed the above procedure, but in this case 15 min binding times were used to better represent the selectivity of the peptide. As shown in Figure 3A, the intensities of the 1007 cm$^{-1}$ peak for *B. cereus*, *B. megaterium* and *B. subtilis* are ~10%, 6%, and 5%, respectively, of that for *B. anthracis*-Sterne. The presence of DPA in the spectra for these bacilli is not surprising, since it is known that some non-specific binding occurs [27]. The lower discrimination against *B. cereus* has been attributed to its closer relation to *B. anthracis* [26]. It should be noted that the inability to completely flush these spores out of the porous sol-gel structure could also contribute to the lack of specificity. While this level of discrimination by itself may not suffice

for positive *B. anthracis*, the use of a second assay, such as one for *B. cereus*, could be used to rule out false positives.

Finally, *B. anthracis*-Sterne was measured at $10^4$ and $10^3$ spores/mL to determine the limits of detection (Figure 3B). At these lower concentrations, other weak spectral features become noticeable, such as a peak at 930 cm$^{-1}$ due to acetic acid (AA), a peak at 1044 cm$^{-1}$ due to nitrate ($NO_3^-$) from the silver nitrate used to prepare the colloid, and a broad feature from 1200 to 1800 cm$^{-1}$ due to glass luminescence from the capillary. It is worth noting that the actual number of spores in the $10^3$ spores/mL sample is ~10 spores, since only 10 μL of sample was introduced into the capillary. Furthermore, the signal-to-noise ratio (S/N) is 28 using the baseline corrected 1007 cm$^{-1}$ peak height and the standard deviation noise between 1700 and 1800 cm$^{-1}$. Based on a limit of detection defined as a S/N of 3, this suggests that a single spore could be detected, or 100 spores/mL assuming the SERS signal intensity is relatively linear in this range. Finally, both spectra presented in Figure 3B were acquired in 1 minute, so that the entire measurement time was ~6.5 minutes (5 min to load the sample, 0.5 min to load the reagents, and 1 min to measure the sample).

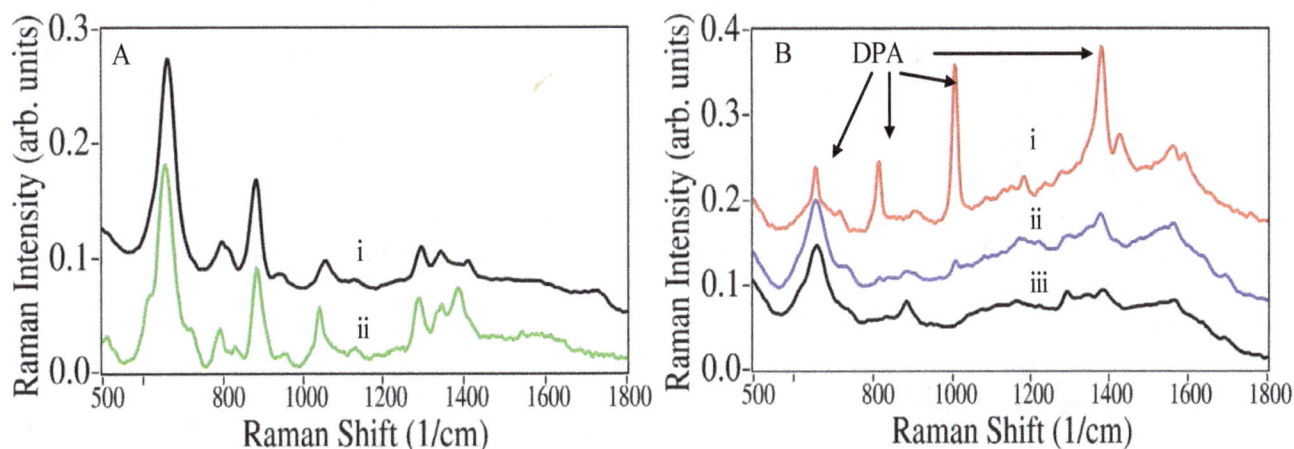

**Figure 1:** SERS of A) 1 mg/mL i) ATYPLPIRC peptide specific to *B. anthracis* functionalized in a SER-active capillary and ii) cysteine in a SER-active capillary. B) 100 μg/mL DPA added to SER-active capillaries treated with i) 5, ii) 10, and iii) 50 μg/mL peptide. Sample conditions: 0.5 hr peptide reaction time, DPA measured immediately. Spectral conditions: 75 mW of 785 nm laser excitation, 9 min acquisition (nine 1-min points along capillary averaged), 8 cm$^{-1}$ resolution using RTA's FT-Raman analyzer. Spectra are displayed on the same intensity scale, but offset for clarity.

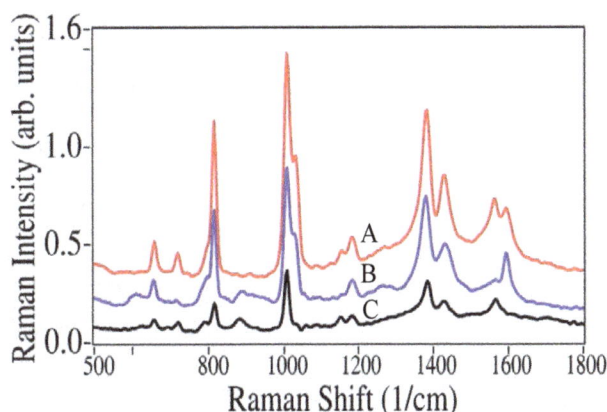

**Figure 2:** SERS of $10^5$ *B. anthracis*-Sterne spores/mL for binding times of A) 30, B) 15, and C) 5 min. Sample conditions: 10 μL sample, 50 μL water wash, 10 μL silver colloid addition, and 10 μL acetic acid addition. Spectral conditions as in Figure 1.

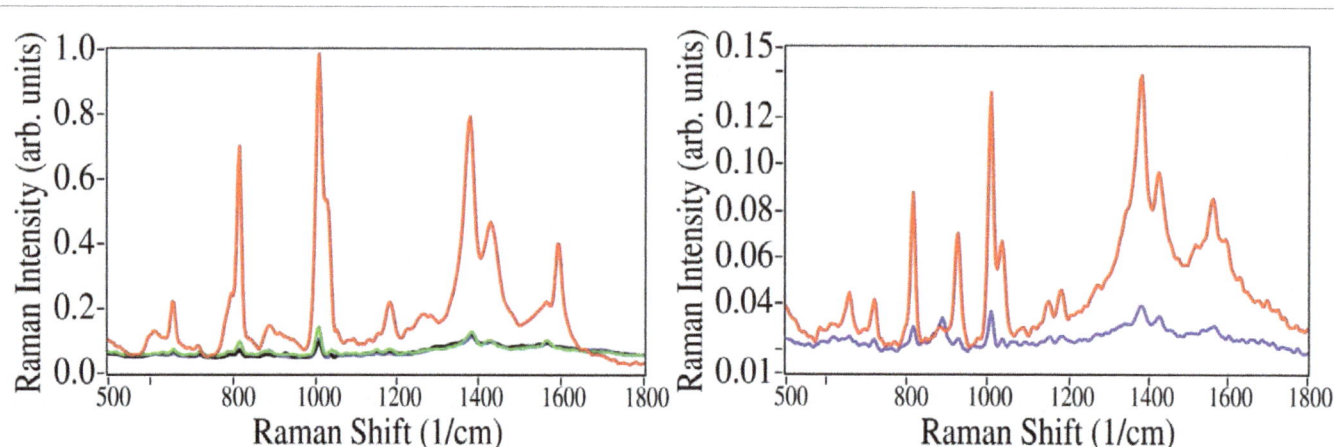

**Figure 3:** SERS of A) i) *B. anthracis*-Sterne, ii) *B. megaterium*, iii) *B. cereus*, and iv) *B. subtilis*, all at $10^5$ spores/mL, conditions: 15 min binding time, 9-min spectral acquisition; and B) *Bacillus anthracis*-Sterne at i) $10^4$ and ii) $10^3$ spores/mL, conditions: 5 min binding time, 1-min spectral acquisition. Other conditions as in Figure 2.

## Conclusion

A method was successfully developed to selectively detect 10 *B. anthracis*-Sterne spores from a $10^3$ spores/mL sample in 6.5 minutes. This was accomplished by functionalizing silver nanoparticles within a porous sol-gel with a *B. anthracis* specific peptide, and then adding acetic acid and silver colloids to release and detect, respectively, DPA by SERS as a biomarker. The assay provided a 10- to 20-fold discrimination against *B. cereus*, *B. megaterium* and *B. subtilis* at $10^5$ spores/mL. Current research is aimed at developing an automated flow system coupling the output of an aerosol collector to the capillaries. The technology, once developed should prove invaluable for real-time monitoring of biological agents at potential terrorist targets.

### Acknowledgements

The authors are grateful for funding from the National Science Foundation (DMI-0349687, IIP-0810335, and II-0956170) and the Defense Advanced Research Projects Agency (N10PC2077 and D11PC20171).

### References

1. Hess G (2012) Biosecurity: an evolving challenge. Chem Eng News 90: 30-32.

2. Wilkening DA (2006) Sverdlovsk revisited: modeling human inhalation anthrax. Proc Natl Acad Sci USA 103: 7589-7594.

3. Hathout Y, Setlow B, Cabrera-Martinez RM, Fenselau C, Setlow P (2003) Small, acid-soluble proteins as biomarkers in mass spectrometry analysis of Bacillus spores. Appl Environ Microbiol 69: 1100-1107.

4. Nudelman R, Bronk BV, Efrim S (2000) Fluorescence emission derived from dipicolinate acid, its sodium, and its calcium salts. Appl Spectrosc 54: 445-449.

5. Pelligrino PM, Fell Jr NF, Gillespie JB (2002) Enhanced spore detection using dipicolinate extraction techniques. Anal Chim Acta 455: 167-177.

6. Manning C, Gross M, Hanshaw T, Kirlin RL, Samuels A (2003) Compact interferometers for chemical and biological agent detection. Proc SPIE 5268: 125-136.

7. Szymanski G, Józwicki R, Wawrzyniuk L, Rataj M, Józwik M (2012) Novel FTIR Spectrometer for the Biological Agent Detection. In: Jabłoński R, Březina T (Eds.) Mechatronics: Recent Technological and Scientific Advances, Springer Berlin Heidelberg pp: 685-689.

8. Farquharson S, Grigely L, Khitrov V, Smith WW, Sperry JF, et al. (2004) Detecting Bacillus cereus spores on a mail sorting system using Raman Spectroscopy. J Raman Spectrosc 35: 82-86.

9. Bell CA, Uhl JR, Hadfield TL, David JC, Meyer RF, et al. (2002) Detection of Bacillus anthracis DNA by LightCycler PCR. J Clin Microbiol 40: 2897-2902.

10. Mabry R, Brasky K, Geiger R, Carrion R Jr, Hubbard GB, et al. (2006) Detection of anthrax toxin in the serum of animals infected with Bacillus anthracis by using engineered immunoassays. Clin Vaccine Immunol 13: 671-677.

11. Tang S, Moayeri M, Chen Z, Harma H, Zhao J, et al. (2009) Detection of anthrax toxin by an ultrasensitive immunoassay using europium nanoparticles. Clin Vaccine Immunol 16: 408-413.

12. Dhayal B, Henne WA, Doorneweerd DD, Reifenberger RG, Low PS (2006) Detection of Bacillus subtilis spores using peptide-functionalized cantilever arrays. J Am Chem Soc 128: 3716-3721.

13. Shekhawat G, Tark SH, Dravid VP (2006) MOSFET-Embedded microcantilevers for measuring deflection in biomolecular sensors. Science 311: 1592-1595.

14. Metfies K, Huljic S, Lange M, Medlin LK (2005) Electrochemical detection of the toxic dinoflagellate Alexandrium ostenfeldii with a DNA-biosensor. Biosens Bioelectron 20: 1349-1357.

15. Connolly J, St. Pierre TG (2001) Proposed biosensors based on time-dependent properties of magnetic fluids. J Magn Magn Mater 225: 156-160.

16. Zhihong M, Xiaohuia L, Weiling F (1999) A new sandwich-type assay of estrogen using piezoelectric biosensor immobilized with estrogen response element. Anal Commun 36: 281-283.

17. Uchida H, Fujitani K, Kawai Y, Kitazawa H, Horii A, et al. (2004) A new assay using surface plasmon resonance (SPR) to determine binding of the Lactobacillus acidophilus group to human colonic mucin. Biosci Biotechnol Biochem 68: 1004-1010.

18. King D, Luna V, Cannons A, Cattani J, Amuso P (2003) Performance assessment of three commercial assays for direct detection of Bacillus anthracis spores. J Clin Microbiol 41: 3454-3455.

19. Radnedge L, Agron PG, Hill KK, Jackson PJ, Ticknor LO, et al. (2003) Genome differences that distinguish Bacillus anthracis from Bacillus cereus and Bacillus thuringiensis. Appl Environ Microbiol 69: 2755-2764.

20. Jeanmaire DL, Van Duyne RP (1977) Surface Raman Spectroelectrochemistry Part I. Heterocyclic, aromatic, and aliphatic amines adsorbed on the anodized silver electrode. J Electroanal Chem Interfacial Electrochem 84: 1-20.

21. Farquharson S, Smith WW, Elliott S, Sperry JF (1999) Rapid biological agent identification by surface-enhanced Raman spectroscopy. Proc SPIE 3855: 110-116.

22. Nelson WH, Dasari R, Feld M, Sperry JF (2004) Intensities of calcium dipicolinate and Bacillus subtilis spore Raman spectra excited with 244 nm light. Appl Spectrosc 58: 1408-1412.

23. Farquharson S, Gift AD, Maksymiuk P, Inscore FE (2004) Rapid dipicolinic acid extraction from Bacillus spores detected by surface-enhanced Raman spectroscopy. Appl Spectrosc 58: 351-354.

24. Farquharson S, Gift AD, Inscore FE (2010) Method for affecting the rapid release of a signature chemical from bacterial endospores, and for detection thereof. US Patent No 7713914.

25. Farquharson S, Inscore F (2008) Detection of invisible Bacilli spores on surfaces using a portable SERS-based analyzer. Int J Hi Spe Ele Syst 18: 407-416.

26. Brigati J, Williams DD, Sorokulova IB, Nanduri V, Chen IH, et al. (2004) Diagnostic probes for Bacillus anthracis spores selected from a landscape phage library. Clin Chem 50: 1899-1906.

27. Williams DD, Benedek O, Turnbough CL Jr (2003) Species-specific peptide ligands for the detection of Bacillus anthracis spores. Appl Environ Microbiol 69: 6288-6293.

28. Sengupta A, Shende C, Farquharson S, Inscore F (2012) Detection of Bacillus anthracis spores using peptide functionalized SERS-active substrates. Spectrosc Intl J ID 176851.

29. Farquharson S, Lee YH, Nelson C (2003) Material for SERS and SERS sensors and method for preparing the same. US Patent No 6623977.

30. Farquharson S, Gift AD, Inscore FE, Shende CS (2008) SERS method and apparatus for rapid extraction and analysis of drugs in saliva. US Patent No 7393691.

31. Lee PC, Meisel D (1982) Adsorption and surface-enhanced Raman of dyes on silver and gold sols. J Phys Chem 86: 3391-3395.

32. Stewart S, Fredericks PM (1999) Surface-enhanced Raman spectroscopy of amino acids adsorbed on an electrochemically prepared silver surface. Spectrochimica Acta Part A 55: 1641-1660.

33. Farquharson S, Shende C, Gift A, Inscore F (2012) Detection of Bacillus spores by surface-enhanced Raman spectroscopy. Bioterrorism 17-40.

34. Carrera M, Zandomeni RO, Sagripanti JL (2008) Wet and dry density of Bacillus anthracis and other Bacillus species. J Appl Microbiol 105: 68-77.

# 19

# Quantification of Sugar Epimers in Polygalactomannans by ESI-MS/MS

Erika Ponzini[1], Greta Borgonovo[1], Luca Merlini[2], Yves M Galante[2], Carlo Santambrogio[1*] and Rita Grandori[1*]

[1]*Department of Biotechnology and Biosciences, University of Milano-Bicocca, Piazza della Scienza 2, 20126, Milan, Italy*
[2]*Istituto di Chimica del Riconoscimento Molecolare, CNR, Via Mario Bianco 9, 20131, Milan, Italy*

## Abstract

Polygalactomannans (PGMs) represent an important family of polysaccharides. They are obtained from the endosperm of leguminous plant seeds and are employed in a growing number of industrial applications as thickening agents and rheology modifiers. Each PGM has a typical mannose/galactose (M/G) ratio (from ~1 to ~5), which determines its solubility, viscosity and other physico-chemical properties.

Complex polysaccharides are commonly characterized by proton nuclear magnetic resonance ([1]H-NMR), whereas mass spectrometry (MS) has not yet seen wide application. In this work, a new quantification method for sugar epimers in PGMs is presented, based on the characteristic fragmentation patterns of mannose and galactose in tandem MS (MS/MS) analyses.

Standard galactose and mannose have been analyzed and fragmented in negative-ion mode on a hybrid quadrupole/time-of-flight (qTOF) mass spectrometer equipped with a nano electrospray ionization (ESI) source. The MS/MS spectra indicate accumulation of the same fragments for the two monosaccharides, but significant and reproducible differences in the relative intensities of the product ions. Known mixtures of mannose and galactose have been analyzed by the same procedure to test the applicability of the method for quantification purposes. The resulting peak intensities over the entire MS/MS spectrum can be deconvoluted as a linear combination of the signals from pure mannose and galactose standards, obtaining reliable and reproducible quantification of the epimers.

The method has been applied to the characterization of hydrolysis products of PGMs from different species of leguminous plants, such as guar (*Cyamopsis tetragonolobus*), sesbania (*Sesbania bispinosa*) and tara (*Caesalpinia spinosa*), in order to assess specific susceptibility to hydrolysis conditions.

**Keywords:** Polygalactomannans; Electrospray-ionization mass spectrometry; Monosaccharides quantification; Collision-induced dissociation; Acidic hydrolysis

## Introduction

Polysaccharides from animal (e.g., chitin, chitosan and hyaluronan), vegetal (e.g., cellulose, starch and pectin) and microbial (e.g., xanthan, gellan and curdlan) organisms are emerging as attractive materials in several industrial fields. They are employed as emulsion stabilizers, rheology modifiers, coating agents etc. in food, textile, cosmetic, biomedical and pharmaceutical branches [1-5]. Thanks to their non-toxic, biodegradable and renewable origin and their amenability to chemical and biochemical modifications, these compounds represent a promising alternative to synthetic polymers for the development of new functionalized materials [6-8].

An important group of polysaccharides is represented by polygalactomannans (PGMs), mainly obtained from the endosperm of the seeds of leguminous plants, where they represent a reserve source for carbon and energy upon germination [9]. Nowadays, the most widely employed PGM for industrial purposes is obtained from the guar plant (*Cyamopsis tetragonolobus*), mainly cultivated in India [10]. Other important, but commercially less exploited PGMs are extracted from locust bean (*Ceratonia siliqua*) and fenugreek (*Trigonella foenum-graecum*), spontaneous plants of the Mediterranean regions, from tara (*Caesalpinia spinosa*), native to South America, and from sesbania (*Sesbania bispinosa*), native to Asia and Africa [9,11]. PGMs are high-molecular weight polymers of about 1-2 MDa [12,13]. They are normally composed of a β(1→4)-D-mannan backbone with several α(1→6) branches of single D-galactose units (Figure 1), although more complex ramifications have been described [6]. The interest in these polymers is mainly due to the possibility of obtaining solutions with unique rheological properties, like extreme viscosities and non-

Newtonian behavior [13-15]. PGMs are often subjected to chemical reactions (e.g., hydroxyalkylation, carboxymethylation) [16,17] or enzymatic treatments (e.g., depolymerization by β-mannanase, oxidation by laccase or galactose oxidase) [18-20], in order to modify their solubility and viscoelastic properties, or to obtain hybrid, functionalized materials [21,22].

PGMs from distinct leguminous species generally differ in structural properties, like the amount and distribution of galactose ramifications along the backbone, resulting in different physico-chemical properties of the PGMs solutions [13]. Mannose/galactose (M/G) ratio can vary from ~1 to ~5 [6,9,14,23], affecting structural properties and solubility of the polymer. The abundance of cis-OH groups in the mannan backbone induces a strong tendency to form inter-chain hydrogen bonds, promoting precipitation and poor solubility in polar solvents. The presence of galactose ramifications enhances solubility preventing such interactions by steric hindrance. Therefore, PGMs with low

---

**\*Corresponding authors:** Carlo Santambrogio, Department of Biotechnology and Biosciences, University of Milano-Bicocca, Piazza della Scienza 2, 20126, Milan, Italy, E-mail: carlo.santambrogio@unimib.it

Rita Grandori, Department of Biotechnology and Biosciences, University of Milano-Bicocca, Piazza della Scienza 2, 20126, Milan, Italy
E-mail: rita.grandori@unimib.it

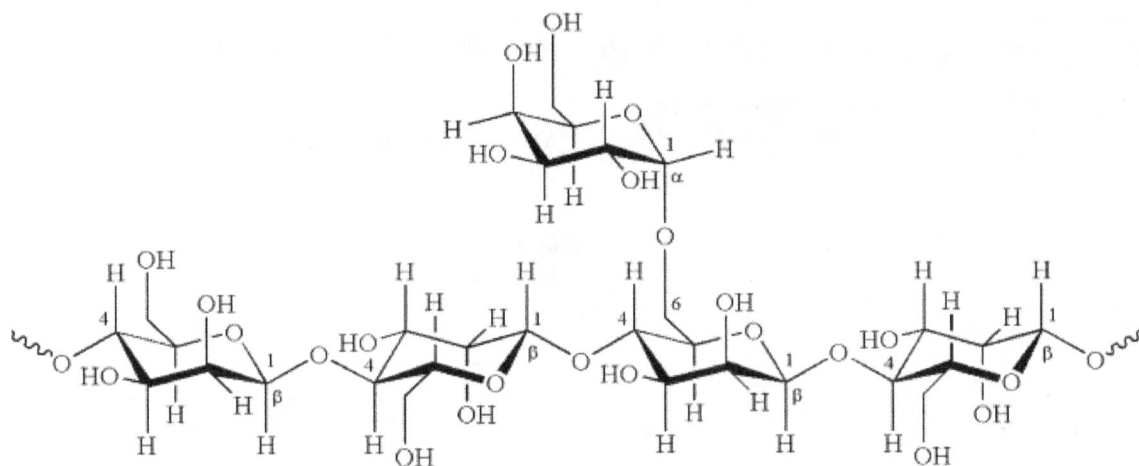

**Figure 1:** Schematic representation of PGM structure, showing a portion of mannan backbone and one galactose ramification.

M/G (i.e., with high galactose content) usually dissolve easier than others, leading to the formation of viscous, non-Newtonian fluids. For example, locust bean gum (M/G ~4) requires a boiling procedure to reach full hydration, while fenugreek (M/G ~1) and guar (M/G ~2) gums are highly soluble even in cold water [9,10]. Thanks to the non-ionic nature of PGMs, the viscosity of the resulting solutions is almost constant over a wide range of temperature and pH [6,10]. The degree of galactose substitutions also affects the propensity of the polymer to form gels, although the relation to the M/G ratio is not straightforward and is strongly dependent on environmental conditions [13,24].

Computational simulations suggest that the M/G ratio is a key factor also for PGMs conformational properties [25-27]. Galactose substitutions promote chain bending *via* intramolecular H-bonding, favoring compact conformations and affecting structural properties like persistence length and gyration radius. As a result, PGMs with low M/G are predicted to be more compact and flexible than the variants with high M/G [27]. Experimental data on the radius of gyration are in agreement with the trend suggested by computational modeling [13,27].

Other factors, such as the pattern of galactose substitutions, have also been suggested to affect the conformational properties of PGMs [25,27]. This "fine structure" of the polymers seems to vary even among variants of the same species [28]. Different patterns have been described, such as periodic [6,28], random [28-30] or hybrid [28,31].

Several experimental techniques have been applied to the study of the physico-chemical properties of PGMs. Size-exclusion chromatography is usually employed for the estimation of the molecular-weight distributions [13,32], while $^{13}$C-NMR combined with enzymatic hydrolysis are used for the determination of the galactose distribution on the backbone [28,33-36]. The M/G ratios are commonly obtained by gas chromatography, liquid chromatography or $^{1}$H-NMR, on intact or hydrolyzed PGMs [13,37-39]. Mass spectrometry (MS) has seen only limited application to PGM analysis [20,40-45], although it has a strong potential as an approach complementary to other techniques for the characterization of complex mixtures.

In this work, a new approach to M/G quantification, based on tandem electrospray-ionization mass spectrometry (ESI-MS/MS), is proposed. The method exploits the distinct fragmentation patterns

of mannose and galactose when subjected to collision-induced dissociation (CID) with an inert gas [46]. The method is applied here to the analysis of different PGMs, in order to characterize the depolymerization products and to assess the specific susceptibility to hydrolysis conditions.

## Materials and Methods

### Analysis of mannose and galactose standard solutions by mass spectrometry

D-(+)-mannose (Carbosynth, Compton, UK) and D-(+)-galactose (Sigma Aldrich, St. Louis, MO, USA) were dissolved in milliQ water at a final concentration of 10 μM. The stock solutions were mixed together obtaining different M/G ratios (0.11, 0.43, 1.00, 2.33 and 9). Pure mannose and galactose solutions and their mixtures were freshly prepared before each ESI-MS analysis. The samples were infused by borosilicate-coated capillaries of 1 μm internal diameter (Thermo Fisher Scientific, Waltham, MA USA) into a QSTAR-Elite hybrid quadrupole/time-of-flight (qTOF) instrument equipped with a nano-electrospray ion source (AB Sciex, ForsterCity, CA, USA). MS and MS/MS measurements were performed both in positive- and negative-ion mode, employing the instrumental settings indicated in the Results section. MS/MS spectra were obtained by CID of singly charged ions with molecular nitrogen. The recorded spectra were averaged over 1 minute acquisition time.

### Quantification of M/G from ESI-MS/MS data

The relative intensity of each peak in MS/MS spectra can be interpreted as the sum of the corresponding peaks generated by pure mannose and galactose, weighted by the fractional amount of the sugars in the sample:

$$I_{iX} = f_M I_{iM} + f_G I_{iG} \tag{1}$$

$$f_M + f_G = 1 \tag{2}$$

where $I_{iX}$ is the intensity of the generic fragment peak from the mixture, $I_{iM}$ and $I_{iG}$ are the intensities of the same peak in pure mannose and galactose samples, and $f_M$ and $f_G$ are the sugar fractional amounts. The above model is based on the assumption that mannose and galactose ions from the same sample have identical fragmentation propensities. The $f_G$ value was estimated by combining equations (1) and (2) and minimizing the following function:

$$\sum_i \left\{ I_{iX} - \left[ f_G I_{iG} + (1 - f_G) I_{iM} \right] \right\}^2 \tag{3}$$

The sum was extended over the 10 most intense peaks of the MS/MS spectrum. Setting the first derivative of the function (3) to zero, the fraction $f_G$ can be expressed as:

$$f_G = \frac{\sum_i (I_{iX} - I_{iM})(I_{iG} - I_{iM})}{\sum_i (I_{iG} - I_{iM})^2} \tag{4}$$

The M/G ratio is then given by:

$$M/G = \frac{f_M}{f_G} = \frac{1 - f_G}{f_G} = \frac{1}{f_G} - 1 \tag{5}$$

## PGMs chemical hydrolysis

Flours from guar (*Cyamopsis tetragonolobus*), tara (*Caesalpinia spinosa*) and sesbania (*Sesbania bispinosa*) (industrial stocks from Lamberti SpA, Albizzate, Italy) employed in this work have a PGM content of 85-90%, and a Brookfield viscosity of ~5000, ~4500 e ~3000 mPa/s, respectively, in 1% (w/v) aqueous solutions. PGMs were purified by dispersing 30 g of flour in a 7:3 ethanol:water mixture (v/v) and vacuum-filtering the solution by a Büchner funnel. The resulting powder was resuspended in acetone, filtered again, and dried for 12 h at 65°C. After purification, 1.5 mg of PGM was dissolved in 300 mL of milliQ water and the pH was lowered below 2 by the addition of an acidic solution (0.5 M HCl, 26.5 M HCOOH or 13.5 M CF$_3$COOH, according to the chosen protocol). Chemical hydrolysis was performed by autoclaving the solution at 121°C for 1.5 h. Right after the incubation

in the autoclave, the solution was cooled on ice for 30 min, and neutralized by the addition of NaOH. Insoluble material was removed by centrifugation before MS analysis.

## Results

### Mannose and galactose fragmentation properties in the gas-phase

Aqueous solutions of standard mannose and galactose were analyzed by ESI-MS in the two different ionization polarities. In positive ion-mode, the spectra of either monosaccharide are dominated by sodiated molecular ions, with only barely visible peaks corresponding to protonated species. CID experiments on sodium adducts do not lead to sugars fragmentation at any value of collision energy (CE), indicating a mere loss of the Na$^+$ ion during the collisions with the inert gas (data not shown), in agreement with previous reports on the MS/MS behavior of pentoses and hexoses in positive ion-mode [46].

In negative ion-mode, the prevalent peak corresponds to the deprotonated form of the sugars (*m/z* 179) (Figure 2A and 2B). Although the isomeric nature of the two monosaccharides gives rise to identical signals, the MS/MS spectra are different for mannose or galactose (Figure 2C and 2D), due to different susceptibility of the sugars to competing fragmentation pathways [46]. The fragments obtained by CID are the same in the two cases, and result from the neutral loss of water and/or formaldehyde (Table 1 and Figure 2D) [46]. The relative intensities of the corresponding peaks, however, are dependent on the sugar nature. The main distinction for the two epimers is represented by the propensity to lose two formaldehyde molecules (producing an ion at *m/z* 119), that is highly pronounced for mannose and quite low for galactose. Besides this major difference, other minor variations in

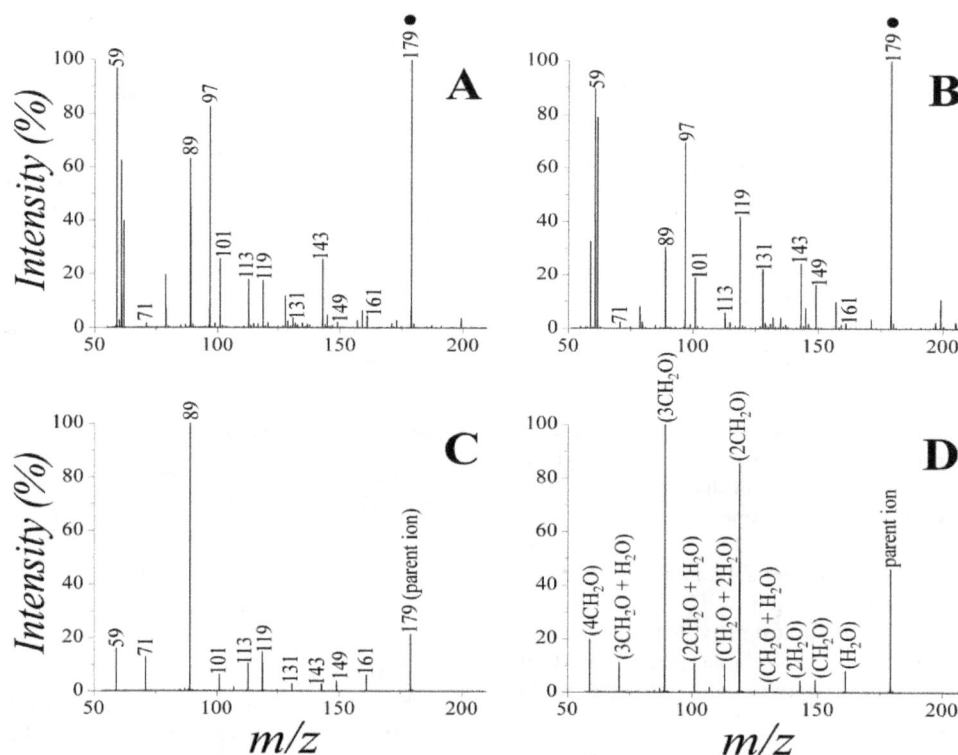

**Figure 2:** ESI-MS (A, B) and ESI-MS/MS (C, D) spectra in negative ion-mode of aqueous solutions of 10 µM galactose (A, C) and mannose (B, D). In (A) and (B) the signal corresponding to the deprotonated monosaccharide (*m/z* 179) is labeled by a black circle. Fragmentation of the 179- ion was performed at DP -80 V and CE -10 V. Peaks are labeled by the *m/z* value (A-C) or by the neutral fragment lost, relative to the parent ion (D).

| $m/z$ | Elemental composition | Neutral loss |
|---|---|---|
| 179 | $C_6H_{11}O_6^-$ | - |
| 161 | $C_6H_9O_5^-$ | $H_2O$ |
| 149 | $C_5H_9O_5^-$ | $CH_2O$ |
| 143 | $C_6H_7O_4^-$ | $2H_2O$ |
| 131 | $C_5H_7O_4^-$ | $CH_2O + H_2O$ |
| 119 | $C_4H_7O_4^-$ | $2CH_2O$ |
| 113 | $C_5H_5O_3^-$ | $CH_2O + 2H_2O$ |
| 101 | $C_4H_5O_3^-$ | $2CH_2O + H_2O$ |
| 89 | $C_3H_5O_3^-$ | $3CH_2O$ |
| 71 | $C_3H_3O_2^-$ | $3CH_2O + H_2O$ |
| 59 | $C_2H_3O_2^-$ | $4CH_2O$ |

**Table 1:** Peak assignment for the MS/MS spectra (parent ion $m/z$ 179).

the fragmentation patterns are also observable. These differences in the fragmentation patterns can be exploited for deconvolution of the spectra of sugar mixtures.

## Quantification of mannose and galactose content in reference mixtures

Known mixtures of mannose and galactose have been analyzed by ESI-MS/MS in the negative ion-mode by fragmentation of the 179⁻ ion. The fractional amount of galactose ($f_G$) has been estimated by applying equation (4) on the 10 most intense fragmentation peaks. The effectiveness of the quantification method has been evaluated by linear regression on measured vs. expected $f_G$ plots (Figure 3).

The influence of instrumental settings has been explored by varying capillary voltage (CV), declustering potential (DP) and collision energy (CE) systematically and independently. The conditions that minimize the quantification error were IS -1.1 kV, DP -80 V and CE -10 V (Figure 3). CV does not have a significant effect on $f_G$ estimation (data not shown). On the contrary, DP influences the quantification results (Figure S1A (supplementary material)), leading to an overestimation of galactose at low voltage values (i.e., -60 V or below). This behavior could be ascribed to different properties of the sugars in desolvation, declustering or in-source dissociation, affecting the transmission efficiency of the ions into the collision cell. Both sugars are prone to in-source dissociation under the here employed experimental conditions, as indicated by the spectra reported in Figure 2A and 2B. The parameter that mostly affects the quantification results is CE (Figure S1B and S1C), inducing biases at either low (i.e., -7.5 V and below) or high (i.e., -15 V and above) values. Low CE values result in poor fragmentation, while high CE values lead to preferential accumulation of the smaller fragments. In either case, quantification is hindered by the lack of several diagnostic peaks. Nevertheless, Figure 3 shows that reliable quantification is obtained upon optimization of the instrumental parameters on standard mixtures. The procedure displays good reproducibility, yielding relative small standard deviations over three independent sets of measurements on freshly prepared samples (Figure 3).

## Analysis of PGMs acidic hydrolysis by mass spectrometry

The method described in the previous sections has been applied to the characterization of the products obtained by chemical hydrolysis of three PGMs: guar, tara and sesbania. Canonical protocols for acidic hydrolysis of polysaccharides at high temperature have been employed [47-50] in order to obtain a complete depolymerization of the PGMs in their monosaccharide constituents.

The analysis of the HCl-hydrolyzed PGMs by ESI-MS in negative

ion-mode (Figure 4) shows the presence of a peak at $m/z$ 179 for all the three samples, in agreement with accumulation of free mannose and/or galactose. However, the peak corresponding to disaccharides ($m/z$ 341) is also present in the spectra (Figure 4A and 4B), suggesting that the hydrolysis process is highly advanced but not fully complete. The MS/MS data for the 179⁻ ion from the three samples are shown in Figure 4D-4F and the quantification results are summarized in Table 2. The results reflect the known ranking in galactose content (tara<guar<sesbania) and display a fairly good agreement, especially for tara, with the M/G values reported in the literature, as assessed by other techniques.

On the other hand, a systematic underestimation of M/G seems to emerge from the data. This bias could be ascribed to incomplete hydrolysis, together with a higher susceptibility of galactose to the depolymerization conditions employed here. Due to exclusive localization of galactose in the ramifications of the PGM structure, it is conceivable that these monomers are released preferentially under suboptimal depolymerization conditions. The discrepancy between expected and measured M/G ratios becomes larger as the galactose content in the PGMs increases (i.e., M/G decreases) (Table 2), consistent with a stronger screening-effect on mannoses and/or higher availability of galactose ramifications.

Guar PGM has also been hydrolyzed by formic or trifluoroacetic acids, in order to compare different depolymerization protocols. The resulting M/G values are in keeping with those obtained by HCl (0.89 ± 0.11 with formic and 0.82 ± 0.16 with trifluoroacetic acid). Chloridric acid seems to be the most suitable acid for PGMs hydrolysis, followed by formic acid and trifluoroacetic acid.

## Discussion

The distinct fragmentation properties of mannose and galactose enable deconvolution of MS/MS spectra to the end of epimer quantification from given mixtures. The method represents an interesting alternative or complementation to the techniques typically employed to this purpose (e.g., NMR and chromatography). MS is attractive in terms of speed and sensitivity and being independent of sugar derivatization. The protocol is not limited to mannose and galactose quantification, and is virtually applicable to any monosaccharides presenting detectable differences in gas-phase fragmentation, like glucose and fructose [46].

The method has been employed to monitor chemical hydrolysis

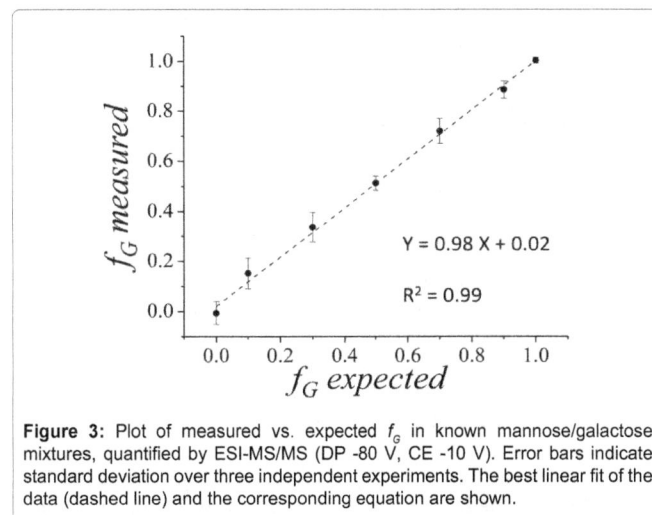

**Figure 3:** Plot of measured vs. expected $f_G$ in known mannose/galactose mixtures, quantified by ESI-MS/MS (DP -80 V, CE -10 V). Error bars indicate standard deviation over three independent experiments. The best linear fit of the data (dashed line) and the corresponding equation are shown.

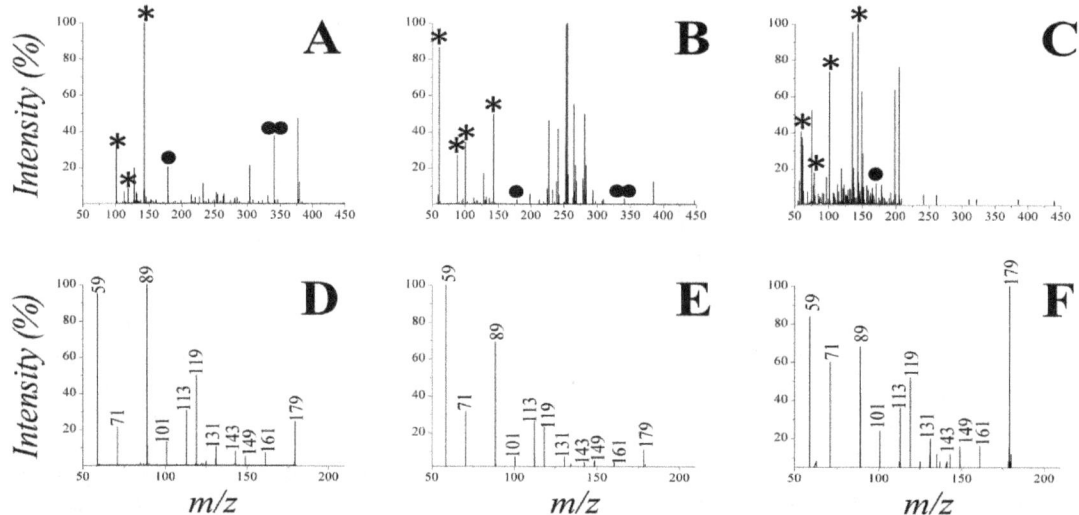

**Figure 4:** ESI-MS (A-C) and ESI-MS/MS of deprotonated monosaccharides (D-F) spectra of HCl-hydrolyzed guar (A, D), sesbania (B, E) and tara (C, F). The peaks corresponding to monosaccharides (•), disaccharides (••) and sugar fragments (*) are labeled in panels (A-C). The peaks in panels (D-F) are labeled by the *m/z* value.

| Leguminous species | Measured M/G | M/G range from literature | References |
|---|---|---|---|
| Guar (*Cyamopsis tetragonolobus*) | 0.96 ± 0.08 | 1.4-1.6 | [14,23] |
| Sesbania (*Sesbania bispinosa*) | 0.79 ± 0.13 | 1.3-1.5 | [14] |
| Tara (*Caesalpinia spinosa*) | 2.33 ± 0.22 | 2.5-3.0 | [14] |

**Table 2:** Estimation of the M/G ratio in different PGMs by HCl hydrolysis and ESI-MS/MS. The range of M/G values from the literature, obtained by different techniques, is given for comparison.

of PGMs from three leguminous species, leading to M/G estimations that are in fair agreement with the values reported in the literature and reflect the different galactose content characterizing the three flours. Systematic discrepancies, however, suggest that the depolymerization process is not complete, as further evidenced by the presence of disaccharides in the hydrolysis products. It is therefore recommendable to systematically optimize the hydrolysis protocol testing the effects of incubation conditions, such as polymer concentration, temperature, stirring, incubation time type and concentration of the acid. ESI-MS and ESI-MS/MS could be further employed to monitor chemical and/or enzymatic depolymerization reactions, in order to optimize hydrolysis conditions and to characterize the response of flours from different species. Another particular issue in the quantitative analysis of hydrolyzed polysaccharides is that mixtures of monosaccharides with vastly different physicochemical properties (i.e., negatively charged, neutral, deoxy etc.) are often obtained [51,52]. This is a challenge for either chromatographic or MS analysis, due to signal dispersion over multiple peaks and/or detection failure for some components of the sample. Further investigation will be needed to assess whether the present method could be extended to monosaccharide degradation products and to test its robustness towards sugar nature.

The applicability of the method is now limited to the quantification of monosaccharides, and does not deliver information on the details of the polysaccharide structure. However, the method might be extended to the analysis of oligosaccharides and the study of galactose-substitution pattern. Differences in the fragmentation properties of disaccharides have been already described [46], and these features could be exploited for structural investigation of PGMs. Further potential applications involve monitoring and mapping PGMs modifications (e.g., oxidation, methylation, etc.), as a first step for the creation of functionalized polymers and nano-engineered materials.

## Acknowledgments

This project was supported by the program "Suschem Lombardia: prodotti e processi chimici sostenibili per l'industria lombarda" Accordo Quadro Regione Lombardia-CNR, 16/07/2012 (Protocol no. 18096/RCC), and by Cariplo Foundation grant 2014-0478.

## References

1. Krishnaiah YS, Karthikeyan RS, Gouri Sankar V, Satyanarayana V (2002) Three-layer guar gum matrix tablet formulations for oral controlled delivery of highly soluble trimetazidine dihydrochloride. J Control Release 81: 45-56.

2. Vieira IGP, Mendes FNP, Gallao MI, Brito ES (2007) NMR study of galactomannans from the seeds of mesquite tree (Prosopis juliflora (Sw) DC). Food Chemistry 101: 70-73.

3. Itzincab-Mejía L, López-Luna A, Gimeno M, Shirai K,Bárzana E (2013) Enzymatic grafting of gallate ester onto chitosan: evaluation of antioxidant and antibacterial activities. Int J Food Sci Technol 48: 2034-2041.

4. Kenawy el R, Worley SD, Broughton R (2007) The chemistry and applications of antimicrobial polymers: a state-of-the-art review. Biomacromolecules 8: 1359-1384.

5. Das D, Ara T, Dutta S, Mukherjee A (2011) New water resistant biomaterial biocide film based on guar gum. Bioresour Technol 102: 5878-5883.

6. Srivastava M, Kapoor VP (2005) Seed galactomannans: an overview. Chem Biodivers 2: 295-317.

7. Rinaudo M (2008) Main properties and current applications of some polysaccharides as biomaterials. Polym Int 57: 397-430.

8. Salwiczek M, Qu Y, Gardiner J, Strugnell RA, Lithgow T, et al. (2014) Emerging rules for effective antimicrobial coatings. Trends Biotechnol 32: 82-90.

9. Prajapati VD, Jani GK, Moradiya NG, Randeria NP, Nagar BJ, et al. (2013) Galactomannan: a versatile biodegradable seed polysaccharide. Int J Biol Macromol 60: 83-92.

10. Mudgil D, Barak S, Khatkar BS (2014) Guar gum: processing, properties and food applications-A Review. J Food Sci Technol 51: 409-418.

11. Pollard MA, Fisher P, Windhab EJ (2011) Characterization of galactomannans derived from legume endosperms of genus Sesbania (Faboideae). Carbohydr Polym 84: 550-559.

12. Cerqueira MA, Pinheiro AC, Souza BWS, Lima AM, Ribeiro C, et al. (2009) Extraction, purification and characterization of galactomannans from non-

traditional sources. Carbohydr Polym 75: 408-414.

13. Merlini L, Boccia AC, Mendichi R, Galante YM (2015) Enzymatic and chemical oxidation of polygalactomannans from the seeds of a few species of leguminous plants and characterization of the oxidized products. J Biotechnol 198: 31-43.

14. Daas PJ, Grolle K, Van Vliet T, Schols HA, De Jongh HH (2002) Toward the recognition of structure-function relationships in galactomannans. J Agric Food Chem 50: 4282-4289.

15. Sittikijyothin W, Torres D, Goncalves MP (2005) Modelling the rheological behaviour of galactomannan aqueous solutions. Carbohydr Polym 59: 339-350.

16. Cheng Y, Prudhomme RK, Chick J, Rau DC (2002) Measurement of forces between galactomannan polymer chains: effect of hydrogen bonding. Macromolecules 35: 10155-10161.

17. Risica D, Dentini M, Crescenzi V (2005) Guar gum methyl ethers. Part I. Synthesis and macromolecular characterization. Polymer 46: 12247-12255.

18. Lavazza M, Formantici C, Langella V, Monti D, Pfeiffer U, et al. (2011) Oxidation of galactomannan by laccase plus TEMPO yields an elastic gel. J Biotechnol 156: 108-116.

19. Cheroni S, Gatti B, Margheritis G, Formantici C, Perrone L, et al. (2012) Enzyme resistance and biostablity of hydroxyalkylated cellulose and galactomannan as thickeners in waterborne paints. Int Biodeterior Biodegrad 69: 106-112.

20. Parikka K, Leppänen AS, Pitkänen L, Reunanen M, Willför S, et al. (2010) Oxidation of polysaccharides by galactose oxidase. J Agric Food Chem 58: 262-271.

21. Gliko-Kabir I, Yagen B, Baluom M, Rubinstein A (2000) Phosphated crosslinked guar for colon-specific drug delivery. II. In vitro and in vivo evaluation in the rat. J Control Release 63: 129-134.

22. Cunha PL, Castro RR, Rocha FA, de Paula RC, Feitosa JP (2005) Low viscosity hydrogel of guar gum: preparation and physicochemical characterization. Int J Biol Macromol 37: 99-104.

23. Crescenzi V, Dentini M, Risica D, Spadoni S, Skjåk-Braek G, et al. (2004) C(6)-oxidation followed by C(5)-epimerization of guar gum studied by high field NMR. Biomacromolecules 5: 537-546.

24. Bresolin TM, Milas M, Rinaudo M, Reicher F, Ganter JL (1999) Role of galactomannan composition on the binary gel formation with xanthan. Int J Biol Macromol 26: 225-231.

25. Petkowicz CLO, Reicher F, Mazeau K (1998) Conformational analysis of galactomannans: from oligomeric segments to polymeric chains. Carbohydr Polym 37: 25-39.

26. Wu Y, Cui W, Eskin NAM, Goff HD (2009) An investigation of four commercial galactomannans on their emulsion and rheological properties. Food Res Int 42: 1141-1146.

27. Wu Y, Li W, Cui W, Eskin NAM,Goff HD (2012) A molecular modeling approach to understand conformation efunctionality relationships of galactomannans with different mannose/galactose ratios. Food Hydrocolloids 26: 359-364.

28. Daas PJ, Schols HA, de Jongh HH (2000) On the galactosyl distribution of commercial galactomannans. Carbohydr Res 329: 609-619.

29. McCleary BV (1979) Modes of action of beta-mannanase enzymes of diverse origin on legume seed galactomannans. Phytochemistry 18: 757-763.

30. McCleary BV, Clark AH, Dea ICM, Rees DA (1985) The fine structures of carob and guar galactomannans. Carbohydr Res 139: 237-260.

31. Hoffman J, Svensson S (1978) Studies of the distribution of the d-galactosylside-chains in guaran. Carbohydr Res 65: 65-71.

32. Pettolino FA, Hoogenraad NJ, Stone BA (2002) Application of a mannan-specific antibody for the detection of galactomannans in foods. Food Hydrocoll 16: 551-556.

33. Gorin PAJ, Iacomini M (1985) Structural diversity of d-galacto-d-mannan components isolated from lichens having ascomycetous mycosymbionts. Carbohydr Res 142: 253-267.

34. Taravel FR, Mazeau K, Tvaroska I (1995) Conformational analysis of carbohydrates inferred from NMR spectroscopy and molecular modeling. Applications to the behavior of oligo-galactomannan chains. Braz J Med Biol Res 28: 723-732.

35. Joshi H, Kapoor VP (2003) Cassia grandis Linn. f. seed galactomannan:

structural and crystallographical studies. Carbohydr Res 338: 1907-1912.

36. Cunha PLR, Vieira IGP, Arriaga AMC, de Paula RCM, Feitosa JPA (2009) Isolation and characterization of galactomannan from Dimorphandra gardneriana Tul seeds as a potential guar gum substitute. Food Hydrocoll 23: 880-885.

37. Ganter JL, Heyraud A, Petkowicz CL, Rinaudo M, Reicher F (1995) Galactomannans from Brazilian seeds: characterization of the oligosaccharides produced by mild acid hydrolysis. Int J Biol Macromol 17: 13-19.

38. Ramesh HP, Yamaki K, Ono H, Tsushida T (2001) Two-dimensional NMR spectroscopic studies of fenugreek (Trigonella foenum-graecum L.) galactomannan without chemical fragmentation. Carbohydr Polymers 45: 69-77.

39. Muschin T, Yoshida T (2012) Structural analysis of galactomannans by NMR spectroscopy. Carbohydr Polymers 87: 1893-1898.

40. Guo S, Mao W, Yan M, Zhao C, Li N, et al. (2014) Galactomannan with novel structure produced by the coral endophytic fungus Aspergillus ochraceus. Carbohydr Polym 105: 325-333.

41. Simões J, Maricato E, Nunes FM, Domingues MR, Coimbra MA (2014) Thermal stability of spent coffee ground polysaccharides: galactomannans and arabinogalactans. Carbohydr Polym 101: 256-264.

42. Chen Y, Mao W, Wang B, Zhou L, Gu Q, et al. (2013) Preparation and characterization of an extracellular polysaccharide produced by the deep-sea fungus Penicillium griseofulvum. Bioresour Technol 132: 178-181.

43. Shang M, Zhang X, Dong Q, Yao J, Liu Q, et al. (2012) Isolation and structural characterization of the water-extractable polysaccharides from Cassia obtusifolia seeds. Carbohydr Polym 90: 827-832.

44. Parikka K, Leppänen AS, Xu C, Pitkänen L, Eronen P, et al. (2012) Functional and anionic cellulose-interacting polymers by selective chemo-enzymatic carboxylation of galactose-containing polysaccharides. Biomacromolecules 13: 2418-2428.

45. Moreira AS, Coimbra MA, Nunes FM, Simões J, Domingues MR (2011) Evaluation of the effect of roasting on the structure of coffee galactomannans using model oligosaccharides. J Agric Food Chem 59: 10078-10087.

46. March RE, Stadey CJ (2005) A tandem mass spectrometric study of saccharides at high mass resolution. Rapid Commun Mass Spectrom 19: 805-812.

47. Laopaiboon L, Nuanpeng S, Srinophakun P, Klanrit P, Laopaiboon P (2009) Ethanol production from sweet sorghum juice using very high gravity technology: effects of carbon and nitrogen supplementations. Bioresour Technol 100: 4176-4182.

48. Sun Y, Cheng J (2002) Hydrolysis of lignocellulosic materials for ethanol production: a review. Bioresour Technol 83: 1-11.

49. Baldaro E, Gallucci M, Formantici C, Issi L, Cheroni S, et al. (2012) Enzymatic improvement of guar-based thickener for better-quality silk screen printing. Color Technol 128: 315-322.

50. Kupiainen L, Ahola J, Tanskanen J (2012) Distinct Effect of Formic and Sulfuric Acids on Cellulose Hydrolysis at High Temperature. Ind Eng Chem Res 51: 3295-3300.

51. Anumula KR (2000) High-sensitivity and high-resolution methods for glycoprotein analysis. Anal Biochem 283: 17-26.

52. Behan JL, Smith KD (2011) The analysis of glycosylation: a continued need for high pH anion exchange chromatography. Biomed Chromatogr 25: 39-46.

# 20

# Separation of Phenol-Containing Pyrolysis Products Using Comprehensive Two-Dimensional Chromatography with Columns Based on Pyridinium Ionic Liquids

**Mikhail V Shashkov[1]\* and Vladimir N Sidelnikov[2]**

[1]*Boreskov Institute of Catalysis SB RAS, Novosibirsk, Russian Federation*
[2]*Novosibirsk State University, Novosibirsk, Russian Federation*

## Abstract

The work describes the application of columns with high-polar stationary liquid phases based on pyridinium ionic liquids for GC × GC separation of bio-oil and product of coal pyrolysis. By using inverse combination columns- the first ionic liquid column, the second is HP-5 the good separation results have been obtained. In the analysis of coal pyrolysis products, the suggested approach provides a much better resolution between components in comparison with a less polar first-dimension column (ZB-WAX). The good selectivity for peaks of phenols is observed, and the group of phenols is well detached and separated from the group of diaromatics. For bio-oil also good separation picture was obtained, the groups of phenols and guaiacol derivatives are distinguished with good resolution of substances within each group.

**Keywords:** Comprehensive two-dimensional chromatography; Ionic liquids; High polar stationary phases; Bio-oil

## Introduction

Although world reserves of conventional hydrocarbon feedstock are big, they are not inexhaustible and are being rapidly depleted [1]. This is why technologies for the production of fuels and valuable chemicals from alternative and renewable sources are under development. Among them are fossil coals, in particular large reserves of low-grade coals, and biomass of different origin. An efficient method for processing of such resources is pyrolysis [2,3]. This process yields a complex mixture of organic compounds that can further be used to produce motor fuels or valuable chemicals.

To choose the optimal conditions of pyrolysis and further processing of pyrolysis liquids, it is necessary to determine their exact chemical composition. Comprehensive two-dimensional chromatography (GC × GC) is the advanced and most informative method for the analysis of multicomponent mixtures of various classes of chemical compounds, specifically those with strongly different polarities [4]. The main advantages of GC × GC are its higher separation capacity in comparison with conventional chromatography, and the possibility to perform group analysis of complex mixtures [4]. However, the best results can be achieved only with the properly chosen separation system (a combination of columns) [5] and separation conditions [6].

It is known that the liquid part of coal pyrolysis products includes various types of aromatic hydrocarbons and substantial amounts of phenols [7]. Another complex mixture is the product of biomass pyrolysis called bio-oil (BO). BO is characterized by a high content of different classes of phenolic compounds, mostly the guaiacol derivatives [8]. In addition, BO contains a wide spectrum of other oxygen-containing compounds, particularly carbohydrates, carbonyl compounds and phenols [8].

At present, publications devoted to GC × GC analysis of the liquid products of pyrolysis of various coals [3,7] can be found in the literature. Due to increased interest in renewable fuel resources, the GC × GC analysis of BO is reported in a greater number of works [9-13]. These works deal with GC × GC analysis of components in the initial BO [9-11] and its fractions [10,13].

However, virtually in all the indicated works, separations of pyrolysis products were carried out on a pair of columns with non-polar (HP-5 and its analogs) and medium-polar (HP-17 and its analogs) liquid phases. Only in some works, columns with phases having a slightly higher polarity, HP-1701 [3] and WAX [12], were employed. Although such a combination of columns gives satisfactory separation results, the separation pattern could be further improved. In some cases, columns with strongly different polarities give better GC × GC separations, which make it possible to find an additional number of components and obtain a more exact boundary between groups of components on separation pictures of complex mixtures [5,14]. In this connection, it seems interesting to perform GC × GC with combinations of non-polar and high-polar columns.

To this end, high-polar columns can be represented by those with stationary liquid phases based on ionic liquids (ILs). ILs form a relatively new class of stationary liquid phases that have some advantages, in particular, high thermal stability in comparison with other types of high-polar phases, polarity, and the possibility to control phase selectivity in a wide range [15,16]. Earlier it was shown that the use of columns with ILs provides good separations of various objects: fatty acids, oil products, objects of plant origin, and pesticides [16,17].

As for the GC × GC analysis of coal and bio-oil pyrolysis products with the use of IL columns, we have found only one work devoted to this issue [18]. However, its authors did not manage to obtain a satisfactory separation on the SP-IL100 column. Also we can found the work devoted GC × GC analysis of sulfur polyaromatics in coal tar on different IL columns [19].

**\*Corresponding author:** Mikhail V. Shashkov, Boreskov Institute of Catalysis SB RAS, Novosibirsk, Russian Federation, E-mail: shashkov11@gmail.com

Our work describes the application of columns with high-polar stationary liquid phases based on pyridinium ILs for GC × GC separation of pyrolysis products. A high contribution of specific interactions is typical of such ILs [16]. Thus, phases with such ILs are highly selective for polar compounds, particularly for phenols [16], which make columns with these phases attractive for separation of objects considered in this work.

## Experimental

### Objects of separation: Sample preparation

A sample of bio-oil, typical flash pyrolysis oil was purchased from Biomass Technology Group BV (Netherlands). The initial bio-oil was extracted with toluene (5 mL of toluene per 1 g of the initial bio-oil). This procedure is required to remove non-volatile components, such as carbohydrates and dicarboxylic acids, which cannot be analyzed by GC [20]. A pyrolizate of bituminous coal from Kuzbass region deposit was obtained by heating at a temperature of 600°C for 3 hours without air access. The liquid fraction of the pyrolizate was analyzed without any preparation steps.

### Columns used in the study

The study was carried out using columns with the non-polar phase HP-5 (Agilent), 30 m × 0.25 mm × 0.25 μm and 5 m × 0.25 mm × 0.15 μm, and a column with the polar phase ZB-WAX (Phenomenex), 30 m × 0.25 mm × 0.25 μm.

Homemade columns prepared by a static high pressure method using the techniques reported in Ref. [21] were employed as high-polar IL columns. ILs were synthesized according to Ref. [15], characterization and purity info also in Ref. [15]. This procedure was used to prepare the (30 m × 0.25 mm × 0.25 μm) columns with pyridine ILs (2MPyC$_9$) and (4MPy), and the (5 m × 0.25 mm × 0.1 μm) column with IL (2MPyC$_9$). Structures of the ILs used in the study are displayed in Figure 1. The efficiency of the prepared columns was ca. 2400 plates per meter.

### GC × GC analysis

All the experiments were carried out using an Agilent 7890A chromatograph and G4513A autosampler. An Agilent flow modulator and ZOEX software (for visualization of 2D samples) were employed in GC × GC experiments.

### Separation conditions

The temperature program of the column oven was as follows: in the case of coal pyrolysis products, 80°C for 3 min followed by programming at a rate of 7°C/min up to a final temperature of 270°C; in the case of bio-oil, 70°C for 3 min followed by programming at 8°C/min up to a final temperature of 270°C. Temperature of inlet and flame-ionization detector was set equal to the final temperature of the analysis. Carrier gas flow rate: for the first column (helium), 0.65 mL/min; for the second column (hydrogen), 23 mL/min. In all cases, modulation period was 1.5 s.

In the experiment with ZB-WAX column, final temperature was 260°C, and flow rate through the first column, 0.9 mL/min. The other conditions were similar to those used for IL columns.

Some components were identified by GC × GC analysis of the mixtures with the test components: naphthalene, phenol, methyl- and dimethylphenols. In other cases, comparative GC-MS analysis of the mixtures was carried out on an Agilent 7000B instrument using a 30 m × 0.25 mm × 0.25 μm column with the liquid phase 2MPyC$_9$ under the conditions similar to those used for the first column in GC × GC.

The retention time of components, which is required to calculate resolution coefficients, was found by means of ZOEX software.

## Discussion

### Analysis of coal pyrolysis products on HP-5/2MPyC$_9$ columns

Let us consider the separation of pyrolysis products using the conventional scheme of columns connection. The commonly accepted sequence of columns (non-polar/polar) is convenient because along the abscissa axis a sample is separated according to the boiling points, and then the axis of ordinates shows the separation that corresponds to increasing polarity. Two-dimensional separation gives the patterns of mixtures (for example, of oil products) that are convenient for interpretation.

The first-dimension column was represented by the most universal HP-5. The second-dimension IL column was 2MPyC$_9$. As was shown in our earlier study, this IL has high polarity and provides good separation of phenols [15].

The optimal separation conditions were chosen for the analysis, which gave the following separation pattern (Figure 2).

It is seen that separation cannot be considered as satisfactory. The peaks corresponding to phenols are strongly broadened on the second column. Although individual zones corresponding to the groups of aromatic hydrocarbons can be distinguished in Figure 1, the pattern is difficult to interpret. As was shown earlier [15], good separations of phenols can be obtained on columns with 2MPyC$_9$ in the case of one-dimensional chromatography, and their peaks have good shape at good

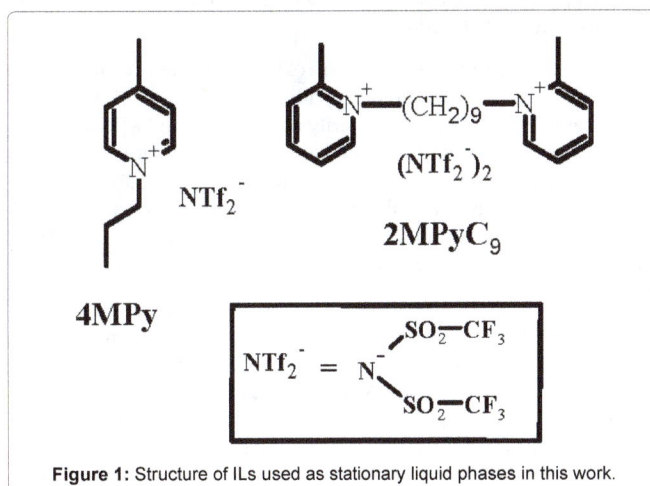

Figure 1: Structure of ILs used as stationary liquid phases in this work.

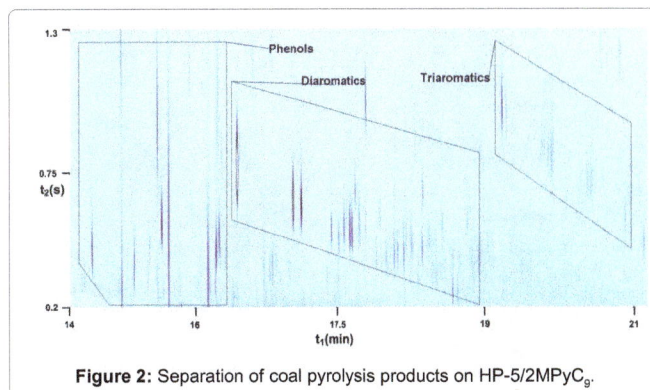

Figure 2: Separation of coal pyrolysis products on HP-5/2MPyC$_9$.

efficiency. Nevertheless, good separations are not observed under the conditions of GC × GC. This effect can be explained as follows. It is known that the second column in two-dimensional chromatography operates at the flow rates strongly exceeding the optimal value. Earlier it was shown that in the case of IL columns the efficiency decreases much faster with increasing the flow rate as compared to the case of columns with polysiloxane stationary liquid phases and phases based on polyethylene glycol [15].

So, further separations were performed with IL column in the first dimension to provide conditions of its operation close to optimal. In some cases, such scheme of columns connection in GC × GC provides a better separation of the groups of polar components [22].

## Analysis of coal pyrolysis products with IL as the first-dimension column

Thus, the conventional scheme of columns connection (non-polar/polar) does not provide a satisfactory separation of a mixture of pyrolysis products. Therefore, the inverse connection of columns (high-polar/non-polar) was employed, and separation conditions (flows through the columns and modulation time) were chosen so as to obtain the best separation of the mixture. An example of such separation is illustrated on Figure 3.

Figure 3 shows that, on the one hand, the group of phenols is well detached and separated from the group of diaromatics. On the other hand, a distinct separation can be observed also between the peaks of phenols in this group. The peaks have good symmetry and are not broadened upon separation on the second column, in distinction to the pattern that is observed on Figure 1. For this reason, the number of the observed peaks of phenols is greater due to a better separation and an increased intensity of minor peaks.

Now let us examine the separation of the same mixture on columns with the phase based on another pyridinium IL, 4MPy. This IL has a somewhat lower McReynolds polarity as compared to 2MPyC$_9$ (Table 1); so, the separation pattern should also be different. However, it was found that separation picture in case of 4Mpy very similar to 2MPyC$_9$, a good resolution between the peaks of phenols within the group is observed, and the group of phenols itself is well separated from diaromatics (due to the similarity the picture is not shown).

Now let us compare the separation results for this mixture with those obtained on a pair of columns where the first column was less polar than ionic liquids. Figure 4 shows the separation obtained on a pair of columns ZB-WAX/HP-5. All separation conditions were similar to those on other columns, except for the carrier flow through the column with ZB-WAX: its value was 0.9 ml/min, which corresponds to the optimal He flow through the column.

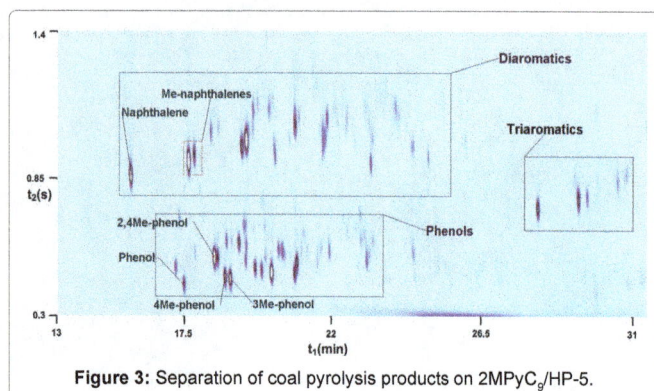

**Figure 3:** Separation of coal pyrolysis products on 2MPyC$_9$/HP-5.

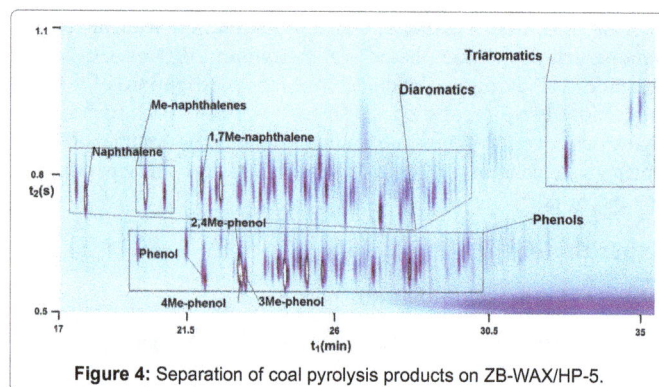

**Figure 4:** Separation of coal pyrolysis products on ZB-WAX/HP-5.

One can see that the general separation pattern is worse than in the case of first columns with the IL phase. The group of phenols has a considerably smaller number of peaks in comparison with chromatograms obtained on IL columns. On Figure 4 the group of phenols is visually more close to the group of diaromatic hydrocarbons than on Figure 3. Such differences in separation patterns in the case of system with the first WAX column can be attributed to its lower polarity as compared to IL. The causes of these differences are considered.

## 2D resolution

To estimate differences in the separation capacity of column systems ZB-WAX/HP-5, 2MPyC$_9$/HP-5 and 4MPy/HP-5, resolution was measured for a series of selected peaks corresponding to this mixture. It is known that the resolution between peaks in GC × GC, $R_{s2D}$ (2D resolution), is calculated by formula (1) [23]:

$$R_{s2D} = \sqrt{R_{s1}^2 + R_{s2}^2} \quad (1)$$

where $R_{s1}$ and $R_{s2}$ are the resolutions along first and second columns, respectively.

For the mixture under analysis, the pairs of peaks were chosen so as to estimate separations: 1) between groups of phenols and diaromatics, and 2) between peaks within the group of phenols. The peak of phenol and the nearest peak from diaromatics group were chosen as a pair to estimate separation between groups of aromatics and phenols. In the case of 2MPyC$_9$, resolution of the Naphthalene/Phenol pair was analyzed; for 4MPy, 1Me-naphthalene/Phenol; and for ZB-WAX, 1,7Me-naphthalene/Phenol.

To estimate peak resolution within the group of phenols, for all columns the pairs of the most close and most intense peaks were chosen: 2,4Me-phenol/4Me-phenol and 4Me-phenol/3Me-phenol. The results obtained are listed in Table 1.

One can see from Table 1 that in all cases except for the 4Me-phenol/3Me-phenol pair, systems with IL columns show better results of separation than the system with the first column based on polyethylene glycol.

The better separation between groups of diaromatics and phenols (column 1 in Table 1) can have the following explanation. Due to high polarity of ILs, and low contribution of dispersive interactions comparing with ZB-WAX [15] compounds retained considerably less, and leaving the column faster in case of ILs. As the result, separation on the second non polar column occurs at a relatively lower temperature, and resolution increases in comparison with ZB-WAX.

A much better separation for the 2,4Me-phenol/4Me-phenol pair on IL can be attributed to a higher polarity of phases with IL (last column in Table 1). Separation for the 4Me-phenol/3Me-phenol pair is slightly better in the case of ZB-WAX column. This may be related to a higher efficiency of this column, because its selectivity for polar compounds was shown to be lower as compared to IL columns [15].

## Separation of the toluene extract of bio-oil on 2MPyC$_9$ column in the first dimension

Now it is clear that the GC × GC separation of a mixture of aromatic and phenolic compounds should be performed using a combination of polar/non-polar columns. Let us consider the separation of bio-oil products.

Bio-oil is a mixture of liquid products that are produced by thermolysis of hemicellulose, cellulose and lignin as the components of biomass. Thermal decomposition products of hemicellulose and cellulose are carbohydrates and carbonyl compounds. Lignin decomposes mostly into phenolic compounds, particularly guaiacol and its derivatives [11]. Among various products constituting BO, only the phenolic compounds can be subjected to GC analysis, because the other compounds are thermally unstable or non-volatile.

Components of the extract were separated using 2MPyC$_9$/HP-5 columns, which showed the best results in separation of coal pyrolysis products. The resulting separation pattern is displayed on Figure 5.

One can see that separation of bio-oil is more difficult as compared to separation of coal pyrolysis products. The reason is that the main substances in the mixture to be separated have close chemical features. Nevertheless, a combination of 2MPyC$_9$/HP-5 columns makes it possible to perform a successful separation. It is seen that the group of phenols is close to the group of guaiacol derivatives. However, phenols are well separated within their group because the IL phase is highly selective for phenols.

## Conclusion

It was shown that in comprehensive two-dimensional chromatography the use of columns with pyridinium ILs in the first

dimension ensures good separations for the analysis of pyrolysis products of coal and biomass. In the analysis of coal pyrolysis products, the suggested approach provides a much better resolution between components in comparison with a less polar first-dimension column (ZB-WAX). Therewith, group analysis can be performed: the regions corresponding to phenolic compounds and diaromatics can be distinguished on a chromatogram. A good selectivity is observed within the group of phenols. In the analysis of bio-oil sample, the groups of phenols and guaiacol derivatives are distinguished with good resolution of substances within each group.

### References

1. Bentley RW (2002) Global oil & gas depletion: an overview. Energy Policy 30: 189-205.

2. Onorevoli B, Machadoa ME, Dariva C, Franceschi E, Krause LC, et al. (2014) A one-dimensional and comprehensive two-dimensional gas chromatography study of the oil and the bio-oil of the residual cakes from the seeds of Crambe abyssinica. Ind Crop Prod 52: 8-16.

3. Rathsack P, Otto M (2014) Classification of chemical compound classes in slow pyrolysis liquids from brown coal using comprehensive gas-chromatography mass-spectrometry. Fuel 116: 841-849.

4. Hantao LW, Najafi A, Zhang C, Augusto F, Anderson JL (2014) Tuning the selectivity of ionic liquid stationary phases for enhanced separation of nonpolar analytes in kerosene using multidimensional gas chromatography. Anal Chem 86: 3717-3721.

5. Omais B, Courtiade M, Charon N, Ponthus J, Thiébaut D (2011) Considerations on orthogonality duality in comprehensive two-dimensional gas chromatography. Anal Chem 83: 7550-7554.

6. Májek P, Krupčík J, Gorovenko R, Špánik I, Sandra P, et al. (2014) Computerized optimization of flows and temperature gradient in flow modulated comprehensive two-dimensional gas chromatography. J Chromatogr A 1349: 135-138.

7. Stihle J, Uzio D, Lorentz C, Charon N, Ponthus J, et al. (2012) Detailed characterization of coal-derived liquids from direct coal liquefaction on supported catalysts. Fuel 95: 79-87.

8. Song Q, Nie JQ, Ren M-G, Guo Q-X (2009) Effective phase separation of biomass pyrolysis oils by adding aqueous salt solutions. Energy Fuels 23: 3307-3312.

9. Marsman JH, Wildschut J, Evers P, de Koning S, Heeres HJ (2008) Identification and classification of components in flash pyrolysis oil and hydrodeoxygenated oils by two-dimensional gas chromatography and time-of-flight mass spectrometry. J Chromatogr A 1188: 17-25.

10. Schneider JK, da Cunha ME, dos Santos AL, Maciel GP, Brasil MC, et al. (2014) Comprehensive two dimensional gas chromatography with fast-quadrupole mass spectrometry detector analysis of polar compounds extracted from the bio-oil from the pyrolysis of sawdust. J Chromatogr A 1356: 236-240.

11. Michailof C, Sfetsas T, Stefanidis S, Kalogiannis K, Theodoridis G, et al. (2014) Quantitative and qualitative analysis of hemicellulose, cellulose and lignin bio-oils by comprehensive two-dimensional gas chromatography with time-of-flight mass spectrometry. J Chromatogr A 1369: 147-160.

12. Hatch LE, Luo W, Pankow JF, Yokelson RJ, Stockwell CE, et al. (2015) Identification and quantification of gaseous organic compounds emitted from biomass burning using two-dimensional gas chromatography–time-of-flight mass spectrometry. Atmos Chem Phys 15: 1865-1899.

13. Cunha ME, Schneider JK, Brasil MC, Cardoso CA, Monteiro LR, et al. (2013) Analysis of fractions and bio-oil of sugar cane straw by one-dimensional and two-dimensional gas chromatography with quadrupole mass spectrometry (GC × GC/qMS). Microchem J 110: 113-119.

14. Cordero C, Rubiolo P, Sgorbini B, Galli M, Bicchi C (2006) Comprehensive two-dimensional gas chromatography in the analysis of volatile samples of natural origin: a multidisciplinary approach to evaluate the influence of second dimension column coated with mixed stationary phases on system orthogonality. J Chromatogr A 1132: 268-279.

15. Shashkov MV, Sidelnikov VN (2013) Properties of columns with several pyridinium and imidazolium ionic liquid stationary phases. J Chromatogr A 1309: 56-63.

| | Diaromatics/ Phenol | 2,4Me-phenol/ 4Me-phenol | 4Me-phenol/ 3Me-phenol | McReynolds polarity [15] |
|---|---|---|---|---|
| 2MPyC$_9$ | 6.18 | 3.80 | 1.55 | 92 |
| 4MPy | 4.78 | 3.58 | 1.58 | 84 |
| ZB-WAX | 3.18 | 0.33 | 1.95 | 51 |

**Table 1:** Resolution of pairs of substances in GC × GC separation using the first column with IL or ZB-WAX and the second HP-5 column.

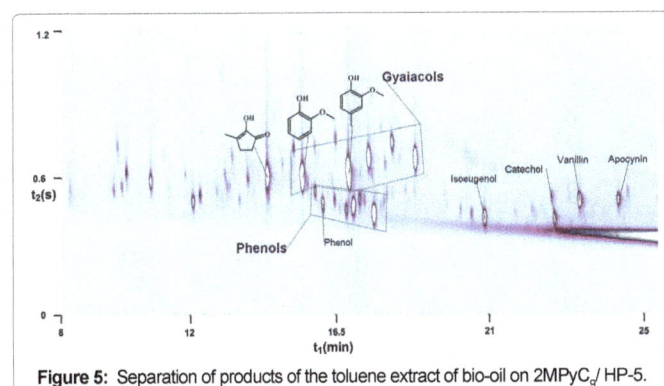

**Figure 5:** Separation of products of the toluene extract of bio-oil on 2MPyC$_9$/ HP-5.

16. Qi M, Armstrong DW (2007) Dicationic ionic liquid stationary phase for GC-MS analysis of volatile compounds in herbal plants. Anal Bioanal Chem 388: 889-899.

17. Vidal L, Riekkola ML, Canals A (2012) Ionic liquid-modified materials for solid-phase extraction and separation: a review. Anal Chim Acta 715: 19-41.

18. Omais B, Courtiade M, Charon N, Thiébaut D, Quignard A, et al. (2011) Investigating comprehensive two-dimensional gas chromatography conditions to optimize the separation of oxygenated compounds in a direct coal liquefaction middle distillate. J Chromatogr A 1218: 3233-3240.

19. Antle P, Zeigler C, Robbat A Jr (2014) Retention behavior of alkylated polycyclic aromatic sulfur heterocycles on immobilized ionic liquid stationary phases. J Chromatogr A 1361: 255-264.

20. Oasmaa A, Kuoppala E, Solantausta Y (2003) Fast pyrolysis of forestry residue. 2. Physicochemical composition of product liquid. Energy Fuels 17: 433-443.

21. Shashkov MV, Sidel'nikov VN (2012) Single cation ionic liquids as high polarity thermostable stationary liquid phases for capillary chromatography. Russ J Phys Chem A 86: 138-141.

22. Seeley JV, Seeley SK, Libby EK, Breitbach ZS, Armstrong DW (2008) Comprehensive two-dimensional gas chromatography using a high-temperature phosphonium ionic liquid column. Anal Bioanal Chem 390: 323-332.

23. Dutriez T, Courtiade M, Thibaut D (2009) High-temperature two-dimensional gas chromatography of hydrocarbons up to $nC_{60}$ for analysis of vacuum gas oils. J Chromatogr A 1216: 2905-2912.

# 21

# Single Drop Microextraction Analytical Technique for Simultaneous Separation and Trace Enrichment of Atrazine and its Major Degradation Products from Environmental Waters Followed by Liquid Chromatographic Determination

**Alula Yohannes[1], Tesfaye Tolesa[1], Yared Merdassa[1,2] and Negussie Megersa[1]\***

[1]*Department of Chemistry, Addis Ababa University, PO Box 1176, Addis Ababa, Ethiopia*
[2]*Department of Chemistry, Jimma University, PO Box 378, Jimma, Ethiopia*

### Abstract

In this work, a method of single drop microextraction (SDME) combined with high performance liquid chromatography (HPLC) with diode array detection (DAD) was studied for trace level enrichment as well as simultaneous determination of atrazine (ATZ) and its major degradation products such as desethylatrazine (DEA) and desisopropylatrazine (DIA) in environmental waters. The main factors influencing the extraction procedure including types and volume of extraction solvent, sample stirring rate, sample solution pH, extraction temperature, extraction time, and salting out effect were optimized. The method detection limits were as low as 0.01 for ATZ and 0.05 for both DIA and DEA, with coefficients of determination better than 0.998 within a linear range of 0.5-150 µg L$^{-1}$. Under the optimal conditions, the proposed method was applied for the analysis of real water samples and good spiked recoveries in the range of 65.6%-96.3% with relative standard deviation of less than 5% were obtained. The results confirmed that the proposed procedure provides reliable precision, linearity and sensitivity and is very effective for analyzing the target compounds in environmental waters. Therefore, the developed SDME method coupled with HPLC-DAD was found to be simple, inexpensive, and environmentally benign sample pretreatment technique.

**Keywords:** SDME technique; Trace enrichment; Environmentally benign; Atrazine; Degradation products; Environmental waters

## Introduction

A pesticide is any substance or mixtures of substances, natural or synthetic, that has been used in agriculture to control or repel any pest that competes for food, affect the qualities and quantities of the yields and spreads disease [1,2]. The effective use of pesticides for agricultural purposes has improved the quality and quantity of food production. It has been estimated that about one-third of the crop production would be lost without the application of pesticide [3]. On the other hand, use of pesticides known to cause pollution in various environmental compartments such as air, soil, ground and surface water as well as serious risks to the aquatic lives and human health [4]. The increasing public concerns, in recent years, associated with the health risks because of the occurrence of pesticide residues led to strict regulations to be issued by the legislative bodies for the maximum residue limits (MRLs) in food commodities and drinking water [5].

Atrazine is one of the most commonly used herbicides, among the symmetrical (*s*-) triazine pesticides applied for the control of annual broadleaf and grassy weeds in maize, sugar cane, pineapples, nuts and non-crop areas [6]. It is also frequently used as a pre-emergent or early post-emergent herbicide and as a result its residues along with its major metabolites have been frequently detected in surface as well as ground waters [7]. Studies on toxicity of atrazine reported its toxic effects across the living species including the decrease in erythrocyte parameters, delayed sexual development and disrupting neuroendocrine system [8]. As a result of pesticide toxicities, the European Union (EU) legislation dictates the maximum admissible concentration of pesticides in drinking water to be 0.5 µg L$^{-1}$ for the total pesticides and 0.1 µg L$^{-1}$ for individual pesticide [9]. Therefore, in order to meet this and other similar regulations, development of highly sensitive and efficient analytical methods is very essential for accurate determination of trace level residues and metabolites of these harmful compounds [10].

Liquid–liquid extraction (LLE) is the classical method traditionally utilized for pesticide extraction from aqueous samples. However, the technique requires large quantities of expensive organic solvents and is time-consuming [11,12]. Although solid phase extraction (SPE) is consuming lesser time than LLE, it still requires some volumes of organic solvents for analytes elution step [13,14]. Besides their undesirable consequences to human health organic solvents are also known for being flammable and pollutant to the environment. These disadvantages have been the driving forces for critical need for developing analytical methods that either eliminate the use of organic solvents or at least minimize their requirements in sample preparation [15].

Solid-phase microextraction (SPME) is a solvent-free sample preparation method which eliminates the disadvantages associated with the use of conventional extraction methods but has inherent drawbacks related to the expensiveness of its fiber and limited lifetime [16]. Recently, different modes of liquid phase microextraction (LPME) were developed as a simple and benign method for sample preparation

**\*Corresponding author:** Negussie Megersa, Department of Chemistry, Addis Ababa University, PO Box 1176, Addis Ababa, Ethiopia
E-mail: megersane@yahoo.com (or) negussie.megersa@aau.edu.et

[17-20]. One mode of LPME is based on the distribution of the analytes between a microdrop of organic solvent at the tip of a microsyringe needle and aqueous sample solution and is termed as single drop microextraction (SDME). The technique combines extraction, preconcentration and sample introduction into a single step [20,21].

The use of a single drop as a collector of analytes in analytical chemistry can be traced back to the work of Liu and Dasgupta [22] in the mid-1990s. At present, there are seven different modes of SDME which could be categorized into either two-phase or three-phase modes; depending on the number of phases co-existing at equilibrium. Two-phase modes include direct immersion, continuous flow, drop-to-drop and directly suspended droplet while three-phase modes consist of a headspace and liquid–liquid–liquid microextraction [23]. Since in direct immersion SDME the microdrop of the extracting solvent is in direct contact with the aqueous sample; the solvent being immiscible with water [24]. This mode of SDME is simple and thus considered to be suitable for quantitative extraction of the target analytes in this study, that have nonpolar or slightly polar properties, from environmental water samples.

The proposed method coupled with HPLC, which has not been considered in any earlier studies for simultaneous determination of atrazine and its major degradations from natural waters, was demonstrated to be a successful and promising sample preparation technique. Quantitative extraction of the degradation products, whose environmental fates are not fully understood, is not easy since the physical and chemical properties of the compounds greatly vary when they lose certain moieties from the parent compound. Even though other sample preparation methods are available, there are no reports for residual extraction of atrazine and its metabolites from environmental sample matrices simultaneously, based on single drop microextraction.

## Experimental

### Chemicals and reagents

Pesticide standards of atrazine, (ATZ) (98.4%); deethylatrazine, (DEA) (96%); and deisopropylatrazine, DIA (99%) used in this study, were purchased from Dr. Ehrenstorfer GmbH (Augsburg, Germany). Selected physicochemical properties of the target analytes are given in Table 1 [25,26]. All the standards were certified reference materials for residue analysis. HPLC grade methanol and acetonitrile, used as mobile phase in chromatographic analysis, were obtained from Acros organics (New Jersey, USA). Other chemicals; including hydrochloric acid, sodium hydroxide, sodium chloride and 1-octanol were the products of Sigma Aldrich Chemie (Steinheim, Japan). Ultrapure water (EASYpure LF) was used throughout and all other chemicals used were of analytical grade reagents, unless otherwise stated.

### Instrumentation

All chromatographic analyses were performed using Agilent Technologies® 1200 series HPLC, equipped with quaternary pump, Agilent 1200 series vacuum degasser, Agilent 1200 series thermostated autosampler and Agilent 1200 series Diode Array Detector (DAD). Chromatographic separations were performed on a reversed phase VP-ODS C$_{18}$ column (150 mm × 4.6 mm I.D., particle size 5 μm) (Agilent Technologies, Germany). The entire units of the HPLC system were interfaced with a computer (Hp, Compaq, Intel® and Pentium 4 HT). The SDME was performed using a bevel shaped 100 μL GC microsyringe (Hamilton, Bonaduz, Switzerland). A S23-2 digital magnetic stirrer (Shanghai Sile Instrument Co., China) and a 5 mm stirring bar were used to stir the solution. All weighing operations were carried out

using Analytical balance (Mettler Toledo, Mettler instrument AG, Switzerland). Ultrasonic bath (Decon F5100b, England) was used for degassing the mobile phases and ensuring complete dissolution of some standards. Water and mobile phases were filtered through 0.22 μm filter paper (Millipore, Germany).

### Preparation of the standard solutions

A stock standard solution with concentration level of 100 mg L$^{-1}$ was prepared by dissolving 2.5 mg of each analyte with methanol and transferred quantitatively into a 25 mL volumetric flask [13]. Finally, it was filled up to the mark with methanol. Intermediate standard solutions were also prepared from the stock solution by diluting with methanol. Working solution of 0.5 μg L$^{-1}$ mixture were prepared daily by appropriate dilution of the stock solutions with ultrapure water. Extraction performances of the method were studied using the aqueous sample solutions that were prepared by spiking a known concentration of the analytes into the reagent water. Stock solutions of the analyte standards were stored in a refrigerator at 4C when not in use.

A series of solutions of the s-triazine standard mixture for calibration were prepared from the intermediate stock solutions in the concentration range of 0.1 mg L$^{-1}$ ppm to 1.0 mg L$^{-1}$, at five points, each day. Evaluation of the analytical results was based on five injections, in most cases, and peak areas were considered as instrumental responses.

### Sample solution pH

The sample solution pH was appropriately varied in order to obtain the target analytes in deionized forms and thus facilitate their transfer into the microdrop. This was achieved by adding drops of 5 mM phosphate buffer into the sample solution containing the model analytes in order to obtain the sample solution pH ranged from 2.0 to 9.0 [25,27]. The phosphate buffers of different pH were prepared by varying the amounts of H$_3$PO$_4$, KH$_2$PO$_4$, and K$_2$HPO$_4$, following the reported literature information [27].

### Procedure for single drop microextraction

For extraction of the target analytes, 5 mL of a salted (7.5% NaCl, w/v) aqueous sample solution spiked at a known concentration with all the target analytes was added to the 6-mL standard vials. The contents were then agitated with a magnetic stirrer by means of a stir bar. Before each extraction, the microsyringe was rinsed three times with methanol to ensure that no air bubbles are available in the barrel and the needle. After uptake of 3 μL of 1-octanol, the microsyringe needle was inserted through the septum and was immersed into the 5 mL sample solution. This was followed by fixing the needle with a stand and clamps assembled at the same height. Then, the microsyringe plunger was depressed to expose a 3 μL drop of 1-octanol to the sample solution. Thereafter, extraction was performed by stirring the sample solution for 20 min. The drop was then retracted back into the microsyringe and the needle was removed from the sample vial (Figure 1). Finally, the extract was directly injected into the HPLC system for quantitative analysis.

### Chromatographic analysis

The mobile phase composed of water, methanol and acetonitrile (45:33:22, v/v/v) was prepared and degassed using ultra sonic bath for about 30 min. The degassed mobile phase was transferred to the HPLC system in isocratic mode and was delivered at a flow rate of 0.8 mL min$^{-1}$. Detection of the analytes was performed at 223 nm. Under these chromatographic conditions, good baseline separation was obtained for all the target compounds. Data acquisition and processing were

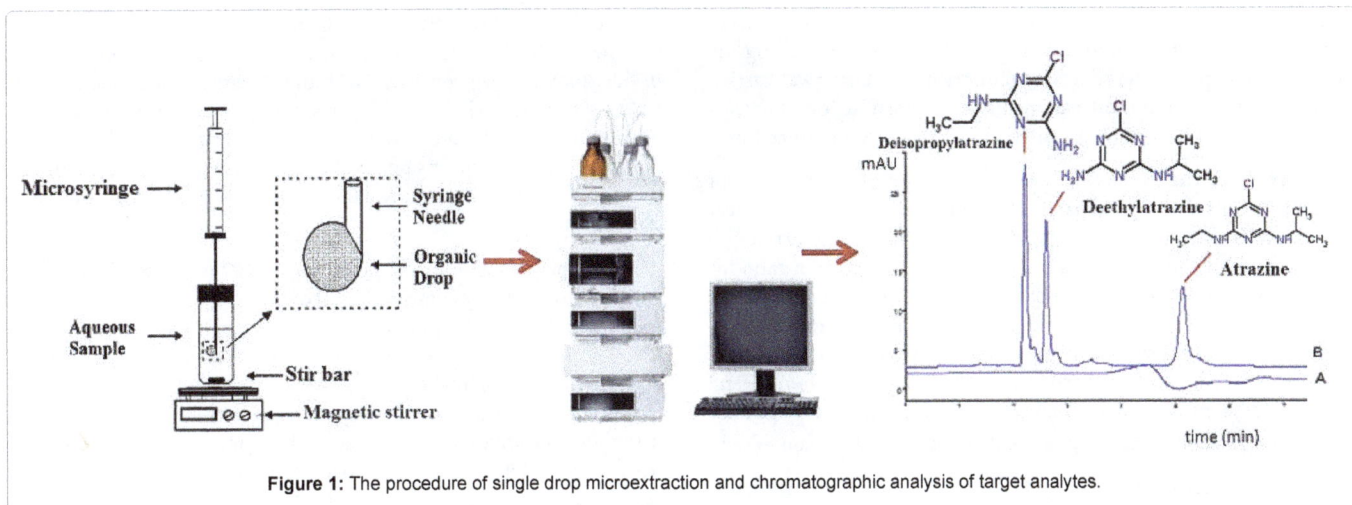

Figure 1: The procedure of single drop microextraction and chromatographic analysis of target analytes.

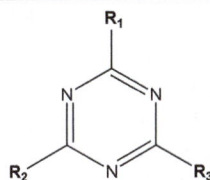

| Analyte | $R_1$ | $R_2$ | $R_3$ | Mol. wt, g mol$^{-1}$ | Solubility, mg L$^{-1}$ (22°C) | Log K$_{ow}$[a] | pK$_a$ |
|---------|-------|-------|-------|-----------------------|-------------------------------|-----------------|--------|
| DIA | -Cl | -NH$_2$ | -NH-CH$_2$-CH$_3$ | 173.60 | 3200 | 1.15 | 1.30-1.58 |
| DEA | -Cl | -NH-CH(CH$_3$)$_2$ | -NH$_2$ | 187.63 | 670 | 1.52 | 1.30-1.65 |
| ATZ | -Cl | -NH-CH(CH$_3$)$_2$ | -NH-CH$_2$-CH$_3$ | 215.7 | 33 | 2.5 | 1.68 |

Table 1: Selected physicochemical properties of the target compounds, from Ref. [25,26].

accomplished with the Chemstation software (Agilent Chemstation, Agilent Technologies®).

## Collection of the water samples

In order to investigate the applicability of the proposed SDME technique, the target analytes were spiked into the water samples obtained from different localities. Tap water samples were obtained from the campus of the College of Natural Sciences, Addis Ababa University; Ababa, Ethiopia, located at latitude of 9°1'48"N 38°44'24"E. Surface water samples were collected from three sites around Ziway Lake (located 160 km south of Addis Ababa; having geographical locations of 8°00'N 38°50'E. The three lake water samples were obtained from the sites locally named: *Cafeteria, Koroconch* and *Wafiko*; that are found on the western, southern and northern extremes of the lake, respectively. Another water sample was also collected from Awash River (located 130 km from the capital, Addis Ababa; with geographical location of 8°55'12"N 40°02'33.65"E), while ground waters were sampled from selected bore holes located around Ziway town. Except for the tap water samples, all the remaining sampling sites are located within the Great Rift Valley of Ethiopia; where extensive agriculture and horticulture activities are taking place, every season.

## Data analysis

In microextraction, the absolute amount of analyte extracted into the organic solvent is usually negligible compared to its total amount in the solution [28]. In other words, the numerical values of extraction efficiency are very small, and thus rather evaluated commonly, and also in this study, in terms of the concentration enrichment factor (EF). EF is defined as the ratio of the initial analyte concentration in the water sample and the enriched final analyte concentration in the organic phase.

Moreover, in order to study the applicability of the microextraction method for real sample analysis relative recovery was calculated. The relative recovery, which indicates the effect of sample matrix on extraction, is defined as the peak area ratio of the natural water sample to the ultrapure water sample spiked with analytes at the same concentration levels [29].

## Results and Discussion

### Optimization of chromatographic separation

In order to obtain the HPLC separation of the three compounds, different HPLC parameters including mobile phase and its flow rate as well as the wavelength of detection were optimized. The results of these experiments have demonstrated that the isocratic elution mode was found to be suitable in terms of analysis time, shape of peaks, and their reproducibility. Under the isocratic elution condition, the optimum separation of the target pesticides was obtained at the mobile phase composition of 45% water, 33% methanol, 22% acetonitrile. UV detection was employed in the range of 200-250 nm to attain the maximum UV absorption spectra of the analytes. It was found that at 223 nm all the target compounds exhibited maximum absorption. Similarly, the flow rate was also varied at 0.5, 0.8 and 1.0 mL min$^{-1}$, and it was noted that 0.8 mL min$^{-1}$ to be the optimum flow rate. Thus, the optimized chromatographic conditions were utilized in all the subsequent experiments.

### Optimization of the SDME parameters

Analytes enrichment based on SDME technique may be influenced by a number of parameters including ionic strength, type and volume of extractant, extraction time, temperature, syringe needle depth, sample agitation rate, sample volume and pH [23,29]. Screening the most important parameters affecting the enrichment factor was carried out and effects of significant parameters, on the extraction efficiency of the analytes, have been investigated in the optimization procedures.

**Selection of the extraction solvent:** The type of organic solvent used in SDME is an essential parameter for establishment of the direct immersion SDME procedure [28,30]. Accordingly, the extraction solvent should be immiscible with water, have high partition ratio for analytes, miscible with the mobile phase of the HPLC and it should further have separate retention time from the analyte peaks in the chromatogram. Moreover, the solvent should ensure drop stability during long periods of stirring [31,32]. For this purpose, Toluene, hexane, cyclohexane and 1-octanol, differing in physicochemical properties; i.e., polarity, surface tension, water solubility, etc were evaluated for use as the extraction solvent.

Toluene, n-hexane and cyclohexane were found to be unsuitable, in this study, because of the difficulty to be held as microdrops at the tip of the microsyringe for the required length of time due to their low viscosity and density. In addition, for ionizable compounds such as atrazine and its degradation products, Table 1, a fairly polar extraction solvent would be the solvent of choice. In a series of experiments that were performed, 1-octanol, which is relatively polar, exhibited relatively stable microdrop at the needle tip during vigorous stirring and demonstrated better extraction efficiency, was the solvent of choice in the current study. Moreover, 1-octanol has also showed no chromatographic interference, i.e., having good chromatographic performances than the other three solvents. Therefore, it was selected to be used as a preferred extraction solvent for all the subsequent works.

**Effect of microdrop volume:** It was described in the SDME theory that the amount of analyte extracted by the drop is related to the volume of the drop, and accordingly the use of a large drop results in an increment of the analytical sensitivity [33]. To this end, the extraction solvent of different microdrop volumes was exposed to the spiked aqueous sample in order to evaluate the effect on the SDME efficiency. The analytes were rapidly extracted to the surface of the drop where the rate of extraction to the surface of the extracting medium is controlled by diffusion to the surface, i.e., the boundary layer, which is governed by the relative solubility of the analyte [34].

The solvent microdrop volumes were varied from 1 μL up to 3 μL (1.0, 2.0, 2.5, and 3.0). It was observed that the amount of analyte extracted increases with the drop volume, however, when the microdrop volume exceeded 3.0 μL, such microdrops were noted by lesser stability and lower reproducibility. The microdrop volume of 3.0 μL was found to be stable for the intended extraction and allowed the highest stirring rate. Therefore, a microdrop volume of 3.0 μL was considered to be the optimal value.

**Effect of stirring rate:** For the drop-based extraction techniques, a fresh interface between the aqueous samples and the extraction solvent can be consecutively generated by stirring [34]. Agitation of the sample solution enhances extraction while reducing the time required for the thermodynamic equilibrium. The effect of stirring rate on the SDME concentration enrichment was evaluated at four levels, i.e., 300, 400, 500, 600 rpm. Apparently, the extraction efficiency increased with the increase in the stirring rate. On the other hand, when the stirring rate exceeded 600 rpm, the organic microdrop was easily lost from the needle tip and affects the extraction performances.

According to the film theory of convective diffusive mass transfer [32], at stable state, the diffusion rate in the aqueous phase increases

with increasing the stirring rate since faster agitation decreases the thickness of the diffusion film in the aqueous phase. The observations in this study are also consistent with the statements of the film theory. It was also observed that at higher stirring speed, the volume of the organic drop was found to reduce, as a result of the possible solvent dissolution in the aqueous phase. Based on the reported facts [35] and the observations of this study, 600 rpm was selected as the optimum stirring rate.

Effect of extraction time: Mass transfer is a time-dependent process and the closer the system is to the equilibrium conditions, may also cause the rate to reduce [36]. For optimum repeatability of the analysis, it is necessary to choose the time at which equilibrium is attained between the microdrop extraction solvent and the sample solution. However, SDME is not an exhaustive extraction technique; although the maximum sensitivity is attained at equilibrium, complete equilibrium need not to be attained provided that extraction conditions are reproduced [37]. In this work, the effect of extraction time was studied at four intervals; i.e., 10, 15, 20, and 25 min. As can be seen from Figure 2, long extraction time could lead to high extraction efficiency of the analytes in the range of 10-20 min, whereas with further prolongation of the extraction time, 20-25 min, no further increment in the peak area was observed, as this may rather result in drop dissolution and high incidences of the drop losses, and thus 20 min was selected as the optimal extraction time.

Effect of salt concentration: The presence of a salt is known to increase the ionic strength of the extraction solution and affect the solubility of organic analyte due to a salting out effect. Therefore, neutral salts are commonly added to the extraction processes primarily to reduce the solubility of the extracted compounds in water [38]. In this work, experiments were conducted to evaluate the effect of salt concentration on the extraction of the analytes by adding different quantities of NaCl varied from 2.5 to 15% (w/v). The results in Figure 3 demonstrated an initial increase, in extraction efficiency, with increased salt concentration, the maximum being reached at 7.5% (w/v), followed by a decrease in the extraction efficiency with further increase in the salt concentration (10-15%). The extra salt added may change the physical properties of the extraction film and thus reduces the diffusion rate of the analytes into the organic phase [39]. Moreover, target analytes

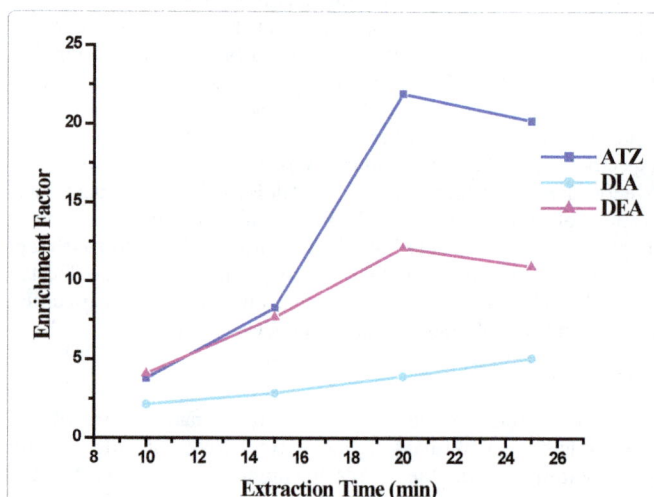

**Figure 2:** Effect of the extraction time on the enrichment factor of SDME. Extraction conditions: extraction solvent, 1-octanol; microdrop volume, 3 μL; stirring rate, 600 rpm; extraction temperature, 25°C; salt concentration, 0%.

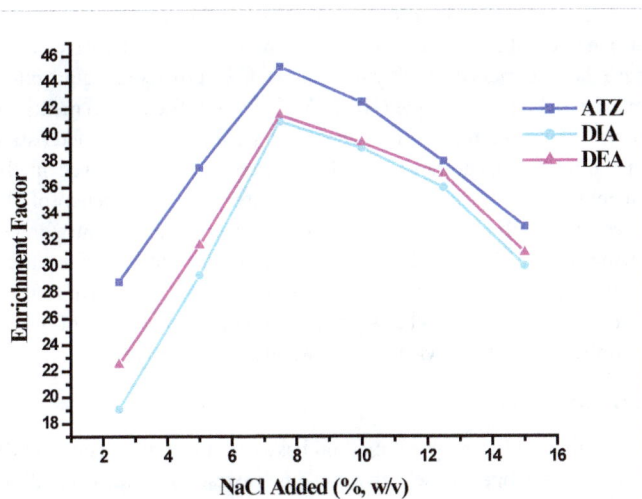

**Figure 3:** Effect of salt addition on the enrichment factor of SDME. Extraction conditions: extraction solvent, 1-octanol; microdrop volume, 3 μL; stirring rate, 600 rpm; extraction time, 20 min; extraction temperature, 25°C.

could also participate in electrostatic interactions with the salt ions in solution, thereby decreasing the tendency for their movement into the organic phase. As a result, 7.5% NaCl (w/v) was chosen as the optimal salt concentration for the developed microextraction technique.

**Effect of temperature:** Generally, increase in temperature is usually associated with increase in extraction efficiency [40], which normally enhances the diffusion coefficient, and therefore affects partition coefficient of the analytes between the aqueous sample and organic extractant phase in SDME. In this experiment, the enrichment factor of the SDME from the spiked reagent water was studied at five different temperatures; namely, 25°C, 30°C, 35°C, 40°C and 45°C. Greater temperatures can promote mass transfer, thereby increasing the extraction efficiency and lowering extraction times. On the other hand, temperatures of 40°C and higher cause the solvent drop to be unstable due to overpressurization and bubble formation in the bulk solution. Therefore, a temperature of 35°C was considered as the optimum temperature for the subsequent extractions.

**Effect of pH:** The pH of the sample solution is one of the most important factors affecting the liquid phase microextraction, as it can significantly influence the form of the analytes in the aqueous phase and consequently its solubility and extractability [41]. In particular, extractability of the weak organic bases such as atrazine and its metabolites, considered in this study, depends on pH of the sample solution, Table 1 [25,42]. In this study, enrichment factor of the considered analytes exhibited increasing tendency with pH ranges of 2-4 and then the increase was noticed to be gradual up to pH 7.0. However, after the neutral pH, the enrichment factor started declining, Figure 4. The decreasing tendencies of the extractability beyond pH 7.0 may be attributed to the degradation of the target analytes [26]. Therefore, pH 7.0, where all the target compounds are deionized for efficient transfer into the extraction solvent, was chosen as the optimum sample solution pH for the subsequent studies. Similar observations have also been noted by other workers for compounds possessing similar chemical natures [25,43,44].

## Validation of the proposed method

To evaluate the proposed direct immersion SDME method, important parameters confirming the performance characteristics such as precision, sensitivity and linearity were determined by

extracting 0.5 μg L-1 spiked reagent water under the optimized conditions. Summary of the optimum conditions selected for the method were as follows: 3.0 μL of 1-octanol as the extraction solvent, 7.5% (w/v) NaCl salt addition, 20 minutes extraction time, stirring rate at 600 rpm, temperature of 35°C and neutral pH. The method validation was carried out at these optimum extraction conditions based on the ICH guidelines [45].

**Precision study:** The precision of the analytical method expresses the amount of scatter in the results obtained from multiple analyses of the homogeneous sample and the results calculated as the relative standard deviation (RSD). The intra-day and inter-day precision of the method was determined under optimal conditions by successive five-time analysis of a 0.5 μg L-1 standard solution of the analytes and the RSDs were below 5% for all the studied target analytes. The figures of merit of the proposed method are given in Table 2.

**Sensitivity:** The sensitivity of the method is usually expressed in terms of the limit of detection (LOD) and limit of quantification (LOQ). The LOD is the lowest analyte concentration that can be detected but not necessarily quantified. The LOQ, on the other hand, is the lowest level or signal of the analyte in sample that can be accurately and precisely measured. The LOD and LOQ are calculated by consecutive analyses of chromatographic sample extracts with decreasing the amounts of the target compounds until a 3:1 and 10:1 signal to noise ratio was achieved, respectively. The LOD and LOQ determined in this study were found to be in the range of 0.01-0.05 μg L-1 and 0.1-0.25 μg L-1, respectively, as indicated in Table 2.

**Linearity:** The linearity of the analytical method is the ability to achieve test results that correspond directly to the concentration of the analyte in the samples within the range of the standard curve. For the

| Analytes | RSD (%, n=5)[a] | RSD (%, n=5)[b] | LR (μg L-1)[c] | r[2, d] | LOD[e] (μg L-1) | LOQ[f] (μg L-1) |
|---|---|---|---|---|---|---|
| ATZ | 4.6 | 4.7 | 0.25–150 | 0.9987 | 0.01 | 0.10 |
| DEA | 4.5 | 4.5 | 0.5–150 | 0.9983 | 0.05 | 0.25 |
| DIA | 4.7 | 4.9 | 0.5–150 | 0.9981 | 0.05 | 0.25 |

[a]Intra-day precision; [b]Inter-day precision; [c]Linear Range; [d]Coefficient of determination; [e]Limit of Detection; [f]Limit of Quantification

**Table 2:** Figures of merit for the proposed SDME method.

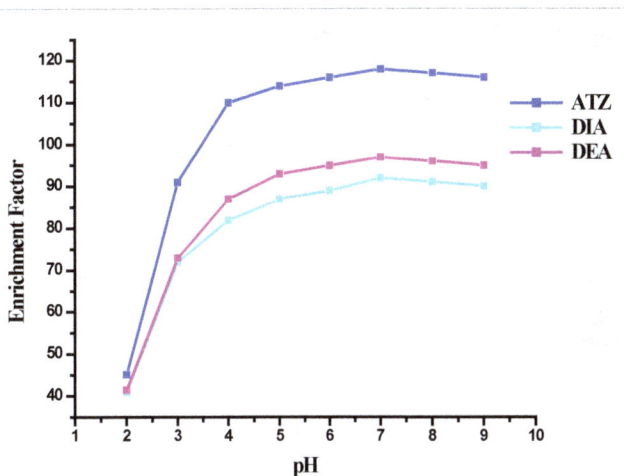

**Figure 4:** Effect of the sample pH on the enrichment factor of SDME. Extraction conditions: extraction solvent, 1-octanol; microdrop volume, 3 μL; stirring rate, 600 rpm; extraction time, 20 min; extraction temperature, 35°C; salt concentration, 7.5 % (w/v).

purpose of quantitative analysis, calibration of the chromatographic system was carried out for each analyte in the linear range from the limit of detection to the highest probable concentration ranging over five orders of magnitudes. For all analytes the coefficients of determinations ($r^2$) of the calibration curves were 0.998 or better, confirming good linearity in the concentration range of 0.25-150 µg L-1 for ATZ and 0.5-150 µg L-1 for DIA and DEA.

## Application to real samples

The developed SDME method was applied for the analysis of target analytes in four environmental water samples including tap water, river water, ground water and lake water. None of the analytes were detected in the real samples. The result indicated that target analytes were not present or they were present at concentration level below method detection limit in the real water samples. The water samples were then spiked with the standards of the target analytes at 0.5 µg L-1 concentration level. The relative recoveries for all these four water samples are listed in Table 3. The relative recovery is defined as the peak area ratio of the natural water sample to the ultrapure water sample spiked with analytes at the same concentration levels [46]. The satisfactory relative recoveries, provided in Table 3, demonstrate that the matrix effect was not significant confirming reliable application of the proposed technique for environment water analysis.

## Comparison with other reported methods

Representative liquid phase microextraction analytical techniques developed by other workers and applied for enrichment of the s-triazine herbicides in water samples were considered for comparison with the analytical technique proposed in this study as shown in Table 4. Two of the reported studies employed the DLLME techniques; in low-density extraction solvent [47] and using solidification of a floating drop [48] for enrichment of the parent s-triazine compounds. Both methods share the advantages of being simple, fast and virtually solvent free with the developed SDME technique. The results reported in these works are comparable with that of the current study, except for the LOD which was much lowered in this study. Moreover, only parent s-triazine analytes were considered in the earlier works [47,48], where simpler quantitative transfer of the analytes into the organic solvents used in DLLME, could be achieved [44].

Further comparison of the results of the current study was also made with another work [49], where the same target analytes have been enriched employing hollow fiber-protected liquid-phase microextraction technique (HF-LPME). Besides the requirements for short extraction time, other analytical performance characteristics, including the linear range and LOD, were greatly improved in the current study. In addition, the simplicity and convenience of the developed method makes it preferable. In general, the high enrichment factor obtained and the corresponding analytical performances confirm the method to be reliable and promising alternative in trace level analysis of the target analytes and other trace pollutants of similar chemical natures in environmental waters.

## Conclusions

In this study, analytical method based on direct immersion SDME has been developed for selective and quick extraction of trace quantities of the most commonly used pesticide; atrazine, and two of its major metabolites frequently, found in the environmental water samples. All the analytes considered were efficiently enriched, exhibiting good linear range which yielded LODs below those established by the European Directive. Both good quantitative performance and absence of matrix effect showed applicability of the method for trace enrichment of the target compounds in real sample analysis.

Therefore, it could be concluded that the developed direct immersion SDME sample preparation method is simple and effective to measure residual amounts of atrazine and its degradation products in environmental water samples. Moreover, it is one of the miniaturized modes of LPME that is virtually solvent free, requiring fraction of a drop of organic solvent, and thus recognized as an environmentally benign. As a result, it can be considered as a preferable method for the analysis of target analytes, and also of other pollutants, occurring in contaminated environmental water samples, possessing similar chemical behaviors.

### Acknowledgments

The authors would like to express their gratefulness to the Department of Chemistry of the Addis Ababa University for provision of the required research facilities. Financial support from the International Science program (ISP) through the "Trace Level Pollutants Analysis" project (ETH:04) is gratefully acknowledged. Fruitful discussions held with the senior research staff of the Analytical lab should also receive our deepest appreciations.

| Methods | Analyte | LR (µg L-1) | LOD (µg L-1) | RSD (%) | $r^2$ | Ref |
|---------|---------|-------------|--------------|---------|-------|-----|
| ST–DLLME-HPLC | Atrazine, Prometryn, Propazine, Terbutryn | 10-400 | 0.6-2.33 | 1.85-8.80 | 0.993-0.997 | [31] |
| SFD-LPME-HPLC | Cyanazine, Simazine, Atrazine | 0.2-200 | 0.01-0.05 | 2.05-8.15 | >0.998 | [32] |
| HF-LPME-HPLC | Deethylatrazine, Deisopropylatrazine, Atrazine | 2.5-200 | 0.5-0.1 | 3.5-4.7 | >0.998 | [33] |
| DI-SDME-HPLC | Deethylatrazine, Deisopropylatrazine, Atrazine | 0.25-150 | 0.01-0.05 | 4.5-4.9 | >0.998 | Current study |

**Table 3:** Comparison of the proposed direct immersed single drop microextraction with various modes of microextraction techniques developed for the determination of s-triazine herbicides and their degradation products in water samples.

| % Recovery | | | | | |
|---|---|---|---|---|---|
| Analytes | Tap Water | Ground Water | River Water | Lake Water | Reagent Water |
| ATZ | 93.3 | 85.8 | 84.3 | 81.1 | 96.3 |
| DEA | 91.8 | 84.4 | 83.3 | 73.3 | 94.8 |
| DIA | 92.6 | 85.1 | 71.5 | 65.6 | 95.1 |

**Table 4:** Summary of results of analysis of target analytes in spiked environmental water samples after SDME.

## References

1. Pinto MI, Sontag G, Bernardino RJ, Noronha JP (2010) Pesticides in water and the performance of the liquid-phase microextractin based techniques. Microchemical Journal 96: 225-237.

2. Fei Y, Zhaoyang B, Xiaoshui C, Sansan L, Yang L, et al. (2014) Analysis of 118 Pesticides in Tobacco after Extraction With the Modified QuEChRS Method by LC-MS-MS. Journal of Chromatographic Science 52: 788-792.

3. Tadeo JL (2008) Analysis of pesticides in Food and Environmental Samples. CRC Press, Boca Raton, NY, USA, pp: 2-34.

4. Megersa N, Solomon T, Jonsson JA (1999) Supported liquid membrane extraction for sample work up and pre-concentration of methoxy-s-triazine herbicides in a flow system. Journal of Chromatography A 830: 203-207.

5. EC (European commission) (2010) Maximum levels for certain pesticides in or on certain products. Official Journal of European Community 129: 3-9.

6. Gast A (1970) Use and performance of triazine herbicides on major crops and major weeds throughout the world. Residue Review 32: 11-18.

7. Michael ES, Susan VO (2000) Extraction of Atrazine and Its Metabolites Using Supercritical Fluids and Enhanced-Fluidity Liquids. Journal of Chromatographic Science 38: 399-408.

8. Jiang H, Adams C, Graziano N, Roberson A, McGuire M (2006) Occurrence and removal of chloro-s-triazines in water treatment plants. Environmental Science and Technology 40: 3609-3616.

9. EC (European commission) (1998) The quality of water intended for human consumption. Official Journal of European Community 330: 32-54.

10. EC (European commission) (2009) Technical specifications for chemical analysis and monitoring of water status. Official Journal of European Community 201: 36-45.

11. Mahara BM, Borossay J, Torkos K (1998) Liquid-liquid extraction for sample preparation prior to gas chromatography and gas chromatography-mass spectrometry determination of herbicide and pesticide compounds. Microchemical Journal 58: 31-38.

12. Majzik-Solymos E, Visi E, Karoly G, Beke-Berczi B, Györfi L (2001) Comparison of extraction methods to monitor pesticide residues in surface water. Journal of Chromatographic Science 39: 325-331.

13. Jiang H, Adams C, Koffskey W (2005) Determination of chloro-s-triazines including didealkylatrazine using solid phase extraction coupled with gas chromatography-mass spectrometry. Journal of Chromatography A 1064: 219-226.

14. do Amaral B, de Araujo JA, Peralta-Zamora PG, Nagata N (2014) Simultaneous determination of atrazine and metabolites (DIA and DEA) in natural water by multivariate electronic spectroscopy. Microchemical Journal 117: 262-267.

15. Ho TS, Pedersen BS, Rasmussen KE (2002) Recovery, enrichment and selectivity in liquid-phase microextraction, comparison with conventional liquid-liquid extraction. Journal of Chromatography A 963: 3-17.

16. François P, Jacques E (2006) Determination of frequently detected herbicides in water by solid-phase microextraction and gas chromatography coupled to ion-trap tandem mass spectrometry. Analytical and Bioanalytical Chemistry 386: 1449-1456.

17. Gang S, Hian KL (2002) Hollow fiber-protected liquid-phase microextraction of triazine herbicides. Analytical Chemistry 74: 648-654.

18. Ru-Song Z, Chun-Peng D, Xia W, Ting J, Jin-Peng Y (2008) Rapid determination of amide herbicides in environmental water samples with dispersive liquid-liquid microextraction prior to gas chromatography-mass spectrometry. Analytical and Bioanalytical Chemistry 391: 2915-2921.

19. dos Anjos JP, de Andrade JB (2015) Simultaneous determination of pesticide multiresidues in white wine and rosé wine by SDME/GC-MS. Microchemical Journal 120: 69-76.

20. Kokosa JM (2015) Recent trends in using single drop microextraction and related techniques in green analytical methods. Trends in Analytical Chemistry 71: 194-204.

21. Bagheri H, Khalilian F (2005) Immersed solvent microextraction and gas chromatography-mass spectrometric detection of s-triazine herbicides in aquatic media. Analytica Chimica Acta 537: 81-87.

22. Liu H, Dasgupta PK (1996) Analytical chemistry in a drop, solvent extraction in a microdrop. Analytical Chemistry 68: 1817-1821.

23. Jeannot MA, Przyjazny A, Kokosa JM (2010) Single drop microextraction development, application and future trends. Journal Chromatography A 1217: 2326-2336.

24. Xu L, Basheer C, Lee HK (2007) Developments in single-drop microextraction. Journal of Chromatography A 1152: 184-192.

25. Megersa N, Jonsson JA (1998) Trace enrichment and sample preparation of alkylthio-s-triazine herbicides in environmental waters using a supported liquid membrane technique in combination with high-performance liquid chromatography. Analyst 123: 225-231.

26. Esser HO, Dupius G, Vogel C, Marco GJ (1976) S-Triazines. In Herbicides: chemistry, degradation and mode of action. Marcel Dekker, New York, NY, USA, pp: 130-208.

27. Christian GD, Purdy WC (1962) The residual current in orthophosphate medium. Journal of Electroanalytical Chemistry 3: 363-367.

28. Wei G, Li Y, Wang X (2008) Comparison of efficiencies between single-drop microextraction and continuous-flow microextraction for the determination of methomyl in natural waters. International Journal of Environmental Analytical Chemistry 88: 397-408.

29. Pinheiro A, da Rocha GO, de Andrade JB (2011) A SDME/GC-MS methodology for determination of organophosphate and pyrethroid pesticides in water. Microchemical Journal 99: 303-308.

30. Jianfeng Y, Cuiying Z, Fayun C, Yingying C, Fazle S, et al. (2015) A Simple, Rapid and Eco-Friendly Approach for the Analysis of Aromatic Amines in Environmental Water Using Single-Drop Microextraction-Gas Chromatography. Journal of Chromatographic Science 53: 360-365.

31. Liang P, Guo L, Liu Y, Liu S, Zhang T (2005) Application of liquid-phase microextraction for the determination of phoxim in water samples by high performance liquid chromatography with diode array detector. Microchemical Journal 80: 19-23.

32. Ye C, Zhoua Q, Wang X (2007) Improved single-drop microextraction for high sensitive analysis. Journal of Chromatography A 1139: 7-13.

33. He Y, Lee HK (1997) Liquid-phase microextraction in a single drop of organic solvent by using a conventional microsyringe. Analytical Chemistry 69: 4634-4640.

34. Zhao L, Lee HK (2001) Application of static liquid-phase microextraction to the analysis of organochlorine pesticides in water. Journal of Chromatography A 919: 381-385.

35. dos Anjos JP, de Andrade JB (2014) Determination of nineteen pesticides residues (organophosphates, organochlorine, pyrethroids, carbamate, thiocarbamate and strobilurin) in coconut water by SDME/GC-MS. Microchemical Journal 112: 119-126.

36. Jeannot MA, Cantwell FF (1997) Mass transfer characteristics of solvent extraction into a single drop at the tip of a syringe needle. Analytical Chemistry 69: 235-239.

37. Jeannot MA, Cantwell FF (1996) Solvent microextraction into a single drop. Analytical Chemistry 68: 2236-2240.

38. Edmar M, Dilma B, Eduardo C (2007) Application of fractional factorial experimental and Box Behnken designs for optimization of single-drop microextraction of 2,4,6-trichloroanisole and 2,4,6-tribromoanisole from wine samples. Journal of Chromatography A 1148: 131-136.

39. Xiao Q, Hu B, Yu C, Xia L, Jiang Z (2005) Optimization of a single-drop microextraction procedure for the determination of organophosphorous pesticides in water and fruit juice with gas chromatography-flame photometric detection. Talanta 69: 848-855.

40. Subhrakanti S, Rajib M, Bidhan CR (2013) A rapid and selective method for simultaneous determination of six toxic phenolic compounds in mainstream cigarette smoke using single-drop microextraction followed by liquid chromatography-tandem mass spectrometry. Analytical and Bioanalytical Chemistry 405: 9265-9272.

41. Chimuka L, Megersa N, Norberg J, Mathiasson L, Jonsson JA (1998) Automated liquid membrane extraction and trace enrichment of triazine herbicides and their metabolites in environmental and biological samples. Analytical Chemistry 70: 3906-3911.

42. Jonsson JA, Lovkvist P, Audunsson G, Nilve G (1993) Mass transfer kinetics for analytical enrichment and sample preparation using supported liquid membranes in a flow system with stagnant acceptor liquid. Analytica Chimica Acta 277: 9-24.

43. Larsson N, Berhanu T, Megersa N, Jonsson JA (2011) An automatic field sampler utilising supported liquid membrane (SLM) for on-site extraction of s-triazine herbicides and degradation products: applied to an agricultural region of Ethiopia. International Journal of Environmental Analytical Chemistry 91: 929-944.

44. Megersa N, Chimuka L, Solomon T, Jonsson JA (2001) Automated liquid membrane extraction and trace enrichment of triazine herbicides and their metabolites in environmental and biological samples. Journal of Separation Science 24: 567-576.

45. ICH (2005) Harmonized Tripartite Guideline Validation of Analytical Procedures. Text and Methodology.

46. Leihong G, Jing Z, Haihong L, Jingbin Z, Yiru W, et al. (2013) Determination of bisphenol A in thermal printing papers treated by alkaline aqueous solution using the combination of single-drop microextraction and HPLC. Journal of Separation Science 361: 298-1303.

47. Tolcha T, Merdassa Y, Megersa N (2013) Low-density extraction solvent based solvent-terminated dispersive liquid-liquid microextraction for quantitative determination of ionizable pesticides in environmental waters. Journal of Separation Science 36: 1119-1127.

48. Yongwei G, Yanbo Z, Liang Z, Lei L (2009) Determination of triazine herbicides in aqueous samples using solidification of a floating drop for liquid-phase microextraction with liquid chromatography. Analytical Letters 42: 1620-1631.

49. Jinfeng P, Jianxia L, Xialin H, Jingfu L, Guibin J (2007) Determination of atrazine, desethyl atrazine and desisopropyl atrazine in environmental water samples using hollow fiber-protected liquid-phase microextraction and high performance liquid chromatography. Microchimica Acta 158: 181-186.

# 22

# Potentiometric and [1]H NMR Spectroscopic Studies of Functional Monomer Influence on Histamine-Imprinted Polymer-Modified Potentiometric Sensor Performance

**Atsuko Konishi\*, Shigehiko Takegami, Shoko Akatani, Rie Takemoto and Tatsuya Kitade**

*Department of Analytical Chemistry, Kyoto Pharmaceutical University, 5 Nakauchicho, Misasagi, Yamashina-ku, Kyoto 607-8414, Japan*

## Abstract

For the development of a histamine (HIS) potentiometric sensor based on molecularly imprinted polymers (MIPs), the effects of four functional monomers, namely acrylamide (AA), atropic acid (AT), methacrylic acid (MAA), and 4-vinylpyridine (4-VP), from which the MIP was synthesized, on the performance of the HIS sensor were examined by potentiometric and [1]H nuclear magnetic resonance (NMR) spectroscopic methods. The intermolecular interactions between HIS as a template molecule and a functional monomer were investigated based on the [1]H NMR spectra of HIS in distilled water in the presence of each functional monomer. Changes to the chemical shift of each HIS proton indicated that HIS typically formed a HIS-functional monomer complex at a ratio of 1:1 via hydrogen bonding with AA, AT and MAA, and interacted with 4-VP between the imidazole ring and pyridine ring of 4-VP. The potential changes of the four HIS sensors were measured in $0.1 \times 10^{-3}$ mol L$^{-1}$ aqueous solution using Ag/AgCl as a reference electrode; the order of the magnitudes of the changes was MAA>AA=4-VP>AT. The potential changes of three non-imprinted polymer-modified potentiometric sensors prepared without HIS were smaller than those of the corresponding HIS sensors, except in the case of AT. The potential response and selectivity of the HIS sensor using MAA were better than those of the other three HIS sensors. The [1]H NMR spectroscopic and potentiometric results showed that the hydrogen bond between HIS and MAA strongly and effectively influenced the potential response of the HIS sensor.

**Keywords:** Potentiometric sensor; Molecularly imprinted polymer; Intermolecular interaction; [1]H NMR spectroscopy; Histamine; Functional monomer

## Introduction

Chemical sensors consist of a recognition element and a transducer. The recognition element recognizes only the analyte, and the transducer converts the recognized information to electrical signals. In the development of chemical sensors, the recognition element is therefore the important factor in detecting an analyte specifically and selectively. Biosensors using biological materials as the recognition element have been developed because enzyme–substrate [1,2] and antigen–antibody [3] reactions are highly specific. Biosensors are widely used in clinical applications such as self-monitoring blood glucose meters [4]. However, there are problems in using biological materials; e.g., they are unstable and expensive.

A molecularly imprinted polymer (MIP) is an artificial tailor-made receptor capable of selectively recognizing and binding target molecules with high affinity. As shown in Figure 1, an MIP is a highly cross-linked polymer. The process usually involves initiating polymerization of a cross-linker with a functional monomer in the presence of a template molecule, which is subsequently extracted, leaving complementary cavities behind [5,6]. The MIP provides binding sites that are complementary in size, shape, and functionality to the template molecule; therefore, the template molecule preferentially rebinds to the cavity. MIPs have many advantages: they are tailor-made, stable, and inexpensive. Recently, MIPs have been studied for many applications such as high-performance liquid chromatography column packings [5,6], solid-phase extraction [7-9], drug-delivery systems [10-13], and sensors [14,15]. Chemical sensors based on MIPs can overcome some of the problems of biosensors.

Histamine (HIS), which is shown in Figure 2A, is important in inflammation, immediate allergic reactions [16], and gastric acid secretion, and acts as a neurotransmitter in the brain [17]. Also, HIS in fish causes allergy-like food poisoning in humans [18]. Because the determination of HIS in blood, urine, and food is therefore important, method developments in this field have been reported, e.g., amperometric determination [19-21] and fluorescent determination using ribonucleopeptide [22]. However, many of the methods employ biosensors.

For the development of MIP-modified potentiometric HIS sensors with good responsivity and selectivity, it is important to investigate the intermolecular interactions between HIS and functional monomers. In this study, we used four functional monomers, i.e., acrylamide (AA), atropic acid (AT) and methacrylic acid (MAA), and 4-vinylpyridine (4-VP), which are neutral, acidic, and basic functional monomers, respectively (Figure 2B-2E). The aims of this study were to elucidate the intermolecular interactions between HIS and each functional monomer and to clarify how each functional monomer affects the responsivity and selectivity of the HIS sensor. First, the intermolecular interactions between HIS and each functional monomer were investigated based on the chemical shift changes of each HIS proton in the presence of

**\*Corresponding author:** Atsuko Konishi, Department of Analytical Chemistry, Kyoto Pharmaceutical University, 5 Nakauchicho, Misasagi, Yamashina-ku, Kyoto 607-8414, Japan, E-mail: konishi@mb.kyoto-phu.ac.jp

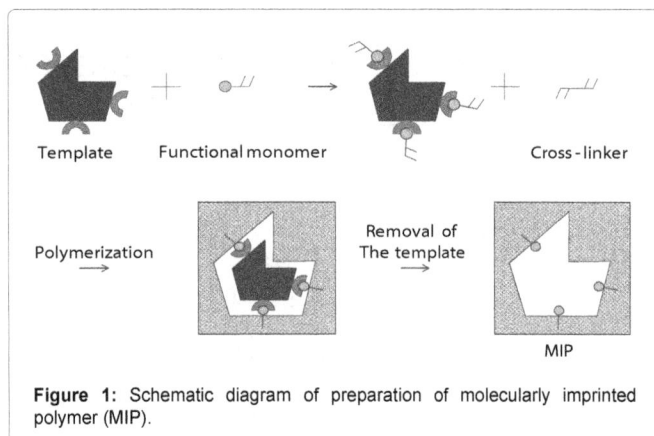

**Figure 1:** Schematic diagram of preparation of molecularly imprinted polymer (MIP).

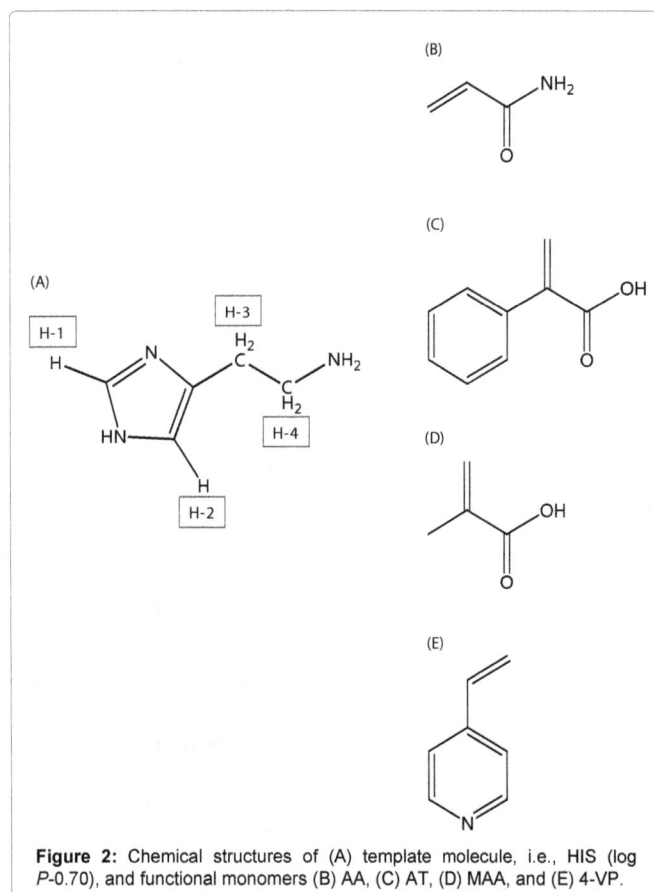

**Figure 2:** Chemical structures of (A) template molecule, i.e., HIS (log P-0.70), and functional monomers (B) AA, (C) AT, (D) MAA, and (E) 4-VP.

Kanto Chemical Co., Inc. (Tokyo, Japan). 2-Aminobenzimidazole and deuterium oxide (99.9%) were purchased from Sigma-Aldrich (St. Louis, MO, USA). HIS dihydrochloride and AT were purchased from the Tokyo Chemical Industry Co., Ltd. (Tokyo, Japan). Sodium dodecyl sulfate (SDS), 2,2'-azobis(2,4-dimethylvaleronitrile) (V-65), dibutyl phthalate, AA, MAA, 4-VP, ethylene dimethacrylate (EDMA), poly(vinyl alcohol) (PVA) 1000 (partially hydrolyzed PVA), methanol, pyrrole, histidine, and L(+)-lysine hydrochloride were purchased from Wako Pure Chemical Industries, Ltd. (Kyoto, Japan). The hydroquinone added to EDMA as a stabilizer was extracted with NaOH before use. Toluene was purchased from Nacalai Tesque, Inc. (Kyoto, Japan).

### $^1$H NMR spectra

Aliquots of an HIS stock solution were placed in 2-mL volumetric flasks. A desired amount of a functional monomer stock solution, i.e., AA, AT, MAA, or 4-VP, was added, and distilled water was added to volume. The final concentrations of HIS and each functional monomer were $1.0 \times 10^{-3}$ mol L$^{-1}$, and 0.0, $1.0 \times 10^{-3}$, $2.0 \times 10^{-3}$, $3.0 \times 10^{-3}$ mol L$^{-1}$ for AA, MAA and 4-VP, and 0.0, $1.0 \times 10^{-3}$, $2.0 \times 10^{-3}$ mol L$^{-1}$ for AT (because of low water solubility), respectively. The flasks were shaken for a short time and then samples were transferred to NMR tubes of diameter 5 mm. A coaxial internal tube containing ca. $2.0 \times 10^{-2}$ mol L$^{-1}$ 3-(trimethylsilyl)propionic-2,2,3,3-$d_4$ acid sodium salt (98 atom% D) in D$_2$O was inserted into the sample tube to provide a reference signal. $^1$H NMR spectra were recorded with a Unity Inova 400NB spectrometer (Agilent Technologies, Inc., CA, USA) operated at 399.97 MHz, using a presaturation method to reduce the signal of protons derived from H$_2$O. The probe temperature was 25 ± 1°C. An accumulation of 200 free induction decays was used to improve the signal-to-noise ratio.

### Sensor synthesis

The HIS sensor was prepared using a modified version of a previously reported procedure [23]. A graphite rod (diameter 3 mm, length 50 mm) was used as a transducer. The graphite rod was polished with sandpaper and then sonicated five times for 5 min in distilled water. A plasma polymerized thin film of ethylbenzene was deposited on the surface of the polished graphite rod using a plasma deposition system (BP-1, Samco Inc., Kyoto, Japan). The first swelling step was performed by immersing the plasma-coated graphite in a suspension of SDS (0.55 mmol) as a surfactant, V-65 (1.37 mmol) as a radical initiator, dibutyl phthalate (14.3 mmol) as a plasticizer, and distilled water (40 mL). The mixture was stirred at room temperature for 24 h. In the second swelling step, the graphite was immersed in PVA solution (10.67 g/500 mL) containing HIS (2 mmol) as a template molecule, a functional monomer (20 mmol), EDMA (25 mmol) as a cross-linker, and toluene (47 mmol) as a porogen. The mixture was stirred at room temperature for 24 h, and then degassed using helium and polymerized by heating at 70-75°C for 12 h. The graphite was then immersed in methanol to remove the HIS. This procedure was repeated until the ultraviolet absorption of HIS in methanol at 210 nm was not observed. The HIS sensor was stored in distilled water. Non-imprinted (NIP) sensors were prepared using the same procedure, but without HIS.

### Sensor measurements

The HIS sensor and an Ag/AgCl reference electrode were immersed in distilled water (100 mL) and the potential response of the HIS sensor against the reference electrode was measured by a potentiometer (pH meter F-52, Horiba Inc., Kyoto, Japan). When the potential response was stable, record was started. After 1 min from the start, $1.0 \times 10^{-2}$ mol L$^{-1}$ stock solution (1 mL) of each chemical substance was injected into

each functional monomer using $^1$H nuclear magnetic resonance (NMR) spectroscopy. Next, the responsivities and selectivities of four HIS sensors containing the functional monomers were assessed by measuring the potential changes in the presence of HIS and other chemical substances. The effects of the intermolecular interactions between HIS and the functional monomer on the performance of the HIS sensor were examined.

## Experimental

### Materials

Graphite rods were purchased from Strem Chemicals Inc. (Newburyport, MA, USA). Ethylbenzene was purchased from the

Potentiometric and 1H NMR Spectroscopic Studies of Functional Monomer Influence on Histamine-Imprinted...

153

the distilled water. The potential response value at the start of record was set to 0 mV and potential change was defined as the difference between the potential response values before and after the addition of each chemical substance. The sensor selectivity was tested by comparing the responses of the template molecule HIS and other chemicals with similar chemical structures or $n$-octanol/water partition coefficients (log $P$), namely histidine, 2-aminobenzimidazole, pyrrole, and lysine.

## Results and Discussion

### ¹H NMR study of intermolecular interactions between HIS and functional monomers

¹H NMR spectra of HIS mixed with each functional monomer, i.e., AA, AT, MAA, and 4-VP, in H₂O were recorded to determine which functional monomer had the highest affinity with HIS. ¹H NMR spectra of HIS-functional monomer systems having a ratio of 1:1 are shown in Figure 3 as an example. The HIS (see structure in Figure 2A) proton signals were observed at 8.61 (H-1), 7.37 (H-2), 3.15 (H-3), and 3.35 (H-4) ppm (Figure 3A). Figure 3B-3D shows that the addition of AA, AT, and MAA shifts the HIS proton signals to slightly lower fields. In contrast, in the presence of 4-VP (Figure 3E), the H-1, H-2 and H-3 signals shifted to higher fields, with the H-4 signal showing almost no change.

The chemical shifts of the HIS protons were measured in the absence and presence of each functional monomer; the results for HIS H1–H4 are shown in Figure 4A–4D. As shown in Figure 4A and 4B, the HIS H-1 and H-2 signals were shifted to lower fields in the presence of AT and MAA at concentrations of $1.0 \times 10^{-3}$ mol L⁻¹. In the presence of AA, small downfield shifts of the HIS H-1 and H-2 signals were also observed, however, the shifts were smaller than those in the presence of AT and MAA. Further increasing the concentration of the monomer above $1.0 \times 10^{-3}$ mol L⁻¹ resulted in no significant change to the chemical shift values. Additionally, Figure 4C-4D shows that regardless of the monomer concentration for AA, AT, and MAA, no significant change to the chemical shift values of the HIS H-3 and H-4 signals were observed. These results indicate that HIS can form a complex with the aforementioned functional monomers at a ratio of 1:1 and the imidazole ring of HIS interacts with the functional monomer. The acid dissociation constants (p$K_a$s) of HIS are 6.15 and 9.84 for the nitrogen atom of the secondary amine on the imidazole ring and the primary amino group on the side chain, respectively [24,25]. The p$K_a$ values of AT [26], MAA [27], and 4-VP [26] are 4.0, 4.66, and 6.0, respectively. The p$K_a$ of AA is above 15 and the proton of the amide does not dissociate under normal conditions. Since the ionization states of HIS and functional monomers depend on the pH value of the sample solution, the pH values of the HIS solution and the mixtures of HIS and each functional monomer used in the ¹H NMR experiments were therefore measured. These results were $4.7 \pm 0.0$ for HIS ($n=3$), $4.6 \pm 0.1$ for HIS/AA, $3.3 \pm 0.1$ for HIS/AT, $3.7 \pm 0.2$ for HIS/MAA and $6.0 \pm 0.2$ for HIS/4-VP ($n=3$ at each functional monomer concentration). Under the experimental conditions in the presence of AA, AT or MAA, the di-cationic form of HIS is protonated at the primary amino group on the side chain and the nitrogen atom of the secondary amine on the imidazole ring. Conversely, AA, AT and MAA are neutral. Previous ¹H NMR investigations have reported that the proton chemical shift is shifted to lower fields by the formation of a hydrogen bond between the template and the functional monomer, and the larger the chemical shift, the stronger the interaction between them [28,29]. Additionally, it has previously been reported that the di-cationic form of HIS forms a complex with MAA more easily than the mono-cationic form in

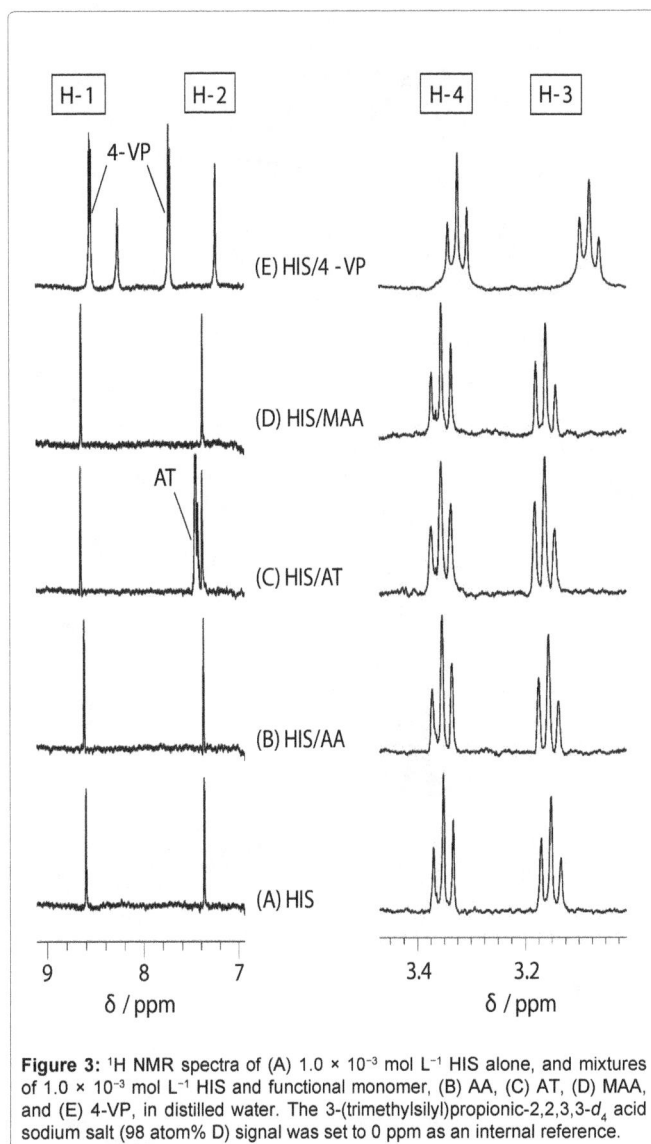

**Figure 3:** ¹H NMR spectra of (A) $1.0 \times 10^{-3}$ mol L⁻¹ HIS alone, and mixtures of $1.0 \times 10^{-3}$ mol L⁻¹ HIS and functional monomer, (B) AA, (C) AT, (D) MAA, and (E) 4-VP, in distilled water. The 3-(trimethylsilyl)propionic-2,2,3,3-$d_4$ acid sodium salt (98 atom% D) signal was set to 0 ppm as an internal reference.

aqueous media [30,31]. Therefore, since the change in chemical shift for the HIS H-1 signal was larger than the corresponding H-2 shift in the presence of AA, AT and MAA, HIS may complex with the functional monomers at a ratio of 1:1 as shown in Figure 4E - an example illustration of the HIS-MAA complex. Furthermore, the strength of the hydrogen bond to form the HIS-functional monomer complex was in the order of HIS-AT>HIS-MAA>HIS-AA.

In contrast, in the presence of 4-VP, all HIS H1–H4 signals shifted to higher fields depending on the 4-VP concentration. Under the experimental conditions in the presence of 4-VP, i.e., pH 6.0, HIS is present in equal quantities of mono-cationic and di-cationic forms with the former protonated only at the primary amino group on the side chain. Meanwhile, 4-VP is 50% neutral and 50% as a mono-cationic species, which is protonated at the nitrogen atom of the tertiary amine on the pyridine ring. The large upfield shifts of the HIS H-1 and H-2 signals in the presence of 4-VP is thought to be the result of the proton exchange between the tertiary and secondary amine on the imidazole ring of HIS and the tertiary amine on the pyridine ring of 4-VP. It has been previously reported that the electrostatic interactions between the protonated amino group and the π-electron-

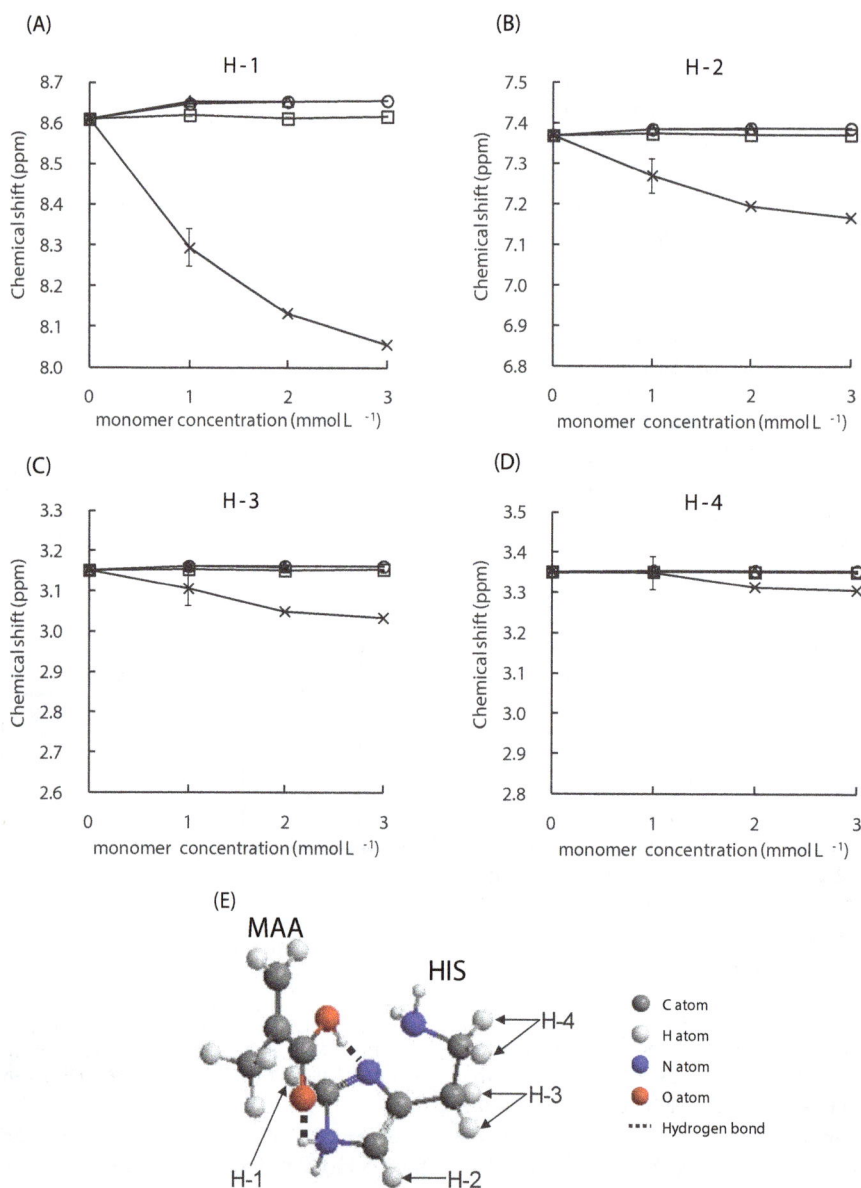

**Figure 4:** Chemical shift of each HIS proton, (A) H-1, (B) H-2, (C) H-3, and (D) H-4, in the absence and presence of the functional monomers as a function of concentration. (□) AA, (Δ) AT, (○) MAA, and (×) 4-VP. (E) Three-dimensional structure of the hydrogen bond between HIS and MAA.

rich pyridine ring can result in chemical shift changes to higher fields [32-34]. If the protonated amino group on the HIS side chain interacts with the pyridine ring of 4-VP, the HIS H-4 shift to higher fields will be greater than the H-3 shift. However, since changes to the chemical shift was in the order of H-1>H-2>H-3>H-4, the results indicate that HIS can interact with 4-VP between the HIS imidazole ring and the pyridine ring of 4-VP, but not between the protonated amino group on the HIS side chain and the pyridine ring of 4-VP. Unfortunately, the results of the $^1$H NMR investigations could not yield details on how HIS complexes with 4-VP. The results of this $^1$H NMR study partly clarify the intermolecular interactions between HIS and the functional monomers in bulk water and show that the order of magnitude of the interactions is: 4-VP>AT>MAA>AA.

## Responsivities of four HIS sensors

The potential response values of four HIS sensors and NIP sensors prepared using each functional monomer on addition of $1.0 \times 10^{-2}$ mol L$^{-1}$ HIS stock solution was recorded; the curves are shown in Figure 5. The potential response value of each HIS sensor increased immediately after addition of HIS stock solution, reached a plateau after 15 s, and then showed a stable potential until 25 min. Potential changes of each HIS sensor at 1500 s after addition of HIS stock solution were shown in Figure 6. As shown in Figure 6, the order of the potential change magnitudes tended to be MAA>AA=4-VP>AT; i.e., the HIS sensor containing MAA showed the largest potential change among the four HIS sensors. These results and those of the $^1$H NMR spectroscopic

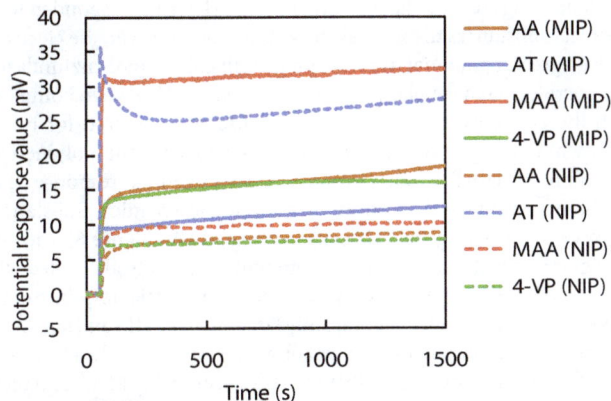

**Figure 5:** Potential response curves of HIS sensors and non-imprinted (NIP) sensors containing different functional monomers after addition of HIS aqueous solution. Final HIS concentration was 0.1 × 10⁻³ mol L⁻¹.

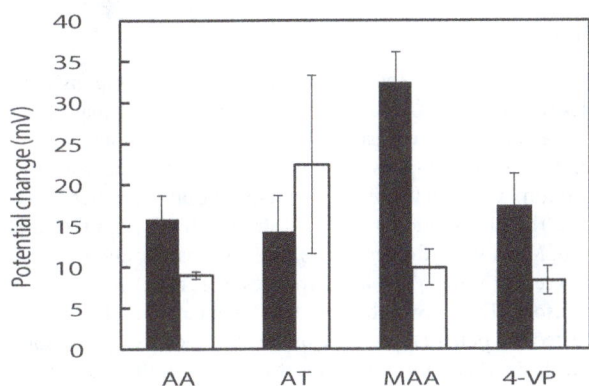

**Figure 6:** Potential changes of (■) HIS sensors and (□) corresponding NIP sensors at 1500 s after addition of HIS aqueous solution. The determination of HIS sensor and NIP sensor using each functional monomer was in triplicate, respectively.

study show that the relationship between the potential responses of the HIS sensors and the interactions between HIS and each functional monomer can be described as follows. The hydrogen bond between HIS and MAA plays an important role in the potential response of the HIS sensor. However, the potential change of the HIS sensor using AT as the functional monomer was smaller than the others, although the ¹H NMR spectroscopic results show that HIS also formed a complex with AT via hydrogen bond. This is considered to be the result of steric hindrance by the AT benzene ring, which is oriented towards the template cavity of the MIP. As the AT benzene ring occupies the cavity void, HIS ingress into the template cavity is hindered because of the aforementioned steric hindrance. As a result, HIS, especially the tertiary and protonated secondary amine on the imidazole ring, cannot bind to the carboxyl group of AT in the template cavity. The ¹H NMR spectra also show that HIS has the strongest and weakest interactions with 4-VP and AA, respectively; the potential changes of the HIS sensors using these monomers were almost the same. Although there are strong interactions between HIS and 4-VP, these interactions did not greatly affect the potential changes of the HIS sensors. The HIS di-cationic form interacts with AA via weak hydrogen bond, but the HIS sensor using AA responded to these hydrogen bonds and showed the same potential response as that of the HIS sensor using 4-VP. Our HIS

sensor therefore mainly responds to surface potential changes caused by hydrogen-bond formation between the template molecule and the functional monomer.

Figure 6 also shows the potential changes for NIP sensors using each functional monomer at 1500 s after addition of HIS stock solution. The potential changes of the NIP sensors using AA, MAA, and 4-VP were significantly smaller than those of the corresponding HIS sensors; however, the potential change of the NIP sensor using AT was larger than that of the corresponding HIS sensor. Usually, the potential change of an NIP sensor is smaller than that of the corresponding HIS sensor because the NIP sensor does not have a HIS template, which provides highly specific HIS-binding sites. The NIP sensors should show no potential response to HIS, but small potential changes were observed for AA, MAA, and 4-VP. This is considered to be the result of non-specific binding and adsorption of HIS on the NIP sensor surface. Both MIP and NIP are hydrophobic polymers prepared using EDMA as a cross-linker. Additionally, the larger numbers of functional groups of each functional monomer may be exposed from the NIP surface when compared with the MIP surface, because the majority of the functional monomers used in the MIP preparations are involved in the construction of the template cavity. Although the di-cationic form of HIS non-specifically binds to the functional group, or adsorbs onto the hydrophobic MIP and NIP surface, the contribution of non-specific binding and hydrophobic adsorption will be larger on NIP than MIP. As a result, the NIP sensor showed a small potential change in response to the non-specific binding and adsorption of HIS on its surface. The surface of the NIP sensor using AT was more hydrophobic than those of the other NIP sensors because AT has a benzene ring in its molecular structure. The increase in non-specific adsorption of HIS on the hydrophobic surface, in addition to the hydrogen bonds between di-cationic HIS and AT, leads to a larger potential change for the NIP sensor using AT than for the other NIP sensors.

The specificities of HIS sensors using each functional monomer were easily compared by calculating the ratio of the potential change of each HIS sensor to that of the corresponding NIP sensor and their ratios were 1.8, 0.6, 3.3, and 2.1 for AA, AT, MAA, and 4-VP, respectively. Therefore, the specificity of the HIS sensor using MAA was also higher than those of the other HIS sensors.

The potentiometric results show that the HIS sensor using MAA gave the best response of the four HIS sensors because of the specific recognition of HIS based on hydrogen bonds between HIS and the carboxylic group of MAA in the MIP template cavity.

### Selectivities of HIS sensors

The selectivities of the HIS sensors using different functional monomers were investigated by comparing their responses to four other chemicals, i.e., histidine, 2-aminobenzimidazole, pyrrole, and lysine (the structures are shown in Figure 7), with their responses to HIS. Histidine is very similar in chemical structure to HIS and has a carboxyl group. The log $P$ values are -0.70, -3.56, 0.91, 0.75 and -3.05 for HIS, histidine, 2-aminobenzimidazole, pyrrole, and lysine respectively [35-38]. Figure 8 shows the potential change curves of the HIS sensor using MAA after addition of each chemical as an example. The potential response curve of HIS differed significantly from those of the other chemicals. For HIS, the HIS sensor showed a potential response value with a large gradient immediately after the addition of HIS stock solution. For the other four chemicals, there was no response or the potential response values gradually decreased with time. These results indicate that this sensor specifically recognized HIS but responded

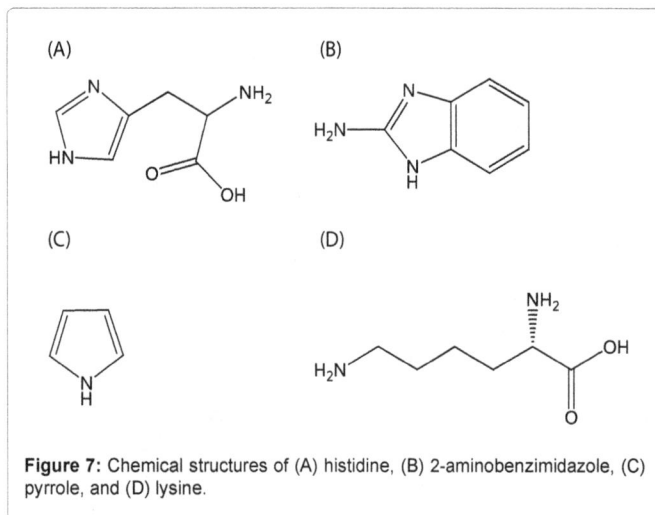

Figure 7: Chemical structures of (A) histidine, (B) 2-aminobenzimidazole, (C) pyrrole, and (D) lysine.

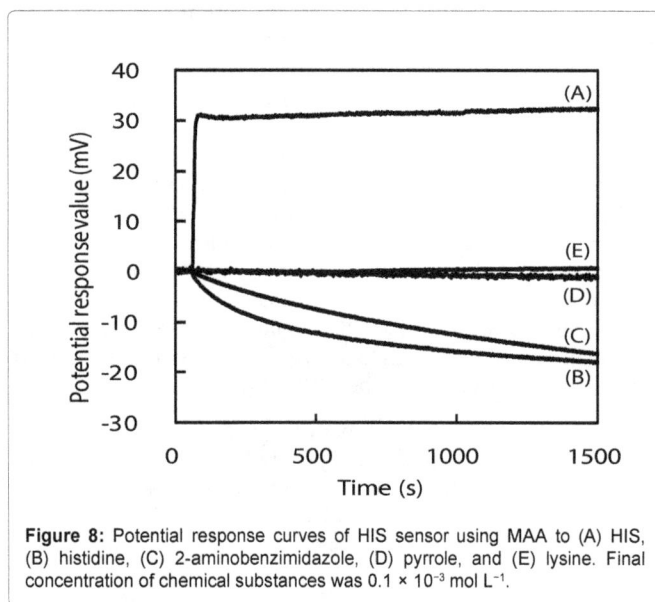

Figure 8: Potential response curves of HIS sensor using MAA to (A) HIS, (B) histidine, (C) 2-aminobenzimidazole, (D) pyrrole, and (E) lysine. Final concentration of chemical substances was $0.1 \times 10^{-3}$ mol L$^{-1}$.

non-specifically to other chemicals adsorbed on the MIP surface of the HIS sensor. The same results were obtained for the other HIS sensors using AA, AT, and 4-VP.

Figure 9 shows the potential changes of the four HIS sensors in the presence of each chemical substance. Three HIS sensors, i.e., those using AA, AT, and 4-VP, showed large potential changes in response to 2-aminobenzimidazole. This large potential change was derived from adsorption of 2-aminobenzimidazole on the hydrophobic surface of MIP because of the large log $P$ value of 2-aminobenzimidazole. In contrast, the HIS sensor using MAA showed a larger potential change to HIS than to the other four chemical substances, because MIP template cavity formed complementary to HIS and hydrogen bonds between HIS and MAA acted effectively to catch HIS in the sample solutions. As a result, the HIS sensor using MAA gave better selectivities than the three HIS sensors using AA, AT, and 4-VP.

To prove that the responsivity and selectivity of the HIS sensor using MAA were as a result of MIP specifically recognizing HIS, the potential change of the bare graphite rod electrode and the graphite rod electrode coated with the polymerized thin film of ethylbenzene

were recorded as a function of sequential addition of each chemical substance (Figure 10). The bare graphite rod electrode responded to all chemical substances the same as the HIS sensor, however, the electrode shows a greater potential response for histidine, 2-aminobenzimidazole and pyrrole, and a smaller potential response for HIS when compared with the HIS sensor. No potential response was observed for lysine. Meanwhile, the graphite rod electrode coated with the polymerized thin film of ethylbenzene shows smaller potential response than the bare graphite rod electrode across all the chemical substances. However, comparing the HIS sensor using MAA (Figure 8) with the bare graphite rod electrode and the graphite rod electrode coated with the polymerized thin film of ethylbenzene, the potential response of the HIS sensor using MAA is significantly larger toward HIS and smaller to other chemical substances than that observed for both of the electrodes. Therefore, these results show that the HIS sensor using MAA enhanced the responsivity and selectivity to HIS as MIP specifically recognized HIS compared with other chemical substances.

Thus, these potentiometric results clearly show that the HIS sensor using MAA had good responsivity to, and selectivity for, HIS as a template molecule.

## Conclusion

An MIP-modified sensor for HIS with good responsivity and selectivity using MAA as a functional monomer was manufactured. Unlike sensors prepared using other functional monomers, i.e., AA, AT, and 4-VP, the HIS sensor using MAA has three-dimensional HIS-recognition cavities in the MIP and can specifically recognize HIS in the cavities. The di-cationic form of HIS strongly binds with the carboxyl group of MAA via hydrogen bonds. The responsivity and selectivity of the HIS sensor using MAA are therefore better than those of the other HIS sensors. This shows that the type of intermolecular interaction influenced the potential response of the MIP-modified potentiometric sensor. It is therefore important to elucidate the intermolecular interactions between the template molecule and functional monomer for fabrication of MIP-modified potentiometric sensors. Based on the evidence obtained in this study, suitable ratios of template molecule, functional monomer, and cross-linker for MIP preparation, and quantification of the HIS sensor are currently investigated. HIS sensor

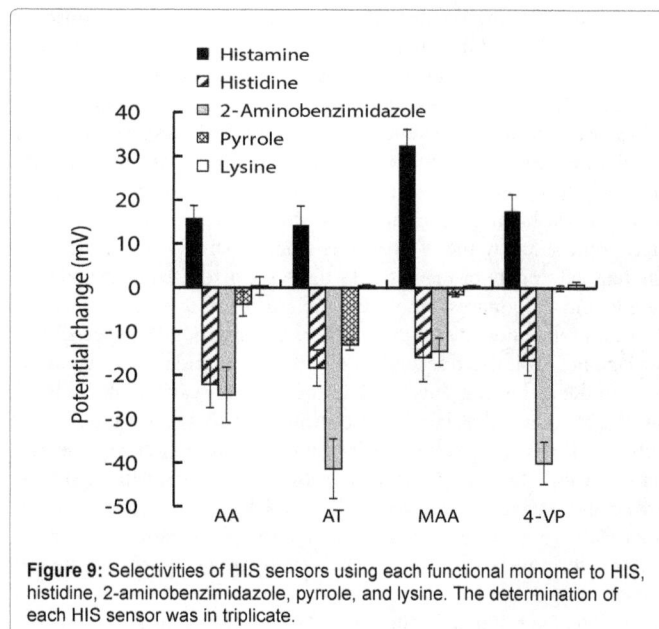

Figure 9: Selectivities of HIS sensors using each functional monomer to HIS, histidine, 2-aminobenzimidazole, pyrrole, and lysine. The determination of each HIS sensor was in triplicate.

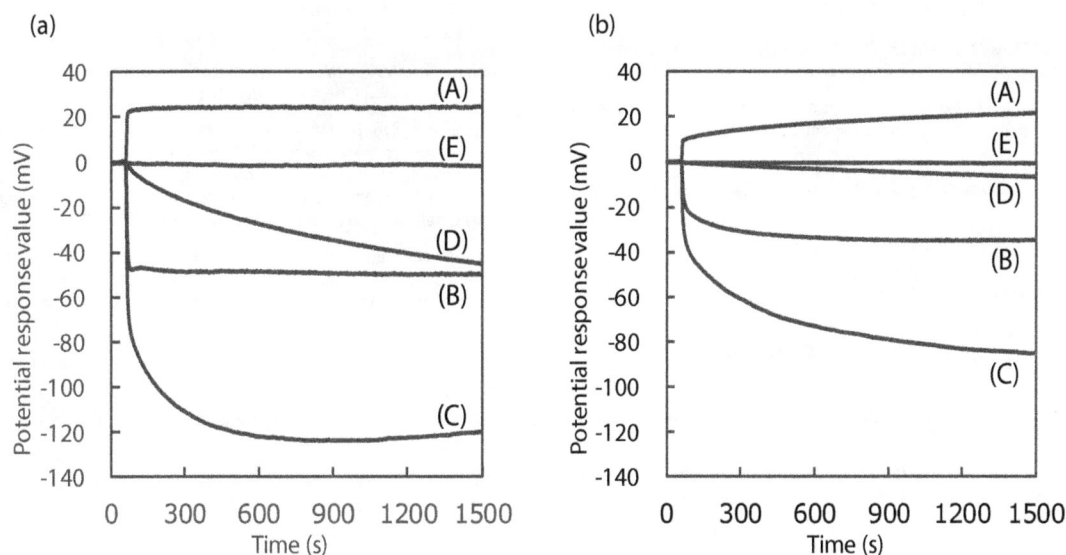

**Figure 10:** Potential response curves of (a) bare graphite rod electrode and (b) graphite rod electrode coated with the polymerized thin film of ethylbenzene to (A) HIS, (B) histidine, (C) 2-aminobenzimidazole, (D) pyrrole, and (E) lysine. Final concentration of chemical substances was $0.1 \times 10^{-3}$ mol L$^{-1}$.

showed a good response to HIS concentrations of $0.1 \times 10^{-3}$ mol L$^{-1}$ and has sufficient capability to sense HIS when compared with other HIS sensors previously reported.

### Acknowledgements

This work was partly supported by JSPS KAKENHI (Grant Number 21590048) for Scientific Research (C).

### References

1. Iyer R, Pavlov V, Katakis I, Bachas LG (2003) Amperometric sensing at high temperature with a "wired" thermostable glucose-6-phosphate dehydrogenase from aquifex aeolicus. Anal Chem 75: 3898-3901.

2. Tani Y, Tanaka K, Yabutani T, Mishima Y, Sakuraba H, et al. (2008) Development of a ᴅ-amino acids electrochemical sensor based on immobilization of thermostable ᴅ-Proline dehydrogenase within agar gel membrane. Anal Chim Acta 619: 215-220.

3. Yu X, Munge B, Patel V, Jensen G, Bhirde A, et al. (2006) Carbon nanotube amplification strategies for highly sensitive immunodetection of cancer biomarkers. J Am Chem Soc 128: 11199-11205.

4. Wang X, Ioacara S, DeHennis A (2015) Long-term home study on nocturnal hypoglycemic alarms using a new fully implantable continuous glucose monitoring system in type 1 diabetes. Diabetes Technol Ther 17: 780-786.

5. Haginaka J, Sanbe H (2001) Uniformly sized molecularly imprinted polymer for (S)-naproxen retention and molecular recognition properties in aqueous mobile phase. J Chromatogr A 913: 141-146.

6. Zhang Y, Lei J (2013) Synthesis and evaluation of molecularly imprinted polymeric microspheres for chloramphenicol by aqueous suspension polymerization as a high performance liquid chromatography stationary phase. Bull Korean Chem Soc 34: 1839-1844.

7. Li H, Li D (2015) Preparation of a pipette tip-based molecularly imprinted solid-phase microextraction monolith by epitope approach and its application for determination of enkephalins in human cerebrospinal fluid. J Pharm Biomed Anal 115: 330-338.

8. Hashemi-Moghaddam H, Ahmadifard M (2016) Novel molecularly-imprinted solid-phase microextraction fibercoupled with gas chromatography for analysis of furan. Talanta 150: 148-154.

9. Svoboda P, Combes A, Petit J, Nováková L, Pichon V, et al. (2015) Synthesis of a molecularly imprinted sorbent for selective solid-phase extraction of β-N-methylamino-L-alanine. Talanta 144: 1021-1029.

10. Sreenivasan K (1999) On the application of molecularly imprinted poly(HEMA) as a template responsive release system. J Appl Polym Sci 71: 1819-1821.

11. Hiratani H, Mizutani Y, Alvarez-Lorenzo C (2005) Controlling drug release from imprinted hydrogels by modifying the characteristics of the imprinted cavities. Macromol Biosci 5: 728-733.

12. Puoci F, Iemma F, Cirillo G, Picci N, Matricardi P, et al. (2007) Molecularly imprinted polymers for 5-fluorouracil release in biological fluids. Molecules 12: 805-814.

13. Men J, Gao B, Yao L, Zhang Y (2014) Preparation and characterization of metronidazole-surface imprinted microspheres MIP-PSSS/CPVA for colon-specific drug delivery system. J Macromol Sci, Part A: Pure Appl Chem 51: 914-923.

14. Sari E, Üzek R, Duman M, Denizli A (2016) Fabrication of surface plasmon resonance nanosensor for the selective determination of erythromycin via molecular imprinted nanoparticles. Talanta 150: 607-614.

15. Florea A, Guo Z, Cristea C, Bessueille F, Vocanson F, et al. (2015) Anticancer drug detection using a highly sensitive molecularly imprinted electrochemical sensor based on an electropolymerized microporous metal organic framework. Talanta 138: 71-76.

16. Lin RY, Schwartz LB, Curry A, Pesola GR, Knight RJ, et al. (2000) Histamine and tryptase levels in patients with acute allergic reactions: An emergency department–based study. J Allergy Clin Immunol 106: 65-71.

17. Schwartz JC, Pollard H, Quach TT (1980) Histamine as a neurotransmitter in mammalian brain: neurochemical evidence. J Neurochem 35: 26-33.

18. Mahmoudi R, Norian R (2014) Occurrence of histamine in canned tuna fish from Iran. J Verbr Lebensm 9: 133-136.

19. Bao L, Sun D, Tachikawa H, Davidson V (2002) Improved sensitivity of a histamine sensor using an engineered methylamine dehydrogenase. Anal Chem 74: 1144-1148.

20. Yamada R, Fujieda N, Tsutsumi M, Tsujimura S, Shirai O, et al. (2008) Bioelectrochemical determination at histamine dehydrogenase-based electrodes. Electrochemistry 76: 600-602.

21. Veseli A, Vasjari M, Arbneshi T, Hajrizi A, Švorc Ľ, et al. (2016) Electrochemical determination of histamine in fish sauce using heterogeneous carbon electrodes modified with rhenium(IV) oxide. Sens Actuators B 228: 774-781.

22. Fukuda M, Hayashi H, Hasegawa T, Morii T (2009) Development of a fluorescent ribonucleopeptide sensor for histamine. Trans Mater Res Soc Jpn 34: 525-527.

23. Kitade T, Kitamura K, Konishi T, Takegami S, Okuno T, et al. (2004)

Potentiometric immunosensor using artificial antibody based on molecularly imprinted polymers. Anal Chem 76: 6802-6807.

24. Sun N, Avdeef A (2011) Biorelevant p$K_a$ (37°C) predicted from the 2D structure of the molecule and its p$K_a$ at 25°C. J Pharm Biome Anal 56: 173-182.

25. Baba T, Matsui T, Kamiya K, Nakano M, Shigeta Y (2014) A density functional study on the p$K_a$ of small polyprotic molecules. Int J Quantum Chem 114: 1128-1134.

26. Calculated using Advanced Chemistry Development (ACD/Labs) Software V11.02

27. Piletska EV, Guerreiro AR, Romero-Guerra M, Chianella I, Turner APF, et al. (2008) Design of molecular imprinted polymers compatible with aqueous environment. Anal Chim Acta 607: 54-60.

28. Dai Z, Liu J, Tang S, Wang Y, Wang Y, et al. (2015) Optimization of enrofloxacin-imprinted polymers by computer-aided design. J Mol Model 21: 290.

29. Dong W, Yan M, Liu Z, Wu G, Li Y (2007) Effects of solvents on the adsorption selectivity of molecularly imprinted polymers; Molecular simulation and experimental validation. Sep Purif Technol 53: 183-188.

30. Shi X, Wu A, Qu G, Li R, Zhang D (2007) Development and characterization of molecularly imprinted polymers based on methacrylic acid for selective recognition of drugs. Biomaterials 28: 3741-3749.

31. Trikka FA, Yoshimatsu K, Ye L, Kyriakidis DA (2012) Molecularly imprinted polymers for histamine recognition in aqueous environment. Amino Acids 43: 2113-2124.

32. Itahara T (1998) NMR study of stacking interactions between adenine and xanthine rings. J Chem Soc, Perkin Trans 2: 1455-1462.

33. Kamiichi K, Doi M, Nabae M, Ishida T, Inoue M (1987) Structural studies of the interaction between indole derivatives and biologically important aromatic compounds. Part 19. Effect of base methylation on the ring-stacking interaction between tryptophan and guanine derivatives: a nuclear magnetic resonance investigation. J Chem Soc, Perkin Trans 2: 1739-1745.

34. Tarui M, Nomoto N, Hasegawa Y, Minoura K, Doi M, et al. (1996) Thermodynamic effect of complementary hydrogen bond base pairing on aromatic stacking interaction in the guanine-X-Trp complex (X=adenine, guanine, cytosine, thymine). Chem Pharm Bull 44: 1998-2002.

35. Perrin D, Dempsey B, Serjeant EP (1981) pKa Prediction for organic acids and bases. Chapman and Hall 107.

36. Charton M, Roche E (1977) Design of biopharmaceutical properties through prodrugs and analogs. Am Pharm Assoc Ch 9: 269.

37. Kabachnik MI, Mastryukova TA (1984) A σρ-analysis of carbon acidity of organophosphorus compounds. Zh Obshch Khim 54: 2161-2169.

38. Hansch C, Leo A, Taft RW (1991) A survey of hammett substituent constants and resonance and field parameters. Chem Rev 91: 165-195.

# 23

# Spectroscopic Characterization of Disulfiram and Nicotinic Acid after Biofield Treatment

**Mahendra Kumar Trivedi[1], Alice Branton[1], Dahryn Trivedi[1], Gopal Nayak[1], Khemraj Bairwa[2] and Snehasis Jana[2]\***

[1]*Trivedi Global Inc., 10624 S Eastern Avenue Suite A-969, Henderson, NV 89052, USA*
[2]*Trivedi Science Research Laboratory Pvt. Ltd., Hall-A, Chinar Mega Mall, Chinar Fortune City, Hoshangabad Rd., Bhopal- 462026, Madhya Pradesh, India*

## Abstract

Disulfiram is being used clinically as an aid in chronic alcoholism, while nicotinic acid is one of a B-complex vitamin that has cholesterol lowering activity. The aim of present study was to investigate the impact of biofield treatment on spectral properties of disulfiram and nicotinic acid. The study was performed in two groups i.e., control and treatment of each drug. The treatment groups were received Mr. Trivedi's biofield treatment. Subsequently, spectral properties of control and treated groups of both drugs were studied using Fourier transform infrared (FT-IR) and Ultraviolet-Visible (UV-Vis) spectroscopic techniques. FT-IR spectrum of biofield treated disulfiram showed the shifting in wavenumber of C-H stretching from 1496 to 1506 $cm^{-1}$ and C-N stretching from 1062 to 1056 $cm^{-1}$. The intensity of S-S dihedral bending peaks (665 and 553 $cm^{-1}$) was also increased in biofield treated disulfiram sample, as compared to control. FT-IR spectra of biofield treated nicotinic acid showed the shifting in wavenumber of C-H stretching from 3071 to 3081 $cm^{-1}$ and 2808 to 2818 $cm^{-1}$. Likewise, C=C stretching peak was shifted to higher frequency region from 1696 $cm^{-1}$ to 1703 $cm^{-1}$ and C-O (COO$^-$) stretching peak was shifted to lower frequency region from 1186 to 1180 $cm^{-1}$ in treated nicotinic acid.

UV spectrum of control and biofield treated disulfiram showed similar pattern of UV spectra. Whereas, the UV spectrum of biofield treated nicotinic acid exhibited the shifting of absorption maxima ($\lambda_{max}$) with respect of control i.e., from 268.4 to 262.0 nm, 262.5 to 256.4, 257.5 to 245.6, and 212.0 to 222.4 nm.

Over all, the FT-IR and UV spectroscopy results suggest an impact of biofield treatment on the force constant, bond strength, and dipole moments of treated drugs such as disulfiram and nicotinic acid that could led to change in their chemical stability as compared to control.

**Keywords:** Disulfiram; Nicotinic acid; Biofield treatment; Fourier transform infrared spectroscopy; Ultraviolet spectroscopy

## Introduction

Disulfiram [bis(diethylthiocarbamoyl)disulphide] is an antabuse drug, being used clinically as an aid to the treatment of chronic alcoholism. It is the first drug approved by US Food and Drug Administration to treat the alcohol addiction [1]. Alcohol (ethanol) transforms into acetaldehyde by alcohol dehydrogenase enzyme, which further oxidized to acetic acid by acetaldehyde dehydrogenase (ADH) enzyme [2]. Disulfiram inhibits the ADH enzyme. As a result, the blood concentration of acetaldehyde increases and causes an unpleasant effect, thus increase the patient's motivation to remain abstinent [3]. In addition to this, disulfiram is reported for protozoacidal effect *in vitro* study [4,5]. Recently, disulfiram has shown the reactivity to latent HIV-1 expression in a primary cell model of virus latency and presently it is assessed in a clinical trial for its potential to diminish the latent HIV-1 reservoir in patients combination with antiretroviral therapy [6].

Nicotinic acid or niacin is one of the B-complex vitamins (Vitamin B₃) that has cholesterol lowering activity. Recent studies showed that therapeutic doses of nicotinic acid induce a profound alteration in plasma concentration of several lipids and lipoproteins, resulting in a greater ability to increase high-density lipoprotein (HDL) cholesterol [7]. Nicotinic acid favorably affects apolipoprotein (apo), very-low-density lipoprotein (VLDL), low-density lipoprotein (LDL) and HDL [7,8].

The exact mechanism of nicotinic acid activity is unknown. However, new findings indicate that nicotinic acid inhibits directly and non-competitively to the triglycerides synthesis enzyme i.e., hepatocyte diacylglycerol acyltransferase-2, which causes acceleration of intracellular hepatic apo B degradation and thus decrease secretion

of VLDL and LDL [9]. Several evidence suggest that nicotinic acid administered either alone or in combination with other cholesterol-lowering medicines can reduce the risk of cardiovascular and atherosclerosis diseases. The clinical uses of nicotinic acid are somewhat limited due to some harmless but unpleasant side effects like cutaneous flushing phenomenon, nausea, vomiting and headache [10]. The chemical and physical stability of pharmaceutical drugs or products are most desired attributes of quality that potentially affect the efficacy, safety and shelf life of drugs [11]. Hence, it is essential to find out an alternate approach, which could enhance the stability of drugs by altering the structural and bonding properties of these compounds.

Contemporarily, biofield treatment is reported to alter the spectral properties of various pharmaceutical drugs like paracetamol, piroxicam, metronidazole, and tinidazole; likewise physical, and structural properties of various metals i.e., tin, lead etc. [12-14]. The conversion of mass into energy is well known in literature for hundreds of years that was further explained by Hasenohrl and Einstein [15,16]. According to Maxwell JC, every dynamic process in the human body had an electrical significance, which generates magnetic field in the human body [17].

**\*Corresponding author:** Snehasis Jana, Trivedi Science Research Laboratory Pvt. Ltd., Hall-A, Chinar Mega Mall, Chinar Fortune City, Hoshangabad Rd, Bhopal-462026, Madhya Pradesh, India, E-mail: publication@trivedisrl.com

This electromagnetic field of the human body is known as biofield and energy associated with this field is known as biofield energy [18,19]. Mr. Trivedi has the ability to harness the energy from environment or universe and can transmit into any living or nonliving object around this Globe. The object(s) always receive the energy and responding into useful way, this process is known as biofield treatment.

Mr. Mahendra Kumar Trivedi's biofield treatment (The Trivedi Effect®) has considerably changed the physicochemical, thermal and structural properties of metals and ceramics [14,20,21]. Growth and anatomical characteristics of some plants were also increased after biofield treatment [22,23]. Further, biofield treatment has showed the significant effect in the field of agriculture science [24,25] and microbiology [26,27].

Considering the impact of biofield treatment on physical and structural property of metals and ceramics, the present study was aimed to evaluate the impact of biofield treatment on spectral properties of disulfiram and nicotinic acid. The effects were analyzed using Fourier transform infrared (FT-IR) and Ultraviolet-Visible (UV-Vis) spectroscopic techniques.

## Materials and Methods

### Study design

The disulfiram and nicotinic acid (Figure 1) samples were procured from Sigma-Aldrich, MA, USA; and each drug was divided into two parts i.e., control and treatment. The control samples were remained as untreated, and treatment samples were handed over in sealed pack to Mr. Trivedi for biofield treatment under laboratory condition. Mr. Trivedi provided this treatment through his energy transmission process to the treated groups without touching the sample [12,13]. The control and treated samples of disulfiram and nicotinic acid were evaluated using FT-IR and UV-Vis spectroscopy.

### FT-IR spectroscopic characterization

FT-IR spectra were recorded on Shimadzu's Fourier transform infrared spectrometer (Japan) with frequency range of 4000-500 cm$^{-1}$. The FT-IR spectroscopic analysis of both control and treated samples of disulfiram and nicotinic acid were carried out to evaluate the impact of biofield treatment at atomic level like force constant and bond strength [28].

### UV-Vis spectroscopic analysis

UV spectra of disulfiram and nicotinic acid were recorded on Shimadzu UV-2400 PC series spectrophotometer with 1 cm quartz cell and a slit width of 2.0 nm. The analysis was carried out using wavelength in the range of 200-400 nm. The analysis was performed to determine the effect of biofield treatment on structural properties of treated drugs [28].

## Results and Discussion

### FT-IR spectroscopic analysis

Vibrational spectral assignment was performed on the recorded FT-IR spectra (Figure 2) based on theoretically predicted wavenumber and presented in Table 1. The FT-IR spectrum of control disulfiram sample (Figure 2a) showed the characteristic vibrational peak at 2975 cm$^{-1}$ that was assigned to C-H (CH$_3$) stretching. Another characteristic peak observed at 1496 cm$^{-1}$ was attributed to C-H symmetrical deformation vibrations. The absorption peaks appeared at 1351-1457 cm$^{-1}$ was assigned to CH$_2$-CH$_3$ deformations. The vibrational peaks

at 1273 cm$^{-1}$ and 1151-1195 cm$^{-1}$ were assigned to C=S stretching and C-C skeletal vibration, respectively. Further, IR peaks observed at 967-1062 cm$^{-1}$ and 818-914 cm$^{-1}$ were attributed to C-N stretching and C-S stretching, respectively. The vibrational peaks appeared at 554-666 cm$^{-1}$ was assigned to S-S dihedral bending. The FT-IR data of control disulfiram was well supported by the literature data [29].

The FT-IR spectrum of biofield treated disulfiram (Figure 2b) showed the vibrational peak at 2975 cm$^{-1}$, which was assigned to CH$_3$ stretching. Vibrational peak appeared at 1506 cm$^{-1}$ was assigned to C-H symmetrical deformation vibrations. Likely, the IR peaks at 1350-1457 cm$^{-1}$ were attributed to CH$_2$-CH$_3$ deformations. The vibrational peaks appeared at 1273 cm$^{-1}$ and 1151-1195 cm$^{-1}$ were assigned to C=S stretching and C-C skeletal vibration, respectively. The IR peaks observed at 967-1056 cm$^{-1}$ and 817-914 cm$^{-1}$ were attributed to C-N stretching and C-S stretching, respectively. The vibrational peaks at 553-665 cm$^{-1}$ were assigned to S-S dihedral bending.

Altogether, the FT-IR data of biofield treated disulfiram (Figure 2b) showed the shifting in frequency of some bonds with respect to control spectra like C-H symmetrical deformation vibrations frequency was shifted from 1496 (control) to 1506 (treated) cm$^{-1}$. The frequency (ν) of vibrational peak depends on two factors i.e., force constant (k) and

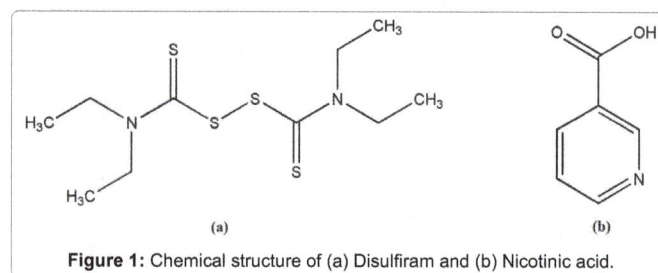

**Figure 1:** Chemical structure of (a) Disulfiram and (b) Nicotinic acid.

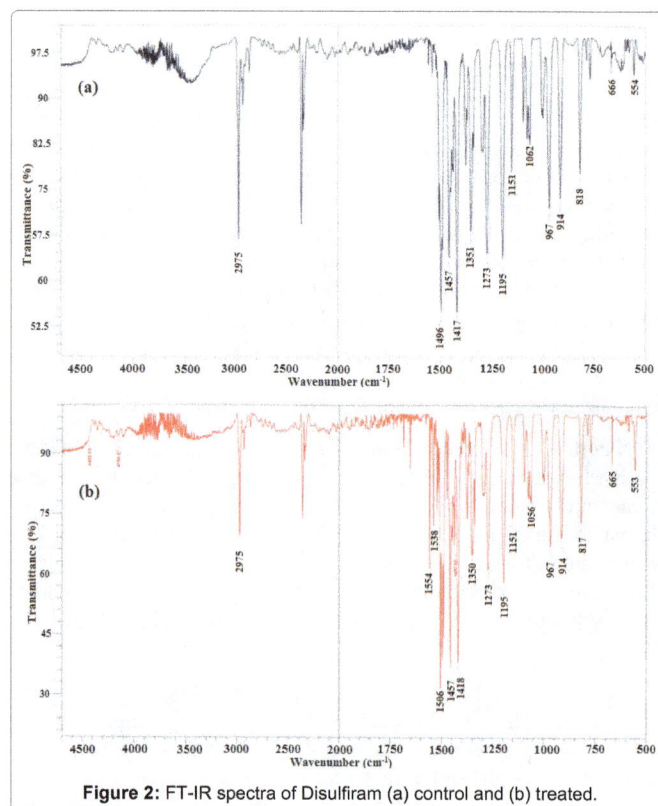

**Figure 2:** FT-IR spectra of Disulfiram (a) control and (b) treated.

| Wave number (cm⁻¹) | | Frequency Assignment |
|---|---|---|
| Control | Treated | |
| 2975 | 2975 | CH₃ stretching |
| 1496 | 1506 | C-H symmetrical deformation vibrations |
| 1351-1457 | 1350-1457 | CH₂ and CH₃ deformation |
| 1273 | 1273 | C=S stretching |
| 1151-1195 | 1151-1195 | C-C skeletal vibrations |
| 967-1062 | 967-1056 | C-N stretching |
| 818-914 | 817-914 | C-S stretching |
| 554-666 | 553-665 | S-S stretching |

**Table 1:** FT-IR vibrational peaks observed in Disulfiram.

reduced mass ($\mu$), which can be explained by following equation [30]

$$\nu = 1/2\pi c \sqrt{(k/\mu)}$$

here, c is speed of light.

If reduced mass is constant, then the frequency is directly proportional to the force constant; therefore, increase in frequency of any bond suggested a possible enhancement in force constant of respective bond and *vice versa* [28]. Based on this it is hypothesized that due to increase in frequency of C-H symmetrical deformation (1496 to 1506 cm⁻¹), the C-H bond strength in treated disulfiram might also be increased with respect of control. Contrarily, the C-N stretching vibration in treated disulfiram was sifted to lower frequency region as compared to control i.e., from 1062 to 1056 cm⁻¹. This could be referred to decrease in C-N bond strength after biofield treatment in compression to control. In addition, the intensity of IR peak appeared at 553-665 cm⁻¹ (S-S dihedral bending) in biofield treated sample was found to be increased, with respect of control peaks in the same frequency region. The intensity of vibrational peaks of particular bond depends on the ratio of change in dipole moment ($\partial\mu$) to change in bond distance ($\partial r$) i.e., the intensity is proportionally change with changes in dipole moment and inversely change with alteration in bond distance [31]. Based on this, it is speculated that ratio of $\partial\mu/\partial r$ might be altered in S-S bonds (appeared in the frequency region of 553-665 cm⁻¹) with the influence of biofield treatment as compared to control.

The vibrational spectral assignment of nicotinic acid was performed on the recorded FT-IR spectra (Figure 3) based on theoretically predicted wavenumber and presented in Table 2. The vibrational peaks appeared at 3071-2808 cm⁻¹ was assigned to C-H stretchings. The IR peaks observed at 1696-1710 cm⁻¹ and 1596 cm⁻¹ were assigned to C=O (COO⁻) asymmetrical stretching and C=C stretching, respectively. Absorption peaks appeared at 1417, 1323, and 1301 cm⁻¹ were attributed to C=N symmetric stretching, C=O symmetrical stretching, and C-N stretching, respectively. The C-O (COO⁻) stretching peak was assigned to IR bend observed at 1186 cm⁻¹. Further, C-H in plane and out plane bending vibrations was assigned to peaks observed in the range of 1033-1114 cm⁻¹ and 642-812 cm⁻¹, respectively. The FT-IR data of control nicotinic acid was well supported by the literature data [32,33].

The FT-IR spectrum of biofield treated nicotinic acid (Figure 3) showed the absorption bands at 2818-3081 cm⁻¹ that were assigned to C-H stretching. Vibrational peaks appeared at 1703-1714 cm⁻¹ and 1594 cm⁻¹ were assigned to C=O (COO⁻) asymmetric stretching and C=C stretching, respectively. Likewise, the IR peaks observed at 1417, 1324, and 1301 cm⁻¹ were assigned to C=N stretching, C=O (COO⁻) symmetric stretching, and C-N stretching, respectively. The IR absorption peak appeared at 1180 cm⁻¹ was attributed to C-O (COO⁻) stretching. Further, the C-H in plane and out plane bending vibrations

was assigned to IR peaks observed at 1037-1116 cm⁻¹ and 642-812 cm⁻¹, respectively.

Overall, the FT-IR data of biofield treated nicotinic acid (Figure 3) showed the shifting in wavenumber of some bonds with respect to control sample. For instance, the C-H stretching towards higher frequency region i.e., from 3071 to 3081 cm⁻¹ and 2808 to 2818 cm⁻¹. This could be due to increase in force constant of C-H bond. Likewise, a slight upstream shifting in C=O stretching peak from 1710 to 1714 cm⁻¹ and 1696 to 1703 cm⁻¹ in treated nicotinic acid also suggests an increase in force constant of C=O bond in treated sample as compared to control. Contrarily, a slight downstream shifting in wavenumber of treated nicotinic acid from 1186 to 1180 cm⁻¹ C-O (COO⁻ stretching); and from 1033 to 1037 cm⁻¹ (=C-H in plane bending) suggests the decrease in force constant of C-O bond and decrease in rigidity of =C-H bond in treated sample as compared to control.

**UV-Vis spectroscopy**

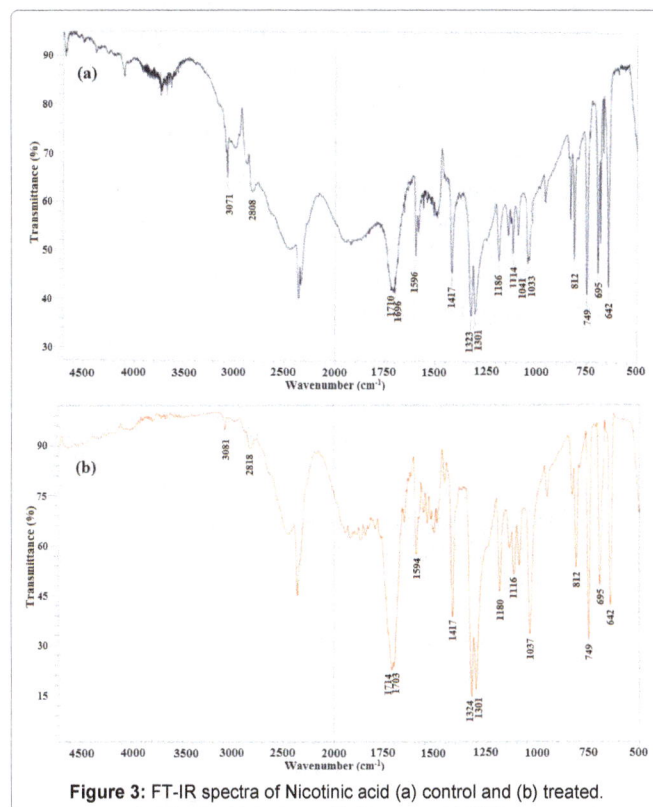

**Figure 3:** FT-IR spectra of Nicotinic acid (a) control and (b) treated.

| Wave number (cm⁻¹) | | Frequency assignment |
|---|---|---|
| Control | Treated | |
| 2808-3071 | 2818-3081 | C-H stretching |
| 1696-1710 | 1703-1714 | C=O (COO-) asymmetric stretching |
| 1596 | 1594 | C=C stretching |
| 1417 | 1417 | C=N stretching |
| 1323 | 1324 | C=O (COO-) symmetric stretching |
| 1301 | 1301 | C-N stretching |
| 1186 | 1180 | C-OH (Ph-OH) stretching |
| 1033-1114 | 1037-1116 | =C-H in-plane bending |
| 642-812 | 642-812 | =C-H out of plane bending |

**Table 2:** FT-IR vibrational peaks observed in Nicotinic acid.

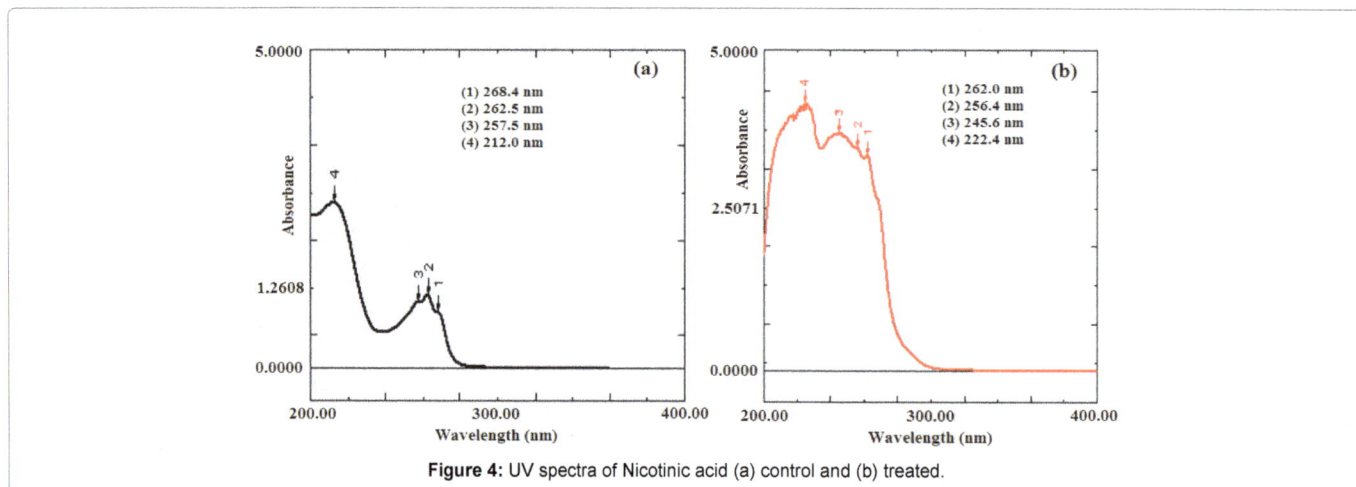

**Figure 4:** UV spectra of Nicotinic acid (a) control and (b) treated.

UV spectra of control and treated disulfiram showed a similar pattern of UV spectra with absorption maxima ($\lambda_{max}$) of 219.8, 250.2, and 281.6 nm in control and 220.8, 249.4, and 281.2 nm in treated sample. This indicates no significant change in the UV spectral property of treated disulfiram with respect to control sample. The UV spectra of control and treated nicotinic acid are showed in Figure 4. The UV spectrum of treated nicotinic acid (Figure 4) exhibited the shifting of absorption maxima ($\lambda_{max}$) from 268.4 to 262.0 nm, 262.5 to 256.4 nm, 257.5 to 245.6 nm, and 212.0 to 222.4 nm. The existing literature on principle of UV spectroscopy suggests that a compound can absorbs UV light due to presence of either or both conjugated pi ($\pi$) -bonding systems ($\pi$-$\pi^*$ transition) and nonbonding electron system (n-$\pi^*$ transition) in the compound. The UV absorption phenomenon occurred when electrons travelled from low energy orbital (i.e., $\sigma$, n, and $\pi$) to high energy orbital (i.e., $\sigma^*$ and $\pi^*$). There is certain energy gap between $\sigma$-$\sigma^*$, $\sigma$-$\pi^*$, $\pi$ -$\pi^*$ and n-$\pi^*$ orbitals. When this energy gap altered, the wavelength ($\lambda_{max}$) was also altered respectively [28]. Based on this, it is speculated that, due to influence of biofield treatment, the energy gap between $\pi$-$\pi^*$ and n-$\pi^*$ transition in nicotinic acid might be altered, which causes shifting of wavelength ($\lambda_{max}$) in treated nicotinic acid as compared to control. To the best of our knowledge, this is the first report showing an impact of biofield treatment on structural properties like force constant, bond strength, dipole moment of disulfiram and nicotinic acid.

## Conclusion

The FT-IR data of biofield treated disulfiram showed an alteration in the wavenumber of C-H and C-N stretching; whereas, wavenumbers of C-H, C=O, and C-O stretching, and =C-H bending were altered in biofield treated nicotinic acid, with respect of control. Also, the peak intensity at 553-665 cm$^{-1}$ (S-S dihedral bending) was increased in biofield treated disulfiram, as compared to control. This alteration in wavenumber referred to alteration in the force constant and bond strength of respective group. The UV spectral data of biofield treated nicotinic acid also support the possible change in the structural property with respect of control.

In conclusion, the results suggest a significant impact of biofield treatment on structural property like force constant, bond strength, dipole moment, and energy gap between bonding and nonbonding orbital of treated drug with respect to control.

## Acknowledgement

The authors would like to acknowledge the whole team of MGV Pharmacy College, Nashik for providing the instrumental facility. Authors would also like to thank Trivedi Science™, Trivedi Master Wellness™ and Trivedi Testimonials for their consistent support during the work.

## References

1. Zindel LR, Kranzler HR (2014) Pharmacotherapy of alcohol use disorders: Seventy-five years of progress. J Stud Alcohol Drugs Suppl 17: 79-88.

2. Svensson S, Some M, Lundsjo A, Helander A, Cronholm T, et al. (1999) Activities of human alcohol dehydrogenases in the metabolic pathways of ethanol and serotonin. Eur J Biochem 262: 324-329.

3. Cederbaum AI (2012) Alcohol metabolism. Clin Liver Dis 16: 667-685.

4. Nash T, Rice WG (1998) Efficacies of zinc-finger-active drugs against *Giardia lamblia*. Antimicrob Agents Chemother 42: 1488-1492.

5. Bouma MJ, Snowdon D, Fairlamb AH, Ackers JP (1998) Activity of disulfiram (bis(diethylthiocarbamoyl)disulphide) and ditiocarb (diethyldithiocarbamate) against metronidazole-sensitive and -resistant *Trichomonas vaginalis* and *Tritrichomonas foetus*. J Antimicrob Chemother 42: 817-820.

6. Doyon G, Zerbato J, Mellors JW, Sluis-Cremer N (2013) Disulfiram reactivates latent HIV-1 expression through depletion of the phosphatase and tensin homolog. AIDS 27: F7-F11.

7. Patel S (2013) A review of available cholesterol lowering medicines in South Africa. S Afr Pharm J 80: 20-25.

8. Gille A, Bodor ET, Ahmed K, Offermanns S (2008) Nicotinic acid: pharmacological effects and mechanisms of action. Annu Rev Pharmacol Toxicol 48: 79-106.

9. Kamanna VS, Kashyap ML (2008) Mechanism of action of niacin. Am J Cardiol 101: 20B-26B.

10. Bodor ET, Offermanns S (2008) Nicotinic acid: An old drug with a promising future. Br J Pharmacol 153: S68-S75.

11. Blessy M, Patel RD, Prajapati PN, Agrawal YK (2014) Development of forced degradation and stability indicating studies of drugs-A review. J Pharm Anal 4: 159-165.

12. Trivedi MK, Patil S, Shettigar H, Bairwa K, Jana S (2015) Effect of biofield treatment on spectral properties of paracetamol and piroxicam. Chem Sci J 6: 98.

13. Trivedi MK, Patil S, Shettigar H, Bairwa K, Jana S (2015) Spectroscopic characterization of biofield treated metronidazole and tinidazole. Med chem 5: 340-344.

14. Dabhade VV, Tallapragada RR, Trivedi MK (2009) Effect of external energy on atomic, crystalline and powder characteristics of antimony and bismuth powders. Bull Mater Sci 32: 471-479.

15. Hasenohrl F (1904) On the theory of radiation in moving bodies. Ann Phys 320: 344-370.

16. Einstein A (1905) Does the inertia of a body depend upon its energy-content? Ann Phys 18: 639-641.

17. Maxwell JC (1865) A dynamical theory of the electromagnetic field. Phil Trans R Soc Lond 155: 459-512.

18. Rubik B (2002) The biofield hypothesis: its biophysical basis and role in medicine. J Altern Complement Med 8: 703-717.

19. Rivera-Ruiz M, Cajavilca C, Varon J (2008) Einthoven's string galvanometer: the first electrocardiograph. Tex Heart Inst J 35: 174-178.

20. Trivedi MK, Patil S, Tallapragada RM (2013) Effect of biofield treatment on the physical and thermal characteristics of vanadium pentoxide powders. J Material Sci Eng S11: 001.

21. Trivedi MK, Patil S, Tallapragada RM (2014) Atomic, crystalline and powder characteristics of treated zirconia and silica powders. J Material Sci Eng 3: 144.

22. Patil SA, Nayak GB, Barve SS, Tembe RP, Khan RR (2012) Impact of biofield treatment on growth and anatomical characteristics of Pogostemon cablin (Benth.). Biotechnology 11: 154-162.

23. Nayak G, Altekar N (2015) Effect of biofield treatment on plant growth and adaptation. J Environ Health Sci 1: 1-9.

24. Lenssen AW (2013) Biofield and fungicide seed treatment influences on soybean productivity, seed quality and weed community. Agricultural Journal 8: 138-143.

25. Shinde V, Sances F, Patil S, Spence A (2012) Impact of biofield treatment on growth and yield of lettuce and tomato. Aust J Basic & Appl Sci 6: 100-105.

26. Trivedi MK, Patil S, Shettigar H, Bairwa K, Jana S (2015) Phenotypic and biotypic characterization of Klebsiella oxytoca: An impact of biofield treatment. J Microb Biochem Technol 7: 203-206.

27. Trivedi MK, Patil S, Shettigar H, Gangwar M, Jana S (2015) Antimicrobial sensitivity pattern of Pseudomonas fluorescens after biofield treatment. J Infect Dis Ther 3: 222.

28. Pavia DL , Lampman GM, Kriz GS (2001) Introduction to spectroscopy. 3rd edn, Thomson learning, Singapore.

29. Marciniec B, Dettlaff K, Naskrent M, Pietralik Z, Kozak M (2012) DSC and spectroscopic studies of disulfiram radiostability in the solid state. J Therm Anal Calorim 108: 33-40.

30. Stuart BH (2004) Infrared Spectroscopy: Fundamentals and applications (analytical techniques in the sciences (AnTs). John Wiley & Sons Ltd, Chichester, UK.

31. Smith BC (1998) Infrared Spectral Interpretation: A systematic approach. CRC Press.

32. Jegannathan S, Mary MB, Ramakrishnan V, Thangadurai S (2014) Vibrational spectral studies of bis (nicotinic acid) hydrogen perchlorate. Asian J Research Chem 7: 67-71.

33. Karabacak M, Kurt M (2008) Comparison of experimental and density functional study on the molecular structure, infrared and Raman spectra and vibrational assignments of 6-chloronicotinic acid. Spectrochim Acta A Mol Biomol Spectrosc 71: 876-883.

# 24

# Shadow Technique Algorithm (STA) Sheds a New Light on Differential Interference Contrast (DIC) Microscopy

Dave Trinel[1], Pauline Vandame[1,2], Magalie Hervieu[3], Emilie Floquet[3], Marc Aumercier[3], Emanuele G Biondi[3], Jean-François Bodart[2] and Corentin Spriet[1]*

[1]TISBio, CNRS, UMR 8576, UGSF, Glycobiology Structural and Functional Unit, Lille University of Science and Technology, Lille, France
[2]Regulation of Signal Division Team, CNRS, UMR 8576, UGSF, Glycobiology Structural and Functional Unit, Lille University of Science and Technology, Lille, France
[3]CNRS, UMR 8576, UGSF, Glycobiology Structural and Functional Unit, Lille University of Science and Technology, France

## Abstract

Diversity of biological samples is still partially considered by conventional Differential Interference Contrast (DIC) microscopy approaches. Here we propose a new algorithm (developed as an ImageJ macro), the STA (Shadow Technique Algorithm), whose originality relies in the 3D/4D visualization of a large range of biological objects, from bacteria, vegetal tissues to living cells in culture. This new approach does not need extensive calculations, systems modifications or in-depth knowledge of acquisition optics. STA, providing 3D DIC reconstruction every hundredth of ms, can be applied to dissect various cellular phenomena. In addition, we propose different methods of graphic representations, which unable to enlighten the specificities of each category of questioning. Specifically we here addressed: i) tissue imaging, ii) cell cycle and cell death imaging iii) vesicle tracking.

**Keywords:** Differential interference contrast; Shadow technique algorithm; Image analysis; ImageJ; Tracking; Cell cycle; 3D reconstruction

## Introduction

Nomarski Differential Interference Contrast (DIC) is a non-invasive photonic microscopy method to gather structural properties at the cellular and sub-cellular levels [1]. In such context, living cells are not damaged, enabling the observation of dynamic processes both *in vitro* and in tissues. From the early fifties, DIC has provided nuances of grey, to detect small structures and vesicles [2,3]. Thus, one shall keep in mind that DIC is intrinsically qualitative, reflecting a difference in both phase and amplitude, and providing a nonlinear response to phase gradient. Furthermore, DIC's images result in a false sense of depth. Indeed, images are produced by interference between a reference and a spatially-sheared light wave, resulting in a differential image having directional contrast, according to the direction of the shear.

DIC microscopy was extensively used when considering objects in 2D, while thick objects have been more difficult to assess. Comparing to fluorescence, which is widely used for 3D and benefits from a black background and grey levels reflecting the intensity of the signal, DIC provides images of structures boundaries exhibiting either bright or dark intensities, while the remainder of the image has an intermediate grey level. Thus, standard image processing methods for DIC image segmentation, such as "thresholding" or edge detection, cannot be employed on DIC's images and limit 3D reconstruction (Figure 1) [4,5].

Several efforts have been made to develop 3D representation of DIC image series. They can be classified following two main approaches, qualitative or quantitative. The quantitative method needs a deep understanding and/or modification of the acquisition system. Rotational diversity is among these techniques, which involve system modifications and extensive data analysis [6]. Rotational diversity consists in taking several images at different rotation with respect to the DIC shear angle and then, combining them by iterative optimization, allowing more quantitative phase gradient maps to be achieved.

However, while the images recorded by DIC are a mixture of non-linear phase and amplitude information, they are intrinsically qualitative, and likely appropriate for 3D visualization of biological objects. Therefore, methods have been developed to achieve a faster preprocessing. Indeed, algorithms can be developed in order to convert classical DIC images into fluorescence-like images. Heise and collaborators [7] provided a critical comparison of most DIC image reconstruction methods and reported advantages and drawbacks of both iterative (iterative line integration [8] and modified iterative Hilbert transform [7]) and non-iterative methods (Hilbert transform [9] and deconvolution [10]). All above-mentioned techniques require precise knowledge of the shear angle of the DIC prism. They may also result in strong visual distortions, implementation complexity or restriction to specific biological objects [7-10]. All together, these considerations have led users to consider DIC as a tough and rough tool for biologist, who are the main potential users for such methods.

In this article, we propose the Shadow Technique Algorithm (STA), a novel and user-friendly algorithm enabling the 3 and 4D visualization of DIC that overcomes the above-mentioned limitations (Figure 1). This method has been applied to representative biological sample ranging from thin and low-contrasted bacteria to thick and dense root slices. We also provide comprehensive methods to extract biological information and to represent sample morphologies.

## Material and Methods

### Sample preparation

*Caulobacter crescentus* bacteria were grown to mid-log phase in PYE (Peptone-Yeast Extract) medium at 30°C [11]. For fixation,

---

*Corresponding author: Corentin Spriet, TISBio, CNRS, UMR 8576, UGSF, Glycobiology Structural and Functional Unit, Lille University of Science and Technology, FRABio EN 3688 CNRS, F 59000 Lille, France
E-mail: corentin.spriet@univ-lille1.fr

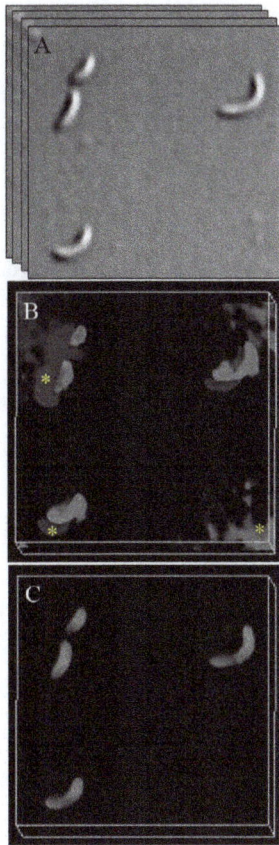

**Figure 1:** DIC imaging of Caulobacter Crescentus. 3D acquisition (A) and corresponding volume reconstruction directly (B) or after processing with the STA (C). While direct 3D reconstruction results in bacteria deformation and artifact (examples highlighted by yellow stars) crucial morphological parameter, like size, shape and curvature are easily retrieved after STA, whatever the original orientation of the bacteria. Stack size: 1004 × 1002 × 52 voxels, voxel size 115 × 115 × 390 nm.

bacteria were centrifuged and placed in 4% paraformaldehyde for 15 min at room temperature followed by 30 min on ice. After washing and resuspension in PBS, 2 μL of suspension were deposited on microscope slides coated with agar pads (1% agar in PBS).

**MDA cells**: Human MDA-MB-231 cells were cultured in Dulbecco's modified Eagle's medium (Invitrogen, Life Technologies, Saint Aubin, France) supplemented with 10% fetal bovine serum and 50 ml gentamycin. 24 hr before observation, cell were seeded on a POC (Perfusion Open and Closed) chamber system (PeCon GmbH) and medium was replaced by Leibovitz's L-15 Medium without phenol red (Sigma-Aldrich Co. LLC) just before observation.

Convallaria root slice is a reference sample kindly provided by Leica Microsystems GmbH.

### Image acquisition

Experiments were carried out with a Leica AF6000 LX microscope (Leica Microsystems GmbH) equipped with a HCX PL APO CS 100X/1.40 DIC (Oil) objective and an Andor iXon DU-885 EM-CCD camera (Andor Technology Ltd). For live MDA acquisitions, the sample temperature was set to 37°C using a PeCon incubator (PeCon GmbH).

### Image processing

**STA:** For details about STA, please see results, and Figure 2.

To use STA:

- Copy the text available as supplementary data in FiJi's macro editor (Plugin/New/Macro).

- Select macro language in the "language" tab

- Optimize the initialization parameters if needed, as explained in the green header

- Open your image or stack

- Run the macro.

**Depth color coding**: We modified the "temporal color code" plugin developed by Kota Miura (Centre for Molecular and Cellular Imaging, EMBL Heidelberg, Germany) to apply a color code on each image depending of its depth. Instead of applying a Maximum Intensity Projection of the stack, we choose to use the ImageJ 3D viewer, and thus obtained a 3D reconstruction of cells with and easily readable depth scale, as shown in Figure 3. The plugin we developed is available upon request.

**Combined volume and slice view:** To compare information usually obtained through 2D DIC imaging and 3D reconstruction along time, we used a mixed representation of the data in Figure 4. First, we performed a 4D reconstruction with high threshold applied on the 4D DIC series processed with STA and color coded it in red.

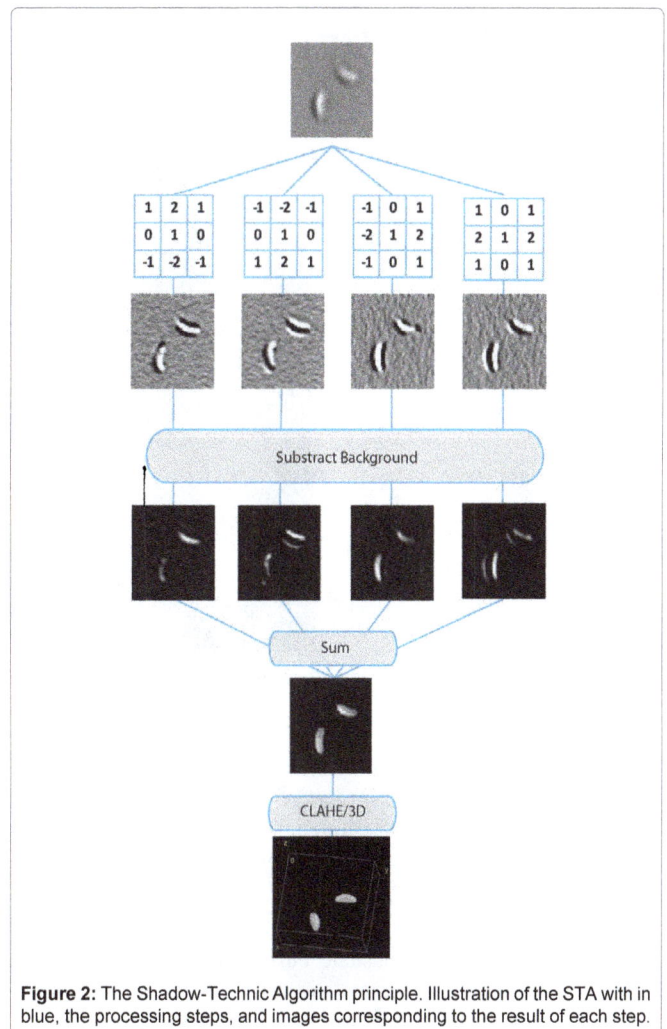

**Figure 2:** The Shadow-Technic Algorithm principle. Illustration of the STA with in blue, the processing steps, and images corresponding to the result of each step.

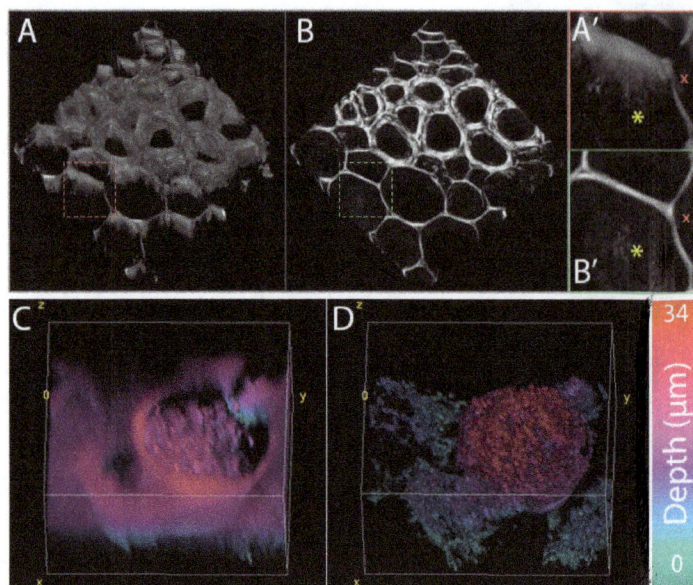

**Figure 3:** STA for thick DIC imaging. 3D representation of Convalaria root slice directly (A, A') or after STA (B, B' and supplementary figure 3B). Direct reconstruction result in a strong loss of detail such has vacuoles envelops (yellow star) or cell wall (red cross) which are realistically preserved after STA. stack size: 1004 × 1002 × 163 voxels, voxel size 115 × 115 × 195 nm. Depth-color-coded 3D representation of membranes of plated MDA cells surrounding an apoptotic one, directly (C) or after STA (D). For such objects, no appropriate threshold can be found to discriminate between cell and background, thus direct 3D reconstruction is less informative than the optimal 2D DIC image. After STA, one can easily identify cell structure at different depth. In this example, membranes from different healthy plated cell (color-coded in blue to purple) are surrounding a thick round cell (color-coded from blue to red). This representation is of major interest for dynamic study of processes strongly affecting cell shape such as cell division or death where traditional 2D acquisition result in out-of-focus images of the studied events. Stack size: 422 × 481 × 62 voxels, voxel size 72 × 72 × 350 nm.

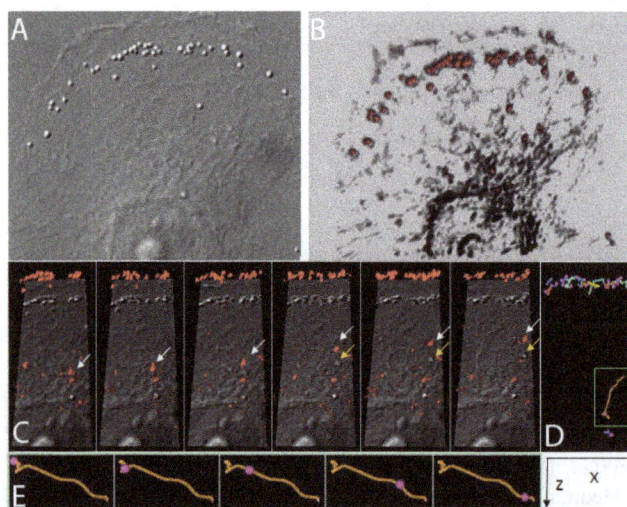

**Figure 4:** STA for 4D DIC imaging: representative DIC image of an *MDA* cell 4D acquisition (A) and associated STA reconstruction (B and supplementary figure 4B) with intensity coded from black to red. C and supplementary figure 4C: mixed representation of high intensity 3D reconstruction (red) with 2D DIC images positioned on the bottom of the volume bounding box (grey). Arrows highlight the displacement of a vesicle in 3D which is visible from the beginning of the experiment in the STA representation (white arrow) and only appears in the 4th time-point on the 2D DIC representation (orange arrow). D represents the result of particle tracking in XY, while (E and supplementary figure 4E) are a zoomed view of YZ particle track.

We then applied on the bottom of the bounding box the best focused unprocessed DIC image. Object correspondence between DIC and STA-reconstructed-DIC can thus be achieved by translation, as seen for vesicles located at the edge of the cell in Figure 4C.

**Particle tracking**: To illustrate our capacity to follow objects in the 4D, we used the trackmate plugin for FiJi as described in ref. [12]. We choose high threshold to segment vesicles only. Vesicles are then represented as pink spheres while their trajectories are shown as colored line, with colors temperature reflecting their velocity.

## Results and Discussion

### STA method

The method has been developed for FiJi [13], a distribution of ImageJ [14] for biological-image analysis. Original stack is duplicated

4 times. On each stacks, either a 3 × 3 north/south/east/west filter is applied (see Figure 2 for filters matrixes) producing a shadow effect on each image. The ImageJ "subtract background" function is then applied, using the "rolling ball" algorithm [15], to subtract local background values from the image. The 4 resulting images are then added up and processed by Contrast Limited Adaptative Histogram Equalization (CLAHE, [16]). 3/4D representations are then obtained on thresholded images either through FiJi's "3D Viewer" or "volume viewer" plugin. STA is available has an ImageJ macro.

## Biological applications

We choose different representative biological samples to illustrate the information that can be extracted by STA: (i) thin and delimitated objects such as fixed bacteria, (ii) thick and structured object like section of complex architecture of *Convalaria* root and (iii) eukaryotic livings cells, such as MDA, to illustrate cellular deformation and intracellular dynamics. These objects would not have been easily and exhaustively considered by conventional DIC approaches.

Figure 1, with bacteria imaging, presents the typical problem encountered after direct 3D DIC reconstruction: object deformation and artifact appearance. After STA, bacteria morphology is faithfully pictured, preserving cell shape, size and curvature. It is thus possible to compare bacteria morphologies whatever their 3D orientation.

In Figure 3A and 3B, we represented the 3D reconstruction of a Convalaria root slice. As observed on such structured object, direct reconstruction was possible with optimized threshold and transparency adjustment. However, while we get a global idea of the biological object, cell walls and vacuoles were not realistically reconstructed (A,A'). After STA, the reconstruction was more consistent with reality, based on previous knowledge we had from fluorescence images. Indeed, the different cells walls were defined on the whole width of the stacks and the vacuole shape was preserved.

On C and D, the same approach was applied to MDA cells. In this case, the contrast was low, and an appropriate threshold could not be found directly from the DIC series (C). After STA, one may clearly visualize both thin plated cells (partial membranes of different cells are visible in blue-purple) and big round cells (coded from blue to red). This visualization could then allow following different cell processes like mitosis, apoptosis or necrosis which is not feasible using 2D DIC microscopy.

We thus extended our approach to 4D imaging (Figure 4). Half of an MDA cell was represented, within which vesicles were easily observed. However, 2D DIC images allowed monitoring only vesicles at the focal plane of an image. It was thus well adapted to the edge of the membrane, which can be considered flat at the optical scale, but we failed to image vesicles on top of the cell. This was exemplified in panel C, a cropped portion of the initial cell over time. The vesicle highlighted by the white arrow was visible from the first time series on the 3D reconstruction, while it appeared in the focal plane of the DIC image only at the fourth time point (orange arrow). STA thus allowed 4D tracking of sub-cellular dynamic as illustrated in D and E.

## Conclusions

### Amenability of STA

We considered several levels of complexity within a biological object: thickness, inner architecture, cellular deformations and intracellular movements of organelles (Figures 3 and 4). Many observations cannot be gathered with conventional DIC use. More complex and time-consuming methods can be used, such as Hilbert transform, rotational diversity, and iterative line integration. Nevertheless, the STA enabled to address these objects, in a user-friendly manner and without requesting either microscopes adaptations or modifications.

In addition, we proposed different methods of graphic representations. Depth scale color coding in Figure 3 provided at glance the relief of membranes: healthy cells plated were visualized in blue-green whereas dying cells were highlighted in red (Figure 3D). Similarly, dividing cells, which are spherical as well, could be easily followed. Thus, morphology of cells can be monitored, with respect of all their aspects of life and death. Regarding organelles tracking, we proposed a mixed representation, merging classical 2D DIC images with 3D volume reconstruction (Figure 4). Such combination provided the familiarity of DIC images with the complementary information of 3D. Furthermore, it can be combined to particles tracking.

In an era that crowns quantitative approaches, still DIC manages to provide valuable and precise qualitative observations through STA. One shall also note that the observations gathered are sufficient to address biological phenomena in their complexity, with an extreme time-resolution without the limits of photo bleaching usually encountered in these contexts. Indeed, in our hands high-resolution images (1000 × 1000 pixels) were gathered every 3 ms and complete stack around 150 ms. Such fast acquisitions, with precise reconstruction of complex structures, provided the opportunity to monitor cellular deformation like the one occurring during cell death (Figure 3) or vesicles movements (Figure 4). Far from remaining an old-grand pa's technique, DIC still provide, through the Shadow Technique Algorithm, a fancy tool for numerous biological applications.

## Acknowledgments

We thank Kota Miura for the temporal color coder plugin for imageJ. Pauline Vandame (2011-2014), Emilie Floquet and Magalie Hervieu (2014-2015) were paid by the University of Lille 1 and the Region Nord-Pas-de-Calais. We thank the personal of the BICeL-Lille1-HB facility for access to the microscopy systems and technical advices.

## References

1. Lang W (1968) Nomarski Differential Interference-Contrast Microscopy.

2. Nomarski G (1955) Microinterférométrie différentielle à onde polarisées. J Phys Radium 16: 9.

3. Obara B, Roberts MA, Armitage JP, Grau V (2013) Bacterial cell identification in differential interference contrast microscopy images. BMC Bioinformatics 14: 134.

4. Simon I, Pound CR, Partin AW, Clemens JQ, Christens-Barry WA (1998) Automated image analysis system for detecting boundaries of live prostate cancer cells. Cytometry 31: 287-294.

5. Obara B, Veeman M, Choi JH, Smith W, Manjunath BS (2011) Segmentation of ascidian notochord cells in DIC timelapse images. Microsc Res Tech 74: 727-734.

6. Preza C (2000) Rotational-diversity phase estimation from differential-interference-contrast microscopy images. J Opt Soc Am A Opt Image Sci Vis 17: 415-424.

7. Heise B, Sonnleitner A, Klement EP (2005) DIC image reconstruction on large cell scans. Microsc Res Tech 66: 312-320.

8. Kam Z (1998) Microscopic differential interference contrast image processing by line integration (LID) and deconvolution. Bioimaging 6: 166-176.

9. Arnison MR, Cogswell CJ, Smith NI, Fekete PW, Larkin KG (2000) Using the Hilbert transform for 3D visualization of differential interference contrast microscope images. J Microsc 199: 79-84.

10. van Munster EB, van Vliet LJ, Aten JA (1997) Reconstruction of optical pathlength distributions from images obtained by a wide-field differential interference contrast microscope. J Microsc 188: 149-157.

11. Fioravanti A, Fumeaux C, Mohapatra SS, Bompard C, Brilli M, et al. (2013) DNA binding of the cell cycle transcriptional regulator GcrA depends on N6-adenosine methylation in Caulobacter crescentus and other Alphaproteobacteria. PLoS Genet 9: e1003541.

12. Jaqaman K, Loerke D, Mettlen M, Kuwata H, Grinstein S, et al. (2008) Robust single-particle tracking in live-cell time-lapse sequences. Nat Methods 5: 695-702.

13. Schindelin J, Arganda-Carreras I, Frise E, Kaynig V, Longair M, et al. (2012) Fiji: an open-source platform for biological-image analysis. Nat Methods 9: 676-682.

14. Schneider CA, Rasband WS, Eliceiri KW (2012) NIH Image to ImageJ: 25 years of image analysis. Nat Methods 9: 671-675.

15. Sternberg SR (1983) Biomedical image processing. Computer 16: 22-34.

16. Pizer SM, Amburn EP, Austin JD, Cromartie R, Geselowitz A, et al. (1987) Adaptative Histogram Equalization and its variations. Computer Vision, Graphics, and Image Processing 39: 355-368.

# 25

# Stability Studies of Ternary Mixtures Containing Fosaprepitant, Dexamethasone, Ondansetron and Granisetron Used in Clinical Practice

Ana Moya Gil[1]*, María Amparo Martínez Gómez[2], Elena Gras Colomer[1], Begoña Porta Oltra[1] and Mónica Climente Martí[1]

[1]Pharmacy Service, Hospital Universitario Doctor Peset, Avda. Gaspar Aguilar 90, Valencia, Spain
[2]Foundation for the Promotion of Health Biomedical Research and Valencia (FISABIO), C/ Micer Masco 31, Valencia, Spain

## Abstract

The use of a combination of $5HT_3$ receptor antagonist, a NK-1 receptor antagonist and dexamethasone has been classified to be state of the art in patients receiving highly as well as moderately emetogenic chemotherapy like cisplatin and anthracyclines. The administration of the ad-hoc admixture of fosaprepitant, dexamethasone and ondansetron (FDO) or granisetron (FDG) in the same IV infusion solution will improve the management of ambulatory procedures related to reducing administration time and number of administered intravenous preparations. All this would improve patient safety and comfort. In order to guarantee security of patients and efficacy of treatment, information about physico-chemical stability of both ternary mixtures at concentrations used in routine clinical practice and at different conditions of storage is needed. In this study, physico-chemical stability of ternary mixtures of fosaprepitant (150 mg), dexamethasone (8 mg) and ondansetron (8 mg) or granisetron (3 mg) in 50, 100 and 250 ml of 0.9 g/dl NaCl at room temperature/refrigerated and protective from/exposed to light has been evaluated. An HPLC method has been developed and validated according to International Conference on Harmonization guidelines to evaluate chemical stability of drugs in mixtures simultaneously. Physical stability study has been carried out by visual inspection, pH measure and gravimetry to control evaporation. The results shown in this paper represent the first evidence of the physico-chemical stability of both ternary mixtures used in clinical practice at different conditions of storage. The ternary mixtures of FDG in 100 and 250 ml of 0.9 g/dl NaCl are physico-chemical stable for 15 days at room temperature and refrigerated and exposed to and protected from light; mixtures in 50 ml are physico-chemical stable for 6 days. The ternary mixtures of FDO in 50, 100 and 250 ml of 0.9 g/dl NaCl are physico-chemical stable for 15 days at both conditions of temperature and light.

**Keywords:** Physico-chemical stability; Fosaprepitant; Dexamethasone; Ondansetron; Granisetron; HPLC

## Introduction

Chemotherapy-induced nausea and vomiting (CINV) is a frequent and potentially treatment-limiting complication of cancer therapy, which is associated with a significant deterioration in quality of life. The emetogenicity depends on factors related to the drug, the combination of antineoplastic drugs administered pharmacotherapy scheme, as well as factors related to the patient. The temporal pattern of appearance of emesis after chemotherapy can be acute or late. Acute emesis occurs within the first 24 hours after chemotherapy; it is the most intense emesis and is related to the release of serotonin and $5\text{-}HT_3$ receptors. The late emesis happens after the first 24 hours post-chemotherapy, there is evidence that the mechanisms involved begin at 8 hours and related to substance P and the NK1 receptors. With the correct use of antiemetic drugs, CINV can be prevented in almost 70%, and even up to 80% of patients [1].

5-hydroxytryptamine 3 ($5\text{-}HT_3$)- receptor antagonists are now the standard therapy for preventing CINV, because emesis is caused by stimulation of $5\text{-}HT_3$ receptors located on vagal afferents by serotonin released from enterochromaffin cells in the small intestine. The first-generation $5\text{-}HT_3$-receptor antagonists, ondansetron (OND), granisetron (GRA), dolasetron, and tropisetron, show considerable efficacy in preventing acute CINV, with acute responses for single agents ranging from 50% to 70%. However, acute responses are further increased when used in combination with the glucocorticoid dexamethasone (DEX) [2]. More recently, understanding the importance of the neurokinin-1 (NK-1) receptor in the emetic pathway in late emesis has led to the development of a new class of effective antiemetics, the NK-1 receptor antagonist (aprepitant, fosaprepitant (FOS)) [3].

According to currently available Multinational Association of Supportive Care in Cancer (MASCC), European Society for Medical Oncology (ESMO) and American Society of Clinical Oncology (ASCO) guidelines, the use of a combination of $5HT_3$ receptor antagonist, a NK-1 receptor antagonist and DEX has been classified to be state of the art in patients receiving highly as well as moderately emetogenic chemotherapy [4,5]. In everyday clinical practice, and ad-hoc admixture of antiemetic drugs in the same IV infusion solution is often highly preferred to accelerate the management of ambulatory procedures, related to reducing administration time and number of administered intravenous preparations. All this would improve patient safety and comfort. So, physico-chemical stability data of ternary mixtures of $5HT_3$ receptor antagonist/NK-1 receptor antagonist/ DEX are needed before to avoid unexpected drug loss or even precipitation.

In data base Stabilis [6], information about physico-chemical stability of single solutions of FOS, DEX, GRA or OND and the binary mixtures DEX/OND and DEX/GRA is available. Furthermore, Sun et al. evaluate physical compatibility of ternary mixtures containing FOS,

---

*Corresponding author: Ana Moya Gil, Pharmacy Service, Hospital Universitario Doctor Peset, Avda. Gaspar Aguilar 90, 46017, Valencia, Spain
E-mail: moya_anagil@gva.es

DEX and GRA (FDG) and FOS, DEX and OND (FDO) [3]. However, until now and to our knowledge, there are no published articles that evaluate chemical and physical stability of both ternary mixtures.

So, the development of an appropriately designed stability study, following the Pharmacopoeia guidelines and the recommendations of the International Committee on Harmonization (ICH), including chemical and physical stability, will allow know stability data of ternary mixtures of FDG or FDO at the concentration levels and storage conditions used in clinical practice [7-9].

Physical stability of drugs in mixture is usually evaluated by measurement of pH of mixture, visual inspection of colour changes, cloudiness (turbidity) and/or precipitation and gravimetry to analyze the water loss measurement [10,11]. Evaluation of chemical stability consists in quantifying concentration of each drug in mixture at different time in order to detect degradation of drugs; in this sense, High Performance Liquid Chromatography (HPLC) has been widely employed due to its high-resolution capacity, sensitivity and specificity [9]. In case of mixtures of more than one drug, the development of a chromatographic method that allows simultaneously quantify all drugs in mixtures should be carried out.

Therefore, the aim of this study was to determine the physico-chemical stability of the ternary mixtures FDG and FDO in 50, 100 and 250 ml of 0.9 g/dl sodium chloride (NaCl) under different storage conditions of light and temperature.

## Experimental

### Instrumentation and chromatographic conditions

An Agilent Technologies 1100 liquid chromatograph with a quaternary pump, a diode array detector (DAD), a thermostated column compartment, an autosampler and a HP Compaq computer equipped with Agilent-Chemstation software was used. 10 µL of each solution was injected, by duplicate, into the chromatograph through a Rheodyne valve (Cotati, CA), with a 20 µ loop. Kromasil C18 column of 5 µm particle size (250 × 4.6 mm inner diameter, Análisis Vínicos, Spain) was used. Mobile phase was ortophosphoric acid (0.1%)-acetonitrile (50:50, v/v); the flow rate was set to 0.8 ml/min, temperature to 20°C and detection to 254 nm. The column was equilibrated with mobile phase for 30 min prior to injection of the drug solution. Ortophosphoric acid and acetonitrile solutions were previously vacuum-filtered through 0.45 µm nylon membranes (Micron Separations, Westboro, MA) and sonicated prior to HPLC analysis.

A pH meter (model 3510, Jenway, UK) connected to a glass pH-electrode and an analytical balance (GF-200, A&D Instruments Ltd, UK) were used to measure the pH and weight, respectively.

### Chemicals

For the preparation of mixtures, FOS (Ivemend 150 mg; Merck Sharp & Dohme, Spain), DEX (Fortecortin 40 mg/5 ml; Merck, Spain), OND (8 mg/4 ml; Normon, Spain) and GRA (3 mg/3 ml; Genéricos Españoles Laboratorios, Spain) were used. 0.9 g/dl NaCl intravenous infusion BP Viaflo 50, 100 and 250 ml were purchased from Baxter (Spain), too.

Acetonitrile (Scharlab SL, Spain), ortophosphoric acid (Fluka Analytical, Sweden) and sterile water for injection (Grifols, Spain) were used to prepare the mobile phase used in chromatographic analysis.

### Solutions preparation for calibration curves

For each drug, calibration curves were done with six standards prepared by making serial dilutions with 0.9 g/dl NaCl from commercial formulations from DEX, OND, GRA and FOS solution obtained after reconstitution of drug powder content in commercially vial (Ivemend) with 5 ml of 0.9 g/dl NaCl. The concentration range assayed for each drug was: FOS, 0.15-4.00 mg/ml; DEX and OND, 0.016-0.300 mg/ml; GRA, 0.005-0.100 mg/ml.

### Mixtures preparation and storage conditions

24 mixtures were prepared in the same way as those prepared for hospital clinical practice following the guidelines of "Pharmaceutical Compounding: Sterile Preparations" of the United States Pharmacopeia (USP) [12-14]. 12 mixtures contained 150 mg of FOS, 8 mg of DEX and 8 mg of OND in 50 ml (mixtures 1-4), 100 ml (mixtures 5-8) and 250 ml (mixtures 9-12) of 0.9 g/dl NaCl. 12 mixtures contained 150 mg of FOS, 8 mg of DEX and 3 mg of GRA in 50 ml (mixtures 13-16), 100 ml (mixtures 17-20) and 250 ml (mixtures 21-24) of 0.9 g/dl NaCl.

For mixtures with the same volume of NaCl, two mixtures were introduced in protective bags for ambient light (PL); one of them was stored at room temperature (27.6°C, IC95% 26.6 to 28.7°C; RT) and the other at 5.8°C (IC95% 3.4 to 8.1°C; F). The other two mixtures were exposed to light (L) and one was stored at RT and the other at F.

Time of study was 15 days, and the assays were performed every 24 hours, and in the first 24 hours were performed at 3 hours, 6 hours and 12 hours, too Tables 1 and 2 summarizes the storage conditions of each mixture assayed.

### Chromatographic method validation

The developed chromatographic method was validated with each drug for linearity, specificity, accuracy, precision, limit of detection and limit of quantification, in accordance with ICH guidelines [15]. The chromatograms were evaluated on the basis of the peak area of each drug. So, for each drug were evaluated the following parameters:

**Linearity:** The graph mean absorbance (y-axis) versus concentration (x-axis) was plotted and correlation coefficient (r), y-intercept and slope of regression line were estimated.

**Specificity:** The specificity of the method was ascertained by evaluating the presence of interferences at the retention time of drug.

| Mixture | Storage condition | | $V_{NaCl}$ (ml) | Concentration (mg/ml) | | |
|---|---|---|---|---|---|---|
| | Light | T | | FOS | DEX | OND |
| 1 | L | F | 50 | 3.000 | 0.160 | 0.160 |
| 2 | L | RT | | | | |
| 3 | PL | F | | | | |
| 4 | PL | RT | | | | |
| 5 | L | F | 100 | 1.500 | 0.080 | 0.080 |
| 6 | L | RT | | | | |
| 7 | PL | F | | | | |
| 8 | PL | RT | | | | |
| 9 | L | F | 250 | 0.600 | 0.032 | 0.032 |
| 10 | L | RT | | | | |
| 11 | PL | F | | | | |
| 12 | PL | RT | | | | |

FOS: Fosaprepitant    DEX: Dexamethasone    OND: Ondansetron
T: Temperature    L: Exposition to ambient light    PL: Protection from light
F: Refrigerated    RT: Room temperature
$V_{NaCl}$: Volume of 0.9 g/dl NaCl

**Table 1:** Concentration and conditions of storage of mixtures of FDO.

| Mixture | Storage condition | | $V_{NaCl}$ (ml) | Concentration (mg/ml) | | |
|---|---|---|---|---|---|---|
| | Light | T | | FOS | DEX | GRA |
| 13 | L | F | 50 | 3.000 | 0.160 | 0.060 |
| 14 | L | RT | | | | |
| 15 | PL | F | | | | |
| 16 | PL | RT | | | | |
| 17 | L | F | 100 | 1.500 | 0.080 | 0.030 |
| 18 | L | RT | | | | |
| 19 | PL | F | | | | |
| 20 | PL | RT | | | | |
| 21 | L | F | 250 | 0.600 | 0.032 | 0.012 |
| 22 | L | RT | | | | |
| 23 | PL | F | | | | |
| 24 | PL | RT | | | | |

FOS: Fosaprepitant   DEX: Dexamethasone       GRA: Granisetron
T: Temperature   L: Exposition to ambient light   PL: Protection from light
F: Refrigerated   RT: Room temperature
$V_{NaCl}$: Volume of 0.9 g/dl NaCl

**Table 2:** Concentration and conditions of storage of mixtures of FDG.

**Accuracy (% Recovery):** The accuracy of the method was determined by calculating recoveries by method of standard additions; known amount of drug (0%, 50%, 100%, 150%) were added to a pre-quantified sample solution and was determined.

**Method precision (Repeatability):** Three standards of drug were analyzed six times and relative standard deviation (%RSD) was calculated for each concentration level.

**Intermediate precision (Reproducibility):** Intra-day precision was determined by analyzing three standards for three times in the same day and inter-day precision, by analyzing three standards daily for five days.

**Limit of Detection (LOD) and Limit of Quantification (LOQ):** LOD and LOQ were calculated using following equation: $LOD = 3 \cdot \sigma / S$, $LOQ = 10 \cdot \sigma / S$; where $\sigma$ is the standard deviation of y-intercepts of regression lines and S is the slope of the calibration curve.

### Physical and chemical stability assessment

Physical compatibility was evaluated daily by: (1) visual inspection of the mixtures for colour changes, cloudiness (turbidity) and/or precipitation. Incompatibility: appearance of some parameter; (2) loss of volume due to evaporation by gravimetry, weighting each mixture before and after extracting aliquot to HPLC analysis. Incompatibility: loss of weight $\geq$ 5%; (3) pH of mixtures, measured each two days in an aliquot of 2.5 ml removed from each mixture by inserting into the bag injection port, previous homogenization by double inversion. Incompatibility: variation of pH > 20%.

Chemical stability was evaluated daily determining the concentration of each drug in the mixture by HPLC. For this purpose, each day, an aliquot of 150 μl was removed from each mixture, in the same way that for the analysis of pH. For each drug, the pair data concentration and time were adjusted, if it was possible, to a zero- (equation 1) or first-order kinetic equation (equation 2):

$C = C_0 - k_0 t$ (equation 1)

$\ln C = \ln C_0 - k_1 t$ (equation 2)

being, C, drug concentration at a specific time; $C_0$, drug concentration at t=0; $K_0$, the zero-order degradation rate constant and $K_1$, the first-order degradation rate constant.

For each drug in mixture, data were reported as a percentage compared with 100% (concentration determined just after preparation); the parameter $T_{90}$ (time at witch remaining drug concentration is of 90%) was calculated by: (1) using equation 3 and 4, depending on the order of reaction or (2) if adjust to kinetic equation was not possible, it was considered the maximum time at which remaining drug concentration was $\geq$ 90%. Caducity of mixture was established by considering the lowest T90 value of the three drugs in the mixture, maximum study time (15 days).

$T_{90} = 0.1 \cdot C_0 / k_0$ (equation 3)

$T_{90} = 0.105 / k_1$ (equation 4)

## Results and Discussion

The present study evaluated the effects of concomitant dilution and storage of FOS with DEX as corticosteroid and OND or GRA as 5-HT$_3$ antagonists. The reason for evaluating physical and chemical stability of the ternary mixtures FDO and FDG was based on the fact that: (1) ad hoc admixtures of these antiemetic drugs in the same IV infusion solution could alleviate the everyday clinical practice particularly in ambulatory settings; (2) both ternary mixtures may represent two potent antiemetic regimens for first-line treatment in case of highly emetogenic chemotherapy as well as salvage regimens in moderately emetogenic chemotherapy; (3) only compatibility information of both mixtures was publicated.

### Optimization and validation of the chromatographic method

To optimize the chromatographic conditions a C18 HPLC column, ortophosphoric acid solution (0.1%) and acetonitrile mixture were found to be the best stationary phase and mobile phase combination to have symmetrical and well-resolved peaks of FOS, DEX and OND or GRA, simultaneously, in 0.9 g/dl NaCl mixtures. The same method was used to analyze both ternary mixtures. The total runtime for the analysis was 5 min and the retention times of drugs were: FOS, 2.4 min; OND, 2.9 min; GRA, 2.9 min and DEX, 4.3 min.

### Chromatographic method validation

In the experimental conditions indicated, the analytical performance parameters suggested by ICH guidelines were evaluated: linearity, specificity, accuracy, method precision, intermediate precision, LOD and LOQ. Because of the simplicity of the procedure, no internal standard was needed. Table 3 shows the values of some of the parameters obtained for each drug. Furthermore, specificity was adequate since no interferences were observed at retention time of drugs. So, since all the criteria were acceptable according to ICH

| Parameters | FOS | DEX | OND | GRA |
|---|---|---|---|---|
| r | 0.9907 | 0.9984 | 0.9991 | 0.9999 |
| Accuracy (%) | 95.1-103.0 | 101.0-103.1 | 101.7-102.7 | 99.3-103.5 |
| Repeatability (%) | $\leq$ 1.9 | $\leq$ 6.0 | $\leq$ 3.2 | $\leq$ 4.1 |
| Intra-day precision (%) | $\leq$ 1.9 | $\leq$ 7.0 | $\leq$ 2.9 | $\leq$ 3.1 |
| Inter-day precision (%) | $\leq$ 4.0 | $\leq$ 6.0 | $\leq$ 5.6 | $\leq$ 6.2 |
| LOD (mg/ml) | 0.097 | 0.011 | 0.009 | 0.002 |
| LOQ (mg/ml) | 0.323 | 0.040 | 0.030 | 0.006 |

FOS: Fosaprepitant                   DEX: Dexamethasone   GRA: Granisetron
OND: Ondansetron   r: correlation coefficient
LOD: Limit of Detection   LOQ: Limit of Quantification

**Table 3:** Validation parameters of chromatographic method.

guidelines, the proposed chromatographic method was adequate to determine simultaneously FOS, DEX and OND or GRA in mixtures. Figures 1 and 2 show the chromatograms of mixtures of FDO and FDG in 50 ml of 0.9 g/dl NaCl, just after preparation.

## Physico-chemical stability assessment of mixtures containing FDO

At the end of the study, none of the mixtures (mixtures 1-12) showed changes in colour, precipitation or measurable losses of volume at the different storage conditions. So, all mixtures were physically stable and so, compatible, during the time of storage.

Table 4 shows initial medium pH values ($pH_0$) of mixtures and variations of pH after 7 and 15 days of storage. In mixtures PL (mixtures 3-4, 7-8, 11-12) mean of pH variations were 0.28% (IC95%, -1.02% to 1.59%) at $7^{th}$ day and 0.017% (IC95%, -1.404% to 1.438%) at $15^{th}$ day; however, in mixtures exposed to L (mixtures 1-2, 5-6, 9-10), mean of pH variations were -6.22% (IC95%, -6.87% to -5.56%) at $7^{th}$ day and -6.07% (IC95%, -6.74% to -5.40%) at $15^{th}$ day, being the lowest pH achieved 7.1 what means a reduction of 3.4% respect to blood pH (7.35-7.45). In a previous study carried out by us about physico-chemical stability of FOS in mixtures [11], a maximum decrease in pH after 15 days of storage of 13.6% was observed, being the lowest pH achieved 7.0, too; it was explained by considering the flow of $CO_2$ through polyolefin bag and the consequent acidification of solutions and no FOS degradation.

Table 5 shows the percentage of remaining drugs concentration (%RC) at the different conditions of storage at day 7 and 15. As can be observed, in all mixtures, %RC of FDO were ≥ 90% after 15 days of storage; so, it seems to be that temperature, light and drug concentration do not affect FDO chemical stability. Table 6 shows the $T_{90}$ values obtained from kinetic adjust and experimental value, being the last one the time at which %RC of all three drugs in the mixture were ≥ 90%; experimental value was in all cases of 15 days.

To sum up, mixtures containing 150 mg of FOS, 8 mg of DEX and 8 mg of OND in 50, 100 and 250 ml of 0.9 g/dl NaCl were physico-chemical stable for 15 days at different conditions of light (PL and L) and temperature (RT and F).

## Physico-chemical stability assessment of mixtures containing FDG

At the end of the study, none of the mixtures (mixtures 13-24) showed changes in colour or measurable losses of volume at the different storage conditions. Only in mixtures 13 and 16, precipitate

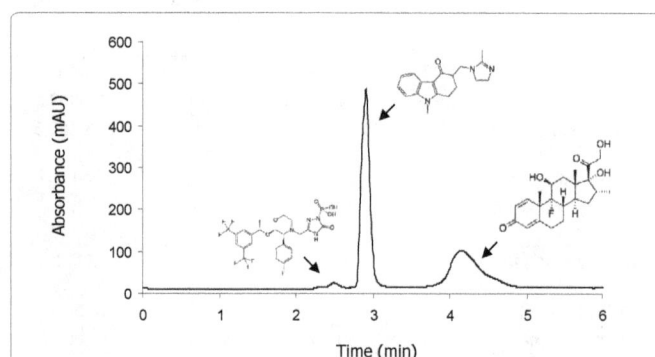

**Figure 2:** Chromatogram of mixture containing 150 mg of FOS, 8 mg of DEX and 3 mg of GRA in 50 ml 0.9 g/dl NaCl, just after preparation, at the chromatographic conditions. Peaks: FOS, 2.4 min; GRA, 2.9 min; DEX, 4.3 min.

| Mixture | $V_{NaCl}$ (ml) | Storage condition | | $pH_0$ | Variation of pH (%) | |
|---|---|---|---|---|---|---|
| | | Light | T | | $t_7$ | $t_{15}$ |
| 1 | 50 | L | F | 7.39 ± 0.27 | -5.7 | -5.1 |
| 2 | | L | RT | | -6.9 | -5.7 |
| 3 | | PL | F | | +1.4 | +2.1 |
| 4 | | PL | RT | | +2.4 | +2.5 |
| 5 | 100 | L | F | 7.32 ± 0.24 | -4.9 | -5.6 |
| 6 | | L | RT | | -6.1 | -5.9 |
| 7 | | PL | F | | +1.3 | -1.1 |
| 8 | | PL | RT | | -1.0 | -1.0 |
| 9 | 250 | L | F | 7.16 ± 0.22 | -6.7 | -7.4 |
| 10 | | L | RT | | -7.0 | -6.7 |
| 11 | | PL | F | | -0.7 | -1.1 |
| 12 | | PL | RT | | -1.7 | -1.3 |
| 13 | 50 | L | F | 8.94 ± 0.05 | -7.9 | -8.1 |
| 14 | | L | RT | | -9.2 | -8.8 |
| 15 | | PL | F | | -8.3 | -7.9 |
| 16 | | PL | RT | | -7.8 | -7.9 |
| 17 | 100 | L | F | 8.89 ± 0.04 | -10.3 | -10.3 |
| 18 | | L | RT | | -10.9 | -11.0 |
| 19 | | PL | F | | -10.7 | -9.6 |
| 20 | | PL | RT | | -11.5 | -12.6 |
| 21 | 250 | L | F | 8.55 ± 0.04 | -14.2 | -16.6 |
| 22 | | L | RT | | -16.1 | -16.5 |
| 23 | | PL | F | | -15.7 | -16.1 |
| 24 | | PL | RT | | -17.8 | -16.6 |

$pH_0$: mean initial pH ± standard deviation
$t_7$: 7 days of storage          $t_{15}$: 15 days of storage

**Table 4:** $pH_0$ and variation of pH of mixtures after 7 and 15 days of storage.

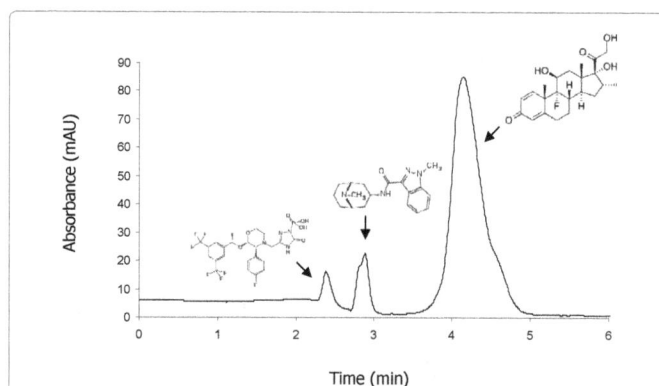

**Figure 1:** Chromatogram of mixture containing 150 mg of FOS, 8 mg of DEX and 8 mg of OND in 50 ml 0.9 g/dl NaCl, just after preparation, at the chromatographic conditions. Peaks: FOS, 2.4 min; OND, 2.9 min; DEX, 4.3 min.

| Mixture | %RC | | | | | |
| --- | --- | --- | --- | --- | --- | --- |
| | FOS | | DEX | | OND | |
| | $t_7$ | $t_{15}$ | $t_7$ | $t_{15}$ | $t_7$ | $t_{15}$ |
| 1 | 99.00 ± 3.00 | 99.00 ± 0.70 | 102.50 ± 0.50 | 100.00 ± 0.06 | 95.10 ± 0.30 | 96.00 ± 0.90 |
| 2 | 100.10 ± 0.30 | 97.80 ± 1.70 | 102.50 ± 1.50 | 100.50 ± 0.13 | 98.00 ± 1.30 | 97.00 ± 0.90 |
| 3 | 98.30 ± 0.60 | 96.40 ± 0.90 | 96.06 ± 1.22 | 98.00 ± 0.50 | 92.50 ± 0.23 | 95.20 ± 0.90 |
| 4 | 93.10 ± 0.50 | 92.30 ± 0.90 | 94.00 ± 5.00 | 98.30 ± 2.50 | 96.70 ± 0.30 | 99.60 ± 1.90 |
| 5 | 100.20 ± 0.60 | 97.60 ± 1.70 | 97.30 ± 2.50 | 99.70 ± 0.14 | 99.20 ± 1.02 | 99.10 ± 0.21 |
| 6 | 97.00 ± 5.00 | 97.50 ± 1.40 | 100.70 ± 0.30 | 99.90 ± 0.30 | 99.30 ± 0.50 | 99.40 ± 1.00 |
| 7 | 97.00 ± 6.00 | 95.00 ± 6.00 | 98.00 ± 0.03 | 96.30 ± 0.40 | 98.70 ± 0.90 | 98.10 ± 0.40 |
| 8 | 99.40 ± 0.15 | 99.30 ± 0.50 | 98.60 ± 0.90 | 97.80 ± 0.50 | 98.00 ± 3.00 | 95.70 ± 0.20 |
| 9 | 98.00 ± 3.00 | 99.10 ± 1.09 | 99.10 ± 0.80 | 98.70 ± 0.90 | 99.10 ± 1.50 | 96.90 ± 1.30 |
| 10 | 100.70 ± 1.00 | 90.00 ± 3.00 | 97.50 ± 0.10 | 97.70 ± 0.30 | 99.70 ± 0.50 | 99.60 ± 0.60 |
| 11 | 96.50 ± 0.90 | 96.60 ± 0.80 | 98.10 ± 1.12 | 96.80 ± 0.80 | 94.80 ± 0.40 | 93.60 ± 0.30 |
| 12 | 96.40 ± 1.15 | 96.30 ± 0.30 | 93.00 ± 1.19 | 94.30 ± 1.17 | 92.00 ± 5.00 | 91.90 ± 0.90 |

FOS: Fosaprepitant          DEX: Dexamethasone          OND: Ondansetron
%RC: Percentage of remaining drug concentration ± standard deviation
$t_7$: 7 days of storage          $t_{15}$: 15 days of storage

**Table 5:** Percentage of remaining concentrations of FDO in mixtures at day 7 and 15 of storage.

| Mixture | $T_{90}$ (days) | | | | | | | | |
| --- | --- | --- | --- | --- | --- | --- | --- | --- | --- |
| | FOS | | | DEX | | | OND | | |
| | Adj | | Exp | Adj | | Exp | Adj | | Exp |
| | $O_0$ | $O_1$ | | $O_0$ | $O_1$ | | $O_0$ | $O_1$ | |
| 1 | - | - | 15 | - | - | 15 | 27 | 27 | 15 |
| 2 | - | - | 15 | - | - | 15 | 53 | 55 | 15 |
| 3 | 28 | 29 | 15 | 45 | 70 | 15 | 15 | 15 | 15 |
| 4 | 12 | 12 | 15 | - | - | 15 | - | - | 15 |
| 5 | - | - | 15 | - | - | 15 | - | - | 15 |
| 6 | - | - | 15 | - | - | 15 | - | - | 15 |
| 7 | - | - | 15 | - | - | 15 | - | - | 15 |
| 8 | - | - | 15 | - | - | 15 | 22 | 22 | 15 |
| 9 | - | - | 15 | - | - | 15 | - | - | 15 |
| 10 | 19 | 20 | 15 | 51 | 53 | 15 | - | - | 15 |
| 11 | - | - | 15 | - | - | 15 | - | - | 15 |
| 12 | - | - | 15 | 17 | 17 | 15 | - | - | 15 |

FOS: Fosaprepitant     DEX: Dexamethasone OND: Ondansetron
Adj: value obtained from kinetic adjust
Exp: experimental time at which %RC was ≥ 90%
$O_0$: zero-order kinetic adjust          $O_1$: first-order kinetic adjust
%RC: Percentage of remaining drug concentration

**Table 6:** $T_{90}$ values for mixtures containing FDO assayed.

was observed by visual inspection after 7 days of storage. So, mixture 13 and 16 were physically compatible for 6 days while the rest of mixtures were compatible for 15 days.

As can be observed in Table 4, $pH_0$ mean values were higher than pH of mixtures containing FDO. Furthermore, variation of pH with time was highest and increased with increasing 0.9 g/dl NaCl volume (Table 7), being the lowest pH achieved 7.0. As has been commented for mixture FDO, acidification could be a consequence of the flow of $CO_2$ through polyolefin bag and in mixtures with highest drugs concentrations (mixtures 13-16) it could have provoked precipitation.

Regards chemical stability, after 15 days of storage, %RC of all drugs were ≥ 93.5% for all mixtures (Table 8). Kinetic adjust was not possible in mixtures and experimental $T_{90}$ values was in all cases of 15 days (Table 9).

So, mixtures containing 150 mg of FOS, 8 mg of DEX and 3 mg of GRA in 100 and 250 ml of 0.9 g/dl NaCl were physico-chemical stable for 15 days at different conditions of light (PL and L) and temperature (RT and F). Mixtures 13 and 16, in 50 ml of 0.9 g/dl NaCl, were physico-chemical stable for 6 days due to appearance of precipitate after day 7.

## Conclusion

The caducity of all-in-one admixtures containing fosaprepitant 150 mg, dexamethasone 8 mg and ondansetron 8 mg or granisetron 3 mg in 50, 100 and 250 ml of 0.9 g/dl NaCl was established at different conditions of light and temperature. The results from this paper represent the first evidence of the physico-chemical stability of both ternary mixtures FDO and FDG used in clinical practice at different drugs concentrations and conditions of storage. These results indicate that advance preparation of these ternary mixtures is possible, reducing

| Mixture | $T_{90}$ (days) | | | | | | | | |
|---|---|---|---|---|---|---|---|---|---|
| | FOS | | | DEX | | | GRA | | |
| | Adj | | Exp | Adj | | Exp | Adj | | Exp |
| | $O_0$ | $O_1$ | | $O_0$ | $O_1$ | | $O_0$ | $O_1$ | |
| 13 | - | - | 15 | - | - | 15 | - | - | 15 |
| 14 | - | - | 15 | - | - | 15 | - | - | 15 |
| 15 | - | - | 15 | - | - | 15 | - | - | 15 |
| 16 | - | - | 15 | - | - | 15 | - | - | 15 |
| 17 | - | - | 15 | - | - | 15 | - | - | 15 |
| 18 | - | - | 15 | - | - | 15 | - | - | 15 |
| 19 | - | - | 15 | - | - | 15 | - | - | 15 |
| 20 | - | - | - | - | - | 15 | - | - | 15 |
| 21 | - | - | 15 | - | - | 15 | - | - | 15 |
| 22 | - | - | 15 | - | - | 15 | - | - | 15 |
| 23 | - | - | 15 | - | - | 15 | - | - | 15 |
| 24 | - | - | 15 | - | - | 15 | - | - | 15 |

FOS:Fosaprepitant    DEX: Dexamethasone GRA: Granisetron
Adj: value obtained from kinetic adjust
Exp: experimental time at which %RC was ≥ 90%
$O_0$: zero-order kinetic adjust          $O_1$: first-order kinetic adjust
%RC: Percentage of remaining drug concentration

**Table 7:** $T_{90}$ values for mixtures containing FDG assayed.

| Time | Mean variation pH (%) (IC95%) | | |
|---|---|---|---|
| | Mixtures 13-16 | Mixtures 17-20 | Mixtures 21-24 |
| $t_7$ | -8.30 (-8.92 to -7.68) | -10.85 (-11.34 to -10.36) | -15.95 (-17.40 to -14.50) |
| $t_{15}$ | -8.18 (-8.59 to -7.76) | -10.88 (-12.13 to -9.62) | -16.45 (-16.68 to -16.22) |

$t_7$: 7 days of storage          $t_{15}$: 15 days of storage
IC95%: Confidence interval at 95%

**Table 8:** Mean variation of pH for mixtures containing FDG after 7 and 15 days of storage.

| Mixture | %RC | | | | | |
|---|---|---|---|---|---|---|
| | FOS | | DEX | | GRA | |
| | $t_7$ | $t_{15}$ | $t_7$ | $t_{15}$ | $t_7$ | $t_{15}$ |
| 13 | 98.9 ± 1.30 | 99.7 ± 0.40 | 99.7 ± 0.30 | 97.8 ± 2.40 | 99.8 ± 0.16 | 98.9 ± 0.90 |
| 14 | 98.4 ± 0.50 | 98.8 ± 1.00 | 100.3 ± 0.80 | 99.8 ± 0.13 | 98.7 ± 0.30 | 98.8 ± 0.70 |
| 15 | 99.4 ± 0.70 | 93.5 ± 0.50 | 99.5 ± 1.21 | 98.7 ± 0.10 | 98.2 ± 2.00 | 98.7 ± 2.30 |
| 16 | 97.1 ± 1.80 | 98.3 ± 0.80 | 102.2 ± 0.70 | 101. ± 0.10 | 100.3 ± 0.70 | 100.5 ± 0.60 |
| 17 | 98.6 ± 0.04 | 99.8 ± 0.30 | 99.8 ± 1.40 | 99.6 ± 0.30 | 98.9 ± 0.70 | 98.5 ± 0.40 |
| 18 | 98.4 ± 0.00 | 97.3 ± 1.50 | 99.7 ± 0.05 | 98.6 ± 0.50 | 98.7 ± 0.30 | 99.5 ± 0.19 |
| 19 | 99.1 ± 1.09 | 99.2 ± 0.21 | 99.4 ± 1.60 | 98.9 ± 0.40 | 94.1 ± 1.30 | 94.8 ± 1.60 |
| 20 | 102.0 ± 3.00 | 98.0 ± 1.02 | 101.4 ± 0.30 | 99.4 ± 1.00 | 100.0 ± 4.00 | 94.2 ± 0.12 |
| 21 | 97.0 ± 6.00 | 99.6 ± 1.50 | 99.8 ± 1.40 | 99.6 ± 0.30 | 98.6 ± 1.70 | 97.9 ± 1.30 |
| 22 | 100.2 ± 0.30 | 99.4 ± 2.00 | 98.0 ± 40 | 101.0 ± 4.0 | 96.0 ± 3.00 | 98.9 ± 1.08 |
| 23 | 95.1 ± 1.20 | 99.7 ± 1.50 | 99.4 ± 0.06 | 99.9 ± 0.30 | 99.9 ± 1.30 | 97.8 ± 0.70 |
| 24 | 100.6 ± 0.05 | 99.6 ± 1.20 | 99.4 ± 0.17 | 99.9 ± 0.20 | 98.0 ± 4.00 | 95.5 ± 0.40 |

FOS: Fosaprepitant    DEX: Dexamethasone GRA: Granisetron
%RC: Percentage of remaining drug concentration (%)
$t_7$: 7 days of storage          $t_{15}$: 15 days of storage

**Table 9:** Percentage of remaining concentrations of FDG in mixtures at day 7 and 15 of storage.

waiting times for patients and that their administration simplify the management of these treatments in terms of reducing number of preparations and improving patient safety and comfort.

### Acknowledgements

The authors acknowledge FISABIO for the financial support.

### References

1. Jordan K, Gralla R, Jahn F, Molassiotis A (2014) International antiemetic guidelines on chemotherapy induced nausea and vomiting (CINV): content and implementation in daily routine practice. Eur J Pharmacol 722: 197-202.

2. Saito M, Aogi K, Sekine I, Yoshizawa H, Yanagita Y, et al. (2009) Palonosetron plus dexamethasone versus granisetron plus dexamethasone for prevention of nausea and vomiting during chemotherapy: a double-blind, double-dummy, randomised, comparative phase III trial. Lancet Oncol 10: 115-124.

3. Sun S, Schaller J, Placek J, Duersch B (2013) Compatibility of intravenous fosaprepitant with intravenous 5-HT3 antagonists and corticosteroids. Cancer Chemother Pharmacol 72: 509-513.

4. MASCC/ESMO (2013) Antiemetic Guideline [http://www.mascc.org/assets/Guidelines-Tools/mascc_antiemetic_english_2014.pdf]    MASCC/ESMO. Accessed on: July 2016.

5. Basch E, Prestrud AA, Hesketh PJ, Kris MG, Feyer PC, et al. (2011) Antiemetics: American Society of Clinical Oncology clinical practice guideline update. J Clin Oncol 29: 4189-4198.

6. Stabilis (2015) Stability and compatibility of drugs [http://www.stabilis.org] Stabilis. Accessed on: July 2016.

7. US Department of Health and Human Services FDA Center for Drug Evaluation and Research; US Department of Health and Human Services FDA Center for Biologics Evaluation and Research; US Department of Health and Human Services FDA Center for Devices and Radiological Health (2006) Guidance for industry: patient-reported outcome measures: use in medical product development to support labeling claims: draft guidance. Health Qual Life Outcomes 4: 79.

8. International Conference on Harmonisation of Technical Requirements for Registration of Pharmaceutical for Human Use (ICH) (2003) ICH Harmonised Tripartite Guideline Stability Testing of New Drug Substances and Products (Q1A-R2) [http://www.ich.org/fileadmin/Public_Web_Site/ICH_Products/ Guidelines/Quality/Q1A_R2/Step4/Q1A_R2__Guideline.pdf] International Conference on Harmonisation of Technical Requirements for Registration of Pharmaceutical for Human Use. Accessed on: February 2016.

9. Bakshi M, Singh S (2002) Development of validated stability-indicating assay methods--critical review. J Pharm Biomed Anal 28: 1011-1040.

10. Gómez MA, Arenas VJ, Sanjuán MM, Hernández MJ, Almenar CB, et al. (2007) Stability studies of binary mixtures of haloperidol and/or midazolam with other drugs for parenteral administration. J Palliat Med 10: 1306-1311.

11. Martinez MA, Moya A, Porta B, Climente M (2014) Physico-Chemical Stability of Mixtures of Fosaprepitant used in Clinical Practice. J Anal Bioanal Tech 5: 2-6.

12. The United States Pharmacopeia (USP) (2015) Pharmaceutical compounding: sterile preparations [http://www.usp.org/sites/default/files/usp_pdf/EN/ USPNF/usp-gc-797-proposed-revisions-sep-2015.pdf] The United States Pharmacopeia (USP). Accessed on: February 2016.

13. The Spanish National Health Service, Ministry of Health, Social Services and Equality (2014) Guide to good practices in preparing medicines in hospital pharmacy services. [http://www.msssi.gob.es/profesionales/farmacia/ documentacion.htm] The Spanish National Health Service. Accessed on: June 2014.

14. Pharmacopoeia Official (2004) Rules of Good Preparation of Medicines in Pharmacy [http://www.fog.it/fogliani/giancarlo/normebp.htm] Pharmacopoeia Official. Accessed on: February 2016.

15. ICH (2005) Harmonised Tripartite Guideline on Validation of analytical procedures: text and methodology ICH Q2(R1) [http://www.ich.org] ICH. Accessed on: June 2014.

# 26

# Raman Microspectroscopy Demonstrates Alterations in Human Mandibular Bone after Radiotherapy

Singh SP[1,2], Parviainen I[1], Dekker H[3], Schulten EAJM[3], Ten Bruggenkate CM[3], Bravenboer N[3], Mikkonen JJ[2], Turunen MJ[4], Koistinen AP[2] and Kullaa AM[1,5,6*]

[1]Institute of Dentistry, University of Eastern Finland, Kuopio Campus, Yliopistonranta 1, Kuopio, Finland
[2]SIB Labs, University of Eastern Finland, Yliopistonranta 1, Kuopio, Finland
[3]Department of Oral and Maxillofacial Surgery and Oral Pathology, VU University Medical Center/Academic Centre for Dentistry Amsterdam (ACTA), Amsterdam, The Netherlands
[4]Department of Applied Physics, Faculty of Science and Forestry, University of Eastern Finland, Kuopio, Finland
[5]Research Group of Oral Health Sciences, Faculty of Medicine, University of Oulu, Oulu, Finland
[6]Educational Dental Clinic, Kuopio University Hospital, Kuopio, Finland

## Abstract

Quality and alterations in the biochemical composition of bones used for dental implantation after radiotherapy in cancer patients is always a critical and debatable factor. Clinically the irradiated bone is similar to control bone. The aim of this study was to verify any compositional alterations in human mandible bone after irradiation using Raman microspectroscopy. A total of 36 bone biopsies (21-control, 4-cancer and 11-irradiated) were investigated. Data acquisition points were determined under histopathological supervision. Both mineral and matrix constituents were analyzed by computing area associated with of phosphate (958 cm$^{-1}$), carbonate (1070 cm$^{-1}$) and matrix (amide I) bands. Unpaired Student's t-test was employed to measure level of significance. Absolute mineral contents (phosphate and carbonate) were highest in cancerous specimens. Spectral profile and band-intensity calculations suggest proximity of irradiated specimens with control specimens. Significant differences in both matrix and mineral contents were observed when control/irradiated samples were compared against cancerous specimens. However, no significant differences were observed between control and irradiated groups. Irradiated bone is similar to control and cause of implant loss could be related to osteocytes of the surrounding tissue.

**Keywords:** Raman spectroscopy; Oral cancer; Mandible; Radiotherapy; Bone

## Introduction

Lower disease free survival rates associated with oral cancers have been primarily attributed to late detection and postoperative complications [1]. Following successful diagnosis treatment of oral cancer subjects often include surgery followed by chemo/radiotherapy or combination of both [1,2]. Oral rehabilitation of cancer patients is a meaningful procedure to increase the quality of life [2]. Implant therapy is widely performed to improve masticatory function after tumour surgery. In oral cancer patients, tumour resection is usually combined with irradiation, which locally impairs bone quality and weakens the bone density [2,3]. Radiotherapy of the planned implant site is known to be an important factor in the etiology of implant loss [2-4]. Bone tissue exhibits numerous changes because of radiation, such as diminished vascularization, reduced remodeling capacity and increased risk of osteoradionecrosis [5-7]. Other side effects of radiation therapy of oral cavity area may include severe complications such as xerostomia (drying of the mucsoa), mucositis, reduced mouth opening etc. [5].

Even though radiotherapy plays an important role in the treatment of head and neck malignancies, its concomitant effects may have a major impact on the final outcome of oral rehabilitation. Placing dental implants in essential irradiated tissue is a clinical challenge: peri-implantitis and osteoradionecrosis (ORN) are frequently observed, resulting in implant loss in up to 22% of patients [5-7]. Therefore it is pertinent to explore possible prognostic factors which can predict the occurrence of peri-implantitis or implant loss preferably in a non-invasive manner.

Variety of invasive and non-invasive techniques can be utilized to assess radiation induced effects. Histopathology is considered as gold standard, but it is invasive and repetitive sampling for response monitoring is a major concern. Micro-computed tomography (micro-CT) is a non-invasive technique that can be used for bone analysis. It can provide superficial information such as micro architecture. However, minor compositional alterations such collagen mineralisation is difficult to monitor. Optical spectroscopy methods offer an alternate approach for bone analysis. Methods based on fluorescence, infra-red and Raman spectroscopy are being widely explored for non-invasive disease diagnosis [8-11]. Greatest benefit of these techniques lies in their ability to provide objective and detailed information about biochemical changes associated with disease onset or therapeutic interventions in a non-invasive manner and within short period of time. Among these, diagnosis based on Raman effect is being considered as clinically implementable as it is not influenced with water and requires no sample preparation [10,11]. Numbers of studies have successfully demonstrated its efficacy in disease diagnosis and therapeutic response monitoring [11]. In case of bone specimens

*Corresponding author: Arja Kullaa, Institute of Dentistry, University of Eastern Finland, Kuopio campus, Yliopistonranta 1, Kuopio, Finland
E-mail: arja.kullaa@uef.fi/Arja.Kullaa@oulu.fi

it offers additional advantages over existing methods as a snapshot of variations in different biochemical parameters such as mineral, crystallinity, carbonate and collagen can be obtained in an objective and non-invasive manner [12-14]. These parameters can serve as direct or indirect representative of the progress of therapeutic interventions and mechanical competence of the bone tissue. Raman spectroscopy has been utilized to study different bone related abnormalities such as osteoporosis, fracture risk, metastasis and drug induced reversal of bone resorption [14,15].

The changes in composition, mechanical properties and the remodelling rate of the bone are critical factors for dental implantation after radiotherapy in cancer patients. An exact understanding of the features of bone quality and bone formation is important in clinical practice for an optimal surgical technique. Irradiated bone presents a challenging environment for implant placement. Implant loss or osteoradionecrosis are common post-treatment complications in oral cancer patients. The reason behind this debatable and can be attributed to either radiation induced changes in the bone or changes in surrounding tissues. The present study aims at determining minor alterations in the irradiated bone using Raman microspectroscopy.

## Materials and Methods

### Bone specimens

A total of 36 bone samples obtained during the dental implant surgery were used. Biopsies of the alveolar bone of the mandible approximately $10 \times 3.5$ mm were obtained with trephine drills from 4 patients with non-radiated oral cancer (OSCC), from 11 patients after radiotherapy and from 21 healthy controls. Bone specimens were fixed in 70% ethanol.

All patients gave their consent to participate in the study, and this work was approved by the ethical committee (Medisch Ethische Toetsingscommissie, VU University medical center, Amsterdam, The Netherlands; 2011/220).

### Preparation of specimens

The undecalcified bone specimens were dehydrated in increasing concentrations of ethanol and embedded into Poly (methyl methacrylate) (PMMA). Thereafter, thin sections were cut using a microtome (Polycut S; Reichert-Jung, Wien, Austria). Fresh surface of bone revealed from PMMA block were used for acquiring Raman spectra. For histological examination 10 μm sections were stained using standard protocols with Masson-Goldner trichrome. An optical microscope (AxioImager M2; Carl Zeiss GmbH, Jena, Germany) was used for histological study.

### Raman microspectroscopy

Remaining bone blocks were used for acquiring spectra with a dispersive Raman microscope (Senterra 200LX, Bruker Optics GmbH, Ettlingen, Germany). The wavelength of 785 nm at 100 mW power (source) was used for excitation. A 20X objective (NA-0.5) was chosen to minimize polarization effects. Three point measurements with an exposure time of 60 s and 5 co-additions were performed. All the data acquisition points were selected in correlation with histopathology and under pathological supervision. The spectra were acquired for 4000-127 cm$^{-1}$ range. On z-axis the beam was focused ~10 μm below the surface to ensure spectral acquisition only from bone sections. A background spectrum of the embedding medium i.e., PMMA was also obtained under similar conditions. Both background and bone spectra were interpolated to the finger-print range (1800-800 cm$^{-1}$)

and cosmic peaks were removed. This was followed by normalization of bone spectra using PMMA spectrum. Subsequently to minimize the influence of the background the embedding medium spectrum was mathematically subtracted from the bone spectra [16-18]. Baseline corrections to remove the fluorescence background were performed by fitting a 5$^{th}$ order polynomial function.

### Data analysis

All the data pre-processing and analysis were performed using MATLAB (MATLAB 7.5.0, The Mathworks, Inc., Natick, MA) using locally written scripts. Mean spectrum for control, irradiated and cancerous specimens was calculated by averaging the variations at X axis keeping the Y axis constant. Curve fitting was performed to compute area associated with Raman bands using a custom MATLAB based in-house program. In this method, first locations of sub-peaks were identified by minima in a second derivative spectrum. The sum of the squared differences between observed and computed spectra are minimized to obtain the best fit. These peaks were modelled using Gaussian function and areas of selected bands were measured for band intensity. Different parameters shown in Table 1 were computed using this algorithm. Briefly, mineral components were computed by intensity of Phosphate (958 cm$^{-1}$) and Carbonate bands (1070 cm$^{-1}$) [14,19]. Mineral crystallinity, inversely proportional to full width at half maximum (FWHM) of the phosphate band was computed [14,19]. Mineral to matrix ratio was calculated by dividing intensity of phosphate band by amide I band (1660 cm$^{-1}$) [14,19]. Carbonate substitution rate was determined by generating carbonate to phosphate ratio [14,19]. The data are expressed as the mean ± standard deviation (SD), and statistical comparisons were performed with unpaired Student's t-test (Graph Pad Prism, version 6.1). $p < 0.05$ was considered significant, $p < 0.01$ as highly significant, and $p < 0.001$ as very highly significant.

## Results and Discussion

Radiation-induced effects on structural and biochemical composition in human mandibular bone was explored with histopathology and Raman spectroscopy. In the following sub-sections a summary of important findings of the study is provided.

### Histological features of control, irradiated and unradiated cancerous bones

Representative histological sections of control, irradiated and cancerous bone specimens is shown in Figure 1. In control bone, osteoblasts and newly formed bone are recognizable (Figure 1A) indicating the capacity of bone remodeling. Previous studies have shown that bone irradiation is characterized by loss of osteocytes and lamellar structure [20]. As can be seen from Figure 1B, irradiated specimens exhibit significant amount of empty lacunae as a sign of

| S No | Raman Parameter | Control (40) | Cancer (19) | Irradiated (26) |
|------|-----------------|--------------|-------------|-----------------|
| 1 | Phosphate content | 4.25 ± 0.09 | 4.32 ± 0.10 | 4.27 ± 0.04 |
| 2 | Carbonate content | 1.00 ± 0.17 | 1.16 ± 0.15 | 1.02 ± 0.17 |
| 3 | Carbonate substitution | 0.23 ± 0.04 | 0.27 ± 0.03 | 0.24 ± 0.04 |
| 4 | Mineral Crystallinity | 0.039 ± 0.001 | 0.038 ± 0.0008 | 0.039 ± 0.001 |
| 5 | Mineral to matrix ratio-Phosphate | 2.49 ± 0.38 | 2.27 ± 0.30 | 2.46 ± 0.31 |
| 6 | Mineral to matrix ratio-Carbonate | 0.59 ± 0.15 | 0.62 ± 0.11 | 0.59 ± 0.10 |

Table 1: Summary of different biochemical parameters analyzed with Raman spectroscopy. Data is presented as mean ± standard deviation. Numbers in bracket are number of spectra used in each case.

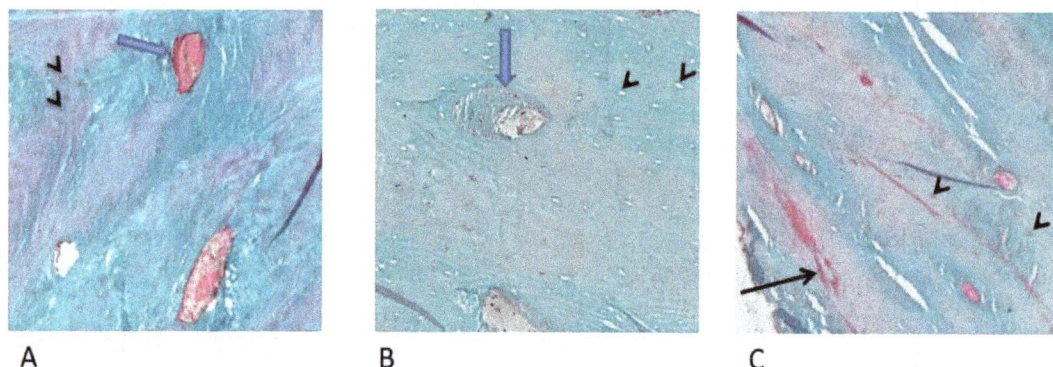

**Figure 1:** Representative photomicrographs showing histological changes of the alveolar bone of the mandible.
A) Control: Arrow heads represent osteocytes in their lacunae. The detection of Haversian canals (arrow) within the newly formed bone indicates a regular bone formation in the bone.
B) Irradiated: Empty osteocytic lacunae are discernible (arrowheads). The diameter of Haversian canals (arrow) smaller than in controls.
C) Cancer: Arrow heads represent osteocytes, but also empty lacunae are discernible.

osteocyte deaths. The bone of cancer patients without radiotherapy shows newly formed bone, but the density of osteocytes has decreased (Figure 1C). Corroborating earlier observations, reduction in vascularization and apoptosis of bone cells were also observed in irradiated specimens [21-24].

## Raman spectral features of control, irradiated and unradiated cancerous bones

Bone is a biological tissue with complex molecular structure. It can be divided into three parts. First is the organic component which mainly consists of collagen and provides bone with its shape and form. The hydroxyapatite (HA) comprises most of the second part which provides bone its strength and rigidity. Third part is comprised of small blood vessels, cells and water etc. Raman spectroscopy can provide detailed information about biochemical composition in a non-destructive manner under a very short span of time. It has been utilized in orthopaedic research for studying changes associated with disease, fracture and aging [12-15]. Mean Raman spectra of control, cancerous and irradiated specimens in 1800-800 cm$^{-1}$ range are shown in Figure 2. Major spectral features can be assigned to bone mineral and matrix components. The major inorganic constituent of bone are present in form of hydroxyapatite ($Ca_{10}[PO_4]_6[OH]_2$) crystals between the collagen fibers. They also contain carbonate and hydrogen phosphate groups (~5-8%). Raman bands around 958 and 1070 cm$^{-1}$ correspond to primary vibrational modes of phosphate ($v_1PO_4^{3-}$) and carbonate ($CO_3^{2-}$), respectively, the major mineral constituent of bone [25]. The origin of phosphate band at 958 cm$^{-1}$ has been assigned to symmetrical stretching of $v1$ band of $PO_4^{3-}$. This band is influenced to a minor extent by environmental factors hence it is considered as the most appropriate band to evaluate phosphate level among the four vibrational modes of $PO_4^{3-}$ [25,26]. Intensity of both carbonate and phosphate bands were highest in cancerous specimens, suggesting higher level of mineralization with respect to control or irradiated specimens. In addition to this, blue-shift was also observed in $PO_4^{3-}$ band of the cancer spectrum with respect to control and irradiated spectrum. Earlier studies have shown that positioning of the phosphate band is influenced by concentrations of carbonate and monohydrogen phosphate ($HPO_4^{2-}$) [27]. In the present study there is a possibility that in cancerous specimens, due to large number of dividing cells, might cause increase in $HPO_4^{2-}$ content by newly deposited mineral which in turn can influence the position of phosphate band. The organic

constituent of the bone mainly consists of collagen in a dense fibrous structural arrangement. Major bands related to the matrix components include 1003 cm$^{-1}$ (ring vibrational modes of amino acid phenylalanine), bands in amide III (collagen and protein structural changes), 1447 cm$^{-1}$ (stretching modes of CH groups of lipids and proteins) and 1660 cm$^{-1}$ (amide I, collagen and protein content) [28]. Variations in term of relative intensity were observed in all these bands. These differences were further explored by computing band intensity and ratios using curve-fitting methods and the findings are discussed below.

## Curve fitting analysis of Raman spectra

Variation among different parameters is shown in Figure 3A-3F and Table 2. Phosphate and carbonate contents in all three conditions i.e., control irradiated and cancers were measured by computing intensity associated with 958 and 1070 cm$^{-1}$ band, respectively. Highest phosphate content was observed in cancerous specimens followed by irradiated and control samples (Table 1). This can be due to high proliferation and deposition of new minerals by cancerous cells. Unpaired Student's t-test was used to evaluate significance of difference between the groups. As shown in Figure 3A and Table 2, differences between control and cancerous specimens and cancerous and irradiated specimens were significant ($p=0.0084$ and $p=0.0225$). However, no significant difference was observed between control and irradiated specimens. Similar trend was observed for carbonate bands (Figure 3B). Intensity of this band was highest in cancerous specimens and it was significantly different from control ($p=0.0009$) and irradiated ($p=0.0064$) samples, Table 2. No significant difference was observed between control and irradiated specimens, Table 2.

In case of bone specimens, absolute intensities of individual bands are influenced by variation in cross-section and losses due to elastic scattering therefore it is advised to study the band ratios [29]. As mentioned earlier, minerals in bone are present in hydroxyapatite crystal form. It has been shown that with time these crystals can undergo carbonate substitutions either by hydroxide (type A) or phosphate (type B) site. This substitution rate has been correlated with bone maturity and aging. Carbonate substitution rates can be computed by peak area ratio of carbonate and phosphate bands [14,19]. Therefore, in the next step phosphate to carbonate ratios were generated. As can be seen from Table 2, the carbonate to phosphate ratio was highest in cancerous specimens. Significant differences were observed when cancerous

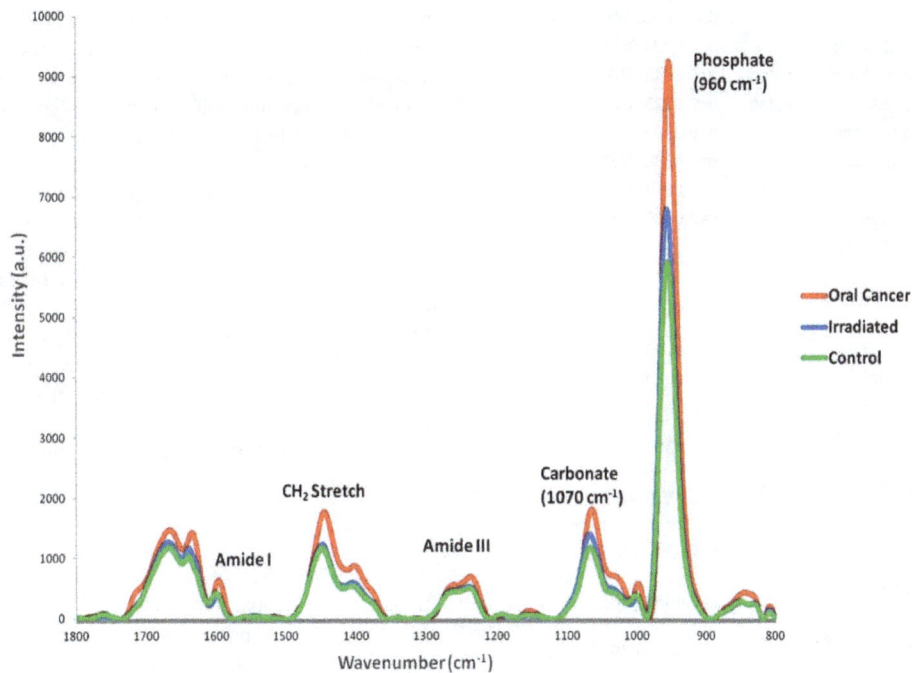

**Figure 2:** Mean baseline corrected spectra from control (40), cancerous (19) and irradiated (26) specimens in fingerprint region.

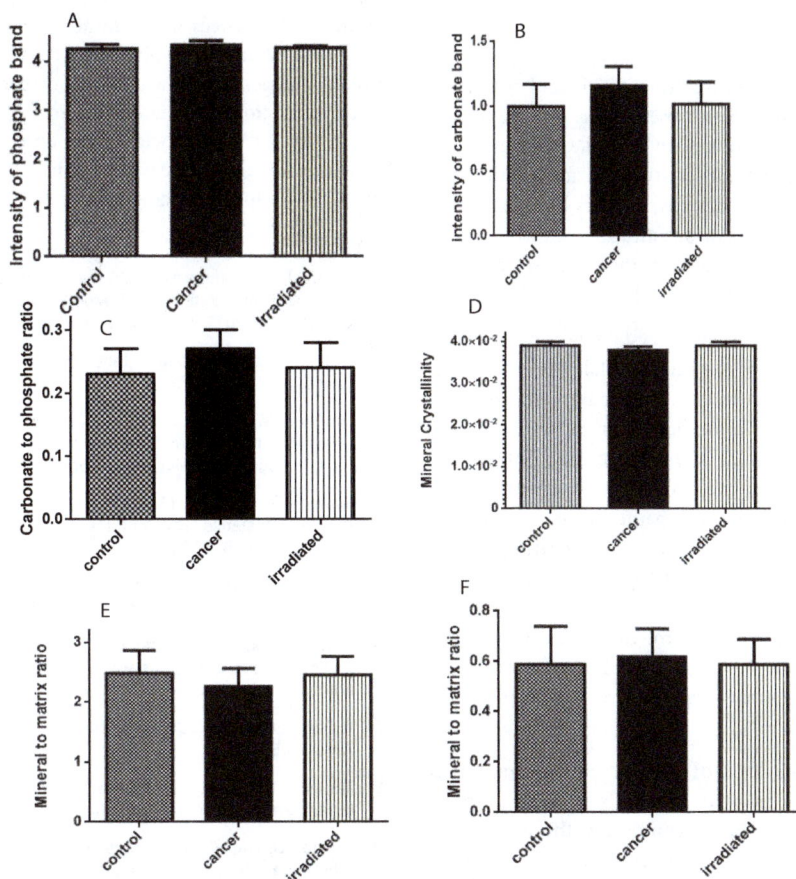

**Figure 3:** Raman band areas and ratios. A: Phosphate; B: Carbonate; C: Carbonate substitution rate; D: Mineral Crystallinity (inverse of FWHM of phosphate peak); E: Mineral to Matrix ratio-phosphate; F: Mineral to matrix ratio- carbonate.

| S No | Raman bands | p-value (Unpaired Student's t-test) |
|------|-------------|-------------------------------------|
| 1 | Phosphate content Control and Cancer | p=0.0084 (VS) |
| 2 | Phosphate content Control and Irradiated | p=0.2749 (NS) |
| 3 | Phosphate content Cancer and Irradiated | p=0.0225 (S) |
| 4 | Carbonate content Control and Cancer | p=0.0009 (VS) |
| 5 | Carbonate content Control and Irradiated | p=0.6588 (NS) |
| 6 | Carbonate content Cancer and Irradiated | p=0.0064 (VS) |
| 7 | Carbonate substitution Control and Cancer | p=0.0006 (VS) |
| 8 | Carbonate substitution Control and Irradiated | p=0.3247 (NS) |
| 9 | Carbonate substitution Cancer and Irradiated | p=0.0163 (S) |
| 10 | Mineral crystallinity Control and Cancer | p=0.0003 (VS) |
| 11 | Mineral crystallinity Control and Irradiated | p=1.00 (NS) |
| 12 | Mineral crystallinity Cancer and Irradiated | p=0.0008 (VS) |
| 16 | Mineral to matrix ratio-phosphate Control and Cancer | p=0.0309 (S) |
| 17 | Mineral to matrix ratio-phosphate Control and Irradiated | p=0.7379 (NS) |
| 18 | Mineral to matrix ratio-phosphate Cancer and Irradiated | p=0.0457 (S) |
| 19 | Mineral to matrix ratio-Carbonate Control and Cancer | p=0.4464 (NS) |
| 20 | Mineral to matrix ratio-Carbonate Control and Irradiated | p=0.9829 (NS) |
| 21 | Mineral to matrix ratio-Carbonate Cancer and Irradiated | p=0.3666 (NS) |

**Table 2:** Statistical analysis (Unpaired Student's t-test) of different biochemical parameters identified by Raman spectroscopy (VS: Very significant; S: Significant; NS: Not significant).

specimens were analyzed against control (p=0.0006) and irradiated (p=0.0163) samples. Similar to earlier observations no significant differences were observed between control and irradiated cases. These findings are in accordance with an earlier study on metastatic bone cancerous specimens [19]. In the cited study, authors have suggested that the increased carbonate content or substitution rates in tumor-bearing bones can be attributed to acid-base imbalance in the bone microenvironment in the extracellular fluid [19].

Mineral crystallinity reflects the mineral crystal size and associated with stoichiometric perfection of apatite crystal [14]. Optimal distribution of crystals can serve as an additional parameter to assess the bone strength. Previous studies have shown a negative correlation between carbonate level and mineral crystal size and distribution [19,29]. With increase in the carbonate content there is replacement of stoichiometric phosphate locations in apatite crystal. It leads to imperfection and decrease in crystallinity. Corroborating earlier observations in the present study the lowest level of crystallinity was observed for cancerous specimens, Table 1 [19]. Very highly significant differences were observed between control-cancerous (p=0.0003) and cancer-irradiated samples (p=0.0008), Table 2. Similar to earlier observations no significant differences were observed between control and irradiated specimens, Table 2.

Mineral to matrix ratio, or level of collagen mineralization, is another major compositional property which is related to bone mechanical strength. It is measured by computing the amount of phosphate and carbonate with respect to collagen [19]. The changes in collagen secondary structure is manifested by deformations in the amide I band (1660 cm$^{-1}$). Both Raman and IR spectroscopy methods have been used to study collagen cross-linking [14]. These

methods work on the principle of measuring two major enzymatic cross links, namely non-reducible pyridinoline (PYD), and reducible dehydrodihydroxynorleucine (deH-DHLNL). As shown in Table 1, collagen mineralization with respect to phosphate content was highest in control specimens followed by irradiated and cancerous samples. This can be attributed to multiple alterations such as loss of lamellar structure or deformation in collagen structure due to cancer and irradiation. The differences between control-cancerous (p=0.0309) and irradiated-cancerous (p=0.0457) specimens was found to be statistically significant, Table 2. No significant difference was observed between control and irradiated specimens. Mineralization level with respect to carbonate content was also computed, however no significant differences were observed. Minor differences between control and irradiated bones are suggestive of reversal of cancer associated changes in the bone composition. These results need to be validated on larger sample size and other gold standard methods. Other aspects such as radiation doses and time lag and their influence on bone quality should also be analyzed.

## Conclusions

Overall findings of the study further support applicability of Raman spectroscopic approaches for non-invasive disease diagnosis and treatment response monitoring. Minor differences between organic and inorganic component of control, cancerous and irradiated mandible bones can be identified. Major differences in both mineral and matrix components were observed between control-cancerous or cancerous-irradiated cases. Phosphate and carbonate content was highest in cancerous specimens. Control and irradiated specimens show no differences in the mineral and organic matrix composition, suggesting cause of implant loss could be primarily associated with minor changes in the vasculature, osteocytes and surrounding tissues. Extrinsic factors such as timing of implant surgery after irradiation could also have influence. Future application of these techniques for routine clinical practice will help in online monitoring of bone quality and could help in reducing post-operative complications.

### References

1. Ferlay J, Soerjomataram I, Dikshit R, Eser S, Mathers C, et al. (2015) Cancer incidence and mortality worldwide: sources, methods and major patterns in GLOBOCAN 2012. Int J Cancer 136: E359-386.

2. Argiris A, Karamouzis MV, Raben D, Ferris RL (2008) Head and neck cancer. Lancet 371: 1695-1709.

3. Brennan MT, Elting LS, Spijkervet FK (2010) Systematic reviews of oral complications from cancer therapies, Oral Care Study Group, MASCC/ISOO: methodology and quality of the literature. Support Care Cancer 18: 979-984.

4. Alsaadi G, Quirynen M, Komárek A, van Steenberghe D (2008) Impact of local and systemic factors on the incidence of late oral implant loss. Clin Oral Implants Res 19: 670-676.

5. Jereczek-Fossa BA, Orecchia R (2002) Radiotherapy-induced mandibular bone complications. Cancer Treat Rev 28: 65-74.

6. Chen JA, Wang CC, Wong YK, Wang CP, et al. (2014) Osteoradionecrosis of mandible bone in patients with oral cancer-associated factors and treatment outcomes. Head Neck.

7. O'Dell K, Sinha U (2011) Osteoradionecrosis. Oral Maxillofac Surg Clin North Am 23: 455-464.

8. Tamura M (1997) Biomedical Optical Spectroscopy and Diagnostics. Measurement Science and Technology 8.

9. Kendall C, Isabelle M, Bazant-Hegemark F, Hutchings J, Orr L, et al. (2009) Vibrational spectroscopy: a clinical tool for cancer diagnostics. Analyst 134: 1029-1045.

10. Hanlon EB, Manoharan R, Koo TW, Shafer KE, Motz JT, et al. (2000) Prospects

for in vivo Raman spectroscopy. Phys Med Biol 45: R1-59.

11. Nijssen A, Koljenovic S, Bakker Schut TC, Caspers PJ, Puppels GJ (2009) Towards oncological application of Raman spectroscopy. J Biophotonics 2: 29-36.

12. Peterson JR, Eboda ON, Brownley RC, Cilwa KE, Pratt LE, et al. (2015) Effects of aging on osteogenic response and heterotopic ossification following burn injury in mice. Stem Cells Dev 24: 205-213.

13. McNerny EM, Gong B, Morris MD, Kohn DH (2015) Bone fracture toughness and strength correlate with collagen cross-link maturity in a dose-controlled lathyrism mouse model. J Bone Miner Res 30: 455-464.

14. Morris MD, Mandair GS (2011) Raman assessment of bone quality. Clin Orthop Relat Res 469: 2160-2169.

15. Balakrishnan B, Indap MM, Singh SP, Krishna CM, Chiplunkar SV (2014) Turbo methanol extract inhibits bone resorption through regulation of T cell function. Bone 58: 114-125.

16. Paschalis EP, Shane E, Lyritis G, Skarantavos G, Mendelsohn R, et al. (2004) Bone fragility and collagen cross-links. J Bone Miner Res 19: 2000-2004.

17. Gadeleta SJ, Boskey AL, Paschalis E, Carlson C, Menschik F, et al. (2000) A physical, chemical, and mechanical study of lumbar vertebrae from normal, ovariectomized, and nandrolone decanoate-treated cynomolgus monkeys (Macaca fascicularis). Bone 27: 541-550.

18. Rieppo J, Hyttinen MM, Jurvelin JS, Helminen HJ (2004) Reference sample method reduces the error caused by variable cryosection thickness in Fourier transform infrared imaging. Appl Spectrosc 58: 137-140.

19. Tchanque-Fossuo CN, Monson LA, Farberg AS, Donneys A, Zehtabzadeh AJ, et al. (2011) Dose-response effect of human equivalent radiation in the murine mandible: part I. A histomorphometric assessment. Plast Reconstr Surg 128: 114-121.

20. Koga DH, Salvajoli JV, Alves FA (2008) Dental extractions and radiotherapy in head and neck oncology: review of the literature. Oral Dis 14: 40-44.

21. Blanco AI, Chao C (2006) Management of radiation-induced head and neck injury. Cancer Treat Res 128: 23-41.

22. Granström G (2005) Osseointegration in irradiated cancer patients: an analysis with respect to implant failures. J Oral Maxillofac Surg 63: 579-585.

23. Ihde S, Kopp S, Gundlach K, Konstantinovic VS (2009) Effects of radiation therapy on craniofacial and dental implants: a review of the literature. Oral Surg Oral Med Oral Pathol Oral Radiol Endod 107: 56-65.

24. Awonusi A, Morris MD, Tecklenburg MM (2007) Carbonate assignment and calibration in the Raman spectrum of apatite. Calcif Tissue Int 81: 46-52.

25. Carden A, Morris MD (2000) Application of vibrational spectroscopy to the study of mineralized tissues (review). J Biomed Opt 5: 259-268.

26. Arnett T (2003) Regulation of bone cell function by acid-base balance. Proc Nutr Soc 62: 511-520.

27. Zanyar M, Shazza R, Rehman IU (2007) Raman Spectroscopy of Biological Tissues. Appl Spectrosc Rev 42: 493-541.

28. Tchanque-Fossuo CN, Gong B, Poushanchi B, Donneys A, Sarhaddi D, et al. (2013) Raman spectroscopy demonstrates Amifostine induced preservation of bone mineralization patterns in the irradiated murine mandible. Bone 52: 712-717.

29. Bi X, Sterling JA, Merkel AR, Perrien DS, Nyman JS, et al. (2013) Prostate cancer metastases alter bone mineral and matrix composition independent of effects on bone architecture in mice-A quantitative study using microCT and Raman spectroscopy. Bone 56: 454-460.

# pKa Determination of a Non-hydro-soluble Chemical Substance, Issued from Chiral Chromatographic Solubility Profiles and Mat-pKa Calculations

**Lionel Vidaud[1]\***, **Antoine Pradines[2]**, **Jérôme Marini[2]**, **Elodie Lajous[2]** and **Patrick Clavières[3]**

[1]*Quality Science and Innovation, EVOTEC Toulouse, France*
[2]*Preparative Chromatography Group, EVOTEC Toulouse, France*
[3]*ADME and Developability Assessment Laboratory, EVOTEC Toulouse, France*

## Abstract

After development and implementation of Mat-pKa software, since 3 years, pKa(s) determinations for approximately 70 compounds were successfully conducted. But a remaining objective was to enlarge the capabilities of this software to resolve cases for non-soluble compounds in aqueous phases. Until now, pKa(s) calculations were based on solubility profiles obtained in aqueous buffered solutions. A new application of this software was successful for solubility profiles obtained in pure water, in which, various concentrations of the substance led to different pH of solutions. A new approach, based on solubility profile elaborated from the Yasuda-Shedlovsky experiment, is presented in this paper, for a non-hydro-soluble chemical substance. More challenging in this case, the pKa value was determined for a chiral compound and, outside of the range 0 to 14. In this third new application, Mat-pKa software confirmed its ability to define pKa values in extended ranges and its independence to diverse techniques (generally the chromatography) delivering solubility profiles, even in very specific and delicate conditions.

**Keywords:** Mat-pKa; pKa determination; pKa of non-hydro-soluble chemical substances; Chromatography; Chiral chromatography; QSAR

## Introduction

### Background of Mat-pKa software creation

Specifically developed 3 years ago, Mat-pKa is a software for calculations of pKa(s) of chemical substances, based on their experimental solubility profiles. Since this time, approximately 70 molecules pre-candidates or candidates to the development have obtained their respective pKa(s) calculated with this software in our R&D environment. These calculations correspond to the resolution of the systems of equations governed by the expressions of the dissociation constants of each entity of the molecule tested, without any approximation, or introduction of additional parameters, or convergence factor, leading finally to the determinations of experimental pKa(s) depending only on the measured solubility and the measured pH of the solutions.

Validation and performances of this software have been provided and tested, as reported in the publication [1] related to the creation and the concept of these calculations. Comparisons with other calculations techniques, or other software have been scrutinized. A large comparative study on 41 molecules, including a predictive software, 2 dedicated experimental equipment and Mat-pKa was also provided.

This approach is of high interest, firstly because solubilities can be measured with different techniques (as Liquid Chromatography, Capillary Electrophoresis, UV spectrometry, or Potentiometry) and secondly because calculations of more than ten pKa(s) can be achieved in few seconds for one molecule, after the insertion of the needed couples of data (solubilities and pH(s) of the solutions). Another interest, is the difference with other software (ACD, VCC, Epik, Marvin, Pallas, Gastro-Plus, Pharma Algorithm and others), based on the structure recognition of moieties and their respective environment in the molecule, to generate empirical estimates (more or less accurate) of these pKa(s).

This new software presents also the advantage to be used on simple computers (PCs) equipped with Microsoft Office™ environment, because the code is written in Visual Basic and of course, don't need

link with experimental equipment.

The Table 1 illustrates the performance of the software compared to the dedicated techniques (Titration, UV, CE) and to ACD v12 software for some drug substances analyzed recently (2013).

This table demonstrates clearly, the better capabilities of Mat-pKa to determine the pKa(s) of these molecules, compared to the dedicated techniques which, most of the time, were not able to provide all the expected values. We can also point out here, the performance of the software when the pKa(s) values are less than 2.5 or more than 9.5. In the light of the example of the Compound 1, Mat-pKa has resolved 2 pKa(s) values, which are close from each other (5.32 and 6.07).

Frequently, results of ACDv12, experimental determinations and Mat-pKa calculations are not too far from each other. Some non-coherences are observed between Mat-pKa and ACDv12, for very low pKa(s) values (considering ACD v12 as the reference).

More recently in 2014, new extensions of Mat-pKa capabilities were evidenced. A simultaneous determination of 4 (or 5) pKas, of one chemical substance and its organic salt was achieved in specific conditions. In this case the solubility profile was determined in pure water, by applying different concentrations of this compound for different generated pH of solutions.

Until now these solubility profiles were managed in aqueous buffered solutions. 2 pKas were attributed to the substance and 2 others

---

**\*Corresponding author:** Lionel Vidaud, Quality Science and Innovation, EVOTEC Toulouse, France, E-mail: lionel.vidaud@evotec.com

| Compounds | Nb of pKas (identified functions) | pKas according to ACD v12 | Experimental pKas | pKas issued from Mat-pKa and solubility profile | Coherence with Mat-pKa |
|---|---|---|---|---|---|
| Compound 1 | 4 | 1.66 | 2.7 | 2.7 | Exp ↔ Mat-pKa# |
| | | 6.28 | | 5.32 | ACD ↔ Mat-pKa' |
| | | 7.76 | 7.6 | 6.07 | Exp ↔ Mat-pKa' |
| | | 12.68 | | 11.69 | ACD ↔ Mat-pKa' |
| Compound 2 | 4 | -1.38 | | -1.86 | ACD ↔ Mat-pKa# |
| | | 2.81 | | 0.71 | No coherence |
| | | 5.17 | 5.2 | 5.44 | ACD ↔ Exp ↔ Mat-pKa# |
| | | 12.21 | | 12.2 | ACD ↔ Mat-pKa# |
| Compound 3 | 1 | 0.42 | 3 | 3.2 | Exp ↔ Mat-pKa# |
| Compound 4 | 4 | -1.52 | | 2.42 | No coherence |
| | | 6.03 | 6.4 | 7.3 | ACD ↔ Exp ↔ Mat-pKa' |
| | | 9.66 | 9.3 | 9.23 | ACD ↔ Exp ↔ Mat-pKa# |
| | | 11.95 | | 11.91 | ACD ↔ Mat-pKa# |
| Compound 5 | 3 | -1.67 | | 3.34 | No coherence |
| | | 8.93 | 8.62 | 8.85 | ACD ↔ Exp ↔ Mat-pKa# |
| | | 11.89 | | 11.3 | ACD ↔ Mat-pKa# |
| Compound 5 (Salified) | 3 | -1.67 | | -0.27 | No coherence |
| | | 8.93 | 8.62 | 8.31 | ACD ↔ Exp ↔ Mat-pKa# |
| | | 11.89 | | 13.1 | ACD ↔ Mat-pKa' |

Results highlighted with# indicates correlation ± 0.5 pKa unit; Results highlighted with' indicates correlation ± 1.5 pKa unit

**Table 1:** Comparative table of pKa(s) determinations for 6 substances in 2013.

were attributed to the organic counter-ion. And finally, a fifth pKa was also determined for one impurity generated in water (and not fully stable in the analytical conditions).

## Objectives of the study

The concept of this study was put in place to resolve the determination of one pKa, for a chemical substance bearing a thiazole ring, as a weak base. Environment of this ring in the chemical structure was also suspected to play a role in the activation or the deactivation of this weak base. Evaluation of pKa was processed through the ACD 2014 software suite (Percepta). The result based on the structure of the compound conducted to the pKa value estimated to 1.9 for the amine ring moiety. This low pKa value suggested that classical dedicated techniques, as for instance, Capillary Electrophoresis coupled to Ultra-Violet detection, or acido-basic titrations by potentiometry [2] or Ultra-Violet detections [3,4], or combined analytical determinations in cases of multi-protic compounds [5] will be very limited to reach an experimental determination of this dissociation constant. This aspect is the first disadvantage for this chemical substance.

Familiarized with the Mat-pKa software over the 3 last years, leading to pKa(s) determinations without any limitation in the pKa range, the idea was to submit the solubility profile of this very weak base to these pKa calculations. But, as a second disadvantage, the chemical structure of this compound reveals 2 chiral centers, leading to 4 possible enantiomers. Consequently, the analytical method, to be used for the measurements of the solubility profile, must be selective for these chiral aspects, to be sure to quantify the contents of the right enantiomer. Of course, the chiral chromatography technique will be the best appropriate method.

Finally, as a third disadvantage, the solubility of this compound in aqueous buffered solutions was very poor, leading to results below the limit of quantitation (LOQ) of the analytical method and non-exploitable by Mat-pKa software. To resolve this last issue, improvements of the solubility in mixtures of organic solvent and aqueous buffered solutions must be studied. This last approach is identical to the one developed by

Yasuda-Shedlovsky [6,7] and applied since a long time by many users.

Consequently the concept of this study was established on the following processes:

- Development and validation (most relevant parameters) of a selective chiral chromatographic method.

- Design of the protocol, for a direct application of the Yasuda-Shedlovsky method, with the selection of different mixtures of organic solvents and water (in this case, buffered solutions), for which dielectric constants are known and pH of the solutions can be measured.

- Calculations of the psKa values (pKa in organic/water solvents) for each mixture, using the Mat-pKa software.

- Extraction of the pKa value of the compound (at 100% of water, $\varepsilon=80.65$), using the linear relation between psKa and the dielectric constant $\varepsilon$, proposed by Yasuda-Shedlovsky:

$$psKa + log([H2O]) = \alpha / \varepsilon + b$$

## Development and validation of the chiral chromatographic method

As chromatographic chiral separation conditions are not predictable, with any specific rules to anticipate if the separation will succeed, a chiral chromatographic analytical screening is necessary. During this screening we test different chiral stationary phases available and usable at the analytical but also at the preparative chromatography scale, combined in a fully automated experimental mode with different mobile phases, leading to a total of more than 60 possible combinations. We perform this screening on analytical HPLC (High Performance Liquid Chromatography) and also on analytical SFC (Supercritical Fluid Chromatography) because these separation techniques are used in a complementary manner in our laboratories.

After this screening step, we chose the best technique HPLC or SFC and the best chiral stationary phase combined with the best mobile phase for chiral analysis or for preparative chromatography. In this paper, only the optimized method is reported for the analytical quantification of the 4 enantiomers, but we have also developed a method and we have performed the separation of each enantiomer using chiral HPLC preparative chromatography.

### Design of the protocol for the Yasuda-Shedlovsky experiments

The protocol was designed according to the following experimental needs:

-4 series of (organic/aqueous buffers) solvent mixtures, respectively, 20/80, 40/60, 60/40 and 80/20 (%v/v). The solution of the chemical substance dissolved in Acetonitrile (~2 mg/mL) was considered as the organic phase.

-6 series of aqueous buffered solutions at expected pHs: 1.3, 3.0, 4.5, 6.4, 7.5 and 9.0. According to the literature (8), solubilities of these reagents at concentrations around 50 mM are compatible with these aqueous / acetonitrile mixtures.

- For each series of the 4 solvent mixtures (combined with each of the 6 buffered solution), ionic strengths to be maintained by addition of KCl solutions prepared at appropriate concentrations, to complete each final mixture to 10 mL.

-The chromatographic method retained to provide the concentrations of each solution was the one used for the chiral purity determination (see experimental section). Potential isomerization of the molecule cannot be excluded and the needs of the strong selectivity of this method might be appreciated in case of impurity formation.

-Sensitive pH-meter and its electrode (713 pH-meter Methrom, calibrated with reference buffered solutions at pH 3.0, 5.0 and 7.0 respectively) to make pH measurements in the mixtures described above.

-Mat-pKa software for the calculation of $psKa + \log([H2O])$

noted **psKa'**

### Experimental Section

### Analytical method for chiral purity determination

The parameters of this method are given bellow:

| Instrument | HPLC 1200 Agilent |
|---|---|
| Software | Agilent Chemstation rev.B.02.03 |
| Column | Chiralpak IB (250 × 4.6 mm, 5 µ) |
| Eluent | Methanol 100% |
| Flow rate | 1 mL/min |
| Detection | 254 nm and 220 nm |
| Column temperature | Ambiant T°C |
| Injection volume | 30 µL |
| Sample preparation | 0.05 to 2 mg/mL, (solubilized in $CH_3CN$) |

As a remark the eluent, the stationary phase (cellulose tris(3,5 dimethylphenycarbamate) as chemical graft) and the detection wavelength are close to the ones used for the preparative HPLC method, developed for the isolation and the characterization of each enantiomer.

### Validation of the analytical method for chiral purity determination

**Selectivity of the method:** The selectivity of the method was validated based on the analysis of the reference sample, corresponding to a mixture of the 4 enantiomers of the chemical substance. As shown in Figure 1, the 4 enantiomers are separated with a high resolution, particularly the last peak eluted corresponding to the substance studied.

**Linearity of the UV response of the method:** Six solutions at 0.05% - 0.1% - 1% - 10% - 50% - 100% of the working concentration (2

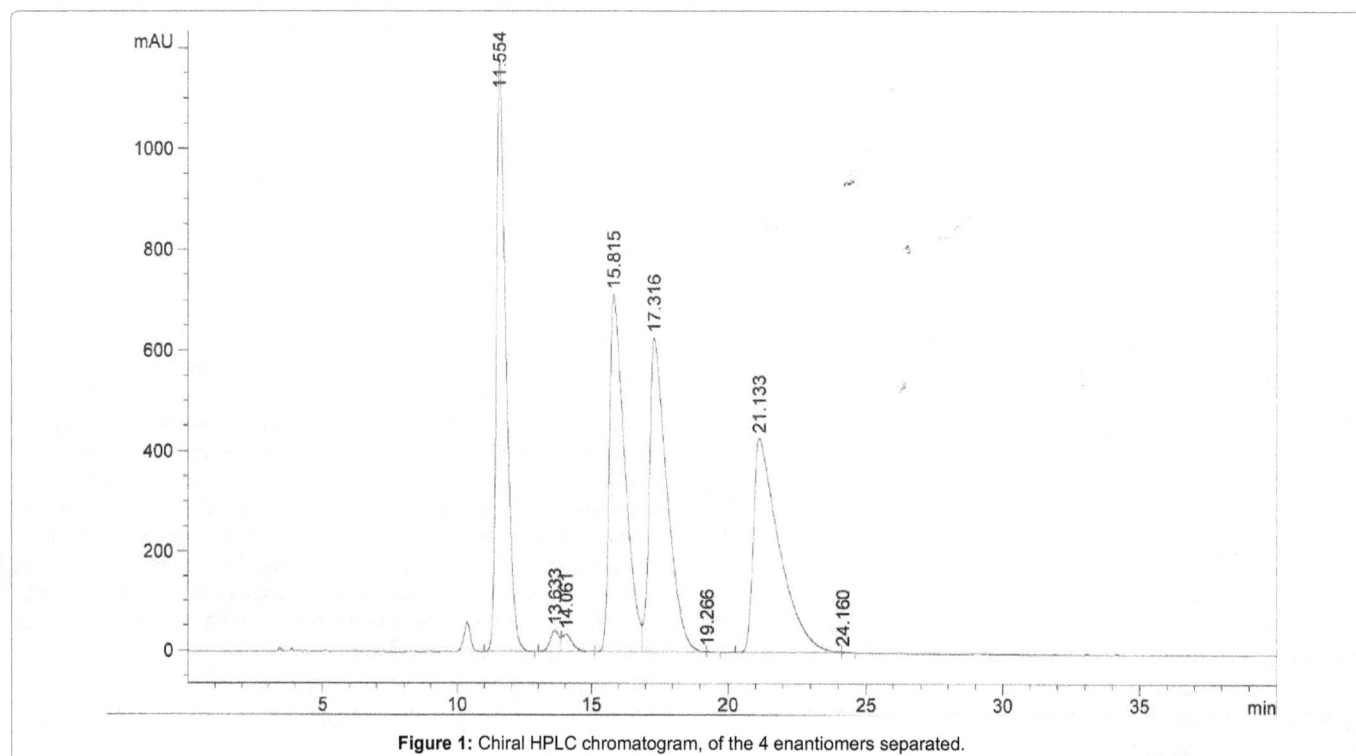

**Figure 1:** Chiral HPLC chromatogram, of the 4 enantiomers separated.

mg/mL) of the substance were prepared by dilution of the 100% stock solution. One chromatogram of each solution was recorded (Table 2). The graphic presentation of the peaks areas obtained for these solutions versus their respective concentrations is represented in Figure 2.

The values obtained for the 2 correlation coefficients (=0.9999) demonstrated the linearity of UV response for the compound in the range of concentrations studied, at 220 nm and at 254 nm.

**Limit of quantification:** Based on the signal to noise ratio (S/N=16.4) obtained for the solution concentrated at 1% (Figure 3) and taking into account the demonstration of the linearity of the method, the limit of quantification (LOQ) was recalculated and was estimated to 12.63 µg/ml (0.6%).

Peaks marked with a retention time value, were not included in the evaluation of the height of the noise of the chromatogram. The height of the peak, at retention time 22.63 min, was taken into account for the calculation of the Signal to Noise ratio (according to the European Pharmacopea 2.2.46).

### Protocols used for solubility measurements

The 6 buffered solutions and the potassium chloride solutions were prepared according the Table 3.

These aqueous solutions were added to the organic solution of the chemical substance studied, according to the Table 4. In which, Ci corresponds to the concentration of each species present in the considered solution and zi the charge of this species. Concentrations of species in buffered solutions (phosphate, acetate, citrate and trishydroxymethylammoniummethane) were extracted from Mat-pKa software according to each measured pH and the respective pKas of each species. As this can be observed in the Table 4 the ionic strength of each mixture of organic solvent /water was maintained approximately at the same level. Before analysis, each solution was filtered on a 0.22 µ GHP filter.

### Results

The Table 5 shows all the couples of data (Calc. concentrations=f(pH)) needed to launch calculation with the Mat-pKa software.

In this table some calculated concentrations are above the maximum values acceptable (bolded). Origins of these abnormalities were investigated and attributed to the non-perfect sealing of the bottles containing the prepared mixtures and or vials for chromatography. In these cases evaporation of the solvents, mainly of the acetonitrile, and consequently increases of concentration of these solutions were observed. Due to the limited quantities of the chemical substance available, these experiments have not been re-processed.

### psKa' determination using Mat-pKa, for 40% of acetonitrile co-solvent

The series of results concerned for these calculations with Mat-pKa, are extracted from the Table 5 corresponding to 40% v/v (2nd series of 6 values in this table).

Mat-pKa calculations were conducted with the couples of data reported in Table 6. Mat-pKa software results are reported in the Table 7.

According to the Table 7, psKa' of the chemical substance, in a mixture of 40% acetonitrile and 60% of water, was calculated to 0.9 by Mat-pKa. Only one psKa' was accessible through these calculations. The entity governing the solubility of the substance in this medium, is clearly the neutral molecule (**A** in the table), which corresponds to a very weak base, according to the low value of this psKa' of the conjugated acid.

The Table 8 and the curve (Figure 4) show the acceptable correlation of the experimental solubility results plotted with red points, with the ones issued from the psKa' calculation (dark blue curve).

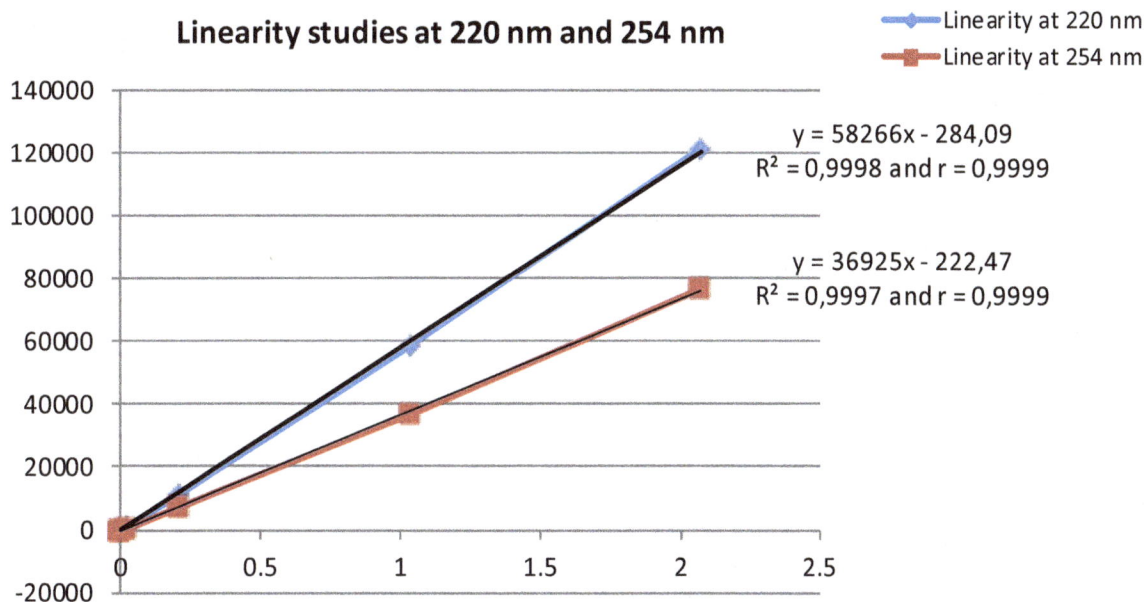

**Figure 2:** Plot of each chromatographic area for different concentrations of solution at 220 nm and at 254 nm.

| Concentration % | 0.05 | 0.1 | 1 | 10 | 50 | 100 |
|---|---|---|---|---|---|---|
| Concentration mg/ml | 0.0001 | 0.0021 | 0.0208 | 0.2071 | 1.0353 | 2.0706 |
| Area (inj 30 µl) at 254 nm | 9.11 | 36.68 | 707.28 | 7345.85 | 37067.1 | 76710.4 |
| Area (inj 30 µl) at 220 nm | 42.2 | 135.91 | 1185.82 | 11681.6 | 58565.2 | 121105 |

**Table 2:** Linearity study results.

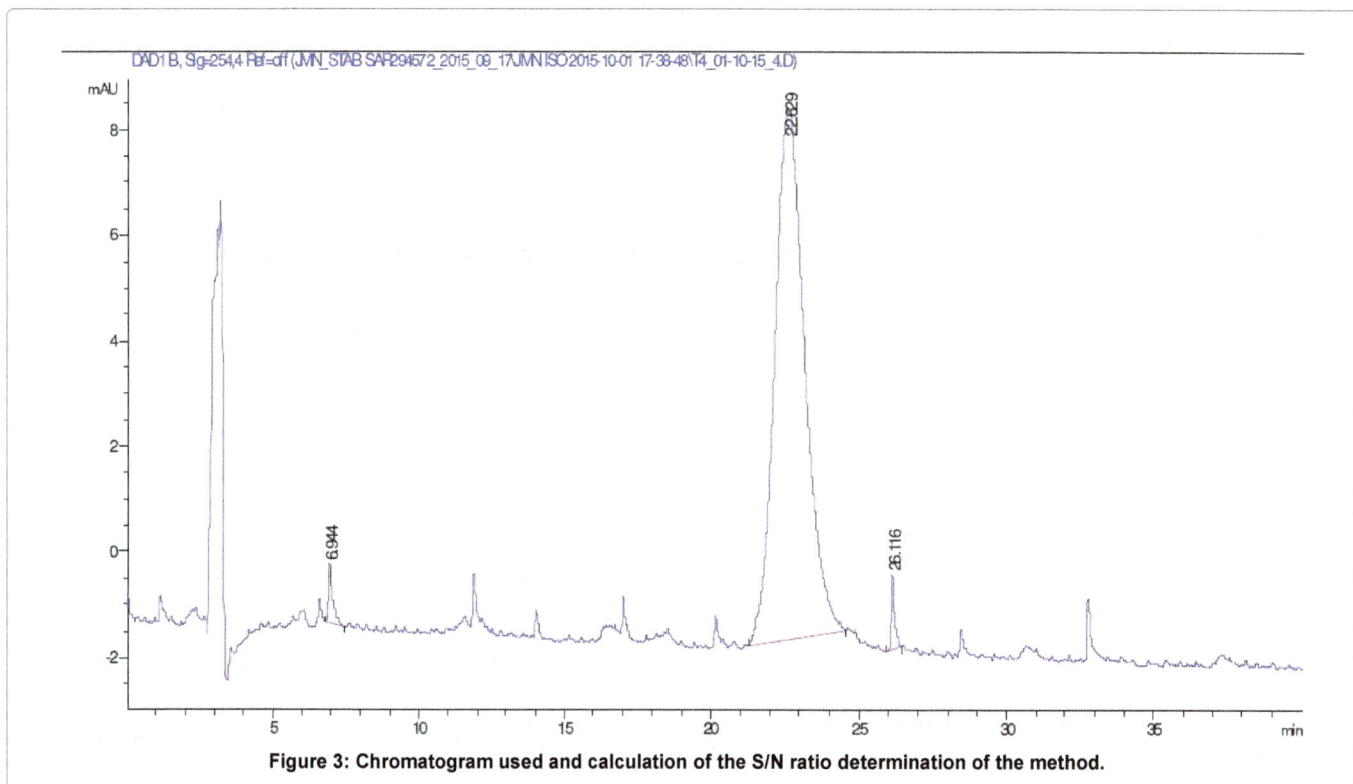

**Figure 3:** Chromatogram used and calculation of the S/N ratio determination of the method.

| Medium | Composition |
|---|---|
| HCl 0.05N (pH 1.3) | 10 mL of HCl 0.1N (Merck) adjusted to 20.0 mL with Millipore water. |
| Citrate buffer pH3 | Solution A is prepared with 267.9 g of citric acid monohydrated (VWR, ref: 20276.292) in 20 mL of Millipore water. Solution B is prepared with 595.9 mg of citric trisodium dihydrated salt (VWR ref: 27830.294) in 40 mL of Millipore water. Solution B is added to solution A to adjust the pH to 3.0. |
| Acetate buffer pH 4.5 | Anhydrous sodium acetate (Fisher ref: S/210/50) 96.88 mg + 1 drop of acetic acid. pH 4.5 is adjusted in this solution by addition of diluted acetic acid (1:10). Solution completed to 20.0 mL with Millipore water. |
| Phosphate buffer pH 6.5 | 3.2 mL $K_2HPO_4$ 0.1M (Merck ref: 1.5099 - 279.8 mg in 20 mL of Millipore water) + 6.8 mL $KH_2PO_4$, $3H_2O$ 0.1M (Fisher ref: P/5320/53 - 484.9 mg in 20 mL of Millipore water) solution adjusted to 20.0 mL with Millipore water. |
| Phosphate buffer pH 7.4 | 1.9 mL of $K_2HPO_4$ 0.1M (Merck ref: 1.5099 - 279.8 mg in 20 mL of Millipore water) + 8.1 mL of $KH_2PO_4$, $3H_2O$ 0.1M (Fisher ref: P/5320/53 - 484.9 mg in 20 mL of Millipore water) solution adjusted to 20.0 mL with Millipore water. |
| Trishydroxymethylaminomethane pH 9.0 buffer | 163.3 mg of Trishyroxymethylaminomethane (Fisher ref: T395.500) in 20 ml of Millipore water. Add 140 µl of HCl 1N and adjust to pH 9.0 with HCl 0.1N. Dilute to 40 ml with Millipore water. |
| Solution of KCl (a) - 0.061M | 15 mL of solution (c) adjusted to 20 mL |
| Solution of KCl (b) - 0.041M | 10 mL of solution (c) adjusted to 20 mL |
| Solution of KCl (c) - 0.082M | 607.2 mg of Potassium Chloride (Merck ref: 1.59707) dissolved in Millipore water and adjusted to 100 mL |

**Table 3:** Preparation of aqueous buffered solutions and potassium chloride solutions.

## psKa' determination using Mat-pKa, for 60% of acetonitrile co-solvent

The series of results concerned for these calculations with Mat-pKa, are extracted from the Table 5 corresponding to 60% v/v (3rd series of 6 values in this table). As it is reported, 4 solubility values were obtained above the maximum possible solubility (in this case, 1.2424 mg/mL).

Consequently these 4 couples of data were not included in the calculations with Mat-pKa. Fortunately the 2 last couples of data (at pH 1.64 and pH 8.12) presented concentration profiles eligible for one psKa' determination, as presented in the Table 9. Mat-pKa software results are reported in the Table 10.

According to the Table 10, psKa' of the chemical substance, in a mixture of 60% acetonitrile and 40% of water, was calculated to -0.34 by Mat-pKa. The entity governing the solubility of the compound in this medium, is again the neutral molecule (**A** in the table), which

| Compound solution, 2 mg/mL in CH₃CN | Aqueous buffered solutions | Aqueous buffered solutions | Aqueous KCl solutions | Aqueous KCl solutions | Organic composition | Dielectric constants of mixtures | Total ionic strengths* |
|---|---|---|---|---|---|---|---|
| Volume (mL) | pH | Volume (mL) | (M) | Volume (mL) | (% v/v) | (ε) | (M) |
| 2.0 | 1.32 | 4.0 | 0.061 | 4.0 | 20 | 73.2 | 0.044 |
| 2.0 | 3.17 | 4.0 | 0.081 | 4.0 | 20 | 73.2 | 0.045 |
| 2.0 | 4.23 | 4.0 | 0.041 | 4.0 | 20 | 73.2 | 0.068 |
| 2.0 | 6.18 | 4.0 | 0.041 | 4.0 | 20 | 73.2 | 0.041 |
| 2.0 | 7.08 | 4.0 | 0 | 4.0 | 20 | 73.2 | 0.040 |
| 2.0 | 8.99 | 4.0 | 0.082 | 4.0 | 20 | 73.2 | 0.037 |
| 4.0 | 1.32 | 3.0 | 0.061 | 3.0 | 40 | 64.75 | 0.033 |
| 4.0 | 3.17 | 3.0 | 0.081 | 3.0 | 40 | 64.75 | 0.034 |
| 4.0 | 4.23 | 3.0 | 0.041 | 3.0 | 40 | 64.75 | 0.051 |
| 4.0 | 6.18 | 3.0 | 0.041 | 3.0 | 40 | 64.75 | 0.031 |
| 4.0 | 7.08 | 3.0 | 0 | 3.0 | 40 | 64.75 | 0.030 |
| 4.0 | 8.99 | 3.0 | 0.082 | 3.0 | 40 | 64.75 | 0.028 |
| 6.0 | 1.32 | 2.0 | 0.061 | 2.0 | 60 | 54.6 | 0.022 |
| 6.0 | 3.17 | 2.0 | 0.081 | 2.0 | 60 | 54.6 | 0.022 |
| 6.0 | 4.23 | 2.0 | 0.041 | 2.0 | 60 | 54.6 | 0.034 |
| 6.0 | 6.18 | 2.0 | 0.041 | 2.0 | 60 | 54.6 | 0.021 |
| 6.0 | 7.08 | 2.0 | 0 | 2.0 | 60 | 54.6 | 0.020 |
| 6.0 | 8.99 | 2.0 | 0.082 | 2.0 | 60 | 54.6 | 0.019 |
| 8.0 | 1.32 | 1.0 | 0.061 | 1.0 | 80 | 45.55 | 0.011 |
| 8.0 | 3.17 | 1.0 | 0.081 | 1.0 | 80 | 45.55 | 0.011 |
| 8.0 | 4.23 | 1.0 | 0.041 | 1.0 | 80 | 45.55 | 0.017 |
| 8.0 | 6.18 | 1.0 | 0.041 | 1.0 | 80 | 45.55 | 0.010 |
| 8.0 | 7.08 | 1.0 | 0 | 1.0 | 80 | 45.55 | 0.010 |
| 8.0 | 8.99 | 1.0 | 0.082 | 1.0 | 80 | 45.55 | 0.009 |

*: total ionic strengths of the media were calculated according to the following formula: $1/2 \times \sum c_i \times (z_i)^2$

**Table 4:** Prepared solutions for solubility measurements.

is a very weak base, according to the low value of this psKa' of the conjugated acid. This observation confirms the previous one, issued from the experiment conducted with 40% (v/v) of acetonitrile. We can observe also that the intrinsic solubility is not too far from the maximum concentration of the solutions prepared for this experiment.

The Table 11 and the curve (Figure 5) show the acceptable correlation of the experimental solubility results plotted with red points and orange points (for those excluded from the calculations), with the ones issued from the psKa' calculation (dark blue curve).

## psKa' determination using Mat-pKa, for 80% of acetonitrile co-solvent

The series of results concerned for these calculations with Mat-pKa, are extracted from the Table 5 corresponding to 80% v/v (4th series of 6 values in this table). As it is reported, 2 solubility values were obtained above the maximum possible solubility (in this case, 1.6565 mg/mL).

Consequently these 2 couples of data were not included in the calculation with Mat-pKa. Fortunately, the 4 last couples of data (at pH 1.42, pH 4.35, pH 6.36 and pH 8.51) were used for psKa' determination, as this is reported in the Table 12. Mat-pKa software results are reported in Table 13.

According to the Table 13, psKa' of the chemical substance, in a mixture of 80% acetonitrile and 20% of water, was calculated to -0.55 by Mat-pKa. The entity governing the solubility of compound in this medium, is again the neutral molecule (**A** in the table), which is a very weak base, according to the low value of this psKa' of the conjugated acid. This last calculation confirms the 2 others previously obtained in 40% and 60% (v/v) of acetonitrile. The intrinsic solubility value is also

comparable to maximum concentration of the solutions prepared for this experiment.

The Table 14 and the curve (Figure 6) present the acceptable correlation of the experimental solubility results plotted with red points and orange points (for those excluded from the calculations), with the ones issued from the psKa' calculation (dark blue curve).

## pKa determination with the series of psKa' issued from the 3 mixtures of acetonitrile with water

Having in hands the 3 psKa' of the chemical substance, obtained in 3 different binary solvents for which each dielectric constant is known, the linear Yasuda-Shedlovsky extrapolation can be tested. The Table 15 summarizes the data that should be taken into account.

The linear regression reported on the curve (Figure 7) was obtained with a correlation coefficient around 0.9, which is probably perfectible, due to the point at ε=54.6 (1/ε=0.0183) for which the psKa' was determined only with 2 couples of experimental points.

According to Yasuda-Shedlovsky method, the negative slope of the curve indicates that we are in presence of a base, as this was also confirmed by Mat-pKa (Tables 7, 10 and 13). Based on the equation of this curve, the extrapolated value at 1/ε=0.0124 (100% of water), leads to a dissociation constant of the conjugated acid function of the molecule:

### pKa=-0.43

*For a simple evaluation, if the second point of this curve is not included in the linearity study, then the pKa value is -0.19, finally not too far from the retained value.*

| Measured pH after filtration | Area at 254 nm | Max conc. (mg/mL) | Calculated conc. (mg/mL)* | Organic Composition (% v/v) |
|---|---|---|---|---|
| | (Reference solution) 76710.4 | 2.0706 | $Calc.C = \dfrac{Area \times 2.0706}{76710.4}$ | 100 |
| 1.67 | N.D. | 0.41412 | N.A. | 20 |
| 3.44 | N.D. | 0.41412 | N.A | 20 |
| 4.59 | N.D. | 0.41412 | N.A | 20 |
| 6.59 | N.D. | 0.41412 | N.A | 20 |
| 7.52 | N.D. | 0.41412 | N.A | 20 |
| 8.87 | N.D. | 0.41412 | N.A | 20 |
| 1.68 | 25174.6 | 0.82824 | 0.67952359 | 40 |
| 3.76 | 21617.7 | 0.82824 | 0.58351423 | 40 |
| 5.09 | 3638.7 | 0.82824 | 0.09821735 | 40 |
| 6.91 | 23964.3 | 0.82824 | 0.64685466 | 40 |
| 7.83 | 27001.6 | 0.82824 | 0.72883876 | 40 |
| 8.79 | 9991.9 | 0.82824 | 0.26970565 | 40 |
| 1.64 | 45766 | 1.24236 | 1.23533549 | 60 |
| 4.01 | 55100.4 | 1.24236 | **1.48729362** | 60 |
| 5.57 | 48839.9 | 1.24236 | **1.31830752** | 60 |
| 7.13 | 56260 | 1.24236 | **1.51859404** | 60 |
| 8.12 | 45295.7 | 1.24236 | 1.22264095 | 60 |
| 8.71 | 54546.4 | 1.24236 | **1.47233981** | 60 |
| 1.42 | 60618.9 | 1.65648 | 1.63625133 | 80 |
| 4.35 | 59978.1 | 1.65648 | 1.61895459 | 80 |
| 6.36 | 60287.6 | 1.65648 | 1.62730874 | 80 |
| 7.35 | 64006.8 | 1.65648 | **1.72769898** | 80 |
| 8.51 | 59231.4 | 1.65648 | 1.59879934 | 80 |
| 8.57 | 84731.6 | 1.65648 | **2.28711167** | 80 |

**Table 5:** Solubility results of the chemical substance in different mixtures of solvents and at different pH.

| Measured pH after filtration | Calculated conc. (mg/mL) |
|---|---|
| 1.68 | 0.67952359 |
| 3.76 | 0.58351423 |
| 5.09 | 0.09821735 |
| 6.91 | 0.64685466 |
| 7.83 | 0.72883876 |
| 8.79 | 0.26970565 |

**Table 6:** Solubility/pH of solutions used for psKa' calculation of the chemical substance, in 40% of acetonitrile.

| Experimental results (Solubility in mg/mL) - Chemical Substance (40% CH$_3$CN) | | | | | | |
|---|---|---|---|---|---|---|
| pH | 1.68 | 3.76 | 5.09 | 6.91 | 7.83 | 8.79 |
| Solubility | 0.6795 | 0.5835 | 0.0982 | 0.6469 | 0.7288 | 0.2697 |
| Nb of acid functions | 1 | | | | | |
| pKas | pKa1 | | | | | |
| ACD (Percepta) | 1.9 | | | | | |
| Calculated results (Solubility in mg/mL and pKas) | | | | | | |
| Entity considered | HA | A | | | | |
| Intrinsic solubility | N.A. | 0.58269481 | | | | |
| pKa1 | Ka <0 | 0.90045749 | | | | |

**Table 7:** Extraction of the Mat-pKa table of results (chemical substance in 40% of acetonitrile).

| Measured pH | 1.68 | 3.76 | 5.09 | 6.91 | 7.83 | 8.79 |
|---|---|---|---|---|---|---|
| Measured solubility | 0.6795 | 0.5835 | 0.0982 | 0.6469 | 0.7288 | 0.2697 |
| Re-calculated solubility (Mat-pKa) | 0.6751# | 0.5834# | *0.5827* | 0.5827* | 0.5827* | *0.5827* |

Results highlighted with # indicates values in the range ± 10%; Results highlighted with * indicates values in the range ± 30%; Results highlighted in *italics* indicates values in the range ± 100%

**Table 8:** Experimental and re-calculated solubilities of the compound in 40% of CH$_3$CN co-solvent.

| Measured pH after filtration | Calculated conc. (mg/mL) |
|---|---|
| 1.64 | 1.23533549 |
| 8.12 | 1.22264095 |

**Table 9:** Solubility/pH of solutions used for psKa' calculation of the chemical substance, in 60% of acetonitrile.

| Experimental results (Solubility in mg/mL) - Chemical Substance (60% CH$_3$CN) | | | |
|---|---|---|---|
| pH | 1.64 | 8.12 | |
| Solubility | 1.2353 | 1.2226 | |
| Nb of acid functions | 1 | | |
| pKas | pKa1 | | |
| ACD (Percepta) | 1.9 | | |
| Calculated results (Solubility in mg/mL and pKas) | | | |
| Entity considered | HA | A | |
| Intrinsic solubility | N.A. | 1.2226 | |
| pKa1 | Ka <0 | -0.34348053 | |

**Table 10:** Extraction of the Mat-pKa table of results (chemical substance in 60% of acetonitrile).

| Measured pH | 1.64 | 4.01 | 5.57 | 7.13 | 8.12 | 8.71 |
|---|---|---|---|---|---|---|
| Measured solubility | 1.2353 | *1.4873* | *1.3183* | *1.5186* | 1.2226 | *1.4723* |
| Re-calculated solubility (Mat-pKa) | 1.2365# | 1.2227˙ | 1.2226# | 1.2226˙ | 1.2226# | 1.2226˙ |

Results highlighted with # indicates values in the range ± 10%; Results highlighted with ˙ indicates values in the range ± 30%; Results highlighted in *italics* indicates measured values over the maximum possible (*indicating their increases of concentration during the experimental process*)

**Table 11:** Experimental and re-calculated solubilities of the chemical substance in 60% of CH₃CN co-solvent.

| Measured pH after filtration | Calculated conc. (mg/mL) |
|---|---|
| 1.42 | 1.63625133 |
| 4.35 | 1.61895459 |
| 6.36 | 1.62730874 |
| 8.51 | 1.59879934 |

**Table 12:** Solubility/pH of solutions used for psKa' calculation of the chemical substance, in 80% of acetonitrile.

| Experimental results (Solubility in mg/mL) - Compound (80% CH₃CN) | | | | |
|---|---|---|---|---|
| pH | 1.42 | 4.35 | 6.36 | 8.51 |
| Solubility | 1.6363 | 1.619 | 1.6273 | 1.599 |
| Nb of acid functions | 1 | | | |
| pKas | pKa1 | | | |
| ACD (Percepta) | 1.9 | | | |
| Calculated results (Solubility in mg/mL and pKas) | | | | |
| Entity considered | HA | A | | |
| Intrinsic solubility | N.A. | 1.61897965 | | |
| pKa1 | Ka <0 | -0.55068474 | | |

**Table 13:** Extraction of the Mat-pKa table of results (chemical substance in 80% of acetonitrile).

| Measured pH | 1.42 | 4.35 | 6.36 | 7.35 | 8.51 | 8.57 |
|---|---|---|---|---|---|---|
| Measured solubility | 1.6353 | 1.619 | 1.6273 | *1.7277* | 1.599 | *2.2871* |
| Re-calculated solubility (Mat-pKa) | 1.6371# | 1.6190# | 1.6190# | 1.6190# | 1.6190# | 1.6190˙ |

Results highlighted with # indicate values in the range ± 10%; Results highlighted with ˙ indicate values in the range ± 100%; Results highlighted in *italics* indicate measured values over the maximum possible (*indicating their increases of concentration during the experimental process*)

**Table 14:** Experimental and re-calculated solubilities of the chemical substance in 80% of CH₃CN co-solvent.

The pKa value (-0.43) is far from the one suggested by ACD 2014 (pKa=1.9). But it confirms the added effects of the environment of the thiazole ring of the molecule, in terms of steric constraints and increase of the delocalization of the free doublet of electrons of the nitrogen atom in the conjugated bonds. On a practical point of view this pKa result confirms also, the difficulties encountered to synthesize a salt, with this very weak base.

## Discussion

### Influence of the organic solvent and of the aqueous buffered pH on the solubility of the chemical substance

Taking into account the assay provided for each solution (Table 5), the following global effects are pointed out through the interpretation of the Figure 8.

It is evidenced that the increase of the organic solvent proportion, increases considerably the solubility of the chemical substance, to be

equal to the maximum when the proportion of acetonitrile starts at 60%. At this level of co-solvent mixture, the influence of pH variation is very limited, suggesting that the non-protonated molecule is preferably solubilized in acetonitrile, consequently away from the effects of protons, which stayed separately in the aqueous solution.

On the other hand, insolubility of the chemical substance is observed between 0% and 20% of acetonitrile mixed with water, whatever the pH value, suggesting its moderate effect. These series of data cannot be exploited with Mat-pKa. Regarding the series of solubilities with 40% of acetonitrile, where values are lower than the maximum, it seems that the pH of the solution has an effect, confirming that, impact of this parameter takes place after the first of the organic organic solvent. In this series, we should notice that the solubility is limited at pH 5.09 and pH 8.79.

Interesting points of enhanced solubility with 40% of acetonitrile are between pH 1.68 and pH 3.76 (compatible with stomach environment) and between pH 6.91 and pH 7.83 (compatible with fasted intestine and plasma environment).

As a first result of this experiment, the drug substance formulated with excipients leading to an equivalent dielectric constant ($\epsilon$) between 56.4<$\epsilon$<64.75 (obtained in our case, with 40% to 60% of acetonitrile in water), should play an important role in the enhancement of the bioavailability of the compound, by increasing its solubility and by maintaining its sensitivity to pH for the transportation. As a second result, we should notice that no racemization and no degradation of the molecule were observed in the solutions tested.

### Mat-pKa software performance for this particular study

The Yasuda-Shedlovsky experiment is extensively used for the determination of pKas for insoluble compounds in aqueous media. Most of the time, it is associated to potentiometric titrations (or pH-metric titrations) of the substance of interest, to determine the inflexion points of the different curves and then the psKa related to each organic/water composition [8-10]. UV titrations are also frequent [11] and psKa are determined based on absorbance responses obtained in different mixtures of organic solvents and pH buffered solutions. Other initiatives were also largely developed to use capillary electrophoresis (CE), by addition of co-solvents in case of insoluble substances [12].

In these cases, psKa(s) are also issued from sigmoidal curves representing the ion mobilities observed versus the pHs applied.

Liquid chromatography technique was also developed to obtain pKas of molecules, but based on the variations of their retention time (as this is also used for the $logD_{pH}$ determination) when submitted to different buffered eluents [13]. Another way was to determine separately logP and $logD_{pH}$, from which pKa can be calculated [14], but apparently limited to one pKa per molecule.

But all of these initiatives are dependent on the capabilities to prepare aqueous buffered solutions (or to obtain them from the market) and of the corrosion resistance of the equipment, which finally are limiting the experimental pH range approximately between pH 2.5 to pH 10. This is a major concern when pKa values must be determined out of this range and particularly in the case of very weak bases, leading to pKas of conjugated acids between -2 and 1.5 for instance.

None of the constraints of methodologies or experiments mentioned above is affecting the pKa(s) determinations with Mat-pKa. And the particular case of the chiral chromatographic method used in this study reveals an extended capability of this software to be

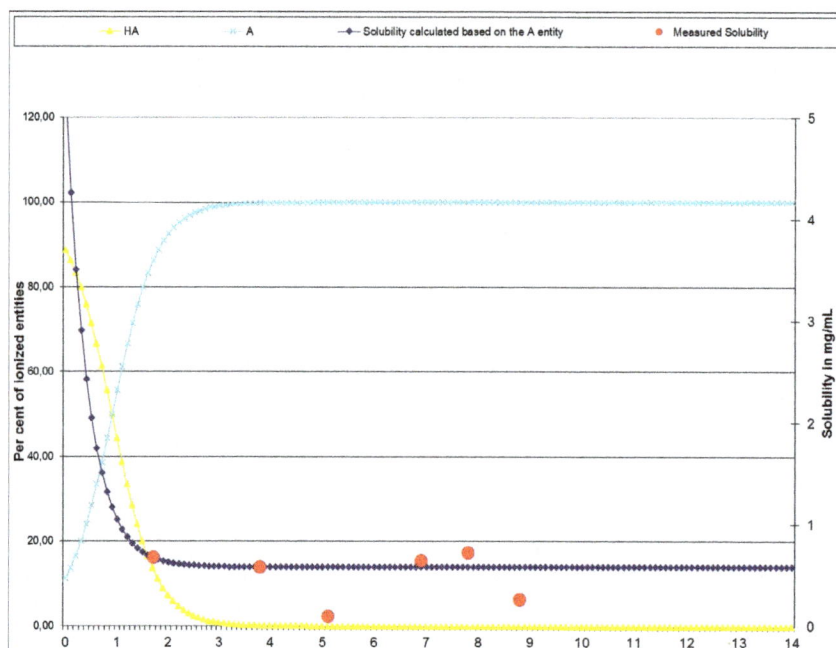

**Figure 4:** Experimental solubility and recalculated one (in 40% of acetonitrile).

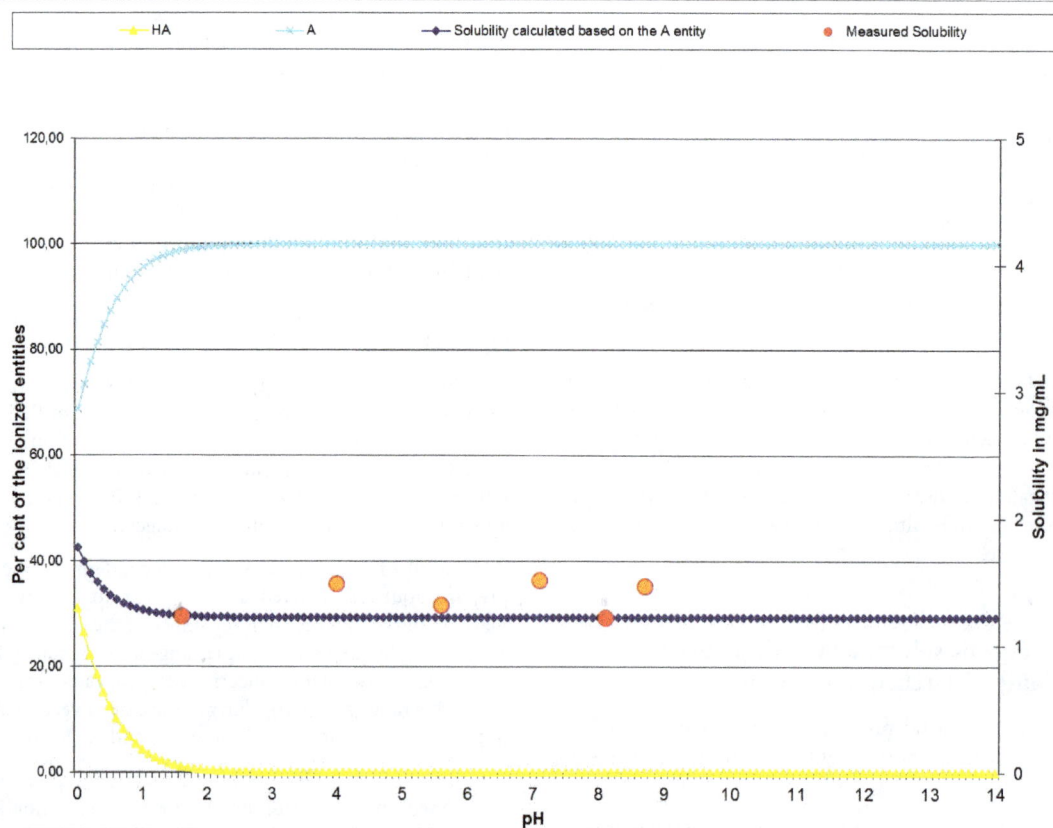

**Figure 5:** Experimental solubility and recalculated one (in 60% of acetonitrile).

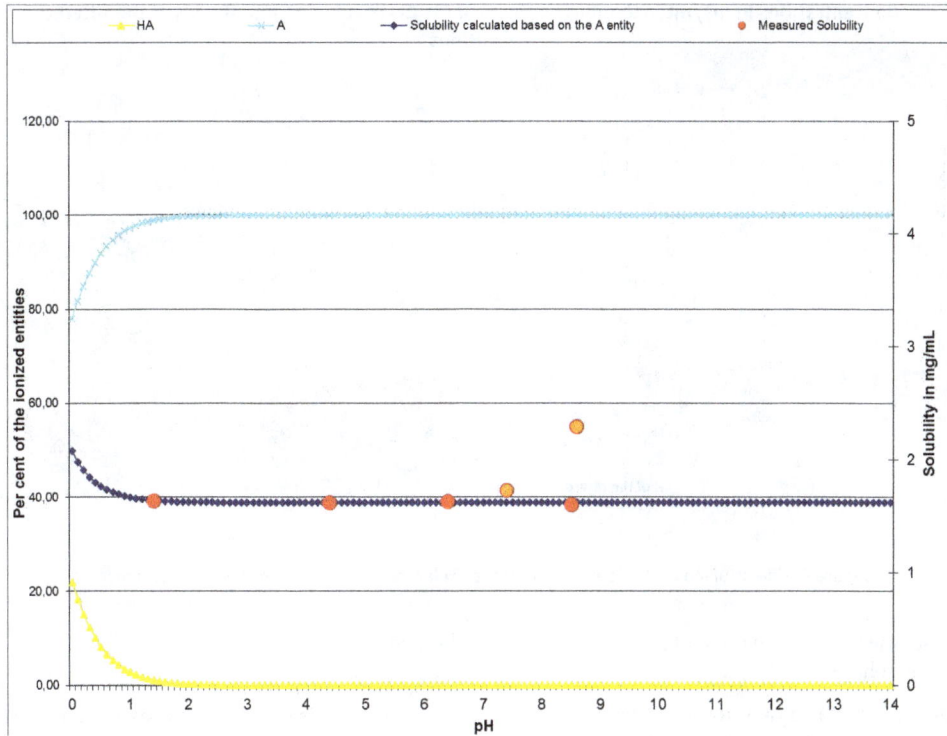

**Figure 6:** Experimental solubility and recalculated one (in 80% of acetonitrile).

| psKa' isuued from Mat-pKa | Organic composition (% V/V) | Dielectric Constant ($\varepsilon$) | 1/$\varepsilon$ | Log ([$H_2O$])=Log (55.508) |
|---|---|---|---|---|
| 0.900 | 40 | 64.75 | 0.0154 | 1.744 |
| -0.345 | 60 | 54.6 | 0.0183 | 1.744 |
| -0.551 | 80 | 45.55 | 0.0220 | 1.744 |

**Table 15:** Summarized data for Yasuda-Shedlovsky linearity calculation.

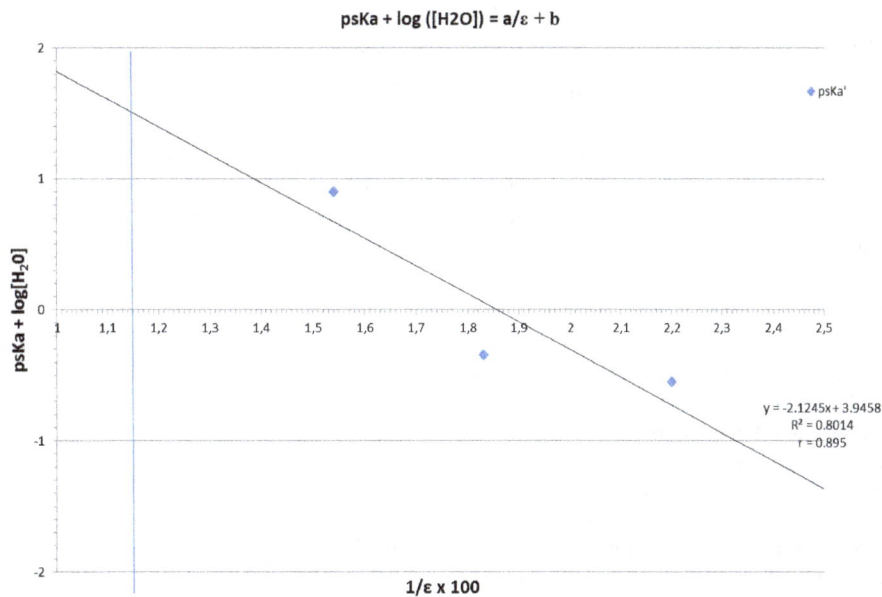

**Figure 7:** Yasuda-Shedlovsky curve based on the 3 psKa' obtained for the Compound.

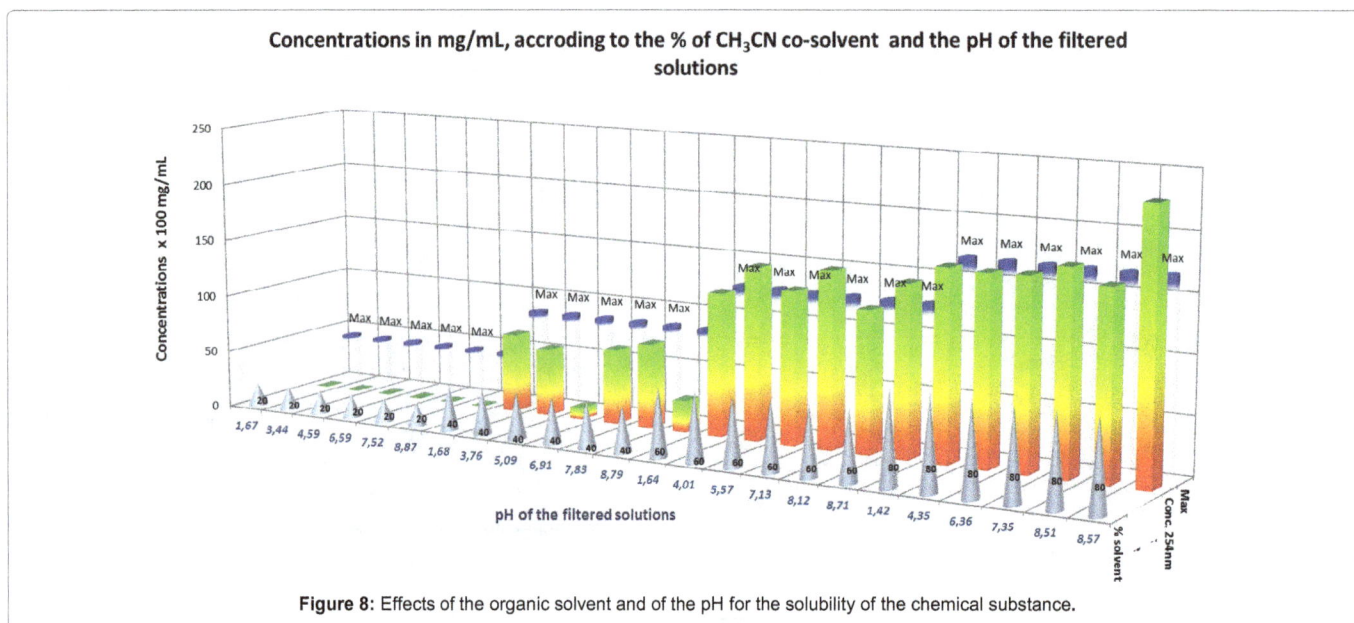

**Figure 8:** Effects of the organic solvent and of the pH for the solubility of the chemical substance.

independent of the technics used, except the needs of concentrations of the substance and pH of the solution.

The principal advantage of Mat-pKa software is based on the determination of unknown pKas driving the solubility curve in extended pH points, avoiding the detection of inflexion points, as this is the case for the major part of the dedicated techniques. This software consists in the resolution of systems of equations of Henderson-Hasselbalch [15], without any approximation. Calculations conducted in few seconds, can deliver a high number (N) of pKas for one molecule, depending on the availability of (N+1) couples of data of solubility/pH. As a result, these calculated values can be determined, outside of the range of experimental pH of buffered solutions. Also, these pKa(s) can be evaluated even if they are spaced or very close to each other.

The particular case of the chemical substance studied here demonstrates clearly the performance of the Mat-pKa software, which generated psKa' values between 0.9 and -0.55. No other techniques mentioned above would have been able to provide such results, either by Titration or UV spectrometry. Titration would have been limited by the needed concentration of the acid reagent to reach the efficient protonation of the amine and the equivalent point determination. UV spectrometry would have also been limited by the pH buffer preparation at negative pH. By the way, it is important to precise here, the prerequisites of the molecule to be sensitive to changes in absorbance when submitted to pHs variations, otherwise the UV metric method cannot be applied [13]. Finally, with these 2 methods, the 4 enantiomers of this chemical substance would have been analyzed as a mixture, which was not the objective of this study.

Considering the Capillary Electrophoresis, based on ions mobilities of the substance depending on the pH applied, we cannot exclude also the loss of the separation and the possible mixture for the 4 enantiomers in these variable and needed conditions to generate the appropriate data, exploited further to calculate the pKa of the molecule of interest. We should notice also, that a specific equipment (for instance, Combisep pKa analyser) and highest number of couples of data are needed (generally 12 different pH buffers, also limited between pH 2.0 to 10.0), compared to the 6 and less (4 and 2, in our non-voluntary cases) for Mat-pKa, with the maintained separation of

the 4 enantiomers.

Similar interpretation can be conducted for some other chromatographic methods, also based on the variation of the retention time of the molecule analyzed, versus the pH of the mobile phase applied.

## Conclusion

After 3 years of repetitive practices of the Mat-pKa software, several objectives have been completed. Now a new extension of Mat-pKa is available for pKa(s) determinations of chemical substances non-soluble in water, through the experiment of Yasuda-Shedlovsky.

This particular study succeeded due to the high selectivity of the chiral chromatographic method developed and validated, to maintain under control the quantitation of the good enantiomer. This is the first time that such a specific chiral chromatographic method is used for pKas determination. In this field of application Mat-pKa is probably the software the most appropriate to avoid the generation of titration curves (either by CE or HPLC) based on the retention time (or mobility) in function of the pH applied, with a high risk to lose the enantiomeric separations. By the way, Mat-pKa software needs less couples of data (meaning also time and quantities of compound) than the other methods, which needs titration curves and inflexion point determinations.

Finally, Mat-pKa has played its new role in these new particular conditions, to calculate the mandatory psKa' needed for the pKa determination through the linear extrapolation of the Yasuda-Shedlovsky experiment. The pKa value of -0.43 determined for this chemical substance is lower than the predicted one (1.9 by ACD Predicta). It demonstrates a highest limitation of access to the free electronic doublet of the nitrogen atom of the molecule, than this was estimated by ACD.

The solubility profile of the chemical substance in different combinations of acetonitrile / water mixture and at different pHs, revealed clearly the following statements:

-between 0% to 20% (v/v) of acetonitrile in water, the chemical substance stayed insoluble

-Above 60% (v/v) of acetonitrile in water, the chemical substance was fully soluble without any effect of the pH

-At 40% (v/v) of acetonitrile in water, the chemical substance was approximately 75% soluble and pHs variations of the solutions have added a supplementary effect on the solubility and potential transport of the molecule in biological media.

These results, issued from the direct exploitation of the Yasuda-Shedlovsky experiment are also indicative for potential adjustments needed for the formulations of the chemical substance.

## References

1. Vidaud L, Kugel C, Boccardi G, Schmidt S, Pommier JY (2012) Mat-pKa calculation tool development for evaluation of acidity constants from solubility profiles- Large study of 41 compounds. Int J Pharm 437: 137-155.

2. Martel AE, Motekaitis RJ (1992) Determination and Use of Stability Constants. In: Experimental procedure for Potentiometric pH measurements of Metal Complex Equilibria.

3. Bates RG, Schwarzenbach G (1955) Über pH-Werte nichtwässeriger Lösungen. Helv Chim Acta 38: 699-716.

4. Allen RI, Box KJ, Comer JEA, Peake C, Tam KY (1998) Multiwavelength Spectrophotometric Determination of Acid Dissociation Constants of Ionizable Drugs. J Pharm Biomed Anal 17: 699-712.

5. Box KJ, Donkor RE, Jupp PA, Leader IP, Trew DF, et al. (2008) The chemistry of multi-protic drugs Part 1: a potentiometric, multi-wavelength UV and NMR pH titrimetric study of the micro-speciation of SKI-606. J Pharm Biomed Anal 47: 303-311.

6. Yasuda M (1959) Dissociation Constants of Some Carboxylic Acids in Mixed Aqueous Solvents. Bull Chem Soc Jpn 32: 429-432.

7. Shedlovsky T (1962) In: Pesce B (Ed) Electrolytes. Pergamon, New York, USA.

8. Schellinger AP, Carr PW (2004) Solubility of Buffers in Aqueous-Organic Eluents for Reversed-Phase Liquid Chromatography. LCGC North America 22: 544-548.

9. Demiralay EC, Yilmaz H (2012) Potentiometric pKa determination of Piroxicam and Tenoxicam in acetonitrile-water binary mixtures. SDU Journal of Science 7: 34-44.

10. Avdeef A, Box KJ, Comer JEA, Gilges M, Hadley M, et al. (1999) pH-metric logP 11. pKa determination of water-insoluble drugs in organic solvent-water mixtures. J Pharm Biomed Anal 20: 631-641.

11. Tam KY, Takacs-Novak K (2001) Multi-wavelength spectrophotometric determination of acid dissociation constant: a validation study. Analitica Chemica Acta 434: 157-167.

12. Poole SK, Patel S, Dehring K, Workman H, Poole CF (2004) Determination of acid dissociation constants by capillary electrophoresis. J Chromatogr A 1037: 445-454.

13. Hossain MF, Obi C, Shrestha A, Faruk Khan MO (2014) UV-metric, pH-metric, RP-HPLC methods to evaluate the multiple pKa values of a ployprotic basic novel antimalarial drug lead, cyclen bisquinoline. Modern Chemistry Application 2: 145.

14. Chaing PC, Hu Y (2009) Simultaneous determination of logD, logP and pKa of Drugs using a Reverse Phase HPLC coupled with a 96-well plate auto injector. Comb Chem High Throughput Screen 12: 250-257.

15. Po HN, Senozan NM (2001) The Henderson-Hasselbach Equation: its History and Limitations. Journal of Chemical Education 78: 1499-1503.

# 28

# Total Reflection X-Ray Fluorescence Spectroscopy to Evaluate Heavy Metals Accumulation in Legumes

**Fabjola Bilo[1], Laura Borgese[1]\*, Annalisa Zacco[1], Pranvera Lazo[2], Claudia Zoani[3], Giovanna Zappa[3], Elza Bontempi[1] and Laura E Depero[1]**

[1]*Department of Mechanical and Industrial Engineering, University of Brescia, Brescia, Italy*
[2]*Faculty of Natural Science, University of Tirana, Tirane, Albania*
[3]*ENEA-UTAGRI, CR Casaccia, Rome, Italy*

## Abstract

This work is to demonstrate the usefulness of total reflection X-ray fluorescence (TXRF) for fast and reliable quantitative analysis of heavy metals in plants used for accumulation studies. A model study of beans germination in lead contaminated environment under controlled laboratory conditions was realized. Metal accumulation in different parts of the plant was evaluated. Two different sample preparation procedures for TXRF analysis were considered: microwave acid digestion and direct analysis of suspended powdered sample. Quantitative determination of macro, micro, and trace elements was performed. Root showed the highest accumulation of lead, followed by stem, leaves and crops. Results showed that direct analysis of suspended powdered samples may be used as a fast and simple method for screening.

**Keywords:** TXRF; Heavy metals; Food; Environment; Accumulation; Beans

## Introduction

Pulses are known for their nutritious composition. They contain a high number of bioactive substances including enzyme inhibitors, oligosaccharides, and phenolic compounds offering beneficial effects [1]. Beans are classified as the main group of pluses used as common food for humans [2] and animals [3,4] due to their composition [5]. According to the 2005 Dietary Guidelines, a frequent consumption of beans (four or more times per week) is recommended [6]. Regarding the mineral composition, elements such as Ca, K, Mg are considered macronutrients, while Fe, Cu, Zn, Mn, Ni micronutrients [7].

Beans are characterized not only by high biomass production but also an intensive heavy metals accumulation [1]. In the case of heavy metals, it is known that both the direct contact with polluted environment and the consumption of contaminated food may cause serious health damages [8-10]. Dramatic effects of heavy metals on growth and development of animals and plants are widely known [11,12], and more recent studies have revealed that even essential elements, such as Mn, may be dangerous if extensive exposure (i.e., from food, work and the environment) occurs [13]. Horticultural plants and cereals are widely produced for human and animal consumption, and they can play an important role in the assumption of potentially toxic elements and heavy metals. The uptake of heavy metals depends on many factors, such as the biological specificity of the plant, the conditions of soil, water, and air in the growing environment [14-16]. However, the correlation between metal contaminants in soil and crops is complex and not obvious [17]. Indeed, the bioavailability of the contaminant is one of the main factors to be considered in order to assess possible effects on the food chain [18-20], but it's not enough. It is crucial to know the elemental composition of food in relation with the estimated amount consumed. In this context, the World Health Organization (WHO) [21], the US- Environment Protection Agency (US-EPA) [22], and the European Commission (EC) [23] have already determined the provisional tolerable daily intake (PTDI) guidelines for potentially toxic elements. In the frame of Surveillance methods for routine monitoring, the association of analytical communities (AOAC) has developed the performance requirements of standard methods for heavy metals determination in a wide diversity range of foods and beverages, comprising plants. The required values for limit of quantification (LOQ), repeatability (r), reproducibility (R), and recovery [24] have been set. Atomic absorption spectroscopy (AAS) and inductively coupled plasma (ICP) based spectroscopies are the reference techniques [25,26].

Total reflection X-ray fluorescence (TXRF) is a technique for elemental analysis which has been recently becoming very attractive in environmental and food fields. Indeed, TXRF offers some advantages compared to AAS or ICP such as the small amount of sample required (few mg or μL), the absence of matrix effects, the possibility to perform direct analysis [27,28], and short measurement times (100-1000 s) for simultaneous multi-elemental analysis. Moreover, the actual commercial bench top systems do not need gasses or water cooling, allowing a very simple instrumental setup and reducing maintenance costs. TXRF is a geometrical modification of conventional X-ray fluorescence (XRF), leading to a substantial improvement of detection limits [29]. In recent years, many studies about TXRF analysis of environmental samples such as water [30-32], soils [33-36], air particulate matter [37-39], bio-monitors [40-42] and plants [43-46] have been published. Recently, applications of TXRF for the analysis of foodstuffs have been also reviewed [47]. An additional interesting aspect to be considered is that TXRF could be used as fast screening tool for simultaneous multi-elemental determination at the very low level.

One of the most successful ways to obtain valuable information about the interaction of heavy metals with plants is the determination of their compositions after growing in controlled environmental

**\*Corresponding author:** Laura Borgese, Dipartimento di Ingegneria Meccanica e Industriale Via Branze 38, 25 123 Brescia, Italy
E-mail: laura.borgese@unibs.it

conditions in presence or absence of heavy metals. The analysis of different parts of the grown plants, such as leaves, stems, roots and crops, figures out composition changes and heavy metals accumulation.

Aim of this work is to demonstrate the usefulness of TXRF for fast screening and reliable quantitative analysis of heavy metals in plants used for accumulation studies. A model study of plant germination and growth in Pb contaminated environment has been carried out in order to evaluate what may happen to plants grown under extremely polluted conditions. A first germination study, TXRF was used to prove its screening capabilities. Then, a second and more extensive germination study TXRF was used to assess the correlation between the content of lead in the environment and in different parts of bean plants germinated modulating the amount of soluble Pb.

## Materials and Methods

Red kidney bean (*Phaseolus vulgaris*) commercially available for human consumption was used. Two germination experiments were performed. In the first germination study, 15 seeds were sown in 10 and 100 mg/L of lead nitrate (Sigma Aldrich) solutions and MilliQ (MQ) water as reference. In the second germination study, seeds were grown in different concentration of $Pb(NO_3)_2$ solution, respectively 2, 4, 10, 50, and 100 mg/L and MQ water. Germination studies were performed in laboratory conditions at 20-22°C and 12 h in artificial light and 12 h in dark. Plants growth was regularly observed. After 12 days the length of stems and roots was measured, and different parts of plant were collected and weighed. In particular roots, leaves, crops and stems were considered. The collected samples were dried at 60°C for 24 hours and weighed as dry mass. A total number of 24 plant samples were analyzed for the determination of macro, micro and trace elements.

The certified reference material (CRM) SRM-1570A (*Trace Elements in Spinach Leaves*) from NIST [48] was considered as reference and used without any further drying or grinding step.

For suspension, the dried sample was ground into fine powder using an agate mortar and sifted to 600 μm. About 10 mg of powdered sample were mixed with 990 μL of water solution containing Triton X-100 1% wt to prepare the suspension. After that, 10 μl of 100 mg/L gallium in nitric acid used as internal standard (IS), (Ga-ICP Standard Solution, Fluka, Sigma Aldrich) were added, in order to obtain a final Ga concentration of 1 mg/L. Samples were vortexed for 1 min at 2500 rpm and homogenized in ultrasonic bath for 15 min.

For digestion, approximately 0.15 g of dried sample were put in Teflon vessels, added with 18 mL of concentrated nitric acid (65% - Sigma Aldrich) and 2 mL of MilliQ (MQ) water [49,50]. Samples were digested using CEM SP-D microwave system, equipped with 24-places auto-sampler closed vessel. Each sample was individually processed. The microwave energy applied was precisely controlled by monitoring temperature and pressure of the sample, to obtain the maximum efficiency. A five steps procedure was automatically performed to have complete digestion: 3 min at 160°C, 5 min at 180°C, 3 min at 200°C, 5 min at 205°C, and 10 min at 210°C. After cooling, the volume of each sample was adjusted to 25 mL adding MQ water. Quantitative analysis was performed using Ga as IS, in concentration 1 mg/L. Therefore, 50 μL of IS solution, with Ga concentration 10 mg/L, were added to 450 μL of digested sample.

Quartz glass reflectors were cleaned, the blank was checked and siliconized, putting a drop of 10 μl of Silicone solution in isopropanol (Serva Electrophoresis, Heidelberg, Germany), to obtain a hydrophobic surface. A drop of 10 μl of the prepared specimens was deposited in

the center of the prepared reflector and dried on a hot plate at 50°C. Three trials were prepared and measured for each specimen. TXRF measurements were carried out by a Bruker S2 Picofox spectrometer (Bruker AXS Microanalysis GmbH, Berlin, Germany), equipped with a Mo tube operating at 750 μA and 50 kV, multilayer monochromator, silicon drift detector (SDD) and energy resolution was 165 eV at 5.9 keV. Samples were irradiated for 600 s live time [50].

## Results and Discussion

A first evaluation of the effect of Pb on the growth of beans was performed measuring the percentage of germination (PG), defined as the ratio between the number of grown seeds with respect to the total germinated seeds. PG was 100% in the reference solution, MQ water, and decreased with increasing the concentration of soluble Pb, as it was expected [2]. PG about 40, 33, 27, 23, and 20 were found for 2, 4, 10, 50 and 100 mg/L of Pb, respectively.

Figure 1 shows the effect of Pb concentration on the length of stems and roots of the second germination study, calculated as the average of three measured samples, highlighting a significant negative relationship.

Elemental analysis of germinated beans was performed by TXRF. The main requirements to perform TXRF analysis are having an X-ray reflector carrier and a sample deposited on it as a thin film [29]. For this reason, most of the literature about TXRF analysis reports the use of pretreatments for solid sample solubilization. A comparison of sample preparation procedures for TXRF analysis of plants is reported in Table

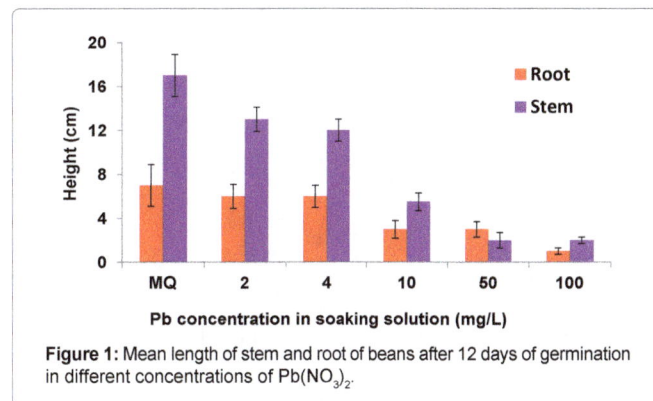

**Figure 1:** Mean length of stem and root of beans after 12 days of germination in different concentrations of $Pb(NO_3)_2$.

| Type of pre-treatment | Type of Sample | Sample amount (g) | Treatment conditions | References |
|---|---|---|---|---|
| Dry Ashing | Leaves | 2 | Heating at 500°C (muffle oven) | [41] |
| Wet Digestion | Leaves | 0.1 | Heating at 120°C in an electronical furnace | [57] |
| Ultrasound-assisted extraction | Spices Leaves Flowers | 0.01 | Sonication using a cup-horn sonoreactor and centrifugation | [51] |
| Microwave acid digestion | Leaves | 0.5 | Digested with $HNO_3$ and $H_2O_2$ in micorwave oven | [46] |
| | Lichens | 0.15 | | [40] |
| | Leaves Root Stem Crop | 0.15 | | This study |
| Suspended Powdered | Root | 0.01 | Suspension of powdered in Triton X-100, 1% solution | This study |

**Table 1:** Sample preparation procedures for elemental chemical analysis of plants by means of TXRF.

1. Dry ashing and wet digestion are the most widely employed for this kind of analysis. However, both these procedures are time-consuming, they require laboratory instrument and loose of volatile elements. Ultrasound Assisted Extraction is a more rapid sample preparation procedure compared to above mentioned, but parameters such as slurry stabilization and sedimentation errors should be carefully considered. Moreover, each sample should be treated independently from the others. Even known as a critical step, microwave digestion is usually the preferred sample preparation procedure for TXRF analysis, leading to higher sample homogeneity and lower spectral background. Furthermore, direct analysis of suspended powders is also possible. Indeed, suspension is simple and fast, it does not require any additional instrumentation, lowering also the risk of sample contamination. The main drawbacks of suspension are lower homogeneity of the sample and higher spectral background, due to particles scattering.

We have tested digestion and suspension as valuable procedure for sample preparation and TXRF analysis. Accuracy of both the proposed methods was tested with SRM 1570 A, selected for the similarity of the matrix with the tested samples. Figure 2 shows TXRF spectra of digested and suspended CRM. Qualitative analysis of TXRF measurements identifies the presence of Cl, K, Ca, Ti, Cr, Mn, Fe, Ni, Cu, Zn, Pb, Br, Rb and Sr. The intensity of all elements, with respect to the IS, is higher for the digested sample, except for Cl. This may be due both to higher absorption effects and lower homogeneity of the suspended

sample. The different behavior of Cl, having higher signal in the suspended sample, highlights one of the main drawbacks of digestion, the possible loss of volatile elements, which is removed as HCl gas during this process. Quantitative analysis is performed starting from K, because significant absorption effects measuring in air conditions and lower fluorescence yield occur for lighter elements [51]. All results and detection limits (DL) obtained for digested and suspended CRM are reported in Table 2. Certified reference values are reported for comparison. As it was expected, considering what have been previously stated, DL of all the elements is higher for suspension. The comparison between digestion and suspension show that: for K, Ca and Pb results of digested samples are higher with respect to suspended, while the opposite occurs for Mn, Ni, Cu, Zn Rb and Sr. Relative Standard Deviation (RSD) values are comparable and lower than 10% for all the elements with the exception of Pb, where the RSD is 14% and 24% for digested and suspended sample respectively, probably due to the low Pb concentration very near to the detection limit (DL). Statistical analysis, based on student t test, shows that results of TXRF analysis differ significantly from the reference values only for K. In this case, $t_{crit}$=4.30 (P>95%, n-1=2). Results obtained with CRM highlights some critical aspects in TXRF analysis of suspended sample. However, the obtained degree of accuracy suggests that this method can be proposed as a suitable tool for a reliable sample screening.

The first germination study was performed to verify that the

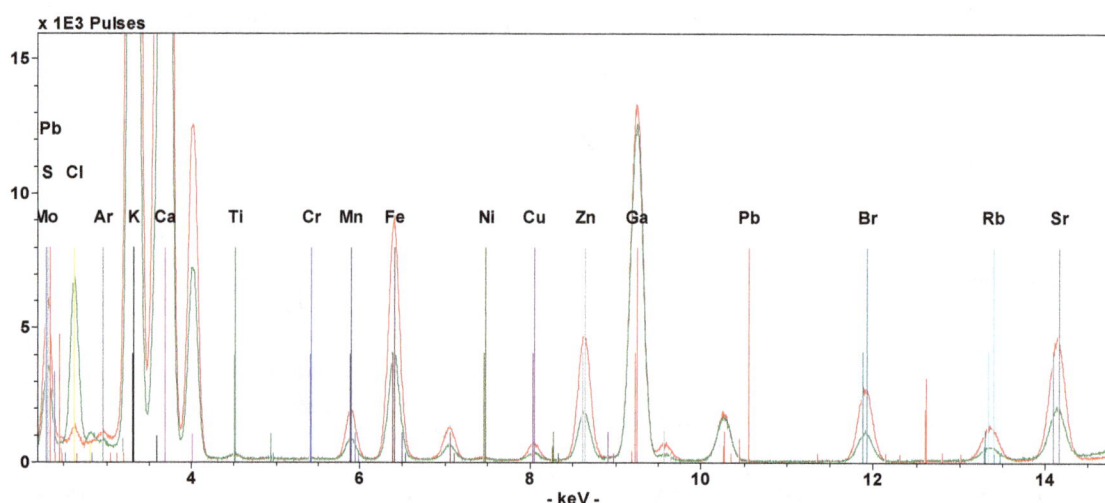

**Figure 2:** TXRF spectra of digested (red) and suspended (green) SRM 1570 A.

| Elements | Certified values (mg/Kg) | TXRF Digested | | | | TXRF Suspended | | | |
|---|---|---|---|---|---|---|---|---|---|
| | | Mean ± CI (mg/Kg) | RSD (%) | $t_{exp}$ | DL (mg/Kg) | Mean ± CI (mg/Kg) | RSD (%) | $t_{exp}$ | DL (mg/Kg) |
| K | 29000 ± 520 | 21227 ± 2499 | 6 | 10.8 | 2.8 | 19879 ± 7173 | 2 | 4.4 | 5.7 |
| Ca | 15300 ± 410 | 13184 ± 2102 | 8 | 3.5 | 2.1 | 16103 ± 5793 | 2 | 0.5 | 4.4 |
| Mn | 75.9 ± 1.9 | 75.7 ± 16 | 3 | 0.04 | 0.3 | 68.1 ± 24 | 1 | 1.1 | 0.8 |
| Ni | 2.14 ± 0.1 | 2.57 ± 1 | 6 | 1.6 | 0.2 | 2.1 ± 1 | 9 | 0.1 | 0.4 |
| Cu | 12.2 ± 0.6 | 13.2 ± 2 | 1 | 1.3 | 0.14 | 11.4 ± 4 | 3 | 0.8 | 0.4 |
| Zn | 82 ± 3 | 85 ± 17 | 0.5 | 0.7 | 0.14 | 70 ± 25 | 1 | 1.6 | 0.3 |
| Rb | 12.7 ± 1.6 | 11.6 ± 2 | 2 | 1.6 | 0.1 | 8.9 ± 3 | 3 | 4.1 | 0.3 |
| Sr | 55.6 ± 0.8 | 57.2 ± 11 | 0.6 | 0.5 | 0.14 | 51.8 ± 19 | 3 | 0.7 | 0.4 |
| Pb | 0.2 ± | 0.20 ± 0.1 | 14 | | 0.1 | 0.25 ± 0.1 | 24 | 1.3 | 0.15 |

`Mean is the average of three measurements and CI is the confidence interval

**Table 2:** Results of TXRF analysis performed on digested and suspended SRM 1570 A in comparison with reference values. Precision is expressed as RSD.

model experiment would have leaded to a substantial and measurable accumulation of Pb in beans germinated in contaminated environment. Roots of 15 seeds germinated in MQ, and 10, 100 mg/L lead nitrate solution were measured by TXRF after suspension in a solution containing MQ water and Triton X-100, as stabilizing agent to prevent particles sedimentation and provide reproducible thin layers on sample carrier after drying [52]. The TXRF spectrum of suspended bean roots germinated in 10 mg/L Pb solution is reported in Figure 3. Qualitative analysis shows the presence of Cl, K, Ca, Ti, Mn, Fe, Ni, Cu, Zn, Pb, Br, Rb and Sr. The comparison of suspended samples spectra reported in Figure 2 and Figure 3 show a higher background in the case of roots. Indeed, the homogeneity of the CRM, with particle size less than 75 μm, is much higher with respect to the sifted root beans powder, leading to lower repeatability and less accurate quantitative analysis in the latter case [52]. The obtained results are reported in Table 3, where higher RSD values are calculated with respect to CRM. Despite of the lower accuracy these results allow to verify a significant accumulation of Pb in germinated beans proportional to the concentration of the growth solution, as well as the unexpected presence of measurable quantities of Pb in beans from uncontaminated environment.

On the basis of these preliminary results, we have performed a second and more specific germination study to analyze Pb accumulation in different parts of plants grown in solutions with modulated content of Pb. The TXRF spectrum of digested bean root germinated in 10 mg/L Pb solution is reported in Figure 3. Comparison with the spectrum of suspended root shows the same composition, highlighting the repeatability of the two germination study and TXRF measurements. As it was already observed for CRM (Figure 2), the background is higher for suspension. Results of quantitative analysis were reported in Table 4. Composition data obtained by both sample preparation

procedures agree with the average mineral composition of beans in all the analyzed parts of the plant: K and Ca (2-5%) are the major elements, Fe, Zn and Cu (2-50 mg/Kg) are minor elements, while Mn, Ni, Rb and Sr (<10 mg/Kg) are in traces. [7] Linear discriminant analysis (LDA) was applied to evaluate elements contribution to the differentiation of the four groups of samples: leaves, crops, stems and roots. LDA plot of two canonical functions is reported in Figure 4. Results of LDA show that the total variance is explained by three discriminant factors. The first factor is responsible for 67.4% of the total variance, and the largest absolute correlation is found for Mn, Ni, and Rb. The second factor accounts for 29.2% of total variance, with Fe and Cu having the largest correlation. The third factor explains 3.4% of the total variance and it includes K and Ca. LDA shows that only seven elements are sufficient to discriminate different parts of plants.

Pb is present in all the analyzed samples, with concentration higher than that reported in other similar studies [14-16,53]. This is probably due to the longer germination period, 12 days in our case compared to 5-7 days of the other studies, and absence of chelators in the growth solution [54]. Pb uptake of plants grown in MQ water (reference) and those grown in all the considered concentrations of lead nitrate solutions shows significant differences. Unexpectedly, Pb was detected also in reference samples, strongly suggesting the unwanted presence of Pb in seeds. The relation between Pb content in plant samples and in the growing solution is shown in Figure 5 for roots, stems, leaves, and crops. A positive correlation is present in all cases, even if a poor linearity is observed for roots and stems. Pb accumulated in root is almost two orders of magnitude higher than in all the other parts of plant. This can be explained by the uptake mechanism of the plant, where metal ions penetrate through the roots. Roots act as a sort of barrier for Pb transfer into the plant [54], because it is known that

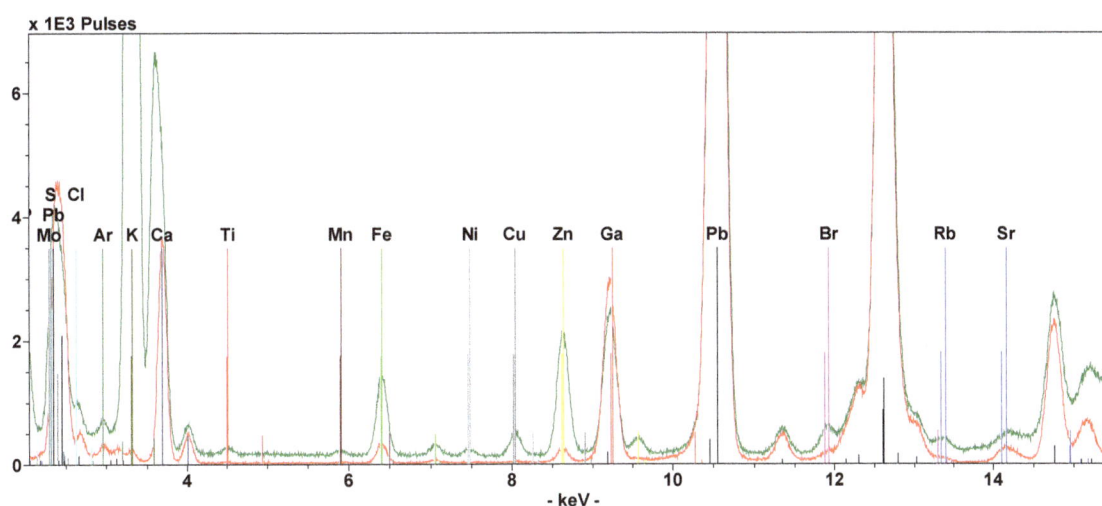

**Figure 3:** TXRF spectra of digested (red) and suspended (green) root beans germinated in 10 mg/L Pb(NO₃)₂.

| Part of plant | Environment Pb concentration (mg/L) | Elemental concentration (mg/Kg) | | | | | | | | | |
|---|---|---|---|---|---|---|---|---|---|---|---|
| | | K | Ca | Mn | Fe | Ni | Cu | Zn | Rb | Sr | Pb |
| Root | 0 | 20471 ± 3559 | 7342 ± 1291 | 14 ± 3 | 273 ± 49 | 15 ± 3 | 44 ± 7 | 244 ± 42 | 3.1 ± 0.6 | 21 ± 4 | 62 ± 13 |
| | 10 | 34260 ± 5947 | 2099 ± 372 | 12 ± 2 | 194 ± 36 | 9 ± 2 | 33 ± 6 | 161 ± 30 | 4.1 ± 0.9 | 3.5 ± 1 | 2759 ± 476 |
| | 100 | 9616 ± 2344 | 869 ± 146 | 3 ± 2 | 60 ± 12 | 2.1 ± 0.4 | 13 ± 3 | 41 ± 7 | n.d. | n.d. | 17615 ± 2964 |

*n.d. = less than detection limit

**Table 3:** Elemental concentration of suspended samples expressed as the average and 95% confidence range.

| Part of plant | Environment Pb Concencentration (mg/L) | Elemental concentration (mg/Kg) | | | | | | | | | |
|---|---|---|---|---|---|---|---|---|---|---|---|
| | | K | Ca | Mn | Fe | Ni | Cu | Zn | Rb | Sr | Pb |
| Leaves | 0 | 23850 ± 8605 | 483 ± 181 | 15 ± 6 | 128 ± 46 | 6 ± 2 | 23 ± 8 | 77 ± 29 | 10 ± 4 | 0.6 ± 0.5 | 0.30 ± 0.02 |
| | 2 | 25723 ± 9370 | 711 ± 280 | 17 ± 7 | 130 ± 47 | 5 ± 2 | 21 ± 8 | 75 ± 27 | 6 ± 2 | 1.0 ± 0.4 | 3.5 ± 1 |
| | 4 | 24919 ± 9000 | 1417 ± 628 | 18 ± 8 | 145 ± 53 | 8 ± 3 | 27 ± 60 | 91 ± 33 | 3 ± 1 | 3 ± 1 | 11 ± 4 |
| | 10 | 25069 ± 9050 | 753 ± 274 | 13 ± 5 | 140 ± 51 | 9 ± 3 | 25 ± 9 | 98 ± 43 | 4 ± 2 | 0.8 ± 0.4 | 38 ± 14 |
| | 50 | 18599 ± 6723 | 1610 ± 592 | 7 ± 3 | 108 ± 44 | 8 ± 3 | 19 ± 7 | 65 ± 23 | 5 ± 2 | 3 ± 1 | 156 ± 61 |
| | 100 | 29643 ± 10794 | 1617 ± 629 | 13 ± 5 | 155 ± 60 | 13 ± 5 | 31 ± 11 | 113 ± 41 | 10 ± 4 | 2.5 ± 1 | 581 ± 212 |
| Roots | 0 | 18777 ± 6782 | 1230 ± 457 | 5 ± 2 | 76 ± 31 | 4 ± 2 | 11 ± 4 | 51 ± 19 | 6 ± 2 | 5 ± 2 | 0.60 ± 0.05 |
| | 2 | 21484 ± 7814 | 1437 ± 519 | 6 ± 2 | 104 ± 38 | 5 ± 2 | 14 ± 5 | 67 ± 24 | 5 ± 2 | 5 ± 2 | 279 ± 101 |
| | 4 | 22650 ± 8463 | 1795 ± 679 | 8 ± 4 | 85 ± 31 | 7 ± 6 | 15 ± 5 | 79 ± 31 | 3 ± 1 | 4 ± 2 | 484 ± 175 |
| | 10 | 23742 ± 8569 | 1143 ± 414 | 8 ± 4 | 110 ± 40 | 4 ± 2 | 14 ± 5 | 77 ± 28 | 4 ± 2 | 3 ± 1 | 1282 ± 462 |
| | 50 | 22465 ± 8111 | 1223 ± 590 | 6 ± 2 | 80 ± 31 | 7 ± 3 | 16 ± 6 | 62 ± 23 | 6 ± 2 | 1.3 ± 0.6 | 2947 ± 1065 |
| | 100 | 20848 ± 7523 | 1627 ± 588 | 5 ± 2 | 61 ± 22 | 1.5 ± 1 | 15 ± 5 | 69 ± 27 | n.d | n.d | 13920 ± 5024 |
| Crops | 0 | 11355 ± 3712 | 708 ± 440 | 15 ± 5 | 60 ± 20 | 1.1 ± 0.7 | 8 ± 2 | 28 ± 9 | 4 ± 1 | 3 ± 1 | 0.10 ± 0.05 |
| | 2 | 12613 ± 4605 | 568 ± 205 | 16 ± 6 | 61 ± 22 | 0.8 ± 0.3 | 6 ± 2 | 28 ± 10 | 3 ± 1 | 4 ± 1 | 4 ± 1 |
| | 4 | 13107 ± 4738 | 634 ± 244 | 17 ± 6 | 78 ± 28 | 1.4 ± 0.9 | 8 ± 3 | 34 ± 12 | 1.8 ± 0.6 | 4 ± 1 | 7 ± 3 |
| | 10 | 10020 ± 3635 | 523 ± 189 | 16 ± 6 | 69 ± 25 | 0.9 ± 0.3 | 7 ± 3 | 37 ± 13 | 2.1 ± 0.8 | 3 ± 1 | 16 ± 6 |
| | 50 | 7718 ± 2798 | 468 ± 220 | 14 ± 5 | 65 ± 25 | 0.7 ± 0.4 | 6 ± 2 | 27 ± 10 | 3 ± 1 | 3 ± 1 | 65 ± 24 |
| | 100 | 10992 ± 4005 | 464 ± 168 | 14 ± 5 | 55 ± 20 | 0.9 ± 0.3 | 7 ± 3 | 28 ± 10 | 4 ± 1 | 2.4 ± 0.9 | 134 ± 49 |
| Stem | 0 | 20227 ± 1752 | 349 ± 30 | 12 ± 4 | 77 ± 28 | 3 ± 1 | 15 ± 5 | 54 ± 19 | 7 ± 2 | 0.4 ± 0.2 | 0.2 ± 0.09 |
| | 2 | 10798 ± 3901 | 1714 ± 667 | 11 ± 4 | 44 ± 20 | 2.5 ± 1.3 | 12 ± 4 | 46 ± 17 | 5 ± 2 | 5 ± 2 | 21 ± 8 |
| | 4 | 27529 ± 9940 | 1143 ± 571 | 15 ± 5 | 150 ± 56 | 8 ± 3 | 23 ± 8 | 100 ± 36 | 2.5 ± 1 | 2 ± 1 | 23 ± 8 |
| | 10 | 20668 ± 7510 | 481 ± 175 | 12 ± 4 | 91 ± 33 | 5 ± 2 | 15 ± 6 | 66 ± 24 | 2.8 ± 1 | 0.6 ± 0.4 | 61 ± 22 |
| | 50 | 26209 ± 9465 | 749 ± 276 | 10 ± 4 | 119 ± 43 | 10 ± 4 | 23 ± 8 | 80 ± 29 | 6 ± 2 | 1.6 ± 0.7 | 331 ± 120 |
| | 100 | 25750 ± 9329 | 1134 ± 409 | 11 ± 4 | 117 ± 42 | 7 ± 2 | 20 ± 7 | 77 ± 28 | 5.5 ± 2 | 3 ± 1 | 750 ± 271 |

˙n.d.: less than detection limit

**Table 4:** Elemental concentration of digested samples expressed as the average and 95% confidence range.

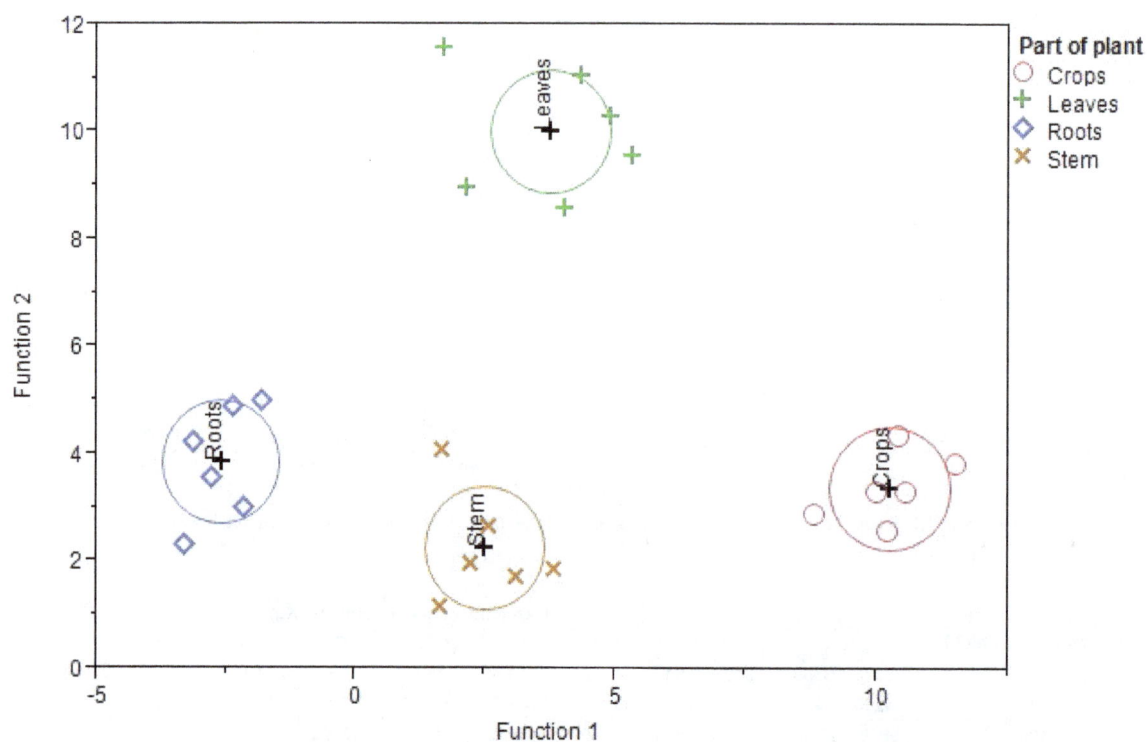

**Figure 4:** Results of LDA for two canonical functions applied to different part of plants.

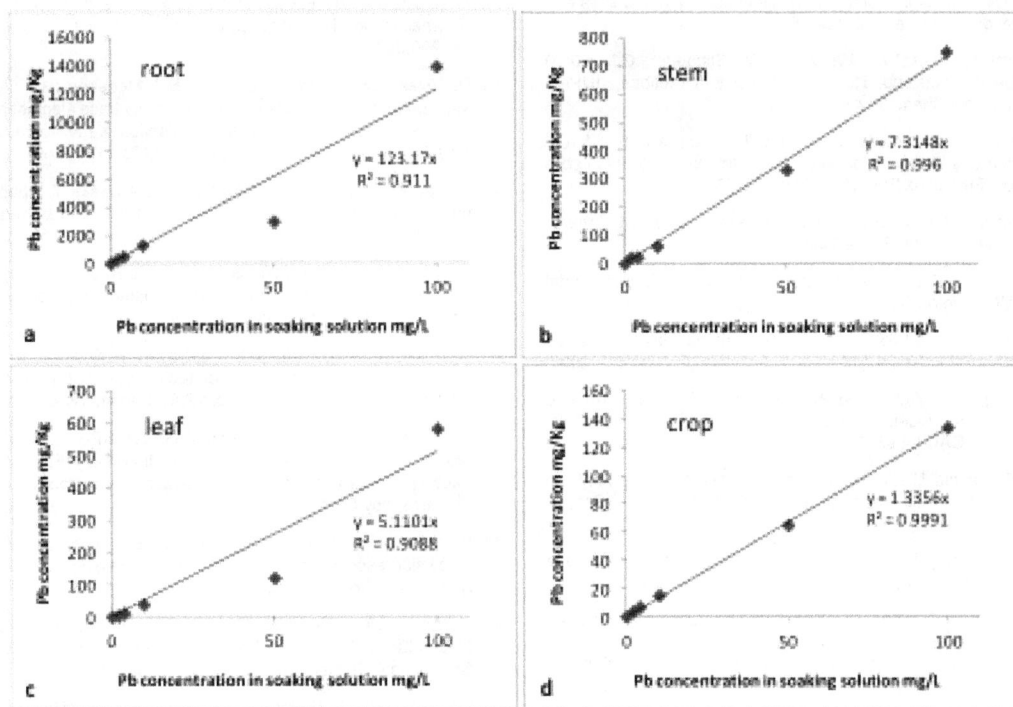

**Figure 5:** Correlation between Pb content in the growing solution and in a) root, b) stem, c) leaf and d) crops.

Pb has lower mobility than other heavy metals, i.e., Cd [53]. Indeed, Pb accumulation decreases from roots, stems, leaves and crops in agreement with literature [53,55,56].

## Conclusion

The composition of plants germinated in polluted environment is fundamental to assess the potential health risk related to the assumption of heavy metals from food. In this work, two germination studies of beans in Pb contaminated controlled conditions are performed to evaluate TXRF as useful method for fast sample screening and accurate quantitative analysis. Two sample preparation procedures are considered: suspension of powdered samples in water and microwave acid digestion. The SRM 1570A is used as CRM to check the accuracy of quantitative analysis. A good correlation is achieved between certified values of SRM NIST 1570 A and those founded from both sample preparation methods. It is highlighted that direct analysis of suspensions gives less accurate results than microwave acid digestion. Our results confirms that root have the highest ability to accumulate Pb, followed by stem, leave and crop. In conclusion, this study demonstrates that TXRF is a suitable analytical technique for reliable quantitative elemental analysis of plant samples with a good accuracy, and direct TXRF analysis is suitable for screening purposes.

## References

1. Campos-Vega R, Loarca-Piña G, Dave Oomah B (2010) Minor components of pulses and their potential impact on human health. Food Res Int 43: 461-482.

2. Yin Y, Fatufe A, Blachier F (2007) Soya Bean Meal and Its Extensive Use in Livestock Feeding and Nutrition.

3. Heuzé V, Tran G, Nozière P, Lebas F (2015) Common bean (Phaseolus vulgaris). Feedipedia, a programme by INRA, CIRAD, AFZ and FAO.

4. Otten JJ, Hellwig JP, Meyers LD (2006) Dietary Reference Intakes: The Essential Guide to Nutrient Requirements. National Academy Press, Washington DC, USA.

5. USDA (2011) Nutrient Database.

6. De la Guardia M, Garrigues S (2015) Handbook of Mineral Elements in Food. John Wiley & Sons.

7. Guerra F, Trevizam AR, Muraoka T, Marcante NC, Canniatti-Brazaca SG (2012) Heavy metals in vegetables and potential risk for human health. Sci Agric 69: 54-60.

8. Puschenreiter M, Horak O, Friesl W, Hartl W (2005) Low-cost agricultural measures to reduce heavy metal transfer into the food chain-a review. Plant Soil Environ 51: 1-11.

9. Jaishankar M, Tseten T, Anbalagan N, Mathew BB, Beeregowda KN (2014) Toxicity, mechanism and health effects of some heavy metals. Interdiscip Toxicol 7: 60-72.

10. Fargasová A (1994) Effect of Pb, Cd, Hg, As, and Cr on germination and root growth of Sinapis alba seeds. Bull Environ Contam Toxicol 52: 452-456.

11. Mahaffey KR, Capar SG, Gladen BC, Fowler BA (1981) Concurrent exposure to lead, cadmium, and arsenic. Effects on toxicity and tissue metal concentrations in the rat. J Lab Clin Med 98: 463-481.

12. Borgese L, Federici S, Zacco A, Gianoncelli A, Rizzo L, et al. (2013) Metal fractionation in soils and assessment of environmental contamination in Vallecamonica, Italy. Environ Sci Pollut Res Int 20: 5067-5075.

13. Liu H, Probst A, Liao B (2005) Metal contamination of soils and crops affected by the Chenzhou lead/zinc mine spill (Hunan, China). Sci Total Environ 339: 153-166.

14. Muchuweti M, Birkett JW, Chinyanga E, Zvauya R, Scrimshaw MD, et al. (2006) Heavy metal content of vegetables irrigated with mixtures of wastewater and sewage sludge in Zimbabwe: Implications for human health. Agric Ecosyst Environ 112: 41-48.

15. Cobb GP, Sands K, Waters M, Wixson BG, Dorward-King E (2000) Accumulation of heavy metals by vegetables grown in mine wastes. Environ Toxicol Chem 19: 600-607.

16. Säumel I, Kotsyuk I, Hölscher M, Lenkereit C, Weber F, et al. (2012) How healthy is urban horticulture in high traffic areas? Trace metal concentrations in vegetable crops from plantings within inner city neighbourhoods in Berlin, Germany. Environ Pollut 165: 124-132.

17. Sahuquillo A, Rigol A, Rauret G (2003) Overview of the use of leaching/ extraction tests for risk assessment of trace metals in contaminated soils and sediments. TrAC Trends Anal Chem 22: 152-159.

18. Hlavay J, Prohaska T, Weisz M, Wenzel WW, Stingeder GJ (2004) Determination of trace elements bound to soils and sediment fractions - (IUPAC technical report). Pure Appl Chem 76: 415-442.

19. Manniello A, Zappa G, Zoani C (2010) Zea mays (L) in areas with different anthropic pollution sources: Relations between toxic element contents in soils and vegetable tissues. Fresenius Envivorn Bull 19: 526-536.

20. World Health Organization (2001) Environmental Health Criteria 224: Arsenic and arsenic compounds. Geneva, Switzerland.

21. USEPA (2010) Risk-based Concentration Table. United State Environmental Protection Agency, Washington DC, USA.

22. Commission Regulation (EC) (2008) No. 1881/2006 setting maximum levels for certain contaminants in foodstuffs. European Regulation (EC) No 629/2008.

23. [No authors listed] (2013) AOAC SMPR 2012.007. Standard method performance requirements for determination of heavy metals in a variety of foods and beverages. J AOAC Int 96: 704.

24. Farooq M, Anwar F, Rashid U (2008) Apraisal of heavy metal contents in different vegetables grown in the vicinity of an industrial area. Pak J Bot 40: 2099-2106.

25. Liu L, Zhang Q, Hu L, Tang J, Xu L, et al. (2012) Legumes can increase cadmium contamination in neighboring crops. PLoS One 7: e42944.

26. Mihucz VG, Tatár E, Varga A, Záray G, Cseh E (2001) Application of total-reflection X-ray fluorescence spectrometry and high-performance liquid chromatography for the chemical characterization of xylem saps of nickel contaminated cucumber plants. Spectrochim Acta Part B 56: 2235-2246.

27. Varga A, Martinez RMG, Zaray G, Fodor F (1999) Investigation of effects of cadmium, lead, nickel and vanadium contamination on the uptake and transport processes in cucumber plants by TXRF spectrometry. Spectrochim Acta Part B 54: 1455-1462.

28. Klockenkämper R, Knoth J, Prange A, Schwenke H (1992) Total Reflection X- ray Flourescence Analysis. Anal Chem 64: 1115A-1123A.

29. Borgese L, Bilo F, Tsuji K, Fernández-Ruiz R, Margui E, et al.(2014) First Total Reflection X-Ray Fluorescence round-robin test of water samples: Preliminary results. Spectrochim Acta Part B 101: 6-14.

30. Moreira S, Ficaris M, Vives AES, Filho VFN, Zucchi OL, et al. (2006) Heavy Metals in Groundwater using Synchrotron Radiation Total Reflection X-Ray Analysis. Instrum Sci Technol 34: 567-585.

31. Stosnach H (2005) Environmental trace-element analysis using a benchtop total reflection X-ray fluorescence spectrometer. Anal Sci 21: 873-876.

32. Bilo F, Borgese L, Cazzago D, Zacco A, Bontempi E, et al. (2014) TXRF analysis of soils and sediments to assess environmental contamination. Environ Sci Pollut Res Int 21: 13208-13214.

33. Marguí E, Floor GH, Hidalgo M, Kregsamer P, Román-Ross G, et al. (2010) Analytical possibilities of total reflection X-ray spectrometry (TXRF) for trace selenium determination in soils. Anal Chem 82: 7744-7751.

34. Stosnach H (2006) On-site analysis of heavy metal contaminated areas by means of total reflection X-ray fluorescence analysis (TXRF). Spectrochim Acta Part B: Atomic Spectroscopy 61: 1141-1145.

35. De Vives AES, Brienza SMB, Moreira S, Zucchi OLAD, Barroso RC, et al. (2007) Evaluation of the availability of heavy metals in lake sediments using SR-TXRF. Nucl Instrum Methods Phys Res : Sect A 579: 503-506.

36. Bontempi E, Zacco A, Benedetti D, Borgese L, Colombi P, et al. (2010) Total reflection X-ray fluorescence (TXRF) for direct analysis of aerosol particle samples. Environ Technol 31: 467-477.

37. Borgese L, Salmistraro M, Gianoncelli A, Zacco A, Lucchini R, et al. (2012) Airborne particulate matter (PM) filter analysis and modeling by total reflection X-ray fluorescence (TXRF) and X-ray standing wave (XSW). Talanta 89: 99-104.

38. Borgese L, Zacco A, Pal S, Bontempi E, Lucchini R, et al. (2011) A new non-destructive method for chemical analysis of particulate matter filters: the case of manganese air pollution in Vallecamonica (Italy). Talanta 84: 192-198.

39. Borgese L, Zacco A, Bontempi E, Colombi P, Bertuzzi R, et al. (2009) Total reflection of x-ray fluorescence (TXRF): a mature technique for environmental chemical nanoscale metrology. Meas Sci Technol 20: 084027.

40. Wannaz ED, Carreras HA, Abril GA, Pignata ML (2011) Maximum values of Ni 2+, Cu 2+, Pb 2+ and Zn 2+ in the biomonitor Tillandsia capillaris (Bromeliaceae): Relationship with cell membrane damage. Environ Exper Bot 74: 296-301.

41. De Vives AES, Moreira S, Brienza SMB, Medeiros JGS, Filho MT, et al. (2006) Monitoring of the environmental pollution by trace element analysis in tree-rings using synchrotron radiation total reflection X-ray fluorescence. Spectrochim Acta Part B: Atomic Spectroscopy 61: 1170-1174.

42. Turnau K, Ostachowicz B, Wojtczak G, Anielska T, Sobczyk L (2010) Metal uptake by xerothermic plants introduced into Zn-Pb industrial wastes. Plant Soil 337: 299-311.

43. Moreira S, Vieira CB, Filho BC, Stefanutti R, Jesus EFO (2005) Study of the Metals Absorption in Culture Corn Irrigated with Domestic Sewage by SR-TXRF. Instrum Sci Technol 33: 73-85.

44. Necemer M, Kump P, Ščancar J, Jacimovic R, Simcic J, et al. (2008) Application of X-ray fluorescence analytical techniques in phytoremediation and plant biology studies. Spectrochim Acta Part B: Atomic Spectroscopy 63: 1240-1247.

45. Martinez T, Lartigue J, Zarazua G, Avila-Perez P, Navarrete M, et al. (2008) Application of the Total Reflection X-ray Fluorescence technique to trace elements determination in tobacco. Spectrochim Acta Part B: Atomic Spectroscopy 63: 1469-1472.

46. Borgese L, Bilo F, Dalipi R, Bontempi E, Depero LE (2015) Total reflection X-ray fluorescence as a tool for food screening. Spectrochim Acta Part B: Atomic Spectroscopy 113: 1-15.

47. Gonzalez CA, Watters RL (2014) Certificate of Analysis Standard Reference Material® 1570a Trace Elements in Spinach Leaves. National Institute of Standards & Technology, Gaithersburg, MD 20899.

48. Varga A, Záray G, Fodor F, Cseh E (1997) Study of interaction of iron and lead during their uptake process in wheat roots by total-reflection X-ray fluorescence spectrometry. Spectrochim Acta Part B: Atomic Spectroscopy 52: 1027-1032.

49. Varga A, Záray G, Fodor F (2002) Determination of element distribution between the symplasm and apoplasm of cucumber plant parts by total reflection X-ray fluorescence spectrometry. J Inorg Biochem 89: 149-154.

50. De La Calle I, Costas M, Cabaleiro N, Lavilla I, Bendicho C (2013) Fast method for multielemental analysis of plants and discrimination according to the anatomical part by total reflection X-ray fluorescence spectrometry. Food Chem 138: 234-241.

51. De La Calle I, Cabaleiro N, Romero V, Lavilla I, Bendicho C (2013) Sample pretreatment strategies for total reflection X-ray fluorescence analysis: A tutorial review. Spectrochim Acta - Part B 90: 23-54.

52. Sekara A, Poniedzialek M, Ciura J, Jedrszczyk E (2005) Zinc and copper accumulation and distribution in the tissues of nine crops: Implications for phytoremediation. Polish J Environ 14: 829-835.

53. Piechalak A, Malecka A (2008) Lead uptake, toxicity and accumulation in Phaseolus vulgaris plants. Biol Plantarum 52: 565-568.

54. Kadhim RE (2011) Effect of Pb , Ni and Co in growth parameters and metabolism of Phaseolus aureus Roxb. Euphrates J Agric Sci 3: 10-14.

55. Tsadilasa C, Shaheen SM, Samaras V, Gizas D, Hu Z (2009) Influence of Fly Ash Application on Copper and Zinc Sorption by Acidic Soil amended with Sewage Sludge. Commun Soil Sci Plan 40: 1-6.

56. Khuder A, Sawan MK, Karjou J, Razouk AK (2009) Determination of trace elements in Syrian medicinal plants and their infusions by energy dispersive X-ray fluorescence and total reflection X-ray fluorescence spectrometry. Spectrochim Acta Part B: Atomic Spectroscopy 64: 721-725.

# 29

# Rapid Analysis of Cocaine in Saliva by Surface-Enhanced Raman Spectroscopy

Kathryn Dana, Chetan Shende, Hermes Huang and Stuart Farquharson*

*Real-Time Analyzers Inc., 362 Industrial Park Road, Unit 8, Middletown, CT 06457, USA*

## Abstract

Increases in illicit drug use and the number of emergency-room visits attributable to drug misuse or abuse highlight the need for an efficient, reliable method to detect drugs in patients in order to provide rapid and appropriate care. A surface-enhanced Raman spectroscopy (SERS)-based method was successfully developed to rapidly measure cocaine in saliva at clinical concentrations, as low as 25 ng/mL. Pretreatment steps comprising chemical separation, physical separation, and solid-phase extraction were investigated to recover the analyte drug from the saliva matrix. Samples were analyzed using Fourier-transform (FT) and dispersive Raman systems, and statistical analysis of the results shows that the method is both reliable and accurate, and could be used to quantify unknown samples. The procedure requires minimal space and equipment and can be completed in less than 16 minutes. Finally, due to the inclusion of a buffer solution and the use of multiple robust pretreatment steps, with minimal further development this method could also be applied to other drugs of interest.

**Keywords:** Raman; SERS; Cocaine; Drug detection

## Introduction

Illicit drug use is a growing problem in the United States, with associated detrimental health effects to users that may cause them to seek urgent care. The number of recent users increased from 8.3 percent in 2002 to 9.4 percent in 2013, an estimated 24.6 million people [1]. This figure includes 1.5 million cocaine users. Within approximately the same period, the number of annual emergency-room visits attributable to drug misuse or abuse rose by 52% (between 2004 and 2011), and cocaine was detected in 40.3% of all illicit-drug-related admissions [2]. In order to accurately diagnose and treat patients in emergency room situations, it is critical to identify the cause of admission, e.g., the type of drug present in the patient's system. The symptoms of illicit drug abuse are often similar and can sometimes be misdiagnosed as physical or psychiatric disorders [3]. In particular, cocaine may cause arrhythmias, myocardial infarction, hypothermia, seizures, and/or hallucinations [4]. Current serum and urine toxicological screening methods can be inconvenient and time-consuming, requiring the collection of samples from potentially uncooperative patients. Thus, there is a need for a rapid and minimally invasive analytical technique to accurately detect illicit substances in emergency room patients. To reduce the invasive nature of sample collection, saliva may be used in lieu of blood or urine. Typically, drugs are represented in saliva at concentrations similar to blood plasma [5,6], while saliva is characterized by better sample integrity than urine and can contain both the parent compound and metabolites [7]. Furthermore, saliva is 99.5% water, making it easy to chemically analyze [8]; and simple saliva collectors are readily commercially available.

In the case of cocaine, previous studies have shown that cocaine and metabolite concentrations in saliva are significantly correlated with blood plasma concentrations and that the metabolite-to-parent ratio detected in oral fluid can potentially be used as an indicator of the time of last use [9,10]. In one study, cocaine was detected in saliva at concentrations 4.9 times higher than in urine and serum, and all saliva samples yielded true positives for cocaine while some matched urine and serum samples had undetectable concentrations of the drug [11].

The current accepted cut-off threshold for the detection of most drugs in saliva falls within the 10-50 ppb (10-50 ng/mL) range, as defined by the U.S. Substance Abuse and Mental Health Services Administration (SAMHSA). For clinical applications, a device that can reliably measure these concentrations in just a few minutes, without false positives or negatives, is desirable [12]. To this end, we have been investigating the potential of surface-enhanced Raman spectroscopy (SERS) to both identify and quantify drugs of interest and their metabolites in an oral fluid matrix [13-15].

In this study, a rapid SERS-based method was developed to measure cocaine in saliva. Various sol-gel chemistries, doped with silver or gold as the SERS-active medium, were investigated. A solid-phase extraction (SPE) pretreatment procedure, including steps to free cocaine from the saliva matrix and concentrate samples prior to measurement, was also developed. Clinically relevant concentrations of cocaine were measured in saliva, and the results were subjected to statistical analysis.

## Materials and Methods

### SERS-active sol-gel capillaries

Gold and silver sol-gels were prepared according to previously published procedures [16,17], with some modifications. Briefly, a metal-ligand precursor (e.g., silver amine or gold chloride) was mixed with a Si-alkoxide precursor (tetramethyl orthosilicate or methyltrimethoxysilane) in methanol. The SERS-active capillaries were prepared by drawing ~20 µL of the resulting gold- or silver-doped sol-gel into a 10 cm long, 1 mm diameter glass capillary (VWR International, Bridgeport, NJ, USA) to form a ~1 cm plug. The plug was

*Corresponding author: Stuart Farquharson, Real-Time Analyzers Inc., 362 Industrial Park Road, Unit 8, Middletown, CT 06457, USA
E-mail: stu@rta.biz

allowed to gel and cure, after which the incorporated gold or silver ions were reduced by flowing 5 mL of 0.1M NaBH$_4$ through the capillary. A sample photograph of a reduced gold sol-gel capillary is included in the Supporting Information (Figure S1).

## Sample pretreatment

SPE cartridges were obtained from United Chemical Technologies (UCT C8+SCX Clean Screen SPE columns; Bristol, PA, USA). All solvents used during the extraction experiments were obtained from Sigma Aldrich (Allentown, PA, USA). For the SPE pretreatment procedure, the column was first conditioned with 1.3 mL of methanol, followed by 1.3 mL of water. Then, 1-2 mL of sample was flowed through the SPE cartridge at a rate of 1 mL/min using a GenieTouch syringe pump (Kent Scientific, Torrington, CT, USA). The column was then dried by flowing air through it for 1 minute using the syringe pump in draw mode, after which analytes were eluted off the column using 100 µL of 50/50 (v/v) hexane/ethyl acetate at a flow rate of 1 mL/min. Finally, 25 µL of the collected eluate was loaded onto a SERS-active capillary and measured using SERS. The recovery efficiency of the SPE method was confirmed by gas chromatography (GC) analysis of samples with known concentration before and after extraction.

For physical separation using centrifugation, a Galaxy Mini centrifuge (VWR, Radnor, PA, USA) was employed for 5 minutes at 6000 rpm. For filtration steps, a 0.2-micron filter (Grace Davison Discovery Science, Deerfiled, IL, USA) was connected to a Luer-Lok plastic syringe (Becton-Dickinson, Franklin Lakes, NJ, USA), and the sample was passed manually through the filter.

## Saliva samples

Saliva sample controls were collected from volunteers using an oral swab (Medimpex United Inc., PA, USA). The swabs were wiped along all inner mouth surfaces for 1 minute, collecting ~1 mL of saliva. Samples were collected throughout the day, then combined and stored in a freezer until use. Varying amounts of a standard cocaine solution were added to saliva samples to yield artificially spiked samples at a desired range of concentrations. These samples were loaded directly onto SERS-active capillaries for SERS measurement.

The SPE pretreatment method described above was modified for saliva samples as follows: A 0.5 mL saliva sample was collected using a swab and transferred to a centrifuge tube containing 0.5 mL of 0.1M pH 6 acetate buffer. A C8+SCX SPE cartridge was preconditioned by sequentially drawing 1 mL each of methanol and 0.1M acetate buffer through it at 2 mL/min. Then, the 1 mL buffered sample was drawn through the column at 1 mL/min. The column was washed by sequentially drawing 1 mL each of water, 0.1M HCl, and methanol through it at 2 mL/min, and then dried by drawing air through it for 1 min. The analyte was eluted from the column using 1 mL of 2% NH$_4$OH in methanol at a flow rate of 0.5 mL/min. All fluid flow through the SPE cartridge was accomplished using a syringe. The eluate was evaporated to dryness under nitrogen flow and reconstituted using 10 µL of 50/50 (v/v) methanol/water, which was then drawn into a SERS-active capillary for measurement.

Analyte concentrations were verified using gas chromatography with flame ionization detection (GC-FID) (Model 17 A, Shimadzu, Columbia, MD, USA). The instrument was fitted with a split-splitless injector and an RTx-1 capillary column (30 m × 0.53 mm, 100% dimethylpolysiloxane; Restek, Bellefonte, PA, USA), and helium was used as the carrier gas. The injector was set to 300°C and a split ratio of 1. The oven temperature program was as follows: initial temperature

150°C; ramp to 305°C at 10°C/min; hold at 305°C for 4 min. The FID temperature was maintained at 300°C. Samples were injected at a volume of 1 µL.

## SERS analysis

For all samples, a 20 µL aliquot was manually drawn into a prepared SERS-active capillary using a syringe. The capillary was mounted on an XY stage above a fiber-optic probe for immediate analysis. SERS measurements were performed using FT-Raman and dispersive Raman analyzers, both employing laser excitation at 785 nm and a power of 80 mW (models RamanID and SERS-Lab, respectively, Real-Time Analyzers, Middletown, CT, USA). Spectra were collected at 10 spots along the sol-gel plug, at 1 mm intervals, for 1 minute each using the FT-Raman analyzer and 10 seconds each using the dispersive Raman analyzer, unless otherwise noted; the results for each capillary were averaged using Raman VISTA software (Real-Time Analyzers).

## Results

Throughout the development of this method, various aspects were investigated to achieve the detection of 10-50 ng/mL cocaine in saliva. This included silver vs. gold as the SERS-active metal; methods to preconcentrate the samples, such as SPE, sample pH, improvements in Raman analyzer technology, and methods to extract the cocaine from saliva.

To determine the optimal sol-gel chemistry, cocaine in solution was measured using multiple variants of gold- and silver-doped sol-gels. The gold-doped sol-gels yielded greater sensitivity than the silver-doped sol-gels, by approximately two orders of magnitude. The lowest measured concentration using the silver-doped sol-gel capillaries was 100 µg/mL, while the lowest measured concentration for the gold-doped sol-gel capillaries was 1 µg/mL. Representative spectra are shown in Figure 1.

Given the better sensitivity obtained using gold, the gold sol-gel was selected for use in further experiments. For the first experiment, a concentration series of cocaine in water was examined to establish the SERS limit of detection using the gold capillaries. The intensity of the peaks, shown in Figure 2, follows a Langmuir curve (13). Furthermore, cocaine was consistently detected at concentrations as low as 125 ng/mL without any sample pretreatment or spectral manipulation.

While 125 ng/mL is adequate for the federal mandatory workplace drug testing cutoff concentration of 150 ng/ mL [18], for clinical applications a lower detection limit is needed. In order to achieve the desired goal of 10-50 ng/mL in saliva, the possibility of a pretreatment step was investigated. In particular, SPE has been shown to be a reliable and inexpensive method for recovering low-concentration analytes from a matrix [19,20], and the authors have successfully used this method previously to recover drugs from saliva [21]. Because cocaine is a basic drug, it was necessary to select an SPE material that is appropriate for basic compounds. The UCT C8+SCX column, which contains monomerically bonded octyl and benzene sulfonic acid on silica and retains analytes with both neutral and amine functional groups, but passes analytes with carboxylate functional groups, was therefore used for pretreatment in all subsequent experiments.

Since the pH of saliva can vary, the effect of pH on SERS measurements of cocaine in water was investigated. The pH of 1 µg/mL cocaine samples was varied by adding dilute HCl or NaOH to yield pH values of 3, 7, and 11. As shown in Figure 3, the characteristic cocaine peaks are relatively stable at acidic pH, but disappear at basic pH. This result was as expected, given that cocaine has a pKa of 8.6. Due to the

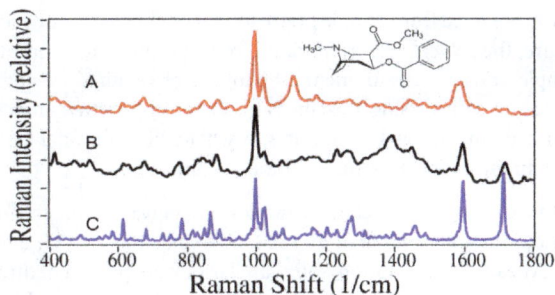

**Figure 1:** a) SERS of 10 µg/mL cocaine on gold-doped sol-gel capillary; b) SERS of 100 µg/mL cocaine on silver-doped sol-gel capillary; and c) normal Raman spectrum (NRS) of pure solid cocaine. Note that the signal is strongest in spectrum (a). Conditions: FT-Raman spectrometer at 8 cm⁻¹ resolution. SERS: 80 mW at 785 nm, ten 1 minute spectra averaged. NRS: 500 mW at 785 nm.

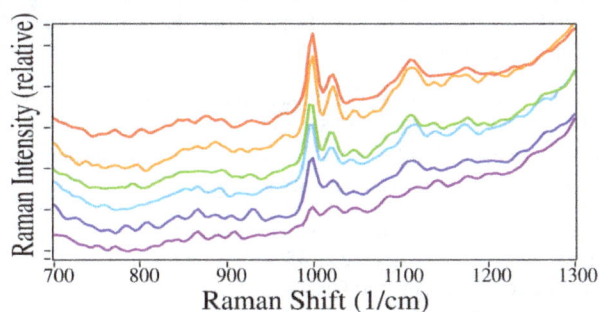

**Figure 2:** SERS of cocaine at 0.125, 0.250, 0.500, 1, 5, and 10 µg/mL on modified gold sol-gel capillaries. Conditions: FT-Raman spectrometer, 80 mW at 785 nm, 8 cm⁻¹ resolution, ten 1 minute spectra averaged. The spectra are on the same intensity scale, but offset for clarity.

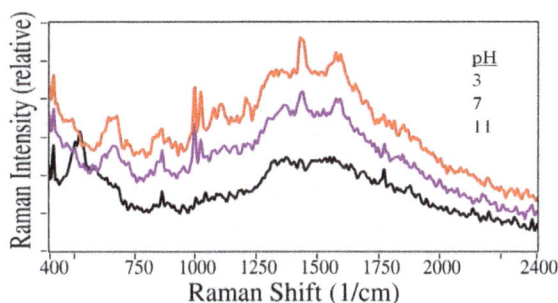

**Figure 3:** SERS of 1 µg/mL cocaine at pH 3, 7 and 11. Conditions: 1 min scan.

demonstrated pH sensitivity, it was concluded that the saliva samples would have to be buffered in order to produce accurate results. For this purpose, acetate buffer was selected to mitigate the effects of pH in subsequent experiments.

After collecting saliva from volunteers using an oral swab, the untreated samples were subjected to SERS analysis to ensure that the background spectrum of the saliva matrix would not interfere with cocaine measurements. First, a saliva blank was compared with 0.5 µg/mL cocaine in saliva using the FT-Raman spectrometer. As shown in Figure 4 (a), the 1002 cm⁻¹ cocaine peak is distinctly visible against the saliva background and the broad feature caused by luminescence of the glass capillary between 1200 and 2000 cm⁻¹. A characteristic peak

at 1020 cm⁻¹ is also visible, attributed to the asymmetric phenyl ring breathing mode [22].

Since the FT-Raman spectra were relatively noisy, a dispersive Raman system was built employing a newly available low-noise charge-coupled-device 2-D array detector. While repeated measurements of the same cocaine sample yielded near-identical-intensity spectra with the same features, the dispersive spectrometer provided a 12-fold improvement in noise reduction based on the signal-to-noise ratio for the 1002 cm⁻¹ peak relative to the spectrum baseline (Figure 4 (b)). It is also worth noting that the comparison of spectra from cocaine in a capillary, saliva in a capillary, and an empty capillary shows that saliva causes very little interference with the cocaine spectrum. Due to the improved signal quality, the dispersive Raman system was selected for use in further measurements.

Cocaine-spiked saliva samples containing 1000, 500, 250, 100, 50, 25, 10, 5, and 1 µg/mL cocaine were loaded into SERS-active capillaries without pretreatment and measured using the dispersive Raman system. Spectral features specific to cocaine were detected at concentrations as low as 25 µg/mL (Figure 5). However, this is 3 orders of magnitude higher than the SAMHSA cut-off threshold.

In order to improve sensitivity, a series of solvents, specifically acetic acid, acetonitrile, chloroform, and hexane, were investigated to chemically separate cocaine from the saliva mucans. The best performance was observed for acetic acid. As noted above, due to the need to control the sample pH, acetate buffer was selected to meet both chemical separation and buffering requirements. Next, filtration and centrifugation were investigated to physically separate the cocaine from the saliva matrix. Both methods yielded similar results. Due to the ease of use and minimal equipment required for manual filtration compared with centrifugation, the former was selected for all further experiments.

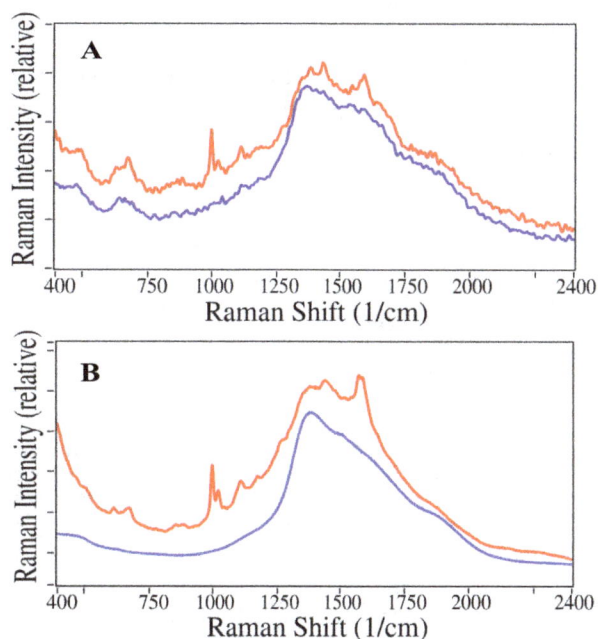

**Figure 4:** SERS of a) 0.5 µg/mL cocaine and saliva blank, collected using FT-Raman, 60 sec; and b) 0.5 µg/mL cocaine and blank glass capillary, collected using dispersive Raman, ten 10-sec spectra averaged. Conditions ((a) and (b)): 80 mW of 785 nm laser excitation.

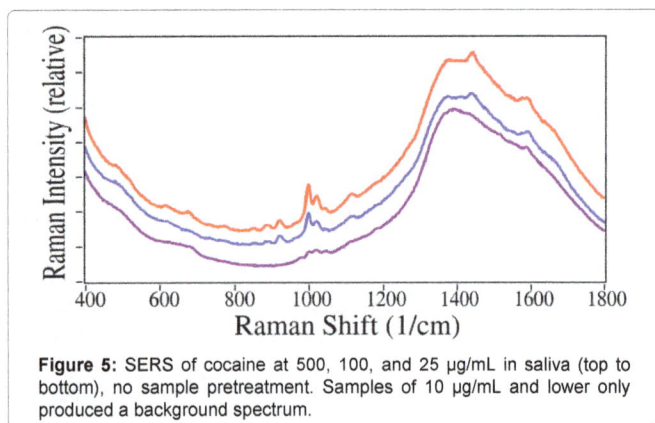

**Figure 5:** SERS of cocaine at 500, 100, and 25 µg/mL in saliva (top to bottom), no sample pretreatment. Samples of 10 µg/mL and lower only produced a background spectrum.

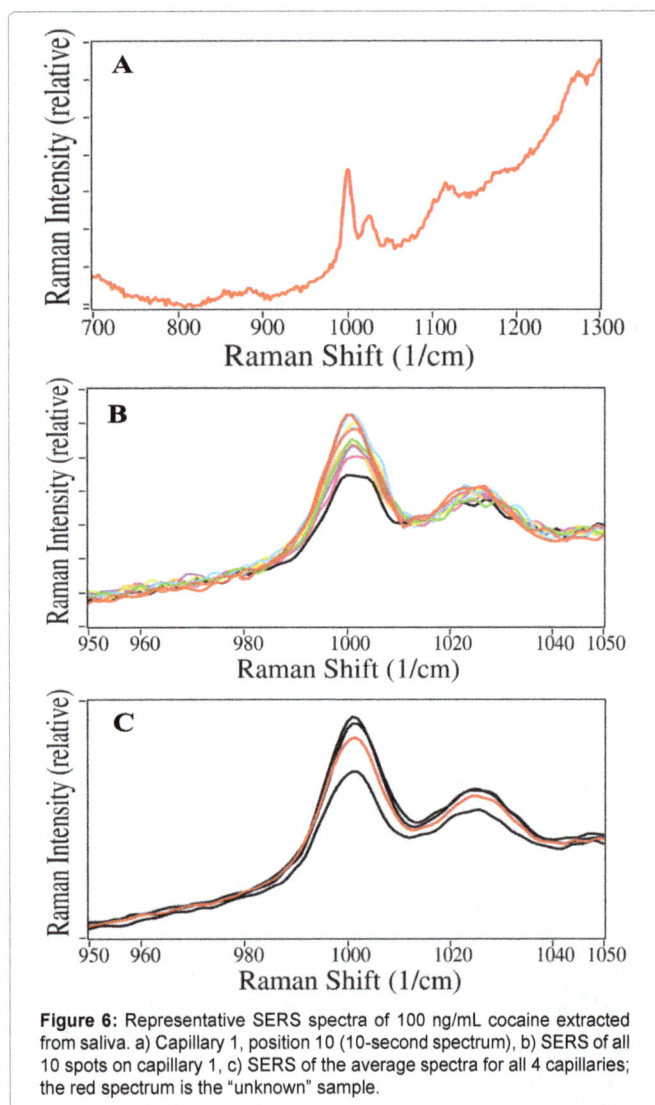

**Figure 6:** Representative SERS spectra of 100 ng/mL cocaine extracted from saliva. a) Capillary 1, position 10 (10-second spectrum), b) SERS of all 10 spots on capillary 1, c) SERS of the average spectra for all 4 capillaries; the red spectrum is the "unknown" sample.

Based on the results obtained for cocaine in water, an SPE pretreatment method was then developed for the spiked saliva samples, using a C8+SCX column as before. Several eluents were investigated, including acetate buffer, acetonitrile, ammonia, ethanol, hexane, hydrochloric acid, methanol, and water. Flow rates were varied from 0.1 to 5 mL/min. In addition to varying the solvent and flow rate, an

additional step was added after the SPE process to further improve the sensitivity obtained from sample pretreatment. Following the extraction procedure, the solvent was evaporated from the eluate to concentrate the sample prior to measurement. The total time required for analysis of cocaine in saliva using this method, including pretreatment and SERS measurement, was 15.5 minutes. A schematic of the final method is shown in the Supporting Information (Figure S2).

Artificially spiked cocaine samples in saliva at much lower concentrations were prepared at 1000, 500, 250, 100, 50, 25, and 10 ng/mL. At each concentration, four capillaries were prepared, three to be analyzed in triplicate and a fourth to be used later as an unknown; all capillaries were measured by SERS. This procedure yielded both sensitive and repeatable results, as illustrated in Figure 6.

The raw averaged spectra collected for spiked saliva samples were overlaid by setting the spectral intensity at 810 cm$^{-1}$ to zero. The cocaine intensity was calculated as the peak height at 1002 cm$^{-1}$ minus a baseline drawn from 950 to 1070 cm$^{-1}$. The average standard deviation for the four capillaries individually was 17.1%, while the standard deviation between capillaries was 21.7%.

Figure 7 (a) shows the 3-capillary averaged spectra for cocaine in saliva at 1000, 500, 250, 100, 50, and 25 ng/mL (10 ng/mL was not observed). In Figure 7 (b), the peak heights at 1002 cm$^{-1}$ are plotted as a function of concentration and fit with a Langmuir curve.

The Langmuir equation is $\Theta=[kC/(1+kC)]S$, where $\Theta$ is the surface coverage (here expressed as the SERS peak height), k is a constant, C is the sample concentration and S is a scaling constant added to produce the correct SERS peak height. Here, k=0.007, C ranges from 0 to 1000 ng/mL, and S=2900. The Langmuir fit then represents a concentration

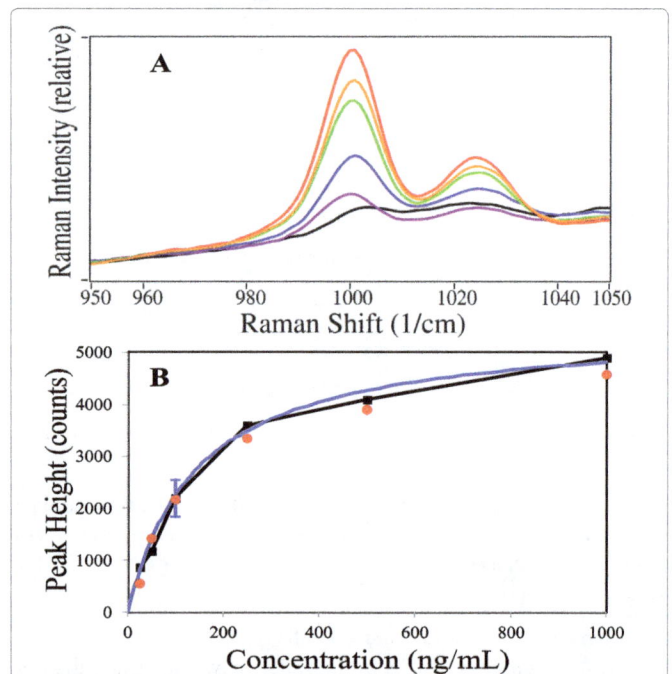

**Figure 7:** a) SERS of concentrated cocaine extracted from saliva containing 25, 50, 100, 250, 500, and 1000 ng/mL. Spectra are scaled to have the same background intensity. b) Plot of 1002 cm-1 peak heights for each concentration (black squares connected by black line for clarity), Langmuir fit to data (blue line), and results for the 4th capillaries as unknowns (red dots). The standard deviation for the 100 ng/mL "unknown" sample is included.

| 100ng/ml | Cap 1 | Cap 2 | Cap 3 | Unknown | 4 Cap Ave |
|---|---|---|---|---|---|
| Spot 1 | 2838 | 2962 | 1662 | 2281 | |
| Spot 2 | 2590 | 2714 | 857 | 1786 | |
| Spot 3 | 2405 | 2405 | 2095 | 1662 | |
| Spot 4 | 2900 | 2466 | 1786 | 2281 | |
| Spot 5 | 2714 | 2157 | 2033 | 2528 | |
| Spot 6 | 2281 | 2219 | 1786 | 1538 | |
| Spot 7 | 2343 | 2405 | 1352 | 2095 | |
| Spot 8 | 1724 | 2219 | 1167 | 2776 | |
| Spot 9 | 2095 | 2714 | 2095 | 2528 | |
| Spot 10 | 2590 | 2095 | 2343 | 2095 | |
| **Average** | 2447.8 | 2435.5 | 1717.4 | 2156.9 | 2189.4 |
| **Stddev** | 340.2 | 268.7 | 443.3 | 382.6 | 475.6 |
| **%dev** | 13.90% | 11.03% | 25.81% | 17.74% | 21.72% |

**Table 1:** Peak height measurements and statistics for 4 capillaries containing 100 ng/mL cocaine extracted from saliva (3 known and 1 unknown, 10 spots/capillary). Values shown are the peak height calculated at $1002\ cm^{-1}$ minus a baseline drawn from 950 to $1070\ cm^{-1}$.

curve for this analysis. For example, the Langmuir curve predicts the following concentrations for the "unknown" samples: 25=15, 50=50, 100=92, 250=225, 500=340, and 1000=750 ng/mL. It is worth noting that smaller errors occur at lower concentrations for a Langmuir curve. However, the SERS measurements produce different amounts of error at each different concentration, as reflected in their standard deviations (Table 1).

## Conclusion

A method was successfully developed to rapidly measure cocaine in saliva at concentrations as low as 25 ng/mL. The procedure requires minimal space and equipment and can be completed in less than 16 minutes, which makes it appropriate for use in settings such as hospital emergency rooms. Due to the inclusion of a buffer solution and the use of the C8+SCX SPE column, this method could also be applied to measure not only other basic drugs, but also acidic drugs in a saliva matrix. Ongoing research using saliva samples from regular drug users is currently being performed to confirm the accuracy of this method and determine its applicability in real-world conditions. Future studies of pretreatment options may allow the detection limit and analysis time to be improved, and experiments using other pharmaceuticals will confirm the applicability of this method to other compounds.

### Acknowledgments

The authors would like to thank the National Institute on Drug Abuse for funding this research (NIH Grant Number 1R43DA032178-01).

### References

1. Substance Abuse and Mental Health Services Administration (SAMHSA) (2014) Results from the 2013 National Survey on Drug Use and Health: Summary of National Findings. NSDUH Series H-48, HHS Publication No. (SMA) 14-4863.

2. SAMHSA (2011) Drug Abuse Warning Network: National Estimates of Drug-Related Emergency Room Visits.

3. Reidy LJ, Junquera P, Van Dijck K, Steele BW, Nemeroff CB (2014) Underestimation of substance abuse in psychiatric patients by conventional hospital screening. J Psychiatr Res 59: 206-212.

4. Rainey PM, Schonfeld DJ (1997) The Yale University School of Medicine Patient's Guide to Medical Tests: Toxicology, Monitoring of Drug Therapy, and Testing for Substance Abuse. Houghton-Mifflin Co., New York, USA. pp: 587-596.

5. Drummer OH (2005) Review: Pharmacokinetics of illicit drugs in oral fluid. Forensic Sci Int 150: 133-142.

6. Crouch D, Day J, Baudys J, Fatah A (2004) Evaluation of Saliva/Oral Fluid as an Alternate Drug Testing Specimen. National Institute of Justice, USA Report Number 605-03.

7. Allen KR (2011) Screening for drugs of abuse: which matrix, oral fluid or urine? Ann Clin Biochem 48: 531-541.

8. Chamberlain J (1995) The Analysis of Drugs in Biological Fluids. 2nd edn. CRC Press, Boca Raton, FL, USA. p: 46.

9. Scheidweiler KB, Kolbrich Spargo EA, Kelly TL, Cone EJ, Barnes AJ, et al. (2010) Pharmacokinetics of Cocaine and Metabolites in Human Oral Fluid and Correlation with Plasma Concentrations following Controlled Administration. Ther Drug Monit 32: 628-637.

10. Dams R, Choo RE, Lambert WE, Jones H, Huestis MA (2007) Oral fluid as an alternative matrix to monitor opiate and cocaine use in substance-abuse treatment patients. Drug Alcohol Depend 87: 258-267.

11. Schramm W, Craig PA, Smith RH, Berger GE (1993) Cocaine and benzoylecgonine in saliva, serum, and urine. Clin Chem 39: 481-487.

12. SAMHSA (2010) Results from the 2009 National Survery on Drug Use and Health, Volume 1. NSDUH Series H-38A, HHS Publication No. SMA 10-4586.

13. Farquharson S, Shende C, Inscore F, Maksymiuk P, Gift A (2005) Analysis of 5-fluoroacil in saliva using surface-enhanced Raman spectroscopy. J Raman Spectrosc 36: 208-212.

14. Shende C, Inscore F, Gift A, Maksymiuk P, Farquharson S (2005) Ten-minute analysis of drugs and metabolites in saliva by surface-enhanced Raman spectroscopy. P Soc Photo-Opt Instrum Eng 6007: 165-171.

15. Farquharson S, Shende C, Sengupta A, Huang H, Inscore F (2011) Rapid detection and identification of overdose drugs in saliva by surface-enhanced Raman scattering using fused gold colloids. Pharmaceutics 3: 425-439.

16. Farquharson S, Lee YH, Nelson C (2003) Material for surface-enhanced Raman spectroscopy, and SER sensors and method for preparing same. US Patent No. 6623977.

17. Farquharson S, Gift AD, Inscore FE, Shende CS (2008) SERS method and apparatus for rapid extraction and analysis of drugs in saliva. US Patent No. 7393691.

18. SAMHSA (2008) Mandatory Guidelines For Federal Workplace Drug Testing Programs. Federal Register 73 FR 71858. p: 71880.

19. Foulon C, Menet MC, Manuel N, Pham-Huy C, Galons H, et al. (1999) Rapid Analysis of Benzoylecgonine, Cocaine, and Cocaethylene in Urine, Serum, and Saliva by Isocratic High-Performance Liquid Chromatography with Diode-Array Detection. Chromatographia 50: 721-727.

20. Hegstad S, Oiestad EI, Johansen U, Christophersen AS (2006) Determination of Benzodiazepines in Human Urine Using Solid-Phase Extraction and High-Performance Liquid Chromatography-Electrospray Ionization Tandem Mass Spectrometry. J Analytical Toxicology 30: 31-37.

21. Inscore F, Shende C, Sengupta A, Huang H, Farquharson S (2011) Detection of drugs of abuse in saliva by surface-enhanced Raman spectroscopy (SERS). Appl Spectrosc 65: 1004-1008.

22. Carter JC, Brewer WE, Angel SM (2000) Raman spectroscopy for the in situ identification of cocaine and selected adulterants. Applied Spectroscopy 54: 1876-1881.

# Permissions

The contributors of this book come from diverse backgrounds, making this book a truly international effort. This book will bring forth new frontiers with its revolutionizing research information and detailed analysis of the nascent developments around the world.

We would like to thank all the contributing authors for lending their expertise to make the book truly unique. They have played a crucial role in the development of this book. Without their invaluable contributions this book wouldn't have been possible. They have made vital efforts to compile up to date information on the varied aspects of this subject to make this book a valuable addition to the collection of many professionals and students.

This book was conceptualized with the vision of imparting up-to-date information and advanced data in this field. To ensure the same, a matchless editorial board was set up. Every individual on the board went through rigorous rounds of assessment to prove their worth. After which they invested a large part of their time researching and compiling the most relevant data for our readers.

The editorial board has been involved in producing this book since its inception. They have spent rigorous hours researching and exploring the diverse topics which have resulted in the successful publishing of this book. They have passed on their knowledge of decades through this book. To expedite this challenging task, the publisher supported the team at every step. A small team of assistant editors was also appointed to further simplify the editing procedure and attain best results for the readers.

Apart from the editorial board, the designing team has also invested a significant amount of their time in understanding the subject and creating the most relevant covers. They scrutinized every image to scout for the most suitable representation of the subject and create an appropriate cover for the book.

The publishing team has been an ardent support to the editorial, designing and production team. Their endless efforts to recruit the best for this project, has resulted in the accomplishment of this book. They are a veteran in the field of academics and their pool of knowledge is as vast as their experience in printing. Their expertise and guidance has proved useful at every step. Their uncompromising quality standards have made this book an exceptional effort. Their encouragement from time to time has been an inspiration for everyone.

The publisher and the editorial board hope that this book will prove to be a valuable piece of knowledge for researchers, students, practitioners and scholars across the globe.

# List of Contributors

Peter Baugh J
The BMSS, C/O 23, Priory Road, Sale, M33 2BU, England, UK

Marjorie Buist, Emy Komatsu, Paul G Lopez, Lauren Girard, Edward Bodnar and Hélène Perreault
Department of Chemistry, University of Manitoba, Canada

Sophie Conchon and Jean-Paul Judor
INSERM UMR 10-64, Institut de Transplantation Urology Nephrology (ITUN), Université de Nantes, France

Apolline Salama
INSERM UMR 10-64, Institut de Transplantation Urology Nephrology (ITUN), Université de Nantes, France
Xenothera, Nantes, France

Jean-Paul Soulillou
INSERM UMR 10-64, Institut de Transplantation Urology Nephrology (ITUN), Université de Nantes, France
Translink Framework Program (FP7), Padova, Italy

David H Sachs
Massachusetts General Hospital, Harvard University, Cambridge, MA, USA

Andrea Perota
Avantea Laboratory of Reproductive Technologies, Cremona, Italy

Giovanna Lazzari
Avantea Laboratory of Reproductive Technologies, Cremona, Italy
Avantea Foundation, Cremona, Italy

Cesare Galli
Avantea Laboratory of Reproductive Technologies, Cremona, Italy
Avantea Foundation, Cremona, Italy
Department of Veterinary Medical Sciences, University of Bologna, Ozzano Emilia, Italy
Translink Framework Program (FP7), Padova, Italy

Jean-Paul Concordet
Université Paris Descartes, Paris, France

Ivkovic B and Vujic Z
Department of Pharmaceutical Chemistry and Drug Analysis, Faculty of Pharmacy, University of Belgrade, Serbia

Karljikovic-Rajic K
Department of Analytical Chemistry, Faculty of Pharmacy, University of Belgrade, Serbia

Ibric S
Department of Pharmaceutical Technology and Cosmetology, University of Belgrade, Serbia

Atilio I Anzellotti, Robert Ylimaki and Alexander Yordanov
IBA Molecular, Virginia Commonwealth University, Richmond, USA

Xiuli Dong, Jessica J. Broglie and Liju Yang
Biomanufacturing Research Institute and Technology Enterprise (BRITE) and Department of Pharmaceutical Sciences, North Carolina Central University, Durham, NC 27707, USA

Yongan Tang
Department of Mathematics and Physics, North Carolina Central University, Durham, NC 27707, USA

Kamal Alizadeh and Nasim Abbasi Rad
Department of Chemistry, Lorestan University, Khorramabad, Iran

Aufried Lenferink TM and Cees Otto
Department of Medical Cell Biophysics, MIRA Institute, University of Twente, Enschede, The Netherlands

Barbara Liszka M
Department of Medical Cell Biophysics, MIRA Institute, University of Twente, Enschede, The Netherlands
Wetsus, European Centre of Excellence for Sustainable Water Technology, Leeuwarden, The Netherlands

Pooja Bansal, Gaurav G, Susheela Rani and Ashok Kumar Malik
Department of Chemistry, Punjabi University, Patiala, Punjab, India

Nidhi N
Department of Chemistry, Atma Ram Sanatan Dharam College, New Delhi, India

Balusamy T and Nishimura T
Corrosion Resistant Steel Group, Research Center for Structural Materials (RCSM), National Institute for Materials Science, Ibaraki, Tsukuba, 305-0047, Japan

Yew-Keong Choong, Nor Syaidatul Akmal Mohd Yousof and Mohd Isa Wasiman
Phytochemistry Unit, Herbal Medicine Research Centre, Institute for Medical Research, Jalan Pahang, 50588 Kuala Lumpur, Malaysia

Jamia Azdina Jamal
Drug and Herbal Research, Faculty of Pharmacy, UKM, Bangi, Selangor, Malaysia

Zhari Ismail
School of Pharmaceutical Science, USM, Gelugor, Penang, Malaysia

Mohammad Mahdi Doroodmand and Fatemeh Ghasemi
Department of Chemistry, College of Sciences, Shiraz University, Shiraz 71454, Iran

Poonam Vats and Tausif Monif
Department of Clinical Pharmacology and Pharmacokinetics, Sun Pharmaceutical Industries Ltd, Gurgaon-122 015, Haryana, India

S Manaswita Verma
Department of Pharmaceutical Sciences, Birla Institute of Technology, Mesra, Ranchi-835 215, Jharkhand, India

Tarin Nimmanwudipong
Department of Environmental Science and Technology, Tokyo Institute of Technology, 4259 Nagatsuta, Midori-ku, Yokohama 226-8502, Japan

Naohiro Yoshida
Department of Environmental Science and Technology, Tokyo Institute of Technology, 4259 Nagatsuta, Midori-ku, Yokohama 226-8502, Japan
Earth-Life Science Institute (WPI-ELSI), Tokyo Institute of Technology, Meguro, Tokyo 152-8551, Japan
Department of Environmental Chemistry and Engineering, Tokyo Institute of Technology, 4259 Nagatsuta, Midori-ku, Yokohama 226-8502, Japan

Alexis Gilbert
Earth-Life Science Institute (WPI-ELSI), Tokyo Institute of Technology, Meguro, Tokyo 152-8551, Japan

Naizhong Zhang and Keita Yamada
Department of Environmental Chemistry and Engineering, Tokyo Institute of Technology, 4259 Nagatsuta, Midori-ku, Yokohama 226-8502, Japan

Magda Akl A, Magdy Bekheit M and Ibraheim Helmy
Faculty of Science, Mansoura University, Mansoura, Egypt

Yuhei Hida
Graduate School of Innovative Life Science for Education, University of Toyama, 3190 Gofuku, Toyama, Japan

Hiroaki Shinohara
Graduate School of Innovative Life Science for Education, University of Toyama, 3190 Gofuku, Toyama, Japan
Graduate School of Science and Engineering for Research, University of Toyama, 3190 Gofuku, Toyama, Japan

Sean Williams, Neeraja Venkateswaran, Travis Del Bonis O'Donnell, Pete Crisalli, Sameh Helmy, Maria Teresa Napoli and Sumita Pennathur
Mechanical Engineering Department, University of California Santa Barbara, Santa Barbara, CA 93106, USA

El-Beshlawy MM
Department of Chemistry, Collage of Girls, Ain Shams University, Egypt

Chetan Shende, Kathryn Dana, Hermes Huang and Stuart Farquharson
Real-Time Analyzers Inc., 362 Industrial Park Road, Unit 8, Middletown, CT-06457, USA
Smiths Detection, 14 Commerce Drive, Danbury, CT 06810, USA

Jay Sperry
University of Rhode Island, 45 Lower College Road, Kingston, RI 02881, USA

Erika Ponzini, Greta Borgonovo, Carlo Santambrogio and Rita Grandori
Department of Biotechnology and Biosciences, University of Milano-Bicocca, Piazza della Scienza 2, 20126, Milan, Italy

Luca Merlini and Yves M Galante
Istituto di Chimica del Riconoscimento Molecolare, CNR, Via Mario Bianco 9, 20131, Milan, Italy

Mikhail V Shashkov
Boreskov Institute of Catalysis SB RAS, Novosibirsk, Russian Federation

Vladimir N Sidelnikov
Novosibirsk State University, Novosibirsk, Russian Federation

Alula Yohannes, Tesfaye Tolesa and Negussie Megersa
Department of Chemistry, Addis Ababa University, Addis Ababa, Ethiopia

**Yared Merdassa**
Department of Chemistry, Addis Ababa University, Addis Ababa, Ethiopia
Department of Chemistry, Jimma University, Jimma, Ethiopia

**Atsuko Konishi, Shigehiko Takegami, Shoko Akatani, Rie Takemoto and Tatsuya Kitade**
Department of Analytical Chemistry, Kyoto Pharmaceutical University, 5 Nakauchicho, Misasagi, Yamashina-ku, Kyoto 607-8414, Japan

**Mahendra Kumar Trivedi, Alice Branton, Dahryn Trivedi and Gopal Nayak**
Trivedi Global Inc., 10624 S Eastern Avenue Suite A-969, Henderson, NV 89052, USA

**Khemraj Bairwa and Snehasis Jana**
Trivedi Science Research Laboratory Pvt. Ltd., Hall-A, Chinar Mega Mall, Chinar Fortune City, Hoshangabad Rd., Bhopal- 462026, Madhya Pradesh, India

**Dave Trinel and Corentin Spriet**
TISBio, CNRS, UMR 8576, UGSF, Glycobiology Structural and Functional Unit, Lille University of Science and Technology, Lille, France

**Pauline Vandame**
TISBio, CNRS, UMR 8576, UGSF, Glycobiology Structural and Functional Unit, Lille University of Science and Technology, Lille, France
Regulation of Signal Division Team, CNRS, UMR 8576, UGSF, Glycobiology Structural and Functional Unit, Lille University of Science and Technology, Lille, France

**Jean-François Bodart**
Regulation of Signal Division Team, CNRS, UMR 8576, UGSF, Glycobiology Structural and Functional Unit, Lille University of Science and Technology, Lille, France

**Magalie Hervieu, Emilie Floquet, Marc Aumercier and Emanuele G Biondi**
CNRS, UMR 8576, UGSF, Glycobiology Structural and Functional Unit, Lille University of Science and Technology, France

**Ana Moya Gil, Elena Gras Colomer, Begoña Porta Oltra and Mónica Climente Martí**
Pharmacy Service, Hospital Universitario Doctor Peset, Avda. Gaspar Aguilar 90, Valencia, Spain

**María Amparo Martínez Gómez**
Foundation for the Promotion of Health Biomedical Research and Valencia (FISABIO), C/ Micer Masco 31, Valencia, Spain

**Parviainen I**
Institute of Dentistry, University of Eastern Finland, Kuopio Campus, Yliopistonranta 1, Kuopio, Finland

**Kullaa AM**
Institute of Dentistry, University of Eastern Finland, Kuopio Campus, Yliopistonranta 1, Kuopio, Finland
Research Group of Oral Health Sciences, Faculty of Medicine, University of Oulu, Oulu, Finland
Educational Dental Clinic, Kuopio University Hospital, Kuopio, Finland

**Singh SP**
Institute of Dentistry, University of Eastern Finland, Kuopio Campus, Yliopistonranta 1, Kuopio, Finland
SIB Labs, University of Eastern Finland, Yliopistonranta 1, Kuopio, Finland

**Mikkonen JJ and Koistinen AP**
SIB Labs, University of Eastern Finland, Yliopistonranta 1, Kuopio, Finland

**Dekker H, Schulten EAJM, Ten Bruggenkate CM and Bravenboer N**
Department of Oral and Maxillofacial Surgery and Oral Pathology, VU University Medical Center/ Academic Centre for Dentistry Amsterdam (ACTA), Amsterdam, The Netherlands

**Turunen MJ**
Department of Applied Physics, Faculty of Science and Forestry, University of Eastern Finland, Kuopio, Finland

**Lionel Vidaud**
Quality Science and Innovation, EVOTEC Toulouse, France

**Antoine Pradines, Jérôme Marini and Elodie Lajous**
Preparative Chromatography Group, EVOTEC Toulouse, France

**Patrick Clavières**
ADME and Developability Assessment Laboratory, EVOTEC Toulouse, France

**Fabjola Bilo, Laura Borgese, Annalisa Zacco, Elza Bontempi and Laura E Depero**
Department of Mechanical and Industrial Engineering, University of Brescia, Brescia, Italy

**Pranvera Lazo**
Faculty of Natural Science, University of Tirana, Tirane, Albania

**Claudia Zoani and Giovanna Zappa**
ENEA-UTAGRI, CR Casaccia, Rome, Italy

# Index

www.ingramcontent.com/pod-product-compliance
Lightning Source LLC
Chambersburg PA
CBHW080640200326
41458CB00013B/4691